senic Exposure
ıd Health Effects V

Arsenic Exposure and Health Effects V

Proceedings of the Fifth International Conference on Arsenic
Exposure and Health Effects, July 14–18, 2002, San Diego, California

Editors:

Willard R. Chappell
University of Colorado at Denver
Denver, CO, USA

Charles O. Abernathy
U.S. Environmental Protection Agency
Washington, DC, USA

Rebecca L. Calderon
U.S. Environmental Protection Agency
Research Triangle Park, NC, USA

David J. Thomas
U.S. Environmental Protection Agency
Research Triangle Park, NC, USA

2003

ELSEVIER

Amsterdam – Boston – Heidelberg – London – New York – Oxford – Paris – San Diego
San Francisco – Singapore – Sydney – Tokyo

ELSEVIER B.V.
Sara Burgerhartstraat 25
P.O. Box 211, 1000 AE Amsterdam
The Netherlands

ELSEVIER Inc.
525 B Street, Suite 1900
San Diego, CA 92101-4495
USA

ELSEVIER Ltd
The Boulevard, Langford Lane
Kidlington, Oxford OX5 1GB
UK

ELSEVIER Ltd
84 Theobalds Road
London WC1X 8RR
UK

First edition 2003

Library of Congress Cataloging in Publication Data
A catalog record is available from the Library of Congress.

British Library Cataloguing in Publication Data
A catalogue record is available from the British Library.

ISBN: 0-444-51441-4

∞ The paper used in this publication meets the requirements of ANSI/NISO Z39.48-1992 (Permanence of Paper).
Printed in The Netherlands.

Preface

The Society of Environmental Geochemistry and Health (SEGH) Fifth International Conference on Arsenic Exposure and Health Effects was held July 14 to 18, 2002 in San Diego, California. Both public and private groups sponsored the conference. In addition to SEGH and the University of Colorado at Denver, other sponsors included the US Environmental Protection Agency (US EPA), the Centers for Disease Control and Prevention (CDC), AWWA Research Foundation (AwwaRF), the Electric Power Research Institute (EPRI), and the National Institute of Environmental Health Sciences (NIEHS).

Over 220 people, including the speakers and poster presenters, attended the 5th conference. Of these, approximately one-third were non-US citizens; the largest groups were from Asia and Europe. The attendees included scientists from academia, industry, local, state, US federal and several foreign governments. The disciplines of geochemistry, chemistry, molecular biology, biochemistry, epidemiology and medicine were well-represented at the meeting. Several outstanding papers and posters presented at the conference generated lively discussion and debate, not only about scientific issues, but also social, public policy and regulatory issues. The interactions between the attendees both during and outside the sessions resulted in many new contacts between scientists.

There were 14 platform sessions and 4 poster sessions with 67 speakers and 130 posters. Panel discussions were held after each session to encourage discussion and debate. Two sessions were devoted to an overview of some US EPA and NIEHS arsenic programs. As in past conferences, the first report of elevated arsenic exposures in a new country was given. This time it was Nepal; Dr. Shrestha reported elevated arsenic concentrations in groundwaters in the Terai region.

The seeds of the conference(s) were sown at a meeting of the SEGH Executive Board in December 1991. They agreed to form an Arsenic Task Force similar to the SEGH Lead in Soil Task Force that had been formed in the 1980s. It was clear that there was a growing controversy regarding the proposed changes in the US EPA Maximum Contaminant Level (MCL) for arsenic in drinking water. This is the enforceable standard for drinking water. In addition to impacting on water utilities, the development of the standard would also have the possibility for significant economic impacts on the cleanup of superfund sites and on the electric power industry (because of arsenic in fly ash).

The Task Force was formed in 1992 and co-chaired by Willard Chappell and Charles Abernathy. An international conference seemed to be an excellent way to begin to compile the data and to determine what needed to be done in this area. The First SEGH International Conference on Arsenic Exposure and Health Effects was held in New Orleans in July, 1993. This conference was successful in attracting the top arsenic researchers in the world. It was followed by the Workshop on Epidemiology and Physiologically-Based Pharmacokinetics that was held in Annapolis, MD in June, 1994. Perhaps the most significant outcome of the 1994 Workshop was the realization that there are many arsenic hot spots in the world. Although not widely recognized by the scientific community at that time, significant public health problems existed in countries such as India, Thailand and China. The Second SEGH International Conference on Arsenic Exposure and Health Effects was held in June, 1995 in San Diego with a primary purpose being to highlight the global aspects of the problem and most of the impacted countries were represented. At that time, the biggest recognized problem area was in West Bengal, India where an estimated 30 million people are at risk from arsenic exposure in the ground water.

The Third SEGH International Conference on Arsenic Exposure and Health Effects (July, 1998) was also held in San Diego and continued the theme of global impact of arsenic. In addition, two new countries with significant arsenic problems, Inner Mongolia and Bangladesh were represented. The attendees were to learn that the Bangladesh problem could be larger than the one in West Bengal with a possible 80 million people (two-thirds of the population) at risk. This situation caught the attention of the media later in 1998 with the publication of a front-page article in the Nov. 16, 1998 New York Times. The article was syndicated and published in newspapers around the world.

The SEGH Fourth International Conference on Arsenic Exposure and Health Effects (June, 2000) continued the focus on the global impact of arsenic, and also featured sessions on mechanisms of cancer carcinogenesis, metabolism, and water treatment technology. New impacted countries represented included Viet Nam.

The Fifth International Conference on Arsenic Exposure and Health Effects (July, 2002) introduced the participants to the problems in Nepal. Numerous speakers discussed advances in understanding the mechanisms of arsenic toxicity and carcinogenicity. The considerable advances in several areas were readily apparent and this monograph represents the state of the art. The organizers are deeply grateful to the authors for their fine work and to the sponsors for the support that made it possible.

We are also deeply appreciative of the fine efforts of Rosemary Wormington of the Environmental Sciences Program of the University of Colorado at Denver who put in long hours as conference coordinator. She kept the entire Conference going and, more than anyone else, is responsible for the success of this and the past conferences.

Contents

List of Contributors

Abernathy, Charles O.
Offices of Science and Technology
Washington, DC

Alam, M.G.M
School of Biological and Chemical
Sciences
Faculty of Science and Technology
Deakin University
221 Burwood Highway
Burwood
Victoria, Australia 3125

Anderson, Henry
Wisconsin Department of Health and
Family Services
Division of Public Health
1 W Wilson St Rm 150
Madison, WI 53701

Andrewes, Paul
Environmental Carcinogenesis
Division
NHEERL, US EPA
Research Triangle Park, NC
USA, 27711

Aposhian, H. Vasken
Department of Molecular and
Cellular Biology
Center for Toxicology
The University of Arizona
Tucson, Arizona, 85721-0106

Aposhian, Mary M.
Department of Molecular and
Cellular Biology
The University of Arizona
Tucson, Arizona, 85721-0106

Arnold, Lora L.
University of Nebraska Medical
Center
Department of Pathology and
Microbiology and the Eppley

Institute for Research on Cancer
983135 Nebraska Medical Center
Omaha, Nebraska 68198-3135

Basu, Arindam
Institute of Post Graduate
Medical Education & Research
Kolkata
India

Basu, Gautam Kumar
School of Environmental Studies
Jadavpur University
Calcutta 700032
INDIA

Bencko, M.D., Ph.D., Vladimír
Charles University of Prague
Czech Republic

Bende, Flóra
Makó Water Works
Hungary

Berg, Michael
Swiss Federal Institute for
Environmental Science and
Technology (EAWAG),
CH-8600
Dubendorf, Switzerland

Beringer, Mike
Solid Waste and Emergency Response
Washington, DC

Board, Philip G.
Molecular Genetics Group
John Curtin School of Medical
Research
Australian National University
Canberra
Australia

Bosland, Maarten C.
The Nelson Institute of
Environmental Medicine and Kaplan

Comprehensive Cancer Center
New York
University School of Medicine

Bruce, Scott L.
National Research Centre for Env.
Toxicology
The Univ. of Queensland
Australia

Burns, Fredric J.
The Nelson Institute of
Environmental Medicine and Kaplan
Comprehensive Cancer Center
New York University School of
Medicine

Calderon, R.L.
Research and Development
Research Triangle Park, NC

Cano, Marty
University of Nebraska Medical
Center
Department of Pathology and
Microbiology and the Eppley
Institute for Research on Cancer
983135 Nebraska Medical Center
Omaha
Nebraska 68198-3135

Carter, Dean E.
Department of Pharmacology and
Toxicology
Center for Toxicology
The University of Arizona
Tucson, Arizona, 85721-0106

Caussy, Deoraj
Department of Evidence for
Information and Policy
World Health Organization Office
of the South East Asia

Chakraborti, Dipankar
Director, School of Environmental
Studies
Jadavpur University
Calcutta 700032
INDIA

Chanda, Chitta Ranjan
School of Environmental Studies
Jadavpur University

Calcutta 700032
INDIA

Chen, Hua
Inorganic Carcinogenesis Section
Laboratory of Comparative
Carcinogenesis
NCI at NIEHS
Research Triangle Park, NC, 27709

Chowdhury, Uttam Kumar
School of Environmental Studies
Jadavpur University
Calcutta 700032
INDIA

Cohen, Samuel M.
University of Nebraska Medical
Center
Department of Pathology
and Microbiology and the Eppley
Institute for Research on Cancer
983135 Nebraska Medical Center
Omaha, Nebraska 68198-3135

Colvile, R. N.
Imperial College of Science
Technology and Medicine
London
UK

Con, Tran Hong
Center for Environmental Technology
and Sustainable Development
Hanoi National University Hanoi
Vietnam

Cornwell, David A.
EE&T, Inc.

Cullen, William R.
Environmental Chemistry Group
University of British Columbia
Vancouver B.C.
Canada

Das, Subhankar
Institute of Post Graduate
Medical
Education & Research
Kolkata
India

Davis, Colin
UNICEF
Dhaka
Bangladesh

Del Razo, Luz María
  CINVESTAV-IPN
  Mexico City
  Mexico

Del Razo, Luz María
  Toxicology Section
  Cinvestav-IPN
  México D.F.
  México

Drobná, Zuzana
  Department of Pediatrics
  University of North Carolina
  Chapel Hill
  North Carolina, 27599-7220
  USA

Dunson, D.
  Biostatistics Branch
  National Institute of Environmental
  Health Sciences (NIEHS)
  Research Triangle Park
  NC 27709
  USA

Easterling, Michael
  Analytical Sciences
  Inc.
  Research Triangle Park, NC

Evans, Marina V.
  U.S. EPA; Research Triangle Park,
  NC

EXPASCAN study group.
  Bencko V., Cordos E., Docx P.,
  Fabianova E., Frank P., Gotzl M.,
  Grellier J., Hong B., Rames J.,
  Rautiu R., Stevens E., Zvarova J.,
  Farago, M. E.
  Imperial College of Science
  Technology and Medicine
  London
  UK

Finnell, Richard H.
  Center for Environmental and
  Genetic Medicine
  Institute of Biosciences and
  Technology
  Texas A&M University System Health
  Science Center
  Houston, Texas 77030

Franěk, Petr
  EuroMISE Centre
  Charles University of
  Prague & Czech
  Academy of Sciences
  Czech Republic

Fukushima, Shoji
  Department of Pathology
  Osaka City
  University Medical School

García-Montalvo, Eliud A.
  Toxicology Section, Cinvestav-IPN
  México D.F.
  México

Germolec, D
  Laboratory of Molecular Toxicology
  National Institute of Environmental
  Health Sciences (NIEHS)
  Research Triangle Park
  NC 27709
  USA

Ghose, Nilima
  Institute of Post Graduate Medical
  Education & Research
  Kolkata
  India

Giger, Walter
  Swiss Federal Institute for
  Environmental Science and
  Technology (EAWAG)
  CH-8600
  Dubendorf, Switzerland

Gong, Zhilong
  Department of Public Health Sciences
  Faculty of Medicine

Götzl, M.D., Miloslav
  Department of Oncology District
  Hospital of Bojnice
  Slovak Republic

Grossman, S.
  Faculty of Life Sciences
  Bar-Ilan University
  Ramat Gan, 52900
  Israel

Gruber, Kornél
  WEDECO Ltd.
  Hungary

Guha Mazumder, D. N.
  Department of Gastroenterology &
  Medicine
  Institute of Post Graduate
  Medical Education & Research
  Calcutta
  India, 700020

Ha, Cao
  The Center for Environmental
  Technology and Sustainable
  Development Hanoi National
  University
  Hanoi, Vietnam

Ha, Hoang Van
  Center for Environmental Technology
  and Sustainable Development
  Hanoi National University
  Hanoi
  Vietnam

Hanh, Nguyen Thi
  Center for Environmental
  Technology and Sustainable
  Development
  Hanoi National University
  Hanoi
  Vietnam

Healy, Sheila M.
  Department of Molecular and
  Cellular Biology
  The University of Arizona
  Tucson
  Arizona

Hlavay, József
  Department of Earth and
  Environmental Sciences
  University of Veszprém
  8201 Veszprém
  P. O. Box 158
  Hungary
  E-mail: hlavay@almos.vein.hu

Hódi, Márta
  Hydra Ltd.
  Hungary

Hong Con, Tran
  Research centre for environmental
  Technology and Sustainable
  Development (CETASD),
  Hanoi University of Science,

Vietnam National University,
  Add: 334 Nguyen Trai,
  Thanh Xuan, Hanoi, Vietnam

Hu, Yu
  School of Biological and Chemical
  Sciences
  Deakin University
  AUSTRALIA
  Nelson Institute of
  Environmental Medicine
  New York University School
  of Medicine
  57 Old Forge Rd
  Tuxedo, NY 10987
  USA

Hughes, Michael F.
  U.S. EPA
  Research Triangle Park
  NC

Hung Viet, Pham
  Research centre for environmental
  Technology and Sustainable
  Development (CETASD),
  Hanoi University of Science,
  Vietnam National University,
  Add: 334 Nguyen Trai,
  Thanh Xuan, Hanoi, Vietnam

IH, Dip.
  Dhaka Community Hospital
  Bangladesh

Jakubis, Marián
  State Institute of Health
  Prievidza
  Slovak Republic

Jakubis, P.
  State Health Institute Prievidza
  Slovakia

Jaspers, Ilona
  Department of Pediatrics and Center
  for Environmental Medicine
  Asthma, and Lung Biology
  University of North Carolina
  Chapel Hill
  North Carolina, 27599-7220
  USA

Jiang, Guifeng
  Department of Public Health
  Sciences
  Faculty of Medicine

Jin, Ximei
School of Biological and Chemical
Sciences
Deakin University
Australia School of Public Health
Xinjiang Medical University
China

Joarder, MBBS, A. I.
Dhaka Community Hospital
Bangladesh

Jones-Lee, Anne
G. Fred Lee & Associates
El Macero, CA

Kadiiska, Maria
Laboratory of Pharmacology and
Chemistry, NIEHS
Research Triangle Park
NC, 27709

Kenyon, Elaina M.
U.S. EPA
Research Triangle Park, NC

Kitchin, Kirk T.
Environmental Carcinogenesis
Division
NHEERL, US EPA
Research Triangle Park, NC
USA, 27711

Klein, Catherine B.
Nelson Institute of Environmental
Medicine
New York University School of
Medicine
57 Old Forge Rd, Tuxedo
NY 10987
USA

Knobeloch, Lynda
Wisconsin Department
of Health and Family
Services
Division of Public Health
1 W Wilson St Rm 150
Madison, WI 53701

Lahiri, Sarbari
Institute of Post Graduate
Medical Education & Research
Kolkata
India

Lai, Vivian W.-M.
Environmental Chemistry Group
University of British Columbia
Vancouver B.C.
Canada

Le, X. Chris
Department of Public Health
Sciences
Faculty of Medicine
University of Alberta
Edmonton, Alberta, T6G 2G3

Lee, G. Fred
G. Fred Lee & Associates
El Macero, CA

Lin, Shan
Curriculum in Toxicology
University of North Carolina
at Chapel Hill
Chapel Hill, North Carolina

Liu, Jie
Inorganic Carcinogenesis Section
Laboratory of Comparative
Carcinogenesis, NCI at NIEHS
Research Triangle Park
NC, 27709

Liu, Zijuan
Department of Biochemistry and
Molecular Biology Wayne State
University
School of Medicine

Lodh, Dilip
School of Environmental Studies
Jadavpur University
Calcutta 700032
INDIA

Lomnitski, L.
Faculty of Life Sciences
Bar-Ilan University
Ramat Gan, 52900
Israel

Lu, Xiufen
University of Alberta
Environmental
Health Sciences
10-110 Clinical
Sciences Building
Edmonton
Canada T6G 2G3

MacPhee, Ph.D., Michael J.
McGuire Environmental
Consultants, Inc.
Denver, CO

Mazumder, Kunal
Institute of Post Graduate Medical
Education & Research
Kolkata
India

McCluskey, Kate L.
School of Biological and Chemical
Sciences
Deakin University
221 Burwood Highway
Victoria 3125
AUSTRALIA

McMahon, T.
Pesticides Program
Washington, DC

Medgyesi, Pál
Makó Water Works
Hungary

Michael Berg
Swiss Federal Institute for
Environmental Science and
Technology (EAWAG),
CH-8600 Dubendorf, Switzerland

Miskovic, P.
State Health Institute
Banska Bystrica
Slovakia

Mollah, MBBS, S. U.
Dhaka Community Hospital
Bangladesh

Molnár, János
WEDECO Ltd.
Hungary

Morimura, Keiichirou
Department of Pathology
Osaka City
University Medical School

Moser, G
Integrated Laboratory Systems
Research Triangle Park
NC 27709

Mukherjee, Subhash Chandra
School of Environmental Studies
Jadavpur University
Calcutta 700032
INDIA

Mukhopadhyay, Rita
Department of Biochemistry and
Molecular Biology
Wayne State University
School of Medicine

Murcott, Susan
Dept. of Civil and Environmental
Engineering
Massachusetts Institute
of Technology

Mutter, P.E., Rodney N.
EE&T, Inc.

Ng, Jack C.
National Research Centre for Env.
Toxicology
The Univ. of Queensland
Australia

Nguyen Thi Hanh
Research Centre for Environmental
Technology and Sustainable
Development (CETASD),
Hanoi University of Science,
Vietnam National University.
Add: 334 Nguyen Trai,
Thanh Xuan, Hanoi, Vietnam

Nieuwenhuijsen, M. J.
Imperial College of Science
Technology and Medicine London
UK

Noller, Barry N.
National Research Centre for Env.
Toxicology
The Univ. of Queensland
Australia

Novak, Ph.D., John T.
Virginia Tech

Nyska, A.
Laboratory of Experimental
Pathology
National Institute of
Environmental Health Sciences
(NIEHS)
Research Triangle Park

NC 27709
USA

Ogra, Yasumitsu
Environmental Carcinogenesis
Division
NHEERL, US EPA
Research
Triangle Park, NC
USA, 27711

Ohmichi, Masayoshi
Chiba City Institute of Health and
Environment
Chiba 261-0001
Japan

Orth, Dip.
Dhaka Community Hospital
Bangladesh

Patterson, R.
Laboratory of Molecular Toxicology
National Institute of Environmental
Health Sciences (NIEHS)
Research Triangle Park
NC 27709

Pesch, B
Institut fuer Umweltmedizinische
Forschung (IUF)
Dusseldorf
Germany

Petrick, Jay S.
Department of Pharmacology and
Toxicology
The University of Arizona
Tucson, Arizona

Pham Hung Viet
Research Centre for Environmental
Technology and Sustainable
Development (CETASD),
Hanoi University of Science,
Vietnam National University.
Add: 334 Nguyen Trai,
Thanh Xuan, Hanoi,
Vietnam

Polishchuk, Elena
Environmental Chemistry Group
University of British Columbia
Vancouver B.C.
Canada

Polyák, Klára
Department of Earth and
Environmental Sciences
University of Veszprém
8201 Veszprém
P. O. Box 158
Hungary
E-mail: hlavay@almos.vein.hu

Quamruzzaman, FRCS, Quazi
Dhaka Community Hospital
Bangladesh

Rahman, FRCP, Mahmuder
Dhaka Community Hospital
Bangladesh

Rahman, Mohammad Mahmudur
School of Environmental Studies
Jadavpur University
Calcutta 700032
INDIA

Rames, D.Sc., Jiří
Charles University of Prague
Czech Republic

Ranft, U.
Institut fuer Umweltmedizinische
Forschung (IUF)
Dusseldorf
Germany

Reimer, Kenneth J.
Environmental Sciences Group
Royal Military College of Canada
Kingston, O.N.
Canada

Rosen, Barry P.
Department of Biochemistry and
Molecular Biology
Wayne State University
School of Medicine

Rossman, Toby G.
The Nelson Institute of
Environmental Medicine and Kaplan
Comprehensive Cancer Center
New York University School of
Medicine

Saha, Kshitish C.
The School of Tropical Medicine
Kolkata
West Bengal
India

Saha, Kshitish Chandra
  School of Environmental
  Studies
  Jadavpur University
  Calcutta 700032
  INDIA

Salam, Ph.D., M. A
  Dhaka Community Hospital
  Bangladesh

Sampayo-Reyes, Adriana
  Department of Molecular and
  Cellular Biology
  The University of Arizona
  Tucson
  Arizona

Sancha F., Ana Maria
  Universidad de Chile

Santra, Amal
  Institute of Post Graduate Medical
  Education & Research
  Kolkata
  India

Schertenleib, Roland
  Swiss Federal Institute for
  Environmental Science and
  Technology (EAWAG),
  CH-8600
  Dubendorf, Switzerland

Schuliga, Michael
  School of Biological and Chemical
  Sciences
  Deakin University
  221 Burwood Highway
  Victoria 3125
  AUSTRALIA

Sengupta, Mrinal Kumar
  School of Environmental Studies
  Jadavpur University
  Calcutta 700032
  INDIA

Shahjahan, MBBS, M.
  Dhaka Community Hospital
  Bangladesh

Shi, Jin
  Department of Biochemistry and
  Molecular Biology
  Wayne State University
  School of Medicine

Smith, Allan H.
  University of California Berkeley
  USA

Snow, Elizabeth T.
  School of Biological and Chemical
  Sciences
  Deakin University
  221 Burwood Highway
  Victoria 3125
  AUSTRALIA

  Nelson Institute of Environmental
  Medicine
  New York University School of
  Medicine
  57 Old Forge Rd
  Tuxedo, NY 10987
  USA

Spiegelstein, Ofer
  Center for Environmental and
  Genetic Medicine
  Institute of Biosciences and
  Technology
  Texas A&M University System Health
  Science Center
  Houston
  Texas 77030

Styblo, Miroslav
  Department of Pediatrics
  School of Medicine
  Department of Nutrition, School of
  Public Health
  University of North Carolina
  at Chapel Hill
  Chapel Hill
  North Carolina 27599

Sun, Yongmei
  Environmental Chemistry Group,
  University of British Columbia
  Vancouver B.C.
  Canada

Suttie, A.
  Integrated Laboratory Systems
  Research Triangle Park
  NC 27709

Suzuki, Kazuo T.
  Environmental Carcinogenesis
  Division NHEERL
  US EPA

Research Triangle Park, NC
USA, 27711

Sykora, Peter
School of Biological and Chemical
Sciences
Deakin University
221 Burwood
Highway
Victoria 3125
AUSTRALIA

Tanaka, A.
Environmental Chemistry Division
National Institute for Environmental
Studies
16-2 Onogawa
Tsukuba, Ibaraki 305-0053
Japan

The Ha, Cao
Research centre for environmental
Technology and Sustainable
Development (CETASD),
Hanoi University of Science,
Vietnam National University,
Add: 334 Nguyen Trai,
Thanh Xuan, Hanoi, Vietnam

Thomas, David J.
Pharmacokinetics Branch
Experimental Toxicology Division
National Health and Environmental
Effects Research Laboratory
Office of Research and Development
U.S. Environmental Protection
Agency
Research Triangle Park
North Carolina

Thornton, I.
Imperial College of Science
Technology and Medicine London
UK

Tomita, Takayuki
Environmental Carcinogenesis
Division
NHEERL, US EPA
Research
Triangle Park, NC
USA, 27711

Tran Hong Con
Research Centre for Environmental
Technology and Sustainable

Development (CETASD),
Hanoi University of Science,
Vietnam National University.
Add: 334 Nguyen Trai,
Thanh Xuan, Hanoi,
Vietnam

Trouba, K.
Laboratory of Molecular Toxicology
National Institute of Environmental
Health Sciences (NIEHS)
Research Triangle Park, NC 27709
USA

Uddin, Ahmed N.
The Nelson Institute of
Environmental Medicine and Kaplan
Comprehensive Cancer Center
New York
University School of Medicine

Valenzuela, Olga L.
Toxicology Section, Cinvestav-IPN
México D.F.
México.

Van Ha, Hoang
Research centre for environmental
Technology and Sustainable
Development (CETASD),
Hanoi University of Science,
Vietnam National University,
Add: 334 Nguyen Trai,
Thanh Xuan, Hanoi, Vietnam

Viet, Pham Hung
Center for Environmental Technology
and Sustainable Development
Hanoi National University
Hanoi
Vietnam

Waalkes, Michael P.
Inorganic Carcinogenesis Section
Laboratory of Comparative
Carcinogenesis
NCI at NIEHS Research
Triangle Park, NC, 27709

Wallace, Kathleen
Environmental Carcinogenesis
Division
NHEERL
US EPA
Research Triangle Park, NC
USA, 27711

Walton, F.
  Department of Pediatrics
  School of Medicine
  Department of Nutrition
  School of Public Health
  University of North Carolina
  at Chapel Hill Chapel Hill
  North Carolina 27599

Wang, Guoquan
  School of Public Health
  Xinjiang Medical University
  China

Wang, Lixia
  Environmental Chemistry Group
  University of British Columbia
  Vancouver B.C.
  Canada

Wanibuchi, Hideki
  Department of Pathology Osaka City
  University Medical School

Waters, Stephen B.
  Curriculum in Toxicology
  University of North Carolina at
  Chapel Hill Chapel Hill
  North Carolina

Wei, Min
  Department of Pathology Osaka City
  University Medical School

Wildfang, Eric
  Department of Pharmacology and
  Toxicology
  The University of Arizona
  Tucson
  Arizona

Winchester, E.
  Research and Development
  Research Triangle Park
  NC

Xie, Yaxiong
  Inorganic Carcinogenesis Section
  Laboratory of Comparative
  Carcinogenesis
  NCI at NIEHS
  Research Triangle Park
  NC, 27709

Ye, Jun
  Department of Biochemistry and
  Molecular Biology
  Wayne State University
  School of Medicine

Zakharyan, Robert A
  Department of Molecular and
  Cellular Biology
  The University of Arizona
  Tucson
  Arizona

PART I
OCCURRENCE & EXPOSURE

Arsenic Exposure and Health Effects V
W.R. Chappell, C.O. Abernathy, R.L. Calderon and D.J. Thomas, editors

3

Chapter 1
Groundwater arsenic exposure in India

Dipankar Chakraborti, Mrinal Kumar Sengupta, Mohammad
Mahmudur Rahman, Uttam Kumar Chowdhury, Dilip Lodh,
Sad Ahamed, Md. Amir Hossain, Gautam Kumar Basu, Subhash
Chandra Mukherjee and Kshitish Chandra Saha

Abstract

The first report on arsenic in hand tubewells, dugwells and spring water was published in 1976 from India. It was reported that people were drinking arsenic-contaminated water in Chandigarh and different villages of Punjab, Haryana, Himachal Pradesh in northern India. High arsenic was found in the liver of those suffering from non-cirrhotic portal fibrosis (NCPF) and drinking arsenic-contaminated water. Arsenic groundwater contamination in the state of West Bengal first came to notice during July 1983. The problem first came to international attention after the international conference held in Calcutta during February 1995. Before Bangladesh's arsenic episode was discovered, West Bengal's arsenic problem was known as the world's biggest arsenic calamity. During July 1983, 16 patients with arsenical skin lesions were identified from one village in the district of 24-Parganas where people were drinking arsenic-contaminated water from their hand tubewells in West Bengal. The present arsenic situation from $38,865 \ km^2$ of affected area with a population of 50 million in West Bengal up to August, 2002 is as follows: 3150 villages from 9 districts, 78 blocks/police stations have been identified where groundwater contains arsenic concentrations above 50 µg/l. On the basis of 125,000 water analyses by a laboratory method from the arsenic-affected areas it was estimated that more than 6 million people are drinking arsenic-contaminated water above 50 µg/l. So far, from our preliminary survey 8500 patients with arsenical skin lesions have been registered from 250 villages, and extrapolation of available data indicates that may be 300,000 people are suffering from arsenical skin lesions from 9 arsenic-affected districts of West Bengal. The source of arsenic is geologic. The mechanism of arsenic contamination from the source to the aquifer has not yet been established.

Groundwater arsenic contamination from industrial effluent discharge by a company producing paris-green (copper-aceto-arsenite) and the suffering of people in Behala – Calcutta came to notice during 1989. The highest arsenic concentration in soil near the effluent discharged point was found to be 10,000 µg/gm and the highest arsenic concentration in hand tubewell water was 38,000 µg/l. The total number of people using arsenic-contaminated water was 7000, and around 200 people were identified with arsenical skin lesions.

In the Rajnandgaon district of the state of Chattisgarh in India, a few villages were found where both dugwells and hand tubewells are arsenic contaminated. The source of arsenic is also geologic. The highest concentrations of arsenic found in the dugwells and hand tubewells were 520 and 880 µg/l, respectively. About 130 people were affected with arsenic poisoning. The number of people estimated to be at risk was 10,000.

About 1000 people are suspected to be suffering from arsenical skin lesions from the Semria Ojha Patty village of Sahapur police station in Bhojpur district of Bihar in the middle Ganga Plain. The magnitude of the problem in Bhojpur district hence in Bihar is unknown.

Keywords: arsenic in Ganga plain, industrial arsenic contamination, arsenic contamination in Bihar, source of arsenic

1. Introduction

Before the onset of the 21st century, groundwater arsenic contamination had already been reported in 20 countries, out of which four major incidents were from Asia. In order of severity of occurrence, these are Bangladesh, West Bengal (WB, India), Inner Mongolia (PR China) and Taiwan. Each year, new arsenic groundwater contamination incidents are being reported from Asian countries. For example, several places in China have been recently reported to be new problem areas (ESCAP–UNICEF–WHO Expert Group Meeting, 2001). Severe arsenic contamination has also been recently reported from Vietnam, where several million people are said to be at considerable risk of chronic arsenic poisoning (Berg et al., 2001). In a recent United Nations Economic and Social Commission for Asia and the Pacific–United Nations International Children's Emergency Fund–World Health Organization (UNESCAP–UNICEF–WHO) expert group meeting (ESCAP–UNICEF–WHO Expert Group Meeting, 2001), arsenic groundwater contamination was also reported from other countries including the Lao People's Democratic Republic, Cambodia, Myanmar and Pakistan. Groundwater arsenic contamination has also been reported from Nepal (Shrestha and Maskey, 2002).

In India, groundwater contamination incidents are increasing with time and the major incidents are in Gangetic Plain.

2. Groundwater arsenic contamination in northern India

A preliminary study on arsenic in dugwells, hand pumps and spring water (Datta and Kaul, 1976) was reported in 1976 from Chandigarh and different villages of Punjab, Haryana and Himachal Pradesh in northern India. Concentrations as high as 545 μg/l of arsenic were obtained in the water sample from a hand pump. Datta (1976) further reported high arsenic contents in the liver of five out of nine patients with non-cirrhotic portal hypertension (NCPH) who had been drinking arsenic-contaminated water. It was further stated (Datta, 1976) that "Cirrhosis (adult and childhood), non-cirrhotic portal fibrosis (NCPF) and extrahepatic portalvein obstruction in adults are very common in India and suggested that consumption of arsenic-contaminated water may have some role in the pathogenesis of these clinical states". To date, no further information on arsenic poisoning from northern India is available.

3. Groundwater arsenic contamination in West Bengal – India

After the incident of groundwater arsenic contamination in northern India in 1976, the next available information on arsenic contamination was from the state of West Bengal in 1983. An arsenic patient with skin lesions was identified by dermatologist Dr K.C. Saha (Docket No. 3/158/33/83) of the School of Tropical Medicine, Calcutta on 6th July 1983. The first arsenic contamination report on West Bengal (Garai et al., 1984) stated that 16 patients in 3 families were identified with arsenical skin lesions on 16th July 1983 and 28th July 1983 from one village in the district of 24-Parganas, who were drinking water from their hand tubewell. The highest arsenic concentration recorded in a tubewell was 1250 μg/l.

The same report stated that the condition of 2 patients was so severe that they needed hospitalisation. They had the symptoms of hyperpigmentation, hyperkeratosis, oedema, ascites, wasting, weakness, pain and a burning sensation in toes and fingers.

In 1984, the government of West Bengal constituted an investigation committee with some research workers of various organizations. In the same year after surveying the villagers, the government of West Bengal was informed by the committee about the extent, severity, possible causes of the disease and measures to prevent this disease. A copy of this report could not be traced; only a summary of the report was available from a published book (Saha, 2002).

During October 1984 Saha (Saha, 1984) reported 127 patients with arsenical skin lesions from 25 families (total members 139) from 5 villages in 3 districts (24 Parganas, Nadia, Bardhaman) of West Bengal. Out of the 127 affected members all had diffuse melanosis, 39% had spotted melanosis, and 37% and 12.6% had palmoplanter keratosis and dorsum keratosis respectively. This was the first report on skin lesions from arsenic patients of West Bengal. Next available report on magnitude of the arsenic problem from West Bengal was in the published paper of Saha & Poddar (Saha and Poddar, 1986) in 1986. The report stated 36 villages from 18 police stations/blocks of six districts are affected. These districts were 24 Parganas, Murshidabad, Nadia, Bardhaman, Malda & Midnapur. Although one patient from Ramnagar police station of Midnapur was reported in 1986 but later on it was found that Midnapur district is not affected and the patient was originally from affected district Nadia but settled in Midnapur. From 36 villages, water samples from 207 hand tubewell were analyzed and 105 (50.7%) showed arsenic concentration above 50 μg/l and highest concentration recorded was 586 μg/l. The report further stated that so far 1000 cases of chronic arsenical dermatosis were recorded from the affected villages and cutaneous malignancy was found in 3. Analysis of arsenic in hair, nail, skin-scale from the people in the affected villages confirmed arsenic toxicity and identified subclinical arsenic toxicity in some of the apparently unaffected members of the affected families.

During 1987 an epidemiological survey in 6 villages from 3 districts (24 Parganas, Bardhaman, and Nadia) revealed 197 patients with arsenical dermatosis in 48 families (Chakraborty and Saha, 1987). In some families all the members had symptom of dermatosis. Liver was enlarged in 34.5% and ascites present in 5.6% of the patients. The youngest patient was 14 months old. One patient developed skin cancer. Three deaths were reported due to chronic arsenic poisoning. Out of 71 water samples collected from tubewells of the affected villages, in 55 (77.5%) concentration of arsenic was higher than the Indian permissible limit (50 μg/l) of arsenic in drinking water. Mean arsenic concentration in 31 water samples collected from tubewells of the affected families were 640 μg/l and that of the 40 water samples collected from tubewells of the unaffected families were 210 μg/l. The study reported that lowest concentration of arsenic in water resulting dermatosis was found to be 200 μg/l.

During 1988 an epidemiological investigation (Guha Mazumder et al., 1988) from an arsenic affected of Ramnagar village, Baruipur Block, 24-Parganas showed evidence of chronic arsenical dermatosis and hepatomegaly in 62 (92.5%) out of 67 members of families who drank contaminated tubewell water (arsenic level 200–2000 μg/l). In contrast, only 6 (6.25%) out of 96 persons from the same area who drank safe water (arsenic level < 50 μg/l) had non-specific hepatomegaly, while none had skin lesions.

Hepatomegaly occurred in all 13 patients who were studied in detail, although 5 had splenomegaly. Biopsies of liver tissue from these patients revealed various degrees of fibrosis and the expansion of the portal zone that resembled NCPF. Datta reported (Datta, 1976) similar incidents in 1976 from Chandigarh and its surroundings from northern India. In 1991, a report from the steering committee, government of West Bengal (Steering Committee Arsenic Investigation Project, 1991) described the regional geology, geomorphology, geohydrology, some borehole sample analyses and the clinical investigation of about 60 arsenicosis patients in arsenic-affected areas of West Bengal. It was also reported that water in the intermediate aquifer is polluted with arsenic. Neither the shallow (first) nor the deep (third) aquifers have been reported to have arsenic contamination. The sand grains in the arsenic-contaminated aquifer are generally coated with iron and arsenic-rich material.

During October 1994, a committee constituted by the government of West Bengal (Committee Constituted by Government of West Bengal, 1994) reported arsenic contamination in 41 blocks in 6 districts of West Bengal. The committee analysed about 1200 water samples for arsenic and other common water quality parameters from these 6 districts, and the highest arsenic concentration reported was 3200 µg/l. The committee recommended alternative water supplies, an arsenic removal plant, an epidemiological study and a survey to discover the magnitude of contamination.

A report by the School of Environmental Studies was published in December 1994 narrating the severity and magnitude of the problem (Das et al., 1994). In this document, it was reported that 312 villages from 37 blocks/police stations in 6 districts were affected by arsenic groundwater contamination, and from the extrapolation of the data it was predicted that more than 800,000 people are drinking arsenic-contaminated water from these affected districts and about 175,000 people may suffer from arsenical skin lesions. The average ratio of arsenite to arsenate in water samples was 1:1.

The groundwater arsenic contamination problem of West Bengal came to limelight after the international conference held in Calcutta during 1995 (International Conference on Arsenic in Groundwater, 1995) and the opinions of experts about the arsenic contamination of West Bengal (Post Conference Report, 1995). Most of the organizations working on arsenic problem in West Bengal reported their findings in this conference. Reported abstracts on West Bengal were on epidemiology (Das Gupta et al., 1995), pathology (Bhattacharyya et al., 1995), health hazard induced by chronic arsenicosis (Guha Mazumder et al., 1995; Saha, 1995a), geology (Saha and Chakrabarti, 1995), hydrogeology (Sinha Ray, 1995), geochemistry (Das et al., 1995a), removal of arsenic from water (Bagchi and Bagchi, 1995; Das et al., 1995b; Nath et al., 1995), analysis of arsenic in biological samples (Samanta et al., 1995) and watershed management (Das et al., 1995c).

The severity and magnitude of the arsenic groundwater contamination in West Bengal were reported by various arsenic experts who attended the conference, in the Post Conference Report published in May 1995. Chappell (1995) narrated the problem as "The chronic arsenic poisonings occurring in the West Bengal area represent the single largest environmental health problem I know of other than that associated with the Chernobyl disaster". Epidemiologist Smith (1995) writes, "The problems are very serious and warrant a very high priority for solutions and further investigations". Seriousness of the problem and the need for its solution have been highlighted by various other experts (Guha Mazumder, 1995; Hering, 1995; Redekopp, 1995; Saha, 1995b).

In a report, Saha (1995c) reported 1214 cases of chronic arsenical dermatosis from drinking arsenic-contaminated tubewell water in 61 villages of 6 districts of West Bengal during 1983–1987. Cutaneous malignancy (SCC) was detected as complication in 6 cases. Liver histology showed NCPF. The author reported that treatment with BAL is superior to penicillamine. The duration of drinking arsenic-contaminated water for symptoms to develop varied from 6 months to 2 years or more depending on the arsenic concentration in the tubewell water and the period of drinking. In describing the arsenic calamity of West Bengal, it was reported (Pearce, 1995) that hunger was an old enemy of poor villagers in West Bengal. But irrigation schemes brought a new and more insidious killer – arsenic poisoning.

Chatterjee et al. (1995) reported on the analysis of a few thousand water samples from six arsenic-affected districts of West Bengal and the study showed that groundwater contains two arsenic species, arsenate and arsenite in the ratio (approximately) 1:1. The highest arsenic concentration in a hand tubewell sample from Ramnagar village of South 24-Parganas district was 3700 μg/l. Urine samples from affected villages were also analysed for As(III), As(V), MMAA and DMAA. DMAA and MMAA were the predominant species. XRF analysis of solid residue after roto-evaporating 5 l of water from contaminated tubewells showed that high concentrations of Fe and Mn were present in the sample along with arsenic. Das et al. (1995d) reported arsenic in the hair, nail, urine, skin-scale and a few liver tissues (biopsy) of people from arsenic-affected villages who had arsenical skin lesions. Results showed elevated levels of arsenic in biological samples. The liver tissue analysis showed high arsenic, but non-detectable selenium suggesting that selenium deficiency might have a relation to arsenic toxicity.

On the basis of analysis of 20,000 water samples from arsenic-affected areas of West Bengal, Mandal et al. (1996) reported that 7 districts (North 24-Parganas, South 24-Parganas, Nadia, Bardhaman, Murshidabad, Malda, Hooghly) are arsenic-affected. Around 45% of these samples have arsenic concentration above 50 μg/l. The average concentration of arsenic in contaminated water was 200 μg/l. Many people have arsenical skin lesions as diffuse melanosis, spotted melanosis, leucomelanosis, keratosis, hyperkeratosis, dorsum, non-pitting oedema, gangrene, skin cancer. In addition, there were reports of arsenic patients suffering from internal cancers, such as bladder, lung etc. A study also reported (Nag et al., 1996) that arsenic is present in the form of arsenite and arsenate in groundwater with low concentration of antimony 0.03–0.9 μg/l. The first report of chronic neuropathy in arsenic patients from West Bengal was in 1996 (Basu et al., 1996). Eight out of 46 patients having arsenical skin lesions and a high arsenic exposure also demonstrated features of chronic peripheral neuropathy. They had features of sensory ataxia due to posterior column affection (87%), distal sensory features (50%) and two patients showed distal motor affection. Water sources were changed and a follow-up study after 5 years found moderate improvements for sensory features and minimal for motor features.

Bagla and Kaiser (1996) reported the magnitude and severity of the problem and the opinion of international experts on the arsenic calamity of West Bengal – India.

More and more information on the groundwater contamination and suffering of people surfaced with time (Roy Chowdhury et al., 1997; Mandal et al., 1997). Guha Mazumder et al. (1997) reported non-cancer effects of chronic arsenicosis with special reference to liver damage. The same report also discussed the features of peripheral vascular disease among some of the patients. A comparative study of the arsenic calamity of Bangladesh with West Bengal – India was also reported by Dhar et al. (1997). World Health Organization consultants visited some of the arsenic-affected districts of West Bengal from 19th to 30th August 1996 and submitted a report (Consultants' Report to the World Health Organization, 1997). Their recommendations were (1) to assess the extent of contamination throughout the state; (2) to develop a master registry of water quality data; (3) to develop an analytical laboratory infrastructure; (4) to replace contaminated sources with safe sources and encourage surface water sources; (5) to develop a programme to ascertain the extent of health problem; (6) to develop local and regional medical programmes to assist in the diagnosis, screening and suggestive treatment; (7) to promote public and professional education on arsenic related health problem; and (8) to promote epidemiological study. World Health Organization also arranged a meeting with international experts to discuss the arsenic problem of West Bengal – India and Bangladesh (Consultation on Arsenic in Drinking Water, 1997). A recommendation for action (Recommendations for Action, 1997) and report of regional consultation were published (Report of a Regional Consultation, 1997).

Clinical and various laboratory investigations were carried out (Guha Mazumder et al., 1998a) on 156 patients to ascertain the nature and degree of morbidity and mortality that occurred due to chronic arsenic toxicity in some affected villages of West Bengal. All the patients studied had arsenical skin lesions. Other features included weakness, gastro-intestinal symptoms, involvement of the respiratory system and the nervous system. Lung function tests showed restrictive lung disease, abnormal electromyography, enlargement of the liver and portal hypertension. Liver biopsy reports of 45 patients showed NCPF in 41, cirrhosis in 2 and normal histology in 2 cases. A prospective, double blind, randomised placebo-controlled trial was carried out on 11 patients (Guha Mazumder et al., 1998b) to evaluate the efficiency and safety of the chelating agent meso-2,3-dimercaptosuccinic acid (DMSA) for chronic arsenicosis due to drinking arsenic-contaminated (≥ 50 μg/l) subsoil water in West Bengal. The other 10 patients were given placebo capsules. The clinical features were evaluated by an objective scoring system before and after treatment. It was concluded from the study that DMSA was not effective in producing any clinical or bio-chemical benefit or any histopathological improvement of skin lesions in patients with chronic arsenicosis. Guha Mazumder et al. (1998c) also reported high arsenic levels in drinking water and the prevalence of skin lesions in West Bengal. The study demonstrates clear exposure–response relationship between prevalence of skin lesions and both arsenic water levels and dose per body weight, with males showing greater prevalence to both keratosis and hyperpigmentation. Based on limited exposure assessment, some cases appear to be occurring at surprisingly low levels of exposure. There is evidence that the risks were somewhat greater for those who might be malnourished. Subramanian and Kosnett (1998) visited the state of West Bengal in August 1996 as consultants to the World Health Organization and reported their overview of the arsenic contamination problem in West Bengal on the basis of field visits, meeting with Indian and West Bengal government officials as well as scientists, engineers and physicians studying various arsenic problems of West Bengal. They made a recommendation for the development of a comprehensive infrastructure and plan of action. Mandal et al. (1998) reported the input of safe water used for drinking and cooking on five arsenic-affected families (17 members) for 2 years in West Bengal. Eight of them with arsenical skin lesions did not recover completely after 2 years of drinking safe water, indicating a long-lasting damage. The investigation also showed that despite having safe water for drinking and cooking, the study group could not avoid an intake of arsenic time and again through edible herbs grown in contaminated water, food materials contaminated through washing and the occasional drinking of contaminated water. After minimizing the level of contamination, a noteworthy declining trend was observed in urine, hair, nail in all cases but not to that level observed in a normal population. Biswas et al. (1998) reported similar results on the village level.

In the international conference on arsenic in Dhaka, Bangladesh in 1998, Dutta, Saha and Guha Mazumder presented their findings on histopathology (Dutta et al., 1998), diagnosis of arsenicosis (Saha, 1998) and clinical manifestation of chronic arsenic toxicity (Guha Mazumder et al., 1998d) from the arsenic-affected population of West Bengal. Samanta et al. (1999) reported analysis of 47,000 and 9640 water samples for arsenic from 8 arsenic-affected districts of West Bengal and 64 districts of Bangladesh. On the basis of about 30,000 biological sample analyses (urine, hair, nail, skin-scale, blood) finding elevated levels of arsenic even in those who have no arsenical skin lesions, it was concluded that many villagers might be subclinically affected. A comparative study was reported (Mandal et al., 1999) between arsenic-affected villages of West Bengal and Bangladesh. Guha Mazumder et al. (1999a) reported about an epidemiological study in West Bengal and treatment with a chelating agent. It was mentioned that chelating agent DMSA was not found to be superior to a placebo, but drinking arsenic safe water, rest, nutritious diet and symptomatic treatment could reduce nearly 40% of the patients' symptoms significantly. While describing the treatment related to arsenic toxicity by Guha Mazumder et al. (1999b), it was reported that chelating agents, like D-penicillamine, DMSA and 2,3-dimercapto-1-propanesulfonate (DMPS), appear to be the rational mode of therapy for chronic arsenicosis, however, their usefulness is yet to be established. In a bird's eye view on the arsenical calamity in West Bengal, Saha (1999) reported that more than 200,000 people in 1206 rural areas of 76 blocks of 9 districts of West Bengal have been found to be affected with arsenicosis.

While presenting a review of arsenic poisoning and its health effect, Saha (1999) described major dermatological signs using photographs of the patients from arsenic-affected districts of West Bengal. Hepatic manifestation in chronic arsenic toxicity from arsenic-affected villages of West Bengal was described by Santra et al. (1999).

Chowdhury et al. (2000a,b) reported groundwater contamination in 985 villages from 69 police stations/blocks in 9 affected districts of West Bengal on the basis of 58,166 water sample analyses. Thousands of hair, nail and urine samples from people living in arsenic-affected villages were analysed and 77% of the samples on the average contained arsenic above normal/toxic levels. From the affected villages at random 29,035 people had been examined and 15% of those examined had skin lesions. Out of the total 6695 children examined 1.7% had arsenical skin lesions. Arsenical neuropathy was found in 37.2% of 413 arsenicosis patients from a few villages. Electrophysiologic studies on 20 patients showed an affliction of the sensory nerves in nine patients (95%) and an affliction of the motor nerves in four patients (25%). After extrapolation of the water analysis data, screening of villagers for arsenical skin lesions and a detailed study of a block, it is estimated that about 5 million people are drinking arsenic-contaminated water above 50 μg/l and around 300,000 people may have arsenical skin lesions. The total population in 9 arsenic-affected districts of West Bengal is about 43 million. This does not mean that all the individuals are drinking arsenic-contaminated water and will suffer from arsenic toxicity, but undoubtedly they are at risk. Guha Mazumder et al. (2000a) made a cross-sectional survey involving 7683 participants of all ages in some arsenic-affected villages of South 24-Parganas between April 1995 and March 1996. The study reported that arsenic ingestion also causes pulmonary effects. Paul et al. (2000) studied skin biopsies of 42 patients suffering from chronic arsenic toxicity. Histological studies of H/E stained sections showed hyperkeratosis in 13, para keratosis in 13, acanthosis in 12, papillo-matosis in 24, elongation of reteridges in 21, increased basal pigmentation in 27 and dysplastic changes in 8 cases. Squamous cell carcinoma was present in 2, basisquamous in 1 and basal cell carcinoma in 1 case. Changes of skin lesions after drug DMSA and DMPS therapy compared to placebo were studied. The results were inconclusive. Proliferative activity of skin lesions in patients with chronic arsenic toxicity was studied with AgNOR (argyrophillic proteins of the nuclear organiser region) stain to assess the biological behaviour.

Samanta et al. (2000) reported high performance liquid chromatography inductively coupled plasma mass spectrometry for speciation of arsenic compounds in urine from some arsenic-affected villages of West Bengal, This study would relate to recent inorganic arsenic exposure. From this study it was concluded that those living in arsenic-affected villages may use safe water from their tubewell but they cannot avoid, from time to time, arsenic contamination as many water sources in the surrounding areas are arsenic conta-minated. In the international workshop on control of arsenic contamination in groundwater, Calcutta, India the 5th and 6th January 2000, De (2000) described the global arsenic scenario with particular reference to West Bengal; Pandey and Raut (2000) described epidemiological study of arsenic contamination in West Bengal; Saha (2000) on malignancy in arsenicosis with respect to West Bengal and Guha Mazumder et al. (2000b) described clinical features and dose related clinical effect. Chappell (2000) described the future danger in West Bengal and Bangladesh about arsenic in food chain as many crops are being irrigated with tubewell water containing elevated level of arsenic. Saha reported arsenicosis and the spread of arsenicosis in West Bengal (Saha, 2001a) and cutaneous malignancy in arsenicosis highlighting West Bengal problem (Saha, 2001b). Saha and Chakraborti (2001) reported their 17 years experience of arsenicosis in West Bengal.

Mazumder (2001) reported clinical aspects of chronic arsenic toxicity with respect to affected areas of West Bengal. A hospital-based study on arsenic and liver disease (Guha Mazumder, 2001) on 248 patients suffering from chronic arsenic toxicity was reported. NCPF is a predominant lesion in this population. An epidemiological study on various non-carcinomatous manifestation of chronic arsenic toxicity in the population of districts of West Bengal was described (Guha Mazumder et al., 2001a). It was reported that chronic arsenic toxicity in man produces protean non-carcinogenic manifestations such as weakness, liver enlargement, chronic lung disease and peripheral neuropathy. Guha Mazumder et al. (2001b) investigated the clinical use of DMPS in some arsenic patients in a randomised placebo-controlled trial. It was concluded from the study that DMPS treatment caused significant improvement in the clinical score of patients suffering from chronic arsenic toxicity. It was further reported that increased urinary excretion of arsenic during the period of therapy is the possible cause of improvement.

Rahman et al. (2001) reported on the basis of 101,934 hand tubewells, approximately 25,000 biological samples analysis and the screening of 86,000 persons in affected villages of West Bengal that 2600 villages are affected with arsenic in groundwater greater than 50 µg/l and that around 6 million people are drinking arsenic-contaminated water above 50 µg/l. It was also reported that children in the affected districts of West Bengal are at a higher risk from arsenic toxicity. The report also stated that while surveying in the affected villages they have found that a significant percentage of the population suffering from arsenic toxicity also suffer from arsenic neuropathy. A critical discussion was made in this paper on metal chelators for treating chronic arsenic toxicity in West Bengal. In addition to reporting groundwater arsenic contamination and suffering of people in West Bengal, Chowdhury et al. (2001) also reported arsenic in the food chain and on socio-economic studies in the affected villages. Arsenic-affected people are facing serious social problems. Mandal et al. (2001) identified the most toxic species, dimethylarsinous and monomethylarsonous acids, along with other arsenic metabolites in urine from the affected areas of West Bengal. In another report, Srivastava et al. (2001) reported arsenicisim in West Bengal – India with dermal lesions and elevated arsenic levels in hair.

The United Nations Industrial Development Organization (UNIDO, Volume I and II, 2001) made reports under a project on concerted action on elimination/reduction of arsenic in groundwater, West Bengal, India. Both the reports are the compilation of the work done by various organizations and assessment of the merits. The issues discussed in these two reports are (1) geohydrogeological issues, (2) water quality and treatment, (3) technology options, (4) community-based units, (5) arsenic toxicity and its effect in health, (6) arsenic in the food chain, (7) socio-economic issues, (8) cost factors, (9) institutional training needs, (10) privatisation and marketing issues and (11) monitoring and evaluation.

Basu et al. (2002) studied the evaluation of the micronuclei (MN) formation in oral mucosa cells, urothelial cells and peripheral blood lymphocytes in symptomatic individuals exposed to arsenic through drinking water in West Bengal. Exposed individuals showed a statistically significant increase in the frequency of MN in oral mucosa, urothelial cells and lymphocytes when compared with the controls. It was concluded from the study that the symptomatic individuals exposed to arsenic through drinking water in arsenic-affected area of West Bengal have significant cytogenetic damage.

Saha (2002) reported the history of clinical features and treatment situation of arsenic calamity in West Bengal. Rahman et al. (2003) and Chowdhury et al. (2003) reported current arsenic groundwater contamination and its magnitude in West Bengal. A detailed study on the growth of arsenic contamination in West Bengal, the present situation, its future danger and the lessons that have been learned from this calamity was described by Chakraborti et al. (2002). Roy Chowdhury et al. (2002) reported total arsenic in food composites collected from a few arsenic-affected villages in the Murshidabad district, West Bengal where arsenic-contaminated groundwater was used for agricultural irrigation. The report shows average daily dietary intake of arsenic from food stuffs for adults and children are 180 and 96.5 µg, respectively. Public health engineering directorate, government of West Bengal published a report (Strategy and Action Taken, 2002) on arsenic contamination in groundwater of West Bengal. The report stated that 75 blocks are arsenic-affected and the total population in affected villages is 6.97 million. The report is based on the analysis of 12,423 hand tubewells.

The World Health Organization and government of West Bengal (WHO–Government of West Bengal, 2002) arranged a workshop to draft a national protocol on case-definition and management of arsenicosis in arsenic-affected districts of West Bengal.

The proper diagnosis of arsenicosis is an important issue and Saha (2003) reported in the context of arsenic-affected population of West Bengal that non-arsenical melanosis, leucomelanosis and keratosis are to be excluded while diagnosing someone suffering from arsenical dermatosis. Liver enlargement, jaundice, ascitis or cancer of arsenic origin must, have features of arsenical dermatosis like melanosis, keratosis in palm, sole or body. Guha Mazumder (2003) reported his experience on clinical features, epidemiology and treatment in some arsenic patients in West Bengal.

Mukherjee et al. (2003) reported neurological involvement in patients of arsenicosis from different districts of West Bengal. Peripheral neuropathy was the predominant neurological complication in patients. Overall sensory features were more common than motor features in the patients of neuropathy.

Rahman et al. (2003), in order to understand the magnitude and suffering of arsenic problem in West Bengal, studied one of the 9 affected districts, North 24-Parganas, of West Bengal for 7 years. On the basis of 48,030 water samples, 21,000 hair, nail, urine samples analyses and screening 33,000 people in North 24-Parganas, it is estimated that about 2.0 million and 1.0 million people are drinking arsenic-contaminated water above 10 and 50 μg/l, respectively, in North 24-Parganas. The extrapolation of the available data indicated that about 0.1 million people may be suffering from arsenical skin lesions from North 24-Parganas alone. A follow-up study indicated that many of the victims suffering from severe arsenical skin lesions for several years are now suffering from cancer or have already died of cancer.

Figure 1 shows the arsenic-affected districts, blocks/police stations in West Bengal. Table 1 shows the database on arsenic-affected areas of West Bengal up to August 2002.

3.1. *Source of arsenic for groundwater contamination in West Bengal – India*

Early on, tubewell strainers, pesticides, insecticides and other anthropogenic sources were considered as the source of groundwater arsenic contamination in West Bengal (Chakraborty and Saha, 1987). Das et al. (1994) first reported that the calculations on the basis of water withdrawal compared to concentrations of arsenic in water showed that a single deep tubewell of rural water supply scheme supplying water to a few villages in Malda, one of the nine arsenic-affected districts, is withdrawing 147.25 kg of arsenic per year indicating the source is geologic. Borehole sediment analyses show high arsenic concentrations in only a few soil layers and in those layers arsenic is found to be associated with iron pyrites. Das et al. (1996) further confirmed the existence of arsenic-rich pyrite by EPMA, LAMMA and XRD studies in borehole sediment. It was stated that heavy groundwater withdrawal and aeration of the aquifer leads to the decomposition of arsenic-rich pyrite leading to groundwater arsenic contamination. Mallick and Rajagopal (1995) also stated that due to heavy groundwater withdrawal, the underground aquifer is aerated and oxygen causes degradation of arsenic-rich source. Ghosh and De (1995) found borehole sediment of North 24-Parganas was arsenic contaminated. The potential source area of arsenicosis sediments was suggested to be the metamorphic terrains of the Bihar Plateau region and dominantly igneous Himalayan.

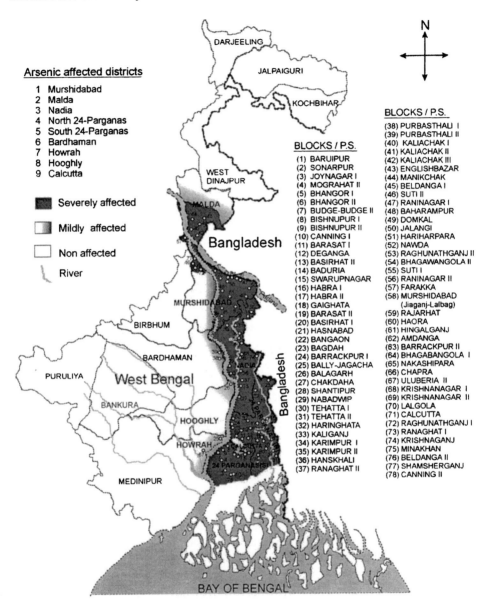

Figure 1. It shows arsenic-affected 9 districts and 78 blocks of West Bengal where groundwater contains arsenic above 50 µg/l.

Bhattacharya et al. (1997, 1998) reported the association of arsenic with hydrated ferric oxide (HFO) and its mobilization to the aquifer due to change of redox conditions during groundwater development. Saha et al. (1997) on the basis of borehole sediment analysis reported that these arsenic-rich sediments were transported from the

Table 1. Present arsenic status in West Bengal, India (up to August 2002).

	West Bengal
Area (km^2)	89,193
Population in million (according to 2001 Census)	80
Total number of districts	18
No. of arsenic-affected districts	9
(groundwater arsenic above 50 μg/l)	
Total number of water samples analysed	115,000
Percentage of samples having arsenic >10 μg/l	50.3
Percentage of samples having arsenic >50 μg/l	25.1
Area of arsenic-affected districts (km^2)	38,865
Population of arsenic-affected districts in million	50
Number of arsenic-affected blocks/police station	78
Number of arsenic-affected villages (approximately)	3,150
where groundwater arsenic above 50 μg/l	
Total hair, nail, urine, skin-scales samples analysed	25,000
Arsenic above normal level (average) in biological samples	89%
People drinking arsenic-contaminated water	6
above 50 μg/l (approximately) in million	
People screened for arsenic patients (preliminary survey)	86,000
Number of registered patients with clinical manifestations	8,500 (9.8%)
Percentage of children having arsenical	1.7
skin lesions of total patients	
People may have arsenical skin lesions[a]	300,000

[a] Extrapolation of the available data on survey, screening, water analysis and biological samples analysis.

Chotanagpur Rajmahal highlands and deposited in sluggish meandering streams under reducing conditions. Ahmed et al. (1998) used various instrumental methods to characterize sediments and suggested arsenic in HFO and its dissolution through reduction. Nickson et al. (1998) also suggested the HFO reduction and mobilization of arsenic from absorbed HFO. Acharyya et al. (1999) suggested that due to the wide use of phosphate fertilizer, desorption of arsenic from sediment took place, and combined microbiological and chemical processes might have increased the natural mobility of arsenic. Roy Chowdhury et al. (1999) reported that arsenic in hand tubewells may be found at any depth, however with depth, concentration of arsenic decreases. Nickson et al. (1999) further suggested the mechanism of arsenic release to groundwater of West Bengal and Bangladesh where it was reported that reduction of Fe is driven by the microbial metabolism of sedimentary organic matter, which is present in a concentration as high as 6%. Acharyya et al. (2000) and Acharyya (2001) made an attempt to explain the arsenic contamination in Bengal basin on the basis of existing data and data which they generated.

Chakraborti et al. (2001) analysed a few thousand borehole sediment samples and characterized arsenic bearing sediments in Gangetic delta.

3.2. Groundwater arsenic contamination in residential area of Behala – Calcutta due to industrial pollution

During July–September 1989, a number of local newspapers in Calcutta and also national papers reported that residents of P.N. Mitra Lane, Behala situated in the southern part of the city of Calcutta, had been suffering from arsenic poisoning, because of drinking arsenic-contaminated ground water. A few died, some of the victims were hospitalised. The symptoms of arsenic poisoning were noticed among many families of the locality. The total number of people using the contaminated water was around seven thousand. The doctors of the local hospitals confirmed that the people were suffering from arsenic poisoning. Pigmentation, i.e. black spots like "rain drops" was noted on the neck, armpits and the trunk of most people using the contaminated water. The familiar arsenical keratosis of the palms and soles, which is a late manifestation of poisoning, was also observed in a large number of people.

A report on chronic arsenicosis from the arsenic-affected area of P.N. Mitra Lane, Calcutta was made by Guha Mazumder et al. (1992). Fifty-three out of 79 members (67%) of 17 families were suffering from chronic arsenicosis. Clinical investigation of these 20 affected persons showed typical skin pigmentation as well as palmar and plantar keratoses while gastro-intestinal symptoms, anaemia and signs of liver disease and peripheral neuropathy were seen in many. However, the source remained unidentified. A detailed study report was made (Chatterjee et al., 1993) on the source of arsenic and the magnitude of the contamination. No arsenic contamination in groundwater was noticed except at the area close to P.N. Mitra Lane of about 2 km diameter. Thus the source in P.N. Mitra Lane seemed to be due to man's activity. The primary suspect for this arsenic episode is the chemical factory located on B.L. Saha Road, since P.N. Mitra Lane happens to be at the back of this factory. This chemical factory was producing several chemical compounds including the insecticide paris-green (acetocopper arsenite). For about twenty years this factory had been producing about 20 tonnes of paris-green per year. Paris-green has been known for almost one hundred years. It is mainly used as insecticide/pesticide, and is one of the most toxic among the arsenic compounds with an LD (mg/kg) to man of 22. Paris-green, $Cu(CH_3COO)_2,3Cu(AsO_2)_2$, itself is not soluble in water but when spread over soil it may undergo degradation with the formation of water-soluble arsenic compounds. Although paris-green was once widely used in agriculture, it is of minor importance today. Its manufacture is restricted in many countries due to its high toxicity, and special permission is needed for its manufacture. The contamination is due to the cumulative effect of the last 20 years dumping of arsenic waste. Although factory authorities say that before dumping they had been removing arsenic, the removal of arsenic from the effluent was not adequate.

Analysis of soil surrounding the dumping ground of the waste showed a very high concentration of arsenic (as high as 10,000 μg/gm). Soil also showed high concentrations of copper and chromium. Nineteen of the hand tubewells people were using for drinking and cooking showed very high concentrations of arsenic. The highest concentration of arsenic recorded in a hand tubewell was 38,000 μg/l and except for 3 samples, all samples contained arsenic above 100 μg/l. Arsenic concentrations in hand tubewells decreased with increasing distance from the dumping ground. A follow-up study in the affected areas was made 8 years after by Chakraborti et al. (1998). Total arsenic concentrations in hand tubewells decreased only 10–15% from what was observed 8 years before. More and more new tubewells, even at longer distances, got arsenic contamination. However, local people are no longer using tubewell water for drinking and cooking. During the last survey in March 1997, 144 people were identified with arsenical skin lesions.

During the last 8 years in the arsenic-affected area of P.N. Mitra lane, 8 people died who had arsenical skin lesions and of those 3 from internal cancer. The study report made after examining 144 people during March 1997 showed (1) patients were feeling weak, (2) acute bronchial problem, (3) liver disorders, (4) those who had diffuse melanosis almost regained their colour after drinking safe water, (5) those who had only light spotted melanosis now apparently do not have such pigmentation, (6) but those who had moderate to heavy spotted melanosis are still showing the symptom with leucomelanosis (spotted pigmentation and depigmentation) and (7) usually patients with moderate to severe keratosis on palm and soles complained their skin lesions were more or less unaltered. In some cases, it was lessened but still persisting.

4. Arsenic contamination in groundwater of Chattisgarh state in India

The state Chattisgarh was within the state of Madhya Pradesh (MP), two years before. Groundwater arsenic contamination from a few villages of Rajnandgaon district of Chattisgarh was reported by Chakraborti et al. (1999). The source of arsenic in groundwater is natural and geologic both for the alluvia Bengal Basin and rocky belt of Dongargarh-Kotri zone of Rajnandgaon district. The total population of the district is 1.5 million. Except for two towns, Rajnandgaon and Khairagarh, the entire district depends on tubewells and dugwells. Water samples ($n = 146$) were collected from 22 villages of Chowki block in Rajnandgaon district. In 8 villages, arsenic in groundwater was found above 10 μg/l and in 4 villages to be above 50 μg/l with the highest concentration being 880 μg/l. The difference between arsenic contamination in West Bengal and Rajnandgaon district is dugwells in West Bengal are not arsenic contaminated but in Rajnandgaon some dugwells are arsenic contaminated and the highest concentration reported from a dugwell was 520 μg/l. Arsenical skin lesions were reported for 42 adults out of 150 examined and 9 children out of 58. About 75% of people have arsenic in hair above toxic level out of 150 hair samples examined. Neurological studies were performed on 61 persons and 34% have shown a positive indication. Pandey et al. (1999) also reported groundwater arsenic contamination from Rajnandgaon district of Chattisgarh. Out of the 390 samples analysed, 26 sites were found to be contaminated with arsenic with the highest concentration being 1010 μg/l. It was further reported that 130 people are critically affected by arsenic poisoning from a single village. The number of people at risk was 10,000. The source and the mobilization process of arsenic from affected areas of Rajnandgaon district Chattisgarh was reported by Acharyya (2002).

5. Groundwater arsenic contamination in Bihar

The School of Environmental Studies, Jadavpur University first discovered the arsenic groundwater contamination in Semria Ojha Patty village of Sahapur Police station in

Bhojpur district of Bihar during June 2002 (Chakraborti et al., 2003). The total population of Semria Ojha Patty village was about 5000 and only 18.4% were drinking safe water ($<$ 10 μg/l), 56.8% were drinking above 50 μg/l and 19.9% above 300 μg/l. The highest concentration of arsenic recorded was 1654 μg/l. Out of 550 people examined for arsenical skin lesions, 60 individuals (10.9%) were registered with arsenical skin lesions. Hair, nail, urine analyses show that about 75% individuals have elevated level of arsenic in biological samples. Thus, many are subclinically affected. Neurological involvements in patients and pregnancy outcome in Semria Ojha Patty village were recorded. Many children have arsenical skin lesions.

Figure 2 shows the arsenic-affected areas in India.

Figure 2. It shows arsenic-affected areas in India.

References

Acharyya, S.K., 2001. Arsenic pollution in groundwater from lower Ganga Plains, Bengal Basin. Indian J. Geol., 73 (1), 1–19.

Acharyya, S.K., 2002. Arsenic contamination in groundwater affecting major parts of southern West Bengal and parts of western Chattisgarh: source and mobilization process. Curr. Sci., 82 (6), 740–744.

Acharyya, S.K., Chakraborty, P., Lahiri, S., Raymahashay, B.C., Guha, S., Bhowmik, A., 1999. Arsenic poisoning in the Ganges delta. Nature, 401, 545.

Acharyya, S.K., Lahiri, S., Raymahashay, B.C., Bhowmik, A., 2000. Arsenic toxicity of groundwater in parts of the Bengal basin in India and Bangladesh: the role of quaternary stratigraphy and holocene sea-level fluctuation. Environ. Geol., 39 (10), 1127–1137.

Ahmed, K.M., Imam, M.B., Akhter, S.H., Hasan, M.A., Alam, M.M., Chowdhury, S.Q., Burgess, W.G., Nickson, R., McArthur, J.J., Hasan, M.K., Ravenscroft, P., Rahman, M., 1998. Mechanism of arsenic release to groundwater: geochemical and mineralogical evidence. International Conference on Arsenic Pollution of Groundwater in Bangladesh: Causes, Effects and Remedies, Dhaka, Bangladesh, Abstr., pp. 125–126.

Bagchi, R.N., Bagchi, S., 1995. Removal of arsenic from the tubewell water for drinking purpose. International Conference on Arsenic in Groundwater: Cause, Effect and Remedy, School of Environmental Studies, Jadavpur University, Calcutta, India, Abstr., pp. 55–56.

Bagla, P., Kaiser, J., 1996. India's spreading health crisis draws global arsenic experts. Science, 274, 174–175.

Basu, D., Dasgupta, J., Mukherjee, A., Guha Mazumder, D.N., 1996. Chronic neuropathy due to arsenic intoxication from geo-chemical source – a 5 year follow-up. JANEI, 1 (1), 45–48.

Basu, A., Mahata, J., Roy, A.K., Sarkar, J.N., Poddar, G., Nandy, A.K., Sarkar, P.K., Dutta, P.K., Banerjee, A., Das, M., Ray, K., Roychaudhury, S., Natarajan, A.T., Nilsson, R., Giri, A.K., 2002. Enhanced frequency of micronuclei in individuals exposed to arsenic through drinking water in West Bengal, India. Mutat. Res., 516, 29–40.

Berg, M., Tran, H.C., Nguyen, T.C., Pham, M.V., Schertenleib, R., Giger, W., 2001. Arsenic contamination of groundwater and drinking water in Vietnam: a human health threat. Environ. Sci. Tech., 35 (13), 2621–2626.

Bhattacharyya, P.K., Poddar, G., Dutta, S.K., 1995. Histological changes in skin with histochemical demonstration of intracellular arsenic crystals in chronic arsenic toxicity from drinking water. International Conference on Arsenic in Groundwater: Cause, Effect and Remedy, School of Environmental Studies, Jadavpur University, Calcutta, India, Abstr., pp. 25–26.

Bhattacharya, P., Chatterjee, D., Jacks, G., 1997. Occurrence of arsenic-contaminated groundwater in alluvial aquifers from delta plains, eastern India: options for safe drinking water supply. Int. J. Water Res. Dev., 13, 79–92.

Bhattacharya, P., Larsson, M., Leiss, A., Jacks, G., Sracek, A., Chatterjee, D., 1998. Genesis of arseniferous groundwater in the alluvial aquifers of Bengal delta plains and strategies for low-cost remediation. International Conference on Arsenic Pollution of Ground Water in Bangladesh: Causes, Effects and Remedies, Dhaka, Bangladesh, Abstr., pp. 120–123.

Biswas, B.K., Dhar, R.K., Samanta, G., Mandal, B.K., Chakraborti, D., Faruk, I., Islam, K.S., Chowdhury, M.M., Islam, A., Roy, S., 1998. Detailed study report of Samta, one of the arsenic-affected villages of Jessore district, Bangladesh. Curr. Sci., 74 (2), 134–145.

Chakraborty, A.K., Saha, K.C., 1987. Arsenical dermatosis from tubewell water in West Bengal. Ind. J. Med. Res., 85, 326–334.

Chakraborti, D., Samanta, G., Mandal, B.K., Roy Chowdhury, T., Chanda, C.R., Biswas, B.K., Dhar, R.K., Basu, G.K., Saha, K.C., 1998. Calcutta's industrial pollution: groundwater arsenic contamination in a residential area and sufferings of people due to industrial effluent discharge – an eight-year study report. Curr. Sci., 74 (4), 346–354.

Chakraborti, D., Biswas, B.K., Roy Chowdhury, T., Basu, G.K., Mandal, B.K., Chowdhury, U.K., Mukherjee, S.C., Gupta, J.P., Chowdhury, S.R., Rathore, K.C., 1999. Arsenic groundwater contamination and sufferings of people in Rajnandgaon district, Madhya Pradesh, India. Curr. Sci., 77 (4), 502–504.

Chakraborti, D., Basu, G.K., Biswas, B.K., Chowdhury, U.K., Rahman, M.M., Paul, K., Roy Chowdhury, T., Chanda, C.R., Lodh, D., Ray, S.L., 2001. Characterization of arsenic-bearing sediments in the Gangetic delta of West Bengal, India. In: Chappell, W.R., Abernathy, C.O., Calderon, R.L. (Eds), Arsenic Exposure and Health Effects IV. Elsevier, New York, pp. 27–53.

Chakraborti, D., Rahman, M.M., Paul, K., Chowdhury, U.K., Sengupta, M.K., Lodh, D., Chanda, C.R., Saha, K.C., Mukherjee, S.C., 2002. Arsenic calamity in the Indian subcontinent what lessons have been learned? Talanta, 58, 3–22.

Chakraborti, D., Mukherjee, S.C., Pati, S., Sengupta, M.K., Rahman, M.M., Chowdhury, U.K., Lodh, D., Chanda, C.R., Chakraborti, A.K., Basu, G.K., 2003. Arsenic groundwater contamination in Middle Ganga Plain, Bihar, India: A future danger. Environ. Health Perspect., 111(9), 1194–1201.

Chappell, W.R., 1995. Impressions of the arsenic situation in West Bengal. Post Conference Report; Experts' Opinions, Recommendations and Future Planning for Groundwater Problem of West Bengal, School of Environmental Studies, Jadavpur University, Calcutta, India, pp. 31–33.

Chappell, W.R., 2000. Concerns regarding elevated arsenic concentrations in irrigation water. International Workshop on Control of Arsenic Contamination in Groundwater, Public Health Engineering Department, Government of West Bengal, Calcutta, India, Abstr., p. 58.

Chatterjee, A., Das, D., Chakraborti, D., 1993. A study of ground water contamination by arsenic in the residential area of Behala, Calcutta due to industrial pollution. Environ. Pollut., 80, 57–65.

Chatterjee, A., Das, D., Mandal, B.K., Roy Chowdhury, T., Samanta, G., Chakraborti, D., 1995. Arsenic in ground water in six districts of West Bengal, India: the biggest arsenic calamity in the world, part I. Arsenic species in drinking water and urine of the affected people. Analyst, 120, 643–650.

Chowdhury, U.K., Biswas, B.K., Roy Chowdhury, T., Samanta, G., Mandal, B.K., Basu, G.K., Chanda, C.R., Lodh, D., Saha, K.C., Mukherjee, S.K., Roy, S., Kabir, S., Quamruzzaman, Q., Chakraborti, D., 2000a. Groundwater arsenic contamination in Bangladesh and West Bengal, India. Environ. Health Perspect., 108 (5), 393–396.

Chowdhury, U.K., Biswas, B.K., Roy Chowdhury, T., Mandal, B.K., Samanta, G., Basu, G.K., Chanda, C.R., Lodh, D., Saha, K.C., Chakraborti, D., Mukherjee, S.C., Roy, S., Kabir, S., Quamruzzaman, Q., 2000b. Arsenic groundwater contamination and sufferings of people in West Bengal – India and Bangladesh. In: Roussel, A.M., Anderson, R.A., Favrier, A.E. (Eds), Trace Elements in Man and Animals 10. Kluwer Academic/Plenum Publishers, New York, pp. 645–650.

Chowdhury, U.K., Rahman, M.M., Mondal, B.K., Paul, K., Lodh, D., Biswas, B.K., Basu, G.K., Chanda, C.R., Saha, K.C., Mukherjee, S.C., Roy, S., Das, R., Kaies, I., Barua, A.K., Palit, S.K., Quamruzzaman, Q., Chakraborti, D., 2001. Groundwater arsenic contamination and human suffering in West Bengal, India and Bangladesh. Environ. Sci., 8 (5), 393–415.

Chowdhury, U.K., Rahman, M.M., Samanta, G., Biswas, B.K., Basu, G.K., Chanda, C.R., Saha, K.C., Lodh, D., Roy, S., Quamruzzaman, Q., Chakraborti, D., 2003. Groundwater arsenic contamination in West Bengal – India and Bangladesh: Case study on bioavailability of geogenic arsenic. In: Naidu, R., Gupta, V.V.S.R., Rogers, S., Kookana, R.S., Bolan, N.S., Adriano, D. (Eds), Bioavailability, Toxicity and Risk Relationships in Ecosystems. Science Publishers Inc., Enfield, USA, 291–329.

Committee Constituted by Government of West Bengal, 1994. Arsenic pollution in groundwater in West Bengal, committee constituted by Government of West Bengal, PHE Department, PHE 1/716/3D-1/88.

Consultants' Report to the World Health Organization, 1997. Arsenic in groundwater in West Bengal, India: past, present and future. Subramanian, K.S. and Kosnett, M.J.: Consultants' Report to the World Health Organization.

Consultation on arsenic in drinking water and resulting arsenic toxicity in India and Bangladesh, World Health Organization, New Delhi, 29 Apr–1 May, 1997.

Das, D., Chatterjee, A., Samanta, G., Mandal, B., Roy Chowdhury, T., Samanta, G., Chowdhury, P.P., Chanda, C., Basu, G., Lodh, D., Nandi, S., Chakraborty, T., Mandal, S., Bhattacharya, S.M., Chakraborti, D., 1994. Arsenic contamination in groundwater in six districts of West Bengal, India: the biggest arsenic calamity in the world. Analyst, 119, 168N–170N.

Das, D., Basu, G., Roy Chowdhury, T., Chakraborti, D., 1995a. Borehole soil-sediment analysis of some arsenic-affected areas. International Conference on Arsenic in Groundwater: Cause, Effect and Remedy, School of Environmental Studies, Jadavpur University, Calcutta, India, Abstr., pp. 44–45.

Das, D., Chatterjee, A., Samanta, G., Chakraborti, D., 1995b. A simple household devise to remove arsenic from groundwater hence making it suitable for drinking and cooking. International Conference on Arsenic in Groundwater: Cause, Effect and Remedy, School of Environmental Studies, Jadavpur University, Calcutta, India, Abstr., pp. 59–61.

Das, D., Chakraborti, D., Santra, S.C., 1995c. Alternative sources other than groundwater in six arsenic-affected districts of West Bengal. International Conference on Arsenic in Groundwater: Cause, Effect and Remedy, School of Environmental Studies, Jadavpur University, Calcutta, India, Abstr., pp. 74–75.

Das, D., Chatterjee, A., Mandal, B.K., Samanta, G., Chakraborti, D., 1995d. Arsenic in groundwater in six districts of West Bengal, India: the biggest arsenic calamity in the world, part 2. Arsenic concentration in drinking water, hair, nails, urine, skin-scale and liver tissue (biopsy) of the affected people. Analyst, 120, 917–924.

Das, D., Samanta, G., Mandal, B.K., Roy Chowdhury, T., Chanda, C.R., Chowdhury, P.P., Basu, G.K., Chakraborti, D., 1996. Arsenic in groundwater in six districts of West Bengal, India. Environ. Geochem. Health, 18, 5–15.

Das Gupta, J., Roy, B., Agarwal, P., Santra, A., Ghose, A., Guha Mazumder, D.N., Samanta, G., Mondal, B., Roy Chowdhuri, T., Chakraborti, D., 1995. Epidemiological study of arsenism in 24-Parganas (South) one of the arsenic-affected six districts. International Conference on Arsenic in Groundwater: Cause, Effect and Remedy, School of Environmental Studies, Jadavpur University, Calcutta, India, Abstr., pp. 21–24.

Datta, D.V., 1976. Arsenic and non-cirrhotic portal hypertension. Lancet, 433.

Datta, D.V., Kaul, M.K., 1976. Arsenic content of tubewell water in villages in northern India. A concept of arsenicosis. J. Assoc. Phys. Ind., 24, 599–604.

De, B.K., 2000. Chronic environmental arsenic toxicity – global scenario with particular reference to West Bengal. International Workshop on Control of Arsenic Contamination in Ground Water, Public Health Engineering Department, Government of West Bengal, Calcutta, India, Abstr., p. 43.

Dhar, R.K., Biswas, B.K., Samanta, G., Mandal, B.K., Chakraborti, D., Roy, S., Jafar, A., Islam, A., Ara, G., Kabir, S., Khan, A.W., Ahmed, S.A., Hadi, S.A., 1997. Groundwater arsenic calamity in Bangladesh. Curr. Sci., 73 (1), 48–59.

Dutta, S.K., Chatterjee, A., Guha Mazumder, D.N., Santra, A., Bhattacharyya, P.K., Banerjee, D., 1998. Histopathology of skin lesions in chronic arsenic toxicity – an approach to grading of the changes and study of proliferative markers. International Conference on Arsenic Pollution of Ground Water in Bangladesh: Causes, Effects and Remedies, Dhaka, Bangladesh, pp. 45–47.

ESCAP–UNICEF–WHO Expert Group Meeting, 2001. Economic and social commission for Asia and the Pacific; geology and health: solving the arsenic crisis in the Asia Pacific Region, Bangkok, May 2–4.

Garai, R., Chakraborty, A.K., Dey, S.B., Saha, K.C., 1984. Chronic arsenic poisoning from tubewell water. J. Ind. Med. Assoc., 82 (1), 34–35.

Ghosh, S., De, S., 1995. Source of the arsenious sediments at Kachua and Itina, Habra Block, North 24-Parganas, West Bengal – a case study. Indian J. Earth Sci., 22 (4), 183–189.

Guha Mazumder, D.N., 1995. Health hazard due to chronic arsenicosis in West Bengal: an action plan for its assessment and remedy. Post Conference Report; Experts' Opinions, Recommendations and Future Planning for Groundwater Problem of West Bengal, School of Environmental Studies, Jadavpur University, Calcutta, India, pp. 51–54.

Guha Mazumder, D.N., 2001. Arsenic and liver disease. J. Indian Med. Assoc., 99 (6), 311–320.

Guha Mazumder, D.N., 2003. Chronic arsenic toxicity: clinical features, epidemiology and treatment: experience in West Bengal. Environ. Sci. Health, Part A, A38 (1), 141–164.

Guha Mazumder, D.N., Chakraborty, A.K., Ghose, A., Gupta, J.D., Chakraborty, D.P., Dey, S.B., Chattopadhyay, N., 1988. Chronic arsenic toxicity from drinking tubewell water in rural West Bengal. Bull. WHO, 66 (4), 499–506.

Guha Mazumder, D.N., Das Gupta, J., Chakraborty, A.K., Chatterjee, A., Das, D., Chakraborti, D., 1992. Environmental pollution and chronic arsenicosis in south Calcutta. Bull. WHO, 70 (4), 481–485.

Guha Mazumder, D.N., Das Gupta, J., Santra, A., Pal, A., Ghose, A., Chattopadhyay, N., Chakraborti, D., 1995. Chronic arsenicosis in West Bengal: a study on health hazard. International Conference on Arsenic in Groundwater: Cause, Effect and Remedy, School of Environmental Studies, Jadavpur University, Calcutta, India, pp. 27–30.

Guha Mazumder, D.N., Das Gupta, J., Santra, A., Pal, A., Ghose, A., Sarkar, S., Chattopadhaya, N., Chakraborti, D., 1997. Non-cancer effects of chronic arsenicosis with special reference to liver damage. In: Abernathy, C.O., Calderon, R.L., Chappell, W.R. (Eds), Arsenic Exposure and Health Effects (SE1 8HN). Chapman and Hall, London, pp. 112–123.

Guha Mazumder, D.N., Das Gupta, J., Santra, A., Pal, A., Ghose, A., Sarkar, S., 1998a. Chronic arsenic toxicity in West Bengal – the worst calamity in the world. J. Ind. Med. Assoc., 96 (1), 4–7 & 18.

Guha Mazumder, D.N., Ghoshal, U.C., Saha, J., Santra, A., De, B.K., Chatterjee, A., Dutta, S., Angle, C.R., Centeno, J.A., 1998b. Randomized placebo-controlled trial of 2,3-dimercaptosuccinic acid in therapy of chronic arsenicosis due to drinking arsenic-contaminated subsoil water. Clin. Toxicol., 36 (7), 683–690.

Guha Mazumder, D.N., Haque, R., Ghosh, N., De, B.K., Santra, A., Chakraborty, D., Smith, A.H., 1998c. Arsenic levels in drinking water and the prevalence of skin lesions in West Bengal, India. Int. J. Epidemiol., 27, 871–877.

Guha Mazumder, D.N., De, B.K., Santra, A., Das Gupta, J., Ghosh, A., Pal, A., Roy, B., Pal, S., Saha, J., 1998d. Clinical manifestations of chronic arsenic toxicity: it's natural history and therapy: experience of study in West Bengal, India. International Conference on Arsenic Pollution of Ground Water in Bangladesh: Causes, Effects and Remedies, Dhaka, Bangladesh, pp. 91–93.

Guha Mazumder, D.N., De, B.K., Santra, A., Dasgupta, J., Ghosh, N., Roy, B.K., Ghoshal, U.C., Saha, J., Chatterjee, A., Dutta, S., Haque, R., Smith, A.H., Chakraborty, D., Angle, C.R., Centeno, J.A., 1999a. Chronic arsenic toxicity: epidemiology, natural history and treatment. In: Chappell, W.R., Abernathy, C.O., Calderaon, R.L. (Eds), Arsenic Exposure and Health Effects. Elsevier Science, UK, pp. 335–347.

Guha Mazumder, D.N., De, B.K., Das Gupta, J., Santra, A., Roy, B., Ghose, M., 1999b. Chronic arsenic toxicity: carcinogenesis and treatment. Med. Update, 9 (1), 463–471.

Guha Mazumder, D.N., Haque, R., Ghosh, N., De, B.K., Santra, A., Chakraborti, D., Smith, A.H., 2000a. Arsenic in drinking water and the prevalence of respiratory effects in West Bengal, India. Int. J. Epidemiol., 29, 1047–1052.

Guha Mazumder, D.N., De, B.K., Santra, A., Ghosh, N., Roy, B., Haque, R., Smith, A.H., Chakraborty, D., 2000b. Chronic arsenic toxicity: clinical features, dose related clinical effect and its natural history. International Workshop on Control of Arsenic Contamination in Ground Water, Public Health Engineering Department, Government of West Bengal, Calcutta, India, Abstr., pp. 40–41.

Guha Mazumder, D.N., Ghosh, N., De, B.K., Santra, A., Das, S., Lahiri, S., Haque, R., Smith, A.H., Chakraborti, D., 2001a. Epidemiological study on various non-carcinomatous manifestations of chronic arsenic toxicity in a district of West Bengal. In: Chappell, W.R., Abernathy, C.O., Calderon, R.L. (Eds), Arsenic Exposure and Health Effects IV. Elsevier, New York, pp. 153–164.

Guha Mazumder, D.N., De, B.K., Santra, A., Ghosh, N., Das, S., Lahiri, S., Das, T., 2001b. Randomized placebo-controlled trial of 2,3-dimercapto-1-propanesulfonate (DMPS) in therapy of chronic arsenicosis due to drinking arsenic-contaminated water. Clin. Toxicol., 39 (7), 665–674.

Hering, J.G., 1995. Concluding remarks and recommendations. Post Conference Report; Experts' Opinions, Recommendations and Future Planning for Groundwater Problem of West Bengal, School of Environmental Studies, Jadavpur University, Calcutta, India, pp. 55–56.

International Conference on Arsenic in Groundwater: Cause, Effect and Remedy, 6–8 February, 1995, School of Environmental Studies, Jadavpur University, Calcutta, India.

Mallick, S., Rajagopal, N.R., 1995. The mischief of oxygen on groundwater. Post Conference Report: Experts' Opinions, Recommendations and Future Planning for Groundwater Problem of West Bengal, School of Environmental Studies, Jadavpur University, Calcutta, India, pp. 71–73.

Mandal, B.K., Roy Chowdhury, T., Samanta, G., Basu, G.K., Chowdhury, P.P., Chanda, C.R., Lodh, D., Karan, N.K., Dhar, R.K., Tamili, D.K., Das, D., Saha, K.C., Chakraborti, D., 1996. Arsenic in groundwater in seven districts of West Bengal, India – the biggest arsenic calamity in the world. Curr. Sci., 70 (11), 976–986.

Mandal, B.K., Roy Chowdhury, T., Samanta, G., Basu, G.K., Chowdhury, P.P., Chanda, C.R., Lodh, D., Karan, N.K., Dhar, R.K., Tamili, D.K., Das, D., Saha, K.C., Chakraborti, D., 1997. Chronic arsenic toxicity in West Bengal. Curr. Sci., 72 (2), 114–117.

Mandal, B.K., Roy Chowdhury, T., Samanta, G., Mukherjee, D.P., Chanda, C.R., Saha, K.C., Chakraborti, D., 1998. Impact of safe water for drinking and cooking on five arsenic-affected families for 2 years in West Bengal, India. Sci. Total Environ., 218, 185–201.

Mandal, B.K., Biswas, B.K., Dhar, R.K., Roy Chowdhury, T., Samanta, G., Basu, G.K., Chanda, C.R., Saha, K.C., Chakraborti, D., Kabir, S., Roy, S., 1999. Groundwater arsenic contamination and sufferings of people in West Bengal, India and Bangladesh. In: Sarkar, B. (Ed.), Metals and Genetics. Kluwer Academic/Plenum Publishers, New York, pp. 41–65.

Mandal, B.K., Ogra, Y., Suzuki, K.T., 2001. Identification of dimethylarsinous and monomethylarsonous acids in human urine of the arsenic-affected areas in West Bengal, India. Chem. Res. Toxicol., 14, 371–378.

Mazumder, D.N.G., 2001. Clinical aspects of chronic arsenic toxicity. JAPI, 49, 650–655.

Mukherjee, S.C., Rahman, M.M., Chowdhury, U.K., Sengupta, M.K., Lodh, D., Chanda, C.R., Saha, K.C., Chakraborti, D., 2003. Neuropathy in arsenic toxicity from groundwater arsenic contamination in West Bengal, India. Environ. Sci. Health, Part A, A38 (1), 165–184.

Nag, J.K., Balaram, V., Rubio, R., Alberti, J., Das, A.K., 1996. Inorganic arsenic species in groundwater: a case study from Purbasthali (Burdwan), India. J. Trace Elem. Med. Biol., 10, 20–24.

Nath, K.J., Majumder, A., Sinha, R.K., Chakravarty, I., 1995. Arsenic contamination in ground water and its removal. International Conference on Arsenic in Groundwater: Cause, Effect and Remedy, School of Environmental Studies, Jadavpur University, Calcutta, India, Abstr., pp. 50–51.

Nickson, R., McArthur, J., Burgess, W., Ahmed, K.M., Ravenscroft, P., Rahman, M., 1998. Arsenic poisoning of Bangladesh groundwater. Nature, 395, 338.

Nickson, R.T., McArthur, J.M., Ravenscroft, P., Burgess, W.G., Ahmed, K.M., 1999. Mechanism of arsenic release to groundwater, Bangladesh and West Bengal. Appl. Geochem., 15 (4), 1–11.

Pandey, G.K., Raut, D.K., 2000. Epidemiological study of arsenic contamination in West Bengal. International Workshop on Control of Arsenic Contamination in Ground Water Calcutta, India, Public Health Engineering Department, Government of West Bengal, Calcutta, India, Abstr., p. 42.

Pandey, P.K., Khare, R.N., Sharma, R., Sar, S.K., Pandey, M., Binayake, P., 1999. Arsenicosis and deteriorating groundwater quality: unfolding crisis in central-east Indian region. Curr. Sci., 77 (5), 686–693.

Paul, P.C., Chattopadhyay, A., Dutta, S.K., Guha Mazumder, D.N., Santra, A., 2000. Histopathology of skin lesions in chronic arsenic toxicity – grading of changes and study of proliferative markers. Indian J. Pathol. Microbiol., 43 (3), 257–264.

Pearce, F., 1995. Death and the devil's water. New Sci., 14–15.

Post Conference Report, 1995. Experts' Opinions, Recommendations and Future Planning for Groundwater Problem of West Bengal, Post Conference Report, School of Environmental Studies, Jadavpur University, Calcutta, India.

Rahman, M.M., Chowdhury, U.K., Mukherjee, S.C., Mondal, B.K., Paul, K., Lodh, D., Biswas, B.K., Chanda, C.R., Basu, G.K., Saha, K.C., Roy, S., Das, R., Palit, S.K., Quamruzzaman, Q., Chakraborti, D., 2001. Chronic arsenic toxicity in Bangladesh and West Bengal, India – a review and commentary. Clin. Toxicol., 39 (7), 683–700.

Rahman, M.M., Paul, K., Chowdhury, U.K., Quamruzzaman, Q., Chakraborti, D., 2003. Groundwater arsenic contamination. The Encyclopedia of Water Science, 324–329.

Rahman, M.M., Mandal, B.K., Roy Chowdhury, T., Sengupta, M.K., Chowdhury, U.K., Lodh, D., Chanda, C.R., Basu, G.K., Mukherjee, S.C., Saha, K.C., Chakraborti, D., 2003. Arsenic groundwater contamination and sufferings of people in North 24-Parganas, one of the nine arsenic-affected districts of West Bengal, India: the seven years study report. Environ. Sci. Health, Part A, A38 (1), 25–59.

Recommendations for Action, 1997. Arsenic in drinking water and resulting arsenic toxicity in India and Bangladesh, Recommendations for Action, World Health Organization, New Delhi, 29 April–1 May.

Redekopp, A., 1995. Post Conference Report; Experts' Opinions, Recommendations and Future Planning for Groundwater Problem of West Bengal, School of Environmental Studies, Jadavpur University, Calcutta, India, p. 89.

Report of a Regional Consultation, 1997. Arsenic in drinking water and resulting arsenic toxicity in India and Bangladesh, Report of a Regional Consultation, World Health Organization, New Delhi, 29 April–1 May.

Roy Chowdhury, T., Mandal, B.K., Samanta, G., Basu, G.K., Chowdhury, P.P., Chanda, C.R., Karan, N.K., Lodh, D., Dhar, R.K., Das, D., Saha, K.C., Chakraborti, D., 1997. Arsenic in groundwater in six districts of West Bengal, India: the biggest arsenic calamity in the world: the status report up to August, 1995. In: Abernathy, C.O., Calderon, R.L., Chappell, W.R. (Eds), Arsenic Exposure and Health Effects (SE1 8HN). Chapman and Hall, London, pp. 93–111.

Roy Chowdhury, T., Basu, G.K., Mandal, B.K., Biswas, B.K., Samanta, G., Chowdhury, U.K., Chanda, C.R., Lodh, D., Roy, S.L., Saha, K.C., Roy, S., Kabir, S., Quamruzzaman, Q., Chakraborti, D., 1999. Arsenic poisoning in the Ganges delta. Nature, 401, 545–546.

Roy Chowdhury, T., Uchino, T., Tokunaga, H., Ando, M., 2002. Survey of arsenic in food composites from an arsenic-affected area of West Bengal, India. Food Chem. Toxicol., 40, 1611–1621.

Saha, K.C., 1984. Melanokeratosis from arsenic contaminated tubewell water. Indian J. Dermatol., 29 (4), 37–46.

Saha, K.C., 1995a. Chronic arsenical dermatoses from tubewell water in West Bengal (India) in eighties. International Conference on Arsenic in Groundwater: Cause, Effect and Remedy, School of Environmental Studies, Jadavpur University, Calcutta, India, Abstr., p. 36.

Saha, K.C., 1995b. Post Conference Report; Experts' Opinions, Recommendations and Future Planning for Groundwater Problem of West Bengal, School of Environmental Studies, Jadavpur University, Calcutta, India, pp. 93–95.

Saha, K.C., 1995c. Chronic arsenical dermatoses from tube-well water in West Bengal during 1983–87. Indian J. Dermatol., 40 (1), 1–12.

Saha, K.C., 1998. Diagnosis of arsenicosis. International Conference on Arsenic Pollution of Ground Water in Bangladesh: Causes, Effects and Remedies, Dhaka, Bangladesh, pp. 94–96.

Saha, K.C., 1999. A birds eye view on arsenical calamity in West Bengal. Indian J. Dermatol., 44 (3), 116–118.

Saha, K.C., 2000. Malignancy in arsenicosis. Abstracts of International Workshop on Control of Arsenic Contamination in Ground Water Calcutta, India, Public Health Engineering Department, Government of West Bengal, Calcutta, India, p. 46.

Saha, K.C., 2001a. Arsenicosis. Indian J. Dermatol., 46 (1), 8–17.

Saha, K.C., 2001b. Cutaneous malignancy in arsenicosis. Br. J. Dermatol., 145, 185.

Saha, K.C., 2002. In: Saha, K.C. (Ed.), Arsenicosis in West Bengal. Sadananda Prakashani, Kolkata, India.

Saha, K.C., 2003. Diagnosis of arsenicosis. Environ. Sci. Health, Part A, A38 (1), 255–272.

Saha, A.K., Chakrabarti, C., 1995. Geological and geochemical background of the arsenic bearing ground water occurrences of West Bengal. International Conference on Arsenic in Groundwater: Cause, Effect and Remedy, School of Environmental Studies, Jadavpur University, Calcutta, India, Abstr., p. 42.

Saha, K.C., Chakraborti, D., 2001. Seventeen years experience of arsenicosis in West Bengal, India. In: Chappell, W.R., Abernathy, C.O., Calderon, R.L. (Eds), Arsenic Exposure and Health Effects IV. Elsevier, New York, pp. 387–395.

Saha, K.C., Poddar, D., 1986. Further studies on chronic arsenical dermatosis. Indian J. Dermatol., 31, 29–33.

Saha, A.K., Chakrabarti, C., De, S., 1997. Studies on genesis of arsenic in groundwater in parts of West Bengal. Indian Soc. Earth Sci., 24 (3–4), 1–5.

Saha, J.C., Dikshit, A.K., Bandyopadhyay, M., Saha, K.C., 1999. A review of arsenic poisoning and its effects on human health. Environ. Sci. Technol., 29 (3), 281–313.

Samanta, G., Samanta, G., Mandal, B., Das, D., Chatterjee, A., Chanda, C.R., Chowdhury, P.P., Chakraborti, D., 1995. Determination of arsenic in water, hair, nail, urine, skin-scale, liver tissue (biopsy) of the affected people and speciation of arsenic in water and urine using FI-HG-QT-AAS. International Conference on Arsenic in Groundwater: Cause, Effect and Remedy, School of Environmental Studies, Jadavpur University, Calcutta, India, Abstr., pp. 69–70.

Samanta, G., Roy Chowdhury, T., Mandal, B.K., Biswas, B.K., Chowdhury, U.K., Basu, G.K., Chanda, C.R., Lodh, D., Chakraborti, D., 1999. Flow injection hydride generation atomic absorption spectrometry for determination of arsenic in water and biological samples from arsenic-affected districts of West Bengal, India and Bangladesh. Microchem. J., 62, 174–191.

Samanta, G., Chowdhury, U.K., Mandal, B.K., Chakraborti, D., Sekaran, N.C., Tokunaga, H., Ando, M., 2000. High performance liquid chromatography inductively coupled plasma mass spectrometry for speciation of arsenic compounds in urine. Microchem. J., 65, 113–127.

Santra, A., Das Gupta, J., De, B.K., Roy, B., Guha Mazumder, D.N., 1999. Hepatic manifestations in chronic arsenic toxicity. Indian J. Gastroenterol., 18, 152–155.

Shrestha, R.R., Maskey, A., 2002. Groundwater arsenic contamination in Nepal: a new challenge for water supply sector. 5th International Conference on Arsenic Exposure and Health Effects, San Diego, CA, USA, Abstr., p. 19.

Sinha Ray, S.P., 1995. On the hydrological aspects of arsenic-rich ground water in West Bengal, India. International Conference on Arsenic in Groundwater: Cause, Effect and Remedy, School of Environmental Studies, Jadavpur University, Calcutta, India, Abstr., pp. 39–40.

Smith, A.H., 1995. Post Conference Report; Experts' Opinions, Recommendations and Future Planning for Groundwater Problem of West Bengal, School of Environmental Studies, Jadavpur University, Calcutta, India, pp. 105–106.

Srivastava, A.K., Hasan, S.K., Srivastava, R.C., 2001. Arsenicism in India: dermal lesions and hair levels. Arch. Environ. Health, 56 (6), 562.

Steering Committee Arsenic Investigation Project, 1991. National drinking water mission submission project on arsenic pollution in groundwater in West Bengal, Final Report, Steering Committee, Arsenic Investigation Project, PHE Department, Government of West Bengal.

Strategy and Action Taken, 2002. Arsenic contamination in groundwater in West Bengal, March 2002, Public Health Engineering Directorate, Government of West Bengal, India.

Subramanian, K.S., Kosnett, M.J., 1998. Human exposures to arsenic from consumption of well water in West Bengal, India. Int. J. Occup. Environ. Health, 4, 217–230.

UNIDO, Volume, I and II, 2001. Study report under the project concerted action on elimination/reduction of arsenic in groundwater, West Bengal, India (Project NC/IND/99/967), Volume I and II, United Nations Industrial Development Organization, New Delhi, September, 2001.

WHO–Government of West Bengal, 2002. Three days workshop to draft to national protocol on case-definition and management of arsenicosis (World Health Organization–Government of West Bengal), Government of West Bengal, Department of Health and Family Welfare, P.H. Branch, Kolkata, 6–8 March, 2002.

Arsenic Exposure and Health Effects V
W.R. Chappell, C.O. Abernathy, R.L. Calderon and D.J. Thomas, editors
© 2003 Elsevier B.V. All rights reserved.

Chapter 2

Groundwater arsenic contamination in Nepal: a new challenge for water supply sector

Roshan R. Shrestha, Mathura P. Shrestha, Narayan P. Upadhyay, Riddhi Pradhan, Rosha Khadka, Arinita Maskey, Sabita Tuladhar, Binod M. Dahal, Sharmila Shrestha and Kabita B. Shrestha

Abstract

Groundwater is the major drinking water source in the Terai population of southern Nepal where 49% of the total population (23 million) lives. About 200,000 shallow tube-wells are installed in the 20 districts that comprise the Terai region serving about 11 million people. Recently, arsenic contamination of the shallow groundwater aquifer has become a big issue in Nepal, although it is not as severe as in Bangladesh or West Bengal of India. To date, water samples from about 15,000 tube-wells have been tested with 23% of the samples exceeding the World Health Organization guideline of 10 µg/l and 5% of the samples exceeding the "Nepal Interim Arsenic Guideline" of 50 µg/l. An estimate of around 0.5 million people are exposed to arsenic in drinking water above 50 µg/l. Some recent studies have reported on the accumulation of arsenic in the human body above toxic levels in arsenic-exposed areas. Government, national and international non-governmental organizations are well aware of the situation now, but to date, the government has not yet been able to come up with concrete prevention or mitigation plans. Nepal still needs more research work on arsenic occurrence, effects and mitigation programs simultaneously. This paper highlights the details of the arsenic studies conducted so far by different agencies and the pros and cons of the current measures to mitigate the arsenic problem in Nepal.

Keywords: arsenic, arsenicosis, Nepal

1. Introduction

Nepal is a small country sandwiched between China and India with an area of 147,181 km^2 and an elevation of 64–8848 m above sea level. Due to its unique range of elevation, the country has three geographical zones – mountains, hills and Terai consisting of distinct ecological, economic and social characteristics. Nepal has a population of 23 million as of the 2001 census. Nepal ranks 144 (out of 174 countries) in the world on the United Nations Development Program (UNDP) Human Development Index. This is comparable to Bangladesh (rank 146) but below India (128) and Pakistan (135) (Water Aid, 2001). The government has estimated that 42% of the population is living below the poverty line with a 40% literacy rate and 57 years life expectancy. The annual gross national product (GNP) per capita is currently $220 (United Nations Children's Fund Statistics, 2002).

2. Drinking water supply and sanitation situation of the country

Nepal is rich in water resources but the water supply system was started only in the 1970s. The national coverage of the water supply system was only about 4% in 1970 and now has increased to 70%. However, there are still several hilly regions where people have to spend hours collecting drinking water. Rural communities continue to use the most convenient sources of water irrespective of quality. Sanitation facilities are still in a very poor condition having only 29% national coverage (Water Aid, 2001). Due to the ineffective institutional programs and the poor economic conditions, water quality issues have not yet been prioritized although water supply coverage has increased. This is one of the major reasons for the high infant mortality rate, 74/1000 live births. Similarly, diarrheal diseases alone kill 44,000 children and thousands of people suffer from water-borne epidemics annually.

During the process of developing the water supply sector, drinking water coverage has been significantly increased in Terai where 47% of country's total population lives. People used to drink water from open dug wells, rivers and ponds, and were, therefore, always at risk of water-borne epidemics. Hence, shallow tube-wells were promoted in the Terai to combat the problem. Now it can be said that the water supply facility is quite good in Terai where about 11 million people drink water from more than 200,000 tube-wells.

Studies on water quality of Terai tube-wells started in 1990. The quality assessment was limited to some physico-chemical parameters (iron, ammonia, nitrate) and coliform tests. Water from most of the tube-wells were reported to have high iron content but were bacteriologically safe (Shrestha, 1998). None of the agencies were thinking about other hazardous chemicals that may occur in the water until 1998 when arsenic-poisoning news from Bangladesh and West Bengal, India, spread over the region and the source of the arsenic in ground water was said to be the alluvial sediments deposited by rivers draining from the Himalayas. Professionals and concerned agencies in Nepal started to consider the possibilities of an elevated arsenic content in Terai groundwater. Rocks and minerals associated with arsenic are widespread in the entire Himalayan Region including Nepal, with flood plain areas in the southern belt of Nepal's Terai region. This indicated a strong possibility for elevated arsenic content in ground water aquifers in Terai.

3. History of arsenic study in Nepal

In 1999, the Department of Water Supply & Sewerage (DWSS) with financial assistance from WHO tested about 268 samples from eastern Nepal (Jhapa, Morong and Sunsari) and indicated the possibility of arsenic contamination in Terai ground water in Nepal (Sharma, 1999). The Nepal Red Cross Society (NRCS), one of the largest implementers of the drinking water supply program in the NGO sector first initiated an arsenic testing program in tube-wells with financial support from the Japanese Red Cross Society. In the beginning of 2000, NRCS tested about 2000 samples from their program areas for arsenic contamination – 3% were found to have more than 50 ppb, and 21% of the samples were above the World Health Organization (WHO) limit of 10 ppb (ENPHO, 2000). This study facilitated the NRCS to test all tube-wells installed under the NRCS program. For this purpose, NRCS launched the Drinking Water Quality Improvement Program under technical collaboration with the Environment and Public Health Organization (ENPHO) with financial support of the Japanese Red Cross Society in mid-2000. The objective of this program was to test all the tube-wells (about 12,000) installed under the NRCS program and to implement mitigation measures in arsenic affected areas. These findings and activities sensitized the decision makers in government and non-government agencies active in the water supply sector. The appointment of the National Steering Committee on Arsenic (NSCA) during early 2000 was the first step taken by the government by involving major stakeholders in the drinking water and sanitation sectors. This committee has prepared an Interim National Policy Document on Arsenic, and proposed an interim arsenic standard as 50 ppb and action guidelines for all concerned agencies dealing with water supply. The document has prepared certain guidelines for health screening, communication, testing, data management and some mitigation measures to be taken by the agencies (NSCA, 2001). Other agencies like Department of Water Supply & Sewerage (DWSS), Rural Water Supply and Sanitation Support Program (RWSSSP), Nepal Water for Health (NEWAH), Rural Water Supply and Sanitation Fund Board, and Nepal Water Supply Corporation have also started testing programs in their program areas.

4. District-wide arsenic contamination level

In total, about 14,932 samples were tested (Fig. 1 and Table 1) by different agencies in 20 Terai districts as of May 2002 (Shrestha and Maskey, 2002, Tandukar, 2001, Personal communication with DWSS, RWSSSP and NEWAH, 2002). More than 90% of the samples were tested in the ENPHO laboratory, Kathmandu. The samples were analyzed through Atomic Absorption Spectrophotometer by using the Continuous Hydride

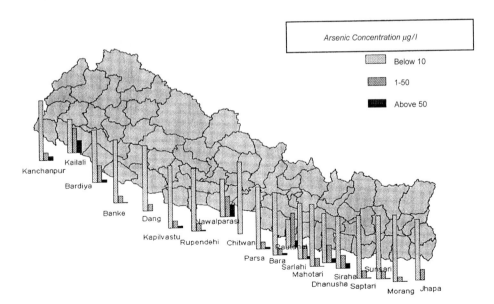

Figure 1. Extent of arsenic contaminated sample throughout the Terai districts.

Table 1. Arsenic concentration in different district.

District	Total samples	Samples with different levels of arsenic (μg/l)						Maximum concentration (μg/l)
		0–10	% 0–10	>10–50	% 10–50	>50	% >50	
Kailali	187	88	47	65	35	34	18	161
Kanchanpur	156	130	83	17	11	9	6	221
Banke	760	671	88	78	10	11	1	270
Bardiya	530	385	73	125	24	20	4	181
Dang	100	91	91	9	9	0	0	50
Rupandehi	1999	1749	87	216	11	34	2	2620
Kapilbastu	2314	2005	87	232	10	77	3	589
Nawalparasi	1164	621	53	340	29	203	17	571
Chitawan	85	85	100	0	0	0	0	8
Parsa	1926	1669	87	204	11	53	3	456
Bara	1850	1612	87	188	10	50	3	254
Rauthat	2007	778	39	968	48	207	10	213
Sarlahi	367	287	78	66	18	14	4	98
Mahottari	67	58	87	8	12	1	1	41
Dhanusha	113	78	69	28	25	7	6	140
Siraha	113	83	73	21	19	9	8	90
Saptari	397	350	88	43	11	4	1	70
Sunsari	200	176	88	22	11	2	1	75
Morong	134	124	93	9	7	1	1	50
Jhapa	463	393	85	69	15	1	0	79
Total	14,932	11,433	77	2708	18	737	5	

Generation Technique. Of the total samples collected, about 5% were reported to have arsenic concentrations more than 50 μg/l and 23% showed more than the WHO limit of 10 μg/l. The results indicate an arsenic problem in Nepal. But it is not as serious as in West Bengal and Bangladesh where 25 and 35% of the samples showed more than 50 μg/l from the total number of samples tested, respectively (Chowdhary et al., 2001).

A maximum concentration of 2620 μg/l is reported in the Rupandehi district, however, only 2% of the total samples contained arsenic concentrations more than 50 μg/l in this district. In Nawalparasi, Rauthat and Kailali districts, the highest arsenic concentrations recorded were 161–571 μg/l and 10–18% of the samples were above 50 μg/l (Table 1). This variation indicates that the arsenic problem is localized rather than uniformly distributed. ENPHO also tested arsenic in tube-well water samples from certain localized (non-project sample) areas where arsenicosis was observed. The arsenic content in water samples varied from < 10 μg/l to more than 1000 μg/l. There are still some districts where only a few samples have been tested. Therefore, the real situation can be known only after testing more samples in all the districts.

The Nepal Red Cross Society is one of the leading agencies working on water quality testing at their program areas. So far, it has tested about 10,000 samples finding arsenic concentrations in the range of 0–456 μg/l with mean value 11.9 μg/l.

There is a very low negative correlation (-0.033) between arsenic concentration and the depth of tube-well. However, this correlation is significant at 1% level of significance. The concentration of arsenic is found to decrease as the depth of the tube-well increases. The high concentrations of arsenic are seen with tube-wells of 25–100 ft depth (Fig. 2). Similarly, a very low correlation (0.023) is seen between arsenic concentrations and the age of the tube-wells. The concentration is significant at 5% level of significance. The concentration of arsenic tends to increase with the age of tube-well (Fig. 2).

5. Arsenic-exposed population and health effects

It has been estimated that about 11 million people in Nepal consume water from about 200,000 tube-wells (National Arsenic Steering Committee, 2001). If this data is considered, then the number of arsenic-exposed population in Nepal can be estimated as follows on the basis of current data on arsenic (Table 2).

A study of health effects due to arsenic contaminated drinking water is yet to be done at the national level. Agencies like NRCS, DWSS and RWSSSP are now carrying out health surveys in some communities exposed to drinking water having arsenic content more than 50 μg/l. The survey includes only the screening of the individuals with visible signs (skin changes) of arsenic poisoning. The symptomatic patients were again confirmed by analyzing either hair or nail samples or both for arsenic content.

In total, 5384 individuals who were exposed to arsenic contaminated water with greater than 50 μg/l were observed by NRCS and DWSS. Overall prevalence of arsenicosis related dermatosis is found to be 2.6% (NRCS/ENPHO, 2001, 2002) ranging from 1.4% in Bara to 5% in Nawalparasi districts. Prevalence tends to increase by age in all study areas. Prevalence is higher among the age group 50 and over. This may be due to the cumulative nature of ingested arsenic. All of the symptomatic patients show early stages of arsenicosis symptoms like Melanosis and or Keratosis on the palm, trunk and sole. The prevalence rate in Nepal is much lower than in Bangladesh where reported figures are 33.1, and 29.0% by different studies; and to West Bengal, India, where the figure is 15.02% (Chowdhury et al., 1999; Tonde et al., 1999; Chowdhary et al., 2001). However, authors recently observed some of the non-study areas in Nawalparasi district and found a very high prevalence of arsenicosis. More than 20% of adult population in Kunwar village and more than 8% in Sarwal village had arsenicosis dermatosis, some with advanced clinical stages. The authors also tested water samples in those areas by using field test kits, which were confirmed in the laboratory. Very high proportions of tube-wells in those villages were contaminated with arsenic with values more than 500 ppb. Therefore, more study is needed to generalize the national situation of arsenic-affected people.

6. Arsenic content in hair and nails

Arsenic levels in nails and hair are considered a reliable indicator of chronic exposure for that which occurred 1–10 months earlier (Shapiro, 1967). The reference value for

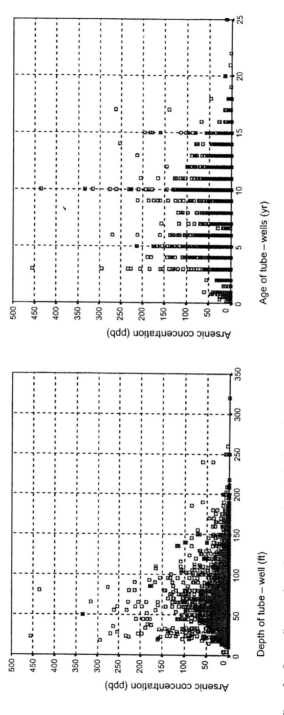

Figure 2. Scatter diagram of arsenic concentration and depth age of tube-well (based on 7886 samples).

Table 2.   Estimation of arsenic-exposed population.

Based on WHO guideline (10 μg/l or higher)		Based on India and Bangladesh guideline (>50 μg/l ppb)	
% Exposed	Population	% Exposed	Population
23	2.53 million	5	555,000

arsenic in nail and hair samples is set at 430–1080 μg/kg (0.43–1.08 mg/kg) and 80–250 μg/kg (0.08–0.25 mg/kg), respectively (Chowdhary et al., 2001). A value of more than 1000 μg/kg in hair is regarded as an indicator of toxicity among acute cases. While conducting a health survey in Nepal, 497 hair samples and 116 nail samples were tested for arsenic content and out of them 95% of the hair samples and 71% of the nail samples contained arsenic concentrations of more than the normal limit. In addition, arsenic content in 62% of the hair samples exceeded the toxic level, i.e. more than 1000 μg/kg. However, the so-called toxic level in chronic arsenicosis does not correlate with the severity of the symptoms. Arsenicosis asymptomatic people exposed or not exposed to arsenic contaminated drinking water but sharing the same aquifer where some samples of tube-well water tested positive for arsenic (>50 μg/l) also show elevated levels of arsenic concentrations in their hair or nail samples. This means that they are sub-clinically affected and may soon develop visible symptoms. Therefore, a low prevalence rate cannot generalize the situation and asymptomatic people are equally at risk of arsenicosis. Thus, even the people drinking water from less contaminated (<50 μg/l) tube-wells but sharing the same aquifer are exposed to arsenic directly or indirectly.

7. Arsenic mitigation in Nepal

The arsenic problem in Nepal is a relatively new issue. Most of the activities conducted by different agencies so far are concentrated on problem identification. However, some organizations have also started to mitigate the problems. The following are some of the major activities going on in Nepal.

7.1. Coordination and networking

For better coordination among the concerned agencies related to water supply, the Government of Nepal has constituted a National Steering Committee on Arsenic (NSCA). The committee, whose members are from all leading government, non-governmental and international agencies working on arsenic, meets once a month to review the progress of different agencies and prepare plans and programs at the national level for arsenic mitigation.

7.2. Human resource development

Most of the leading agencies are now sending their professionals to Bangladesh and West Bengal to share their experiences and to learn.

7.3. Development of IEC materials

The NSCA is finalizing a standard set of information, education and communication (IEC) materials for frontline workers and for distribution in arsenic-affected communities.

7.4. Laboratory testing of arsenic in water and biological samples

There are a few well-equipped water quality testing laboratories in the private sector, and those are only in Kathmandu, the capital of the country. Some of the laboratories are able to detect arsenic in biological samples, using the HG-AAS technique. Recently, NSCA has formulated a subcommittee under the coordination of the Royal Nepal Academy of Sciences (RONAST) to develop standard sampling and testing protocols for arsenic so that all laboratories will have a uniform and standard method of detection.

7.5. Development of arsenic testing kits

ENPHO has recently assembled field level arsenic test kits. It is user-friendly and economical (0.6 USD/test). It is now under field testing with strict monitoring and cross-checking with laboratory tests using the atomic absorption spectrophotometer.

7.6. Arsenic testing and mitigation programs

Arsenic testing programs are now being undertaken by most of the water supply agencies, but they are limited to tube-wells that were installed under their program. Some of the agencies such as the Nepal Red Cross Society (NRCS) and the Rural Water Supply Support Program (RWSSSP) have started some mitigation activities especially to provide safe water options such as household arsenic removal filters and the installation of iron–arsenic removal plants. NRCS has set up an arsenic information center at district level where people can obtain information about arsenic and it also provides a testing facility. Some of the major mitigation activities are as follows.

7.6.1. Marking of tube-wells

NSCA has proposed marking the arsenic contaminated tube-well (>50 µg/l) by painting it white with a black cross (✕) whereas the safe tube-well would be marked with a "✔". These signs should be placed on a part of the tube-well that is easily visible. Soon all the tested tube-wells will be marked this way.

7.6.2. Household arsenic removal filters

Two types of household arsenic removal filters are now in practice: (I) ENPHO arsenic removal powder with filtration device and (II) three pitcher system.

7.6.2.1. ENPHO arsenic removal powder with filtration device

This system has been developed by the School of Environmental Studies, Jadhavpur University, West Bengal, India. In this method, coagulation, co-precipitation followed by filtration occurs to remove arsenic from water (School of Environmental Studies, 2000). It is prepared at ENPHO, Kathmandu and distributed in the NRCS program areas where ground water arsenic content is more than 50 μg/l.

In this process, one packet of arsenic removal powder (mixture of ferric chloride, charcoal powder and sodium hypochloride) is added into a bucket containing 20 l of arsenic contaminated water and then filtered through a ceramic candle filter. Arsenic removal rate of this system was reported as 91% (ENPHO, 2002; Hwang, 2001). Similarly, fecal coliform, iron and turbidity are substantially reduced by 93, 95 and 99%, respectively (Hwang, 2002). As the system removes the iron, water tastes good and stays cool due to the storage of water in earthen pots. This makes the system socially acceptable. The cost for complete sets includes two Ghainto (filtration and collection unit), and the 20-liter capacity plastic bucket is NRs. 425 or US $6. The cost of a year's supply of chemicals is NRs. 500 or US $7.

7.6.2.2. Three pitcher system

This system originated in Bangladesh to treat surface water containing impurities at the household level, but now it is used for household arsenic removal (Ziya et al., 2001). This system is under trial in Nepal. There is some reluctance to its use due to the higher chance of bacterial contamination as no disinfectant materials are used. Another problem might be early clogging of the filter. However, this system is more cost effective than the previous one. Some experiments done by research students on its efficiency showed arsenic removal of more than 98%. It has a relatively low cost of about US $10.00 (Hurd, 2001).

7.6.3. Iron–arsenic removal plant

This technique is based on co-precipitation and adsorption with naturally occurring iron. This method is also widely used in Bangladesh since about 65% of the area of Bangladesh contains dissolved iron concentrations in excess of 2 mg/l (Mamtaz and Bache, 2001). The condition of ground water in Nepal's Terai region is also similar to Bangladesh where many tube-wells' water has a high iron content.

The system consists of three chambers: an aeration chamber to oxidize and precipitate naturally occurring ferrous to ferric compounds. The precipitated ferric-arsenic compounds then flow to the filtration chamber where they are filtered out through a layer of gravel, charcoal and sand. Finally, the clean water flows into the storage chamber. The experiences in Bangladesh show arsenic removal efficiency only up to 70% (Rahman, 2000). Several units of this type of system has been installed by the Nepal Red Cross Society program area and were found to have similar results as in Bangladesh

(ENPHO, 2001). A study found a lower arsenic removal rate where dissolved iron content is only 2 mg/l (Poole, 2002).

7.7. Research undertaken by national and international agencies

The Massachusetts Institute of Technology (MIT) Nepal Water Project has been conducting drinking water quality-related research since 1999. Under this program, twenty Master of Engineering students from the MIT Civil and Environmental Engineering Department have traveled to Nepal to study water quality and household water treatment systems and methods. Some of the students have also conducted research related to arsenic in groundwater and treatment technologies (Hurd, 2001; Hwang, 2002; Murcott, 2002; Ngai, 2002; Poole, 2002). UNICEF Nepal with technical support from the US Geological Survey reviewed the arsenic research activities in Nepal and proposed hydrogeological studies to determine the cause of arsenic contamination and its extent in the shallow and deeper Terai aquifers (Clark and Whitney, 2001). The Royal Nepal Academy of Science & Technology (RONAST) and Tribhuvan University are now starting pure science research projects related to hydrogeology and geochemistry, which was recommended by USGS.

8. National challenges for arsenic mitigation

- Create awareness about arsenic but avoid panic in the communities.
- Country has no standard reference water quality testing laboratory.
- Water testing facilities at the community level with affordable price and reliable result.
- Provide alternative drinking water options in arsenic-affected communities.
- Prepare guideline for installation of new tube-wells.
- Improvement of bacteriological quality after using arsenic removal filters.
- Most of the health personnel are still not aware of or trained about arsenicosis.
- Future monitoring of tube-wells having arsenic content more than 10 ppb.
- Different interests of national and international experts and donor communities.

9. Conclusion and recommendation

Drinking water arsenic contamination in Nepal, though not as widespread as in West Bengal and Bangladesh, is serious as indicated by preliminary studies in some areas of Terai. As the geological source for the Bengal and Indus Basins primarily originated in the Himalayas and geochemical processes are responsible for transforming arsenic from its mineral form into a soluble toxic element, more studies in Nepal can benefit national policymakers and other countries in this region (Clark and Whitney, 2001). In this context international communities should give more attention to studies in Nepal.

Apart from the studies, Nepal itself needs to develop a good strategy to overcome the above-mentioned challenges. All agencies should work together with better coordination. However, the current approach is more agency-oriented. Each agency looks after only the tube-wells installed under its program. This approach causes confusion among the local people. With this fragmentary approach it is not possible to create mass awareness.

The government should act as a facilitator to mitigate the problem. It should be committed to launch programs to control and prevent arsenicosis and arsenic contaminated drinking water through its local authorities, NGOs and CBOs. The formation of National Arsenic Steering Committee by representing all relevant government agencies and concerned INGOs and NGOs is a good step toward developing a uniform and coordinated program. The country needs a huge investment to mitigate this problem. All donor communities working in the water and sanitation sector should allocate funds for arsenic mitigation.

Hydrogeological investigation is one of the urgent studies needed to be conducted to identify the arsenic spatial distribution pattern throughout the country and the pattern of arsenic mobilization in groundwater in Terai. It will help to identify risk areas and at the same time will help to prepare guidelines for future tube-well installation, and preventive/mitigation strategies.

Water sample collection and testing protocols should be standardized to obtain correct data and a central data management system should be established as soon as possible.

A training program for health personnel from the Terai district should be launched so that they will be able to deal with arsenicosis patients. A special course on arsenic should be designed in medical, public health, engineering and other relevant academic institutions.

The international health community can also carry out detailed epidemiological studies. Research in Nepal could be performed to study health effects due to low to moderate levels of arsenic exposure and the effect of nutritional status. It could also help to recommend guideline values of arsenic for developing countries. However, the research should be comprehensive drawing samples from all types of tube-well sources without bias.

More operational research is needed to recommend appropriate arsenic removal techniques at the small community and household levels. Research may be focused on those techniques, which are already approved by other countries. However, research to develop innovative and locally appropriate technologies should also be considered.

As non-symptomatic cases as well as non-exposed cohorts using non-contaminated tube-wells sharing the same aquifer showed high concentrations of arsenic in hair, nails or both, a carefully designed cohort study should be conducted to determine the relative risk of chronic arsenic exposure. Therefore, the control (non-exposed) cohort should be drawn from the population not even sharing the same aquifer.

Steps are to be initiated to develop a national reference laboratory to assure standard laboratory tests on arsenic problems in water and biological samples. With a little support and backup the ENPHO laboratory could be upgraded to a national reference laboratory closely monitored and supervised by a competent expert committee jointly developed by ENPHO and NSCA.

Acknowledgements

The authors are grateful to Dr Willard R. Chappell from University of Colorado for providing opportunity to present this paper in the Fifth International Conference on Arsenic Exposure and Health Effects and to Dr Susan Murcutt, Massachusetts Institute of Technology for her support in preparing this paper. The authors would also like to thank Mr Padam K. Khadka, Nepal Red Cross Society, Mr Madhav Pahari, UNICEF, Ms Kalawati Pokharel, RWSSSP, Mr Anil Pokharel, NEWAH and Mr Robin Lal Chitrakar, DWSS for providing data and necessary information to update the national data on arsenic.

References

Chowdhury, U.K., Biswas, B.K., Dhar, R.K., Samanta, G., Mandal, B.K., Chowdhury, T.R., Chakraborti, D., Kabir, S., Roy, S., 1999. Groundwater arsenic contamination and suffering of people in Bangladesh. In: Chappell, W.R., Abernathy, C.O., Calderon, R.L. (Eds), Arsenic Exposure and Health Effects – Proceedings of the Third International Conference on Arsenic Exposure and Health Effects, July 12–15, 1998, San Diego, California. Elsevier, New York, pp. 165–182.

Chowdhary, U.K., Rahaman, M.M., Mondal, B.K., Paul, K., Lodh, D., Biswas, B.K., Basu, G.K., Chanda, C.R., Saha, K.C., Mukharjee, S.C., Roy, S., Das, R., Kaies, I., Baura, A.K., Palit, S.K., Quamruzzaman, Q., Chakraborti, D., 2001. Groundwater arsenic contamination and human suffering in West Bengal, India and Bangladesh. Environ. Sci., 8 (5), 393–415.

Clark, D., Whitney, J.W., 2001. Options for Hydrogeological Investigations on Arsenic Occurrence in Groundwater in Nepal. UNICEF/National Steering Committee for Arsenic/American Embassy/US Geological Society, Kathmandu, Nepal.

ENPHO, 2000. Ground Water Arsenic Contamination in Terai Tubewells. Nepal Red Cross Society/Environment & Public Health Organization (ENPHO), Kathmandu, Nepal.

ENPHO, 2001. Drinking Water Quality Improvement Program, Annual Report. NRCS/JRCS/ENPHO, Kathmandu, Nepal.

ENPHO, 2002. A study on Health Effects of Arsenic Contaminated Drinking Water in Nawalparasi District, Nepal. Department of Water Supply and Sewerage (DWSS), Government of Nepal.

Hurd, J., 2001. Evaluation of three arsenic removal technologies in Nepal. Submitted to the Department of Civil and Environmental Engineering in partial fulfillment of the requirements for the degree of Master of Engineering in Civil and Environmental Engineering.

Hwang, S.K., 2002. Point-of-use arsenic removal from drinking water in Nepal using coagulation and filtration. Thesis submitted to the Department of Civil and Environmental Engineering in partial fulfillment of the requirements for the degree of Master of Engineering in Civil and Environmental Engineering.

Mamtaz, R., Bache, D.H., 2001. Low-cost Technique of Arsenic Removal from Water and Its Removal Mechanism. Technologies for Arsenic Removal from Drinking Water, Preprints of BUET – UNU International Workshop, May, p. 43.

Murcott, S., 2002. Nepal Water Project. Department of Civil and Environmental Engineering Masters of Engineering Program, Massachusetts Institute of Technology.

NEWAH, 2002. Draft report. Personal communication from Pokhrel, A., Chief, Technical Section, Nepal Water for Health (NEWAH) on June 2002.

Ngai, T.K., 2002. Arsenic speciation and evaluation of an adsorption media in Rupandehi and Nawalparasi districts of Nepal. Thesis submitted to the Department of Civil and Environmental Engineering in partial fulfillment of the requirements for the degree of Master of Engineering in Civil and Environmental Engineering.

NSCA, 2001. Nepal's interim arsenic policy preparation report. Report prepared for the Nepal National Arsenic Steering Committee DWSS, Government of Nepal.

Poole, B.R., 2002. Point-of-use water treatment for arsenic removal through iron oxide coated sand: application for the Terai region of Nepal. Thesis submitted to the Department of Civil and Environmental Engineering in partial fulfillment of the requirements for the degree of Master of Engineering in Civil and Environmental Engineering.

Rahman, Z., 2000. Alternative Water Supply Options. NGO Forum for Drinking Water Supply and Sanitation, Dhaka, Bangladesh.

School of Environmental Studies, 2000. A simple household device to remove arsenic from groundwater hence making it suitable for drinking and cooking. Jadhavpur University, Calcutta, India.

Shapiro, H.A., 1967. Arsenic content in human hair and nails, its interpretation. J. Forensic Med., 14, 65–71.

Sharma, R.M., 1999. Research Study on Possible contamination of groundwater with Arsenic in Jhapa, Morong, and Sunsari districts of Eastern Terai of Nepal. Report of the WHO Project, DWSS Government of Nepal.

Shrestha, R.R., 1998. Drinking Water Crisis in Nepal. Environment & Public Health Organization (ENPHO), Kathmandu, Nepal.

Shrestha, R.R., Maskey, A., 2002. Groundwater arsenic contamination in Terai Region of Nepal and its Mitigation. ENPHO Mag., March, 36–44 (Published by Environment & Public Health Organization, Kathmandu).

Tandukar, N. 2001. Scenario of arsenic contamination in groundwater of Nepal. Paper presented at BUET/ Bangladesh and UNU/Japan. Workshop on Technologies for Arsenic Removal from Drinking Water, Dhaka, May 5–7, 2001.

Tonde, M., Rahman, M., Magnuson, A., Chowdhury, I.A., Faruquee, M.H., Ahmad, S.A., 1999. The relationship of arsenic levels in drinking water and the prevalence rate of skin lesions in Bangladesh. Environ. Health Perspect., 107 (9), 727–729.

United Nations Children's Fund Statistics, 2002. Nepal Country Information. Available: http://www.unicef.org/ statis/Country_1Page122.html, accessed on February 29, 2002.

Water Aid, 2001. Country strategy. Water Aid, Nepal.

Ziya Uddin, M., Islam, B., Reza, T., Ferdous, A.T.M., Rakib, A., 2001. A study to investigate the efficiency and acceptability of arsenic removal 3-pitcher filter in WPP area. CARE, Bangladesh.

Arsenic Exposure and Health Effects V
W.R. Chappell, C.O. Abernathy, R.L. Calderon and D.J. Thomas, editors
© 2003 Published by Elsevier B.V.

Chapter 3

Environmental impacts, exposure assessment and health effects related to arsenic emissions from a coal-fired power plant in Central Slovakia; the EXPASCAN Study

I. Thornton, M.E. Farago, T. Keegan, M.J. Nieuwenhuijsen,
R.N. Colvile, B. Pesch, U. Ranft, P. Miskovic, P. Jakubis
and the EXPASCAN study group[1]

Abstract

The coal-fired power plant at Novaky in the Prievidza district of Central Slovakia has emitted in excess of 3000 tonnes As since commencing operations in the 1950s. This resulted from the combustion of brown coal containing up to 1500 mg/kg arsenic (As). Pollution control measures reduced emissions from around 200 to 2 tonnes per year in the 1980s. This chapter presents (a) the results of environmental monitoring undertaken in 1999–2000, based on the analysis of soils and dusts from 550 households comprising a population-based case-control study of non-melanoma skin cancer (NMSC), (b) data on As levels in coal currently used for power generation and in fly ash dumped within the district, (c) estimates of As exposure in the population in relation to distance from the plant and (d) a summary of health statistics.

Although arsenic levels in soils and house dusts fell with distance from the plant, actual concentrations were low with soils averaging around 40 μg/g within 5 km and 20–25 mg/kg over 5 km from the plant, and dusts 18 and 11 mg/kg As, respectively. Total urine arsenic concentrations were low (median 6.9 μg/l), but showed a correlation with soil arsenic (r = 0.21, p ≤ 0.01). Further, the median urinary As concentrations declined from 7.5 to 5.8 μg/l by increasing distance from the plant (p < 0.05).

Predictions of As deposition rates from atmospheric dispersion modelling system (ADMS) based on current and historical levels of airborne As indicated soil As concentrations considerably in excess of current levels. Possible losses due to leaching and biogeochemical cycling are discussed together with the implications of these losses to human exposure.

NMSC incidence was 21% higher in the vicinity of the power plant, when related to the incidence of the total district, with a decreasing trend by distance from the emission source. The highest exposure level to environmental arsenic in the study area was associated with an increased risk of NMSC, assessed on residential history by distance to the source of the arsenic and annual As emissions (OR 1.90; 95% CI 1.4–2.6) as well as assessed with exposure to arsenic by locally grown food (OR 1.8; 95% CI 1.0–3.4).

Arsenic emissions were highest 20 years ago and current environmental levels might not reflect past exposure. The epidemiological exposure model included historical emissions data but past and present industrial emissions from other sources and lack of accurate exposure assessment of arsenic in the local diet could have confounded the distance- and food-related exposure variables.

Keywords: arsenic, coal-burning, power generation, soil contamination, non-melanoma skin cancer, Slovakia

---

[1] V. Bencko, E. Cordos, P. Docx, E. Fabianova, P. Frank, M. Gotzl, J. Grellier, B. Hong, J. Rames, R. Rautiu, E. Stevens, J. Zvarova.

1. Introduction

Human exposure to elevated levels of arsenic from natural sources is now recognised worldwide. The main exposure route is drinking water associated with natural As-rich geological strata. Exposure to contaminated dust and soil from mining and smelting residues and from the combustion of As-rich coal is also documented (Thornton and Farago, 1997).

The coal-fired power plant at Novaky in the Nitra Valley of the Prievidza district of Central Slovakia is located in a highly industrialised region, known to be heavily polluted. The plant has emitted in excess of 3000 tonnes As since commencing operations in the 1950s as a result of burning locally produced brown coal containing up to 1500 μg/g As. The plant produces around 12% of Slovakia's electricity. These arsenic emissions were identified as a potential health hazard as early as the1970s and links were suggested with elevated rates of skin, lung and bladder cancer.

Arsenic emission statistics are available from 1953 (Fig. 1), ranging from 120 to 230 tonnes As per year in the 1960s and 1970s, falling with pollution abatement measures to around 2 tonnes per year in the 1990s and to almost negligible levels with the introduction of electrostatic precipitators in 1997.

Health statistics show the incidence of non-melanoma skin cancer (NMSC) in this district to be the highest in Slovakia, with 90 male and 50 female cases per 100,000 in 1997 (age adjusted to the world standard population) (National Cancer Institute of Slovakia, 2000).

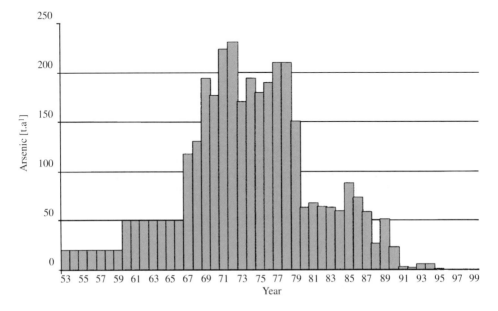

Figure 1. Emission levels of the power plant.

This chapter presents the results of a research project examining the association between the results of programmes of environmental monitoring, prediction of arsenic deposition and a population-based case study of NMSC in the Prievidza district. This work falls within a broader based EU funded study (the EXPASCAN study) to determine (i) if there was a relation between arsenic emissions by the coal-burning power plants and the non-ferrous metal smelting and processing industry in CEE countries, and high rates of cancer of the skin, lung and bladder and (ii) if there were other populations at risk in Central and East Europe, and to identify possible sources of arsenic pollution.

2. Environmental sampling

Methods of sampling garden soils, house dusts, coal and ash residues from the power plant have been described in detail by Keegan et al. (2002). In summary:

(a) 220 garden soils each comprising 20 sub-samples (0–5 cm) were collected from households randomly selected from the population of the case-control study. Where there was no garden, a nearby allotment or amenity area was sampled.
(b) 240 house dust samples were each collected from 1 m^2 of carpet into the Whatman cellulose filter fitted to the entry nozzle of an adapted vacuum cleaner (Watt et al., 1993).
(c) 150 dust samples were taken from the household's vacuum cleaner, providing a composite sample from the house.
(d) 18 samples of coal were taken from storage and transport facilities at the plant.
(e) 22 samples of ash were taken from silos and hoppers at the plant and from nearby dumps.

Soil (<2 mm fraction ground in a Tema mill) and dusts (<1 mm fraction) were digested in nitric and perchloric acids and analysed by ICP-AES. Analytical quality control procedures were based on the analysis of duplicate samples, blanks and reference materials (Thompson and Walsh, 1989).

2.1. Results

The results of soil and house dust analysis are presented in Table 1 in relation to distance from the power station. Arsenic concentrations in soils were relatively low, ranging from 9 to 139 mg/kg with an overall geometric mean value of 27 mg/kg. This compares with a worldwide geometric mean value of 9 mg/kg, with a range in uncontaminated soils of <1–95 mg/kg, reflecting the geological nature of the parent material and soil type (Kabata-Pendias and Pendias, 1984). The level fell from around 40 mg/kg within 5 km of the plant by approximately 40% to 20–25 mg/kg at distances greater than 5 km.

Arsenic concentrations in house dusts ranged from 1 to 170 mg/kg and again decreased in samples collected by both methods of collection from an average of 18 mg/kg within 5 km of the plant to 11 mg/kg at distances greater than 5 km from the plant.

Coal samples of 1998 and 1999 supplies ranged widely in arsenic content up to 1540 mg/kg with a mean value of 690 mg/kg from the main most recent supply (Table 2).

Table 1. Concentrations of arsenic in soil and dust (mg/kg) by distance from the power plant.

	Distance (km)	N	Geometric mean	Geometric standard deviation	Range
Soil (0–15 cm)	<5	40	43	1.7	14–134
	5–10	102	25	1.7	9–139
	>10	68	22	1.5	10–58
	Total	210	27	1.7	9–139
Dust collected	<5	32	18	1.9	6–116
from unit area of	5–10	109	11	2.0	1–170
carpet	>10	70	11	2.0	1–112
	Total	211	12	2.1	1–170
Composite dust	<5	25	19	1.6	7–57
from vacuum	5–10	55	8	2.0	1–22
cleaner	>10	29	8	2.6	1–61
	Total	109	9	2.3	1–61

High concentrations of arsenic were found in ash samples, ranging up to 3030 mg/kg with several exceeding 1000 mg/kg As (Table 2). Arsenic levels were somewhat lower in materials from ash dumps that had been in operation for over 20 years. It is possible that arsenic has been mobilised over time and either leached or lost by volatilisation.

3. Atmospheric dispersion modelling

The short-range atmospheric dispersion model UK-ADMS was applied to calculate ground level concentrations of arsenic in air within 20 km of the Novaky Power plant (Colville et al., 2001). The model was applied for two occasions based on emission data for the periods 1973–1975 and 1991–1993, and was supported by meteorological data, from Prievidza airport, over a period of 4 years.

The arsenic concentration bands, shown in Figure 2 for the periods 1973–1975 and 1991–1993, are elongated along the axis of the Nitra Valley, as the wind, in which the plume is dispersed is channelled by the valley sides. The maximum concentrations are displaced 2 km to the northeast and southwest of the plant and fall off rapidly with distance by a factor of 30 between 2 and 10 km from the plant.

Table 2. Concentrations of arsenic (mg/kg) in coal and ash from the Novaky power station.

	N	Mean (AM)	Range
1998/1999 coal	17	527	37–1540
Ash at plant	9	1453	217–1550
Ash in nearby dumps	13	455	124–1010

a) 1973-1975

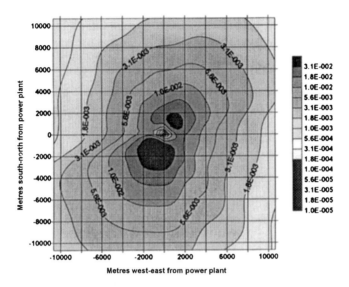

b) 1991-1993

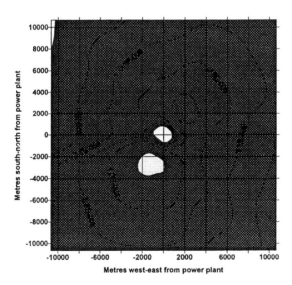

Figure 2. Dispersion profiles of arsenic around the power plant.

By applying the model to the two time periods before and after the application of pollution abatement strategies and an increase in stack height, it is clearly shown that ground level concentrations of arsenic in air will have decreased significantly as will rates of arsenic deposition in the soil (Colville et al., 2001).

4. Case-control study of NMSC

NMSC cases were eligible if (1) they currently resided in this district, (2) they were not older than 80 years and (3) the diagnosis of NMSC as a primary, first tumour was confirmed histologically during 1996–1999. From 374 eligible cases, 328 were randomly contacted and 264 NMSC cases were recruited, together with 280 population controls, frequency-matched to cases on gender and age, as described by Pesch et al. (2002). A questionnaire was used to ascertain demographic characteristics as well as details on the residential and occupational history. Furthermore, questions on dietary habits, outdoor activities, skin type and smoking habits were asked.

4.1. Exposure assessment

Assessment of environmental arsenic exposure was based on the subject's residential history and on the annual arsenic emission. Every place of residence held at least for 1 year since 1953 was classified into three groups: <5, $5-10$ and >10 km of the plant. Historical estimates from the atmospheric dispersion modelling were used to include changes over time for the three groups. Although, for financial reasons, it was not possible to undertake a detailed dietary study, exposure by oral uptake was estimated using the frequency of consumption of the 25 food items from an in-person interview of the study subjects and information on food intake and concentrations of arsenic in food items were obtained from the earlier EU-funded PHARE project (Fabianova and Bencko, 1995). This was assisted by an assessment of nutritional and drinking habits based on data from a semi-quantitative food questionnaire completed by the study population. A second assessment of the nutritional exposure (Asnut2) classified interviewees reporting a relevant contribution of home-grown products to their food consumption during the period of highest As emission (1970–1989) as high exposed. No data were available for As levels in drinking water and an assumed concentration of 1 μg/l was applied. A median value of As uptake from food and water of 13.1 mg/year was estimated, ranging up to 76.5 mg As per year. Occupational exposure was also taken into account (Pesch et al., 2002). In addition to those working in the power plant, a large number of men had been working as coal miners. Further, there are a chemical plant and a plant manufacturing building material from coal ash in the vicinity of the power plant.

Analysis of arsenic species in spot urine samples collected after interview was used as a biomarker of current arsenic exposure and uptake. From the 550 subjects enrolled in the epidemiological case-control study, 548 subjects provided urine samples. In 544 urine samples, at least one arsenic species out of inorganic As, monomethylarsonic acid (MMA) and dimethylarsinic acid (DMA) could be determined using hydride generation atomic absorption spectroscopy. 518 samples have results for all three species, and 411 were statistically analysed after quality control for urine density and creatinine concentrations. Results are summarised in Figure 3. The data show a small but significant decrease in total urinary As and MMA in the population residing over 5 km from the power plant compared with those living within 5 km of the plant (Ranft et al., 2003).

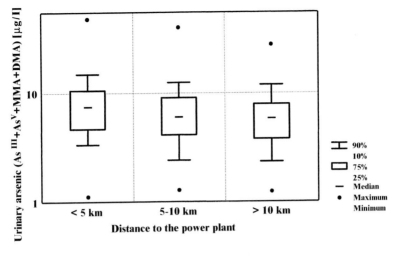

	47.9	40.1	27.7	Maximum
	15.0	12.5	11.9	90th percentile
	10.6	9.1	7.9	75th percentile
Urinary arsenic	7.5	6.0	5.8	Median
	4.7	4.1	3.8	25th percentile
(As[III]+As[V]+	3.4	2.4	2.4	10th percentile
MMA+DMA)	1.1	1.3	1.2	Minimum
[μg/l]	58 (47)	225 (52)	128 (52)	N (% cases)

Figure 3. Urinary arsenic concentrations by distance of places of residence to the power plant, EXPASCAN Study, Prievidza district, Slovakia, 1999. Box-whisker plot in logarithmic scale (from Ranft et al., 2003).

Significant correlations are found between both total As in urine and total inorganic As in urine and As in soil (Table 3). Arsenic in house dust is also related to As in soil, though As in urine is not related to As in dusts.

4.2. Non-melanoma skin cancer risk

Risk of NMSC in the Prievidza district was calculated as odds ratios (OR) with 95% confidence intervals (CI) for populations considered to have had "low", "medium" and "high" exposure to As (Table 4). The methods for exposure assessment and risk estimation are published elsewhere (Pesch et al., 2002). These categories took into account residential history in relation to (a) distance from the plant, (b) estimated annual As emissions from the atmospheric dispersion modelling and (c) nutritional habits and As content of food. This approach took the 30th and 90th percentiles of the distribution of each variable in the total study sample as a cut-off (Pesch et al., 2002).

Significant excess risks were found for both medium- and high-exposure groups. Similar effects were found when environmental exposure to arsenic at the workplaces was included. However, it is recognised that emissions from other industrial plants in the area could have confounded the distance-based exposure variables, and residual confounding

Table 3. Correlations between arsenic in soil, dust and urine.

	Soil	House dust	Total urinary arsenic	Total inorganic urinary arsenic
Soil	–	0.34[**]	0.21[**]	0.15[*]
N		165	159	175
House dust		–	0.13	0.08
N			162	179
Composite house dust	0.40[**]	0.40[**]	0.02	−0.05
N	94	92	80	90

N = number of samples, [**]$p < 0.01$, [*]$p < 0.05$.
Pearson's correlation coefficients based on log-transformed data.

from occupational exposure, although not associated with NMSC with the measures applied, has to be taken into account.

5. Discussion

This study clearly illustrates the problems encountered when attempting to relate well-founded epidemiological data, based on recent and current databases, to exposure data when no reliable information is available for environmental variables at what may have been the relevant time period of critical exposure. Actual levels of As exposure in the Prievidza district from dusts, soils, the atmosphere and food stuffs, some 30–40 years ago

Table 4. Risk of NMSC in the Prievidza district assessed with a distance-based exposure variable for the residences of the study population since 1953.

Exposure to arsenic[a]	Odds ratio[b]	95% CI
Distance-based exposure (places of residence, annual arsenic emissions)[c]		
Low	1	
Medium	1.72	1.42–2.08
High	1.90	1.39–2.60
Trend test	1.50	1.31–1.72
Dietary-based exposure (arsenic in food, consumption of home-grown food)		
Low	1	
Medium	1.12	0.77–1.64
High	1.90	0.98–3.43
Trend test	1.27	0.96–1.69

After Pesch et al. (2002).
[a] Categorisation with the 30th and 90th percentiles of the distribution as cut-off for medium and high.
[b] Adjusted for gender and age.
[c] Using a re-sampling procedure to correct for a selection bias.

are unknown, though emission data from the plant would indicate that they could be expected to be far greater than at the present time.

The emission data for 1960s and 1970s showing some 200 tonnes As per year, mostly falling within a radius of 5 km of the plant, would indicate potentially greater levels of As in surface soils and dusts than found at the time of this study. Within this time frame, As in the surface soil will have been mixed with deeper material as a result of bioturbation through earthworm and other biota activity. Downward migration in the soil may also have resulted through leaching of soluble As species, and losses to the atmosphere may have resulted from volatilisation. Arsenic levels in locally grown food stuffs would also have been potentially larger at the time of maximum emissions, both as a function of uptake from the contaminated surface soil and direct deposition onto crops from the atmosphere. However, the chemical forms of As incorporated into the surface soil from emitted As are unknown and these would have played an important role in controlling As solubility and bioavailability. These forms may well have changed over time as a result of aging or transformation processes in the soil. Mineral and chemical phases of As in both soils and house dusts have been studied using scanning electron microscopy in materials generated through the smelting of copper sulphide ores in Anaconda, Canada showing the presence of oxides, sulphides, silicates and phosphates as well as associations of As with Fe (Davis et al., 1996).

Arsenic concentrations in Prievidza soils and house dusts sampled in 1999 were relatively low and comparable with those of a "control" area near an historical As mining district in southwest England. Median total As concentrations were also similar in both locations. It is noted that median values for arsenic in surface (0–15 cm) soils both within and exceeding 5 km distance from the plant are greater than the soil guideline values of 20 mg/kg As recently published in the UK (Department for Environment, Food and Rural Affairs and Environment Agency, 2002). This value is based on the output from the contaminated land exposure assessment (CLEA) model for assessing the risks to human health from chronic exposure to soil contaminated with arsenic. The small increases in As concentrations in soils, dusts and urine within 5 km of the power plant can be attributed to residues from earlier emissions and possibly dispersion of As-rich ash and slag during transport from the plant. Surface soils and dusts within the UK mining areas were heavily contaminated with As (means 365 and 217 mg/kg, respectively). Even with the possibility of high daily intakes of As by direct ingestion of soil and dust, confirmed by elevated urinary As levels, no measurable health effects were found (Farago et al., 1997; Farago and Kavanagh, 1999). However, possible differences in As speciation and bioavailability in environmental samples may well make a comparison between the two locations invalid. The incidence of NMSC was not assessed in the UK; an ecological study of bladder cancer failed to find any significant relationships with environmental arsenic (Leonardi et al., 1995).

Possible health effects due to exposure to As derived from the combustion of As-rich coals have received detailed assessment in the People's Republic of China. Arsenic levels in coals from Guizhou province range from 20 to over 2000 mg/kg. A study of eight villages showed 860 cases of endemic arsenosis (an incidence rate of 40%) compared to none in a control village (Zheng et al., 1996). However, in Jialin province the burning of high-As coal did not result in any cases of arsenism (Chen et al., 1987). It was speculated that exposure to contaminated air, dusts and foodstuffs would have been much greater in Guizhou as the coal was burned in open stoves within poorly ventilated houses; the domestic circumstances in Jialin were better. Diagnosis of arsenosis was based on a range of symptoms, including neurological and digestive disorders and hyperkeritosis.

It is fully recognised that the routes of exposure to As from coal combustion are less direct in Prievidza than in China.

6. Conclusions

Historic emission data from a coal-fired Slovak power plant point to a high environmental pollution with arsenic especially in the 1970s. Atmospheric dispersion modelling showed decreasing trends of airborne arsenic over time since then by distance to the plant. The special spatial pattern was broadly similar to that of current arsenic concentrations in house dust, soil and urine levels, which all decrease by distance to the emission source, although these current levels of arsenic were generally low. There was a significant correlation between arsenic in soil and urine levels of arsenic, but not between arsenic in house dust and urinary arsenic.

NMSC incidence has been highest in this Slovak district, and there is a 21% higher incidence within 5 km of the plant. A population-based case-control study showed an excess risk of NMSC for those living since 1953 closer to the plant compared to those further away (approximate risk estimates up to about two) suggesting an association with past arsenic emissions from the plant.

Acknowledgements

Funding for the EXPASCAN was provided by European Union contract IC 15 CT98 0525, and financial management by IC Consultants.

References

Chen, B.-R, Yang, Y.-N., et al., 1987. Distribution of arsenic and selenium in coal samples in China. Proceedings of the Third National Symposium on Environmental Geochemistry and Health, Earthquake Publication, P.R. China (in Chinese) cited in Wing-Sun Chow, 1999. Review of Investigations in Arsenic Occurrence, Exposure and Health Effects in the People's Republic of China. Unpublished MSc Thesis, Imperial College of Science, Technology and Medicine, London.

Colville, R.N., Stevens, E.C., Niewenhuijsen, M.J., Keegan, T., 2001. Atmospheric dispersion modelling for assessment of exposure to arsenic for epidemiological studies in the Nitra Valley, Slovakia. J. Geophys. Res – Atmos., 106, 17421–17432.

Davis, A., Ruby, M.V., Bloom, M., Schoof, R., Freeman, G., Bergstrom, P.D., 1996. Mineralogic constraints on the bioavailability of arsenic in smelter-impacted soils. Environ. Sci. Technol., 30, 392–399.

Department for Environment, Food and Rural Affairs and the Environment Agency, 2000. Soil Guideline Values for Arsenic Contamination. R&D Publication SGVI, Environment Agency, Bristol, UK.

Fabianova, E., Bencko, V., 1995. Central European Study on Health Impact of Environmental Pollution. Final report, Project PHARE EC/91/HEA/18. European Union, Brussels, Belgium.

Farago, M.E., Kavanagh, P., 1999. Proportions of arsenic species in human urine. In: Chappell, W.R., Abernathy, C.O., Calderon, R.L. (Eds), Arsenic Exposure and Health Effects. Elsevier, Amsterdam, pp. 325–334.

Farago, M.E., Thornton, I., Kavanagh, P., Elliott, P., Leonardi, G.S., 1997. Health aspects of human exposure to high arsenic concentrations in soil in South West England. In: Abernathy, C.O., Calderon, R.L., Chappell, W.R. (Eds), Arsenic Exposure and Health Effects. Chapman & Hall, London, pp. 210–226.

Kabata-Pendias, A., Pendias, H., 1984. Trace Elements in Soils and Plants. CRC Press, Boca Raton, FL.

Keegan, T., Bing Hong, Thornton, I., Farago, M.E., Jakubis, M., Pesch, B., Ranft, U., Niewenhuijsen, M.J., the EXPASCAN Study Group, 2002. Assessment of arsenic levels in Prievidza district. J. Exp. Anal. Environ. Epidemiol., 12, 179–185.

Leonardi, G.S., Elliott, P., Thornton, I., Farago, M., Grundy, C., Shaddick, G., 1995. Arsenic contamination and bladder cancer: an ecological study. Presented at the Annual Conference of the International Society for Environmental Epidemiology and International Society for Exposure Analysis, August, 1995. Abstract PO 71, Epidemiology 6, Supplement to No 4.

National Cancer Institute of Slovakia (SK NCI), 2000. Cancer incidence in Slovakia 1997, Bratislava, Slovakia (in Slovak).

Pesch, B., Ranft, V., Jakubis, P., Niewenhuijsen, M.J., Hergemoller, A., Unfried, K., Jakubis, M., Miskovic, P., Keegan, T., the EXPASCAN Study Group, 2002. Environmental arsenic exposure from a coal-burning power plant as a potential risk factor for non-melanoma skin cancer: results from a case-control study in the district of Prievidza, Slovakia. Am. J. Epidemiol., 155, 798–809.

Ranft, V., Miskovic, P., Pesch, B., Jakubis, P., Fabianova, E., Nieuwenhuijsen, M.J., Hergemoller, A., Jakubis, M., Keegan, T., EXPASCAN Study Group, 2003. Association between arsenic exposure from a coal-burning power plant and urinary arsenic concentrations in Prievidza district, Slovakia. Environ. Health Perspectives, 2003, 111, 889–894.

Thompson, M., Walsh, J.M., 1989. Handbook of Inductively Coupled Plasma Spectroscopy. Blackie, Glasgow.

Thornton, I., Farago, M.E., 1997. The geochemistry of arsenic. In: Abernathy, C.O., Calderon, R.L., Chappell, W.R. (Eds), Arsenic Exposure and Health Effects. Chapman & Hall, London, pp. 1–16.

Watt, J., Moorcroft, S., Brooks, K., Culbard, E., Thornton, I., 1993. Metal contamination of dusts and soils in urban and rural housedusts in the United Kingdom: 1. Sampling and analytical techniques for household and external dusts. In: Hemphill, D. (Ed.), Trace Substances in Environmental Health, Vol. 17, University of Missouri, Columbia, pp. 229–235.

Zheng, B.-S., Zhang, J., et al., 1996. Geochemical investigation on arsenism caused by high-arsenic coal. Chin. Med. J., 15, 44–48 (in Chinese) cited in Wing-Sun Chow, 1999. Review of Investigations in Arsenic Occurrence, Exposure and Health Effects in the People's Republic of China.Unpublished MSc Thesis, Imperial College of Science, Technology and Medicine, London.

Arsenic Exposure and Health Effects V
W.R. Chappell, C.O. Abernathy, R.L. Calderon and D.J. Thomas, editors
© 2003 Published by Elsevier B.V.

Chapter 4

Trivalent arsenic species: analysis, stability, and interaction with a protein

Guifeng Jiang, Xiufen Lu, Zhilong Gong, William R. Cullen
and X. Chris Le

Abstract

Methods for speciation analysis of arsenite (As^{III}), monomethylarsonous acid (MMA^{III}), and dimethylarsinous acid (DMA^{III}) were summarized. Separation of these trivalent arsenic species was carried out by high performance liquid chromatography, solvent extraction, or selective hydride generation under controlled reaction pH. Hydride generation atomic fluorescence, inductively coupled plasma mass spectrometry (ICPMS), and hydride generation atomic absorption were used for detection. The methods have been used for determination of arsenic species in water and urine samples. To improve the stability of MMA^{III} and DMA^{III}, diethylammonium diethyldithiocarbamate was used to complex with the trivalent arsenic species. Diethylammonium diethyldithiocarbamate was found to be a promising preservative to stabilize MMA^{III} and DMA^{III}. Interactions of As^{III}, MMA^{III}, and DMA^{III} with metallothionein were demonstrated. Size exclusion chromatography separation with ICPMS analysis showed that As^{III}, MMA^{III}, and DMA^{III} could readily bind with metallothionein. Hybrid quadrupole time-of-flight mass spectrometry analysis provided information on the binding stoichiometry. Metallothionein was able to bind with a maximum of 6 As^{III}, 10 MMA^{III}, and 20 DMA^{III} when these arsenic species were present in excess. No binding was observed between metallothionein and the pentavalent arsenic species. Studies of arsenic interaction with proteins are useful to a better understanding of arsenic health effects.

Keywords: arsenic speciation, stability, protein interaction, metallothionein, methylation, metabolism, analytical technique

1. Introduction

Biomethylation is the major metabolic process for inorganic arsenic (Cullen and Reimer, 1989; Yamauchi and Fowler, 1994; Styblo et al., 1995; Goyer, 1996; Aposhian, 1997; NRC, 1999; Vahter, 1999). The methylation process involves stepwise two-electron reduction of arsenic followed by oxidative addition of a methyl group (Cullen et al., 1984, 1989; Cullen and Reimer, 1989). Methyltransferases are responsible for methyl transfer and *S*-adenosyl-methionine (SAM) is the methyl donor (Zakharyan et al., 1995, 2001; Aposhian, 1997; Lin et al., 2002). Trimethylarsine oxide (TMAO^{V}) and trimethylarsine (TMA^{III}) are the end products formed by some microorganisms. Dimethylarsinic acid (DMA^{V}) is the usual end product detected in humans, although dimethylarsinous acid (DMA^{III}) has recently been detected in human urine (Aposhian et al., 2000a,b; Le et al., 2000a; Del Razo et al., 2001; Mandal et al., 2001).

The determination of the trivalent arsenic metabolites, monomethylarsonous acid (MMA^{III}) and dimethylarsinous acid (DMA^{III}), is important to the understanding of the arsenic methylation pathway (Cullen and Reimer, 1989; Aposhian et al., 1999; Styblo et al., 1999; Zakharyan et al., 1999; Sampayo-Reyes et al., 2000). The first component of this paper summarizes the methods that have been used for the speciation of MMA^{III} and DMA^{III}.

An important issue regarding the speciation of MMA^{III} and DMA^{III} is the stability and preservation of these species. We have reported that MMA^{III} and DMA^{III} were less stable than the other arsenic species, suggesting that new strategies for sample handling are needed for analysis of these trivalent arsenic species (Gong et al., 2001). We have recently developed a technique that is able to preserve the speciation of MMA^{III} and DMA^{III} in human urine samples. A metal complexing agent (diethylammonium diethyldithiocarbamate) was added to urine samples as a preservative, and the pH of the samples was adjusted to 6.0. Samples were stored at $-20°C$ and arsenic species were found to be stable for up to 3 weeks.

Toxicological evaluations of these trivalent arsenic methylation metabolites show that MMA^{III} and DMA^{III} are as toxic as, or more toxic than, inorganic arsenite (As^{III}) (Styblo and Thomas, 1995; Styblo et al., 1997; Lin et al., 1999; Petrick et al., 2000, 2001; Mass et al., 2001; Nesnow et al., 2002). However, mechanisms of action responsible for the toxicity of these arsenicals are not clear. Binding of arsenic to proteins (enzymes) has been presumed to be a cause of toxicity. However, no arsenic-containing protein has been identified. Little is known about how these arsenic compounds interact with biological molecules, such as proteins. We report here binding of trivalent arsenical with metallothionein (MT), MT is a small protein (6000–8000 Da) that is rich in cysteine. Several heavy metals and stress factors can induce this protein. MT is induced as a response to toxicity. Trivalent arsenic is presumed to bind with sulfhydryl groups. However, arsenic binding to MT has not been studied. Therefore, we have examined arsenic binding to MT as an example of arsenic–protein interactions.

2. Methodology

2.1. HPLC separation with hydride generation atomic fluorescence detection (HPLC-HGAFS)

Separation of various arsenic species was carried out using a high performance liquid chromatography (HPLC) system that consisted of an HPLC pump (Model 307, Gilson, Middletone, WI), a 6-port sample injector with a 20-μl sample loop (Rheodyne, Rohnet Park, CA), and an appropriate column (Le and Ma, 1998; Le et al., 2000a,b). The column was mounted inside a column heater (Model CH-30, Eppendorf, Westbury, NY) that was controlled by a temperature controller (Model TC-50, Eppendorf). The column temperature was maintained at $50 \pm 1°C$. Mobile phase was pre-heated to the temperature of the column by using a precolumn coil of 50 cm stainless steel capillary tubing, which was also placed inside the column heater. A reversed-phase column (ODS-3, 150 mm × 4.6 mm, 3 μm particle size, Phenomenex) was used for separation. A mobile phase solution (pH 5.9) contained 5 mM tetrabutylammonium hydroxide, 3 mM malonic acid, and 5% methanol, and its flow rate was 1.2 ml/min. A hydride generation atomic fluorescence spectrometer (HGAFS) (Model Excalibur 10.003, P.S. Analytical, Kent, UK) was used for the detection of arsenic as described previously (Le et al., 1996; Le and Ma, 1998; Ma and Le, 1998; Le et al., 2000a). For hydride generation, continuous flows of hydrochloric acid (1.2 M, 10 ml/min) and sodium borohydride (1.3%, 3 ml/min) were introduced to react with arsenic in the HPLC effluent. A continuous flow of argon (250 ml/min) was used to carry arsenic hydrides to the atomic fluorescence detector.

2.2. HPLC separation with inductively coupled plasma mass spectrometry detection (HPLC-ICPMS)

Another HPLC separation system consisted of a Perkin-Elmer (Norwalk, CT) 200 Series pump, autosampler, column oven, and an analytical column. For the separation of As^{III}, As^{V}, MMA^{V}, DMA^{V}, MMA^{III}, and DMA^{III}, a reversed-phase column (ODS-3, 150 mm × 4.6 mm, 3 μm particle size, Phenomenex) was used. A mobile phase solution (pH 5.9) contained 5 mM tetrabutylammonium hydroxide, 3 mM malonic acid, and 5% methanol, and its flow rate was 1.2 ml/min.

For separation of MT-bound from the unbound arsenicals, a size exclusion column (300 mm × 4.6 mm, Phenomenex) was used. Separation was carried out using isocratic elution with a mobile phase of 5 mM sodium phosphate (pH 7.3) and a flow rate of 0.8 ml/min. The injection volume was 20 μl.

The effluent coming from HPLC was directly detected using an Elan 6100 DRC^{plus} ICPMS (PE Sciex, Concord, Ont., Canada). A standard liquid sample introduction system consisted of a Meinhard nebulizer coupled to a standard cyclonic spray chamber. HPLC column outlet was directly connected to the nebulizer and the spray chamber using a short (20 cm) PEEK tubing (0.3 mm i.d.). The instrument settings were optimized for sensitivity with a solution containing 1 μg/l of each of Mg, In, Pb, Ce, Ba, and Th. The ICPMS instrumental operating conditions are summarized in Table 1. Turbochrom workstation software (Perkin-Elmer/Sciex) and IgorPro (WaveMetrics, Lake Oswego, OR) were used to convert ICPMS files and to plot HPLC data as chromatograms.

Table 1. ICPMS operating conditions.

RF power	1100 W
Ar auxiliary gas flow rate	1.2 l/min
Ar plasma gas flow rate	15 l/min
Ar nebulizer gas flow rate	0.90 l/min
Dwell time	200 ms
Detector mode	Dual mode
Scan mode	Peak hopping
Lens	Autolens On
Mass (*m/z*)	75 (As) and 114 (Cd)

2.3. Hybrid quadrupole time-of-flight mass spectrometry

An Applied Biosystem/MDS Sciex QSTAR Pulsar i mass spectrometer (Concord, Ont., Canada) was equipped with a Turbo Ionspray ionization source. Analyte solutions were introduced into the source by a 1-ml gas tight syringe (Hamilton, Reno, NV) and an integrated syringe pump. The flow rate was 10 μl/min. The mass spectrometer was operated in the positive ion mode. The instrument was calibrated daily with a commercial calibration standard. Analyst QS software (Applied Biosystems, Foster City, CA) was used for the spectrum acquisition and data analysis. IgorPro software (WaveMetrics) was used to plot the mass spectra.

Mass spectra were acquired with an electrospray voltage of 5500 V, first declustering potential (DP1) of 65 V, second declustering potential (DP2) of 15 V, and focusing potential (FP) of 215 V. The resolution was 10,000 (fwhm) at m/z 850 and the mass accuracy was 10 ppm using internal standard.

In the tandem MS/MS mode, the parent ion was selected by the first quadrupole (Q1) with low mass resolution, and fragmented in the second quadrupole (Q2) by collision-induced dissociation (CID) with a collision energy of 90 eV and collision gas setting of 12. The resulting product ions were analyzed by the time-of-flight (TOF) analyzer. In assigning the product ions, the theoretical fragmentation pattern of MT was generated for comparison using the MS-product tool in Protein Prospector (http://prospector.ucsf.edu).

2.4. Standards and reagents

An atomic absorption arsenic standard solution containing 1000.0 mg As/l as arsenite (Sigma, St. Louis, MO) was used as the primary arsenic standard. Arsenic trioxide (99.999%), sodium arsenate, $As(O)OH(ONa)_2 \cdot 7H_2O$ and sodium cacodylate, $(CH_3)_2 As(O)ONa$ were obtained from Sigma/Aldrich. Monomethylarsonate, $CH_3As(O)OHONa$, was obtained from Chem Service (West Chester, PA). The source of MMA^{III} was the solid oxide (CH_3AsO), and DMA^{III} was the iodide $[(CH_3)_2AsI]$, which was prepared following literature procedures (Cullen et al., 1989, Burrows and Turner, 1920). Trimethylarsine oxide $[(CH_3)_3AsO, TMAO]$ was synthesized following the procedures of Merijanian and Zingaro (1966). Solutions of standard arsenic compounds were prepared by appropriate dilutions with deionized water from 1000 mg/l stock solutions. Solutions of As^V, MMA^V, MMA^{III}, DMA^{III}, DMA^V and TMAO were standardized against a primary As^{III} standard using an inductively coupled plasma mass spectrometer.

Metallothionein II (Cd_4Zn_3MT) was purchased from Sigma (St. Louis, MO) and was used without further purification. The protein stock solution (5 mg/ml) was prepared by dissolving 5 mg MT in 1 ml water. Working solutions were prepared by serial dilutions of the stock solution with water (or methanol as required).

Diethylammonium diethyldithiocarbamate (DDDC), tetrabutylammonium hydroxide (TBAH), malonic acid, and disodium hydrogen phosphate (Na_2HPO_4) were obtained from Aldrich (Milwaukee, WI). HPLC grade methanol and formic acid were from Fisher Scientific (Fair Lawn, NJ). All reagents used were of analytical grade or better.

2.5. Sample storage and preservation

Deionized water samples containing 100 µg/l of either MMA^{III} or DMA^{III} were prepared fresh by serial dilutions of the stock solutions with deionized water. Aliquots (1 ml) of these samples were placed in 1.5-ml plastic tubes, and separately stored at 25°C (room temperature), 4°C (refrigerated), and -20°C (frozen), respectively.

A first-morning void urine was collected from three volunteers who had not eaten any seafood for 2 weeks prior to the sample collection. Measured amounts of MMA^{III} and DMA^{III} stock solutions were spiked into the urine so that the urine samples contained 100 µg/l of either MMA^{III} or DMA^{III}. Aliquots (1 ml) of the urine samples were placed in 1.5-ml plastic tubes, and separately stored at 25°C, 4°C, and -20°C, respectively.

The water and urine samples containing added MMA^{III} and DMA^{III} were stored for up to 5 months at the specified temperatures. They were periodically analyzed for arsenic speciation by using HPLC separation with hydride generation atomic fluorescence spectrometry detection (HPLC-HGAFS). Triplicate analyses were carried out for each sample aliquots.

For the preservation of MMA^{III} and DMA^{III} in urine samples, DDDC was added to urine samples to a concentration of $5-10$ mM. Aliquots (1 ml) of the urine samples with DDDC were placed in 1.5-ml plastic tubes, and separately stored at -20°C for up to 4 months. Arsenic species were analyzed weekly to examine the stability of MMA^{III} and DMA^{III} in the urine samples.

2.6. Arsenic binding with MT

As^{III}, MMA^{III}, and DMA^{III} (1 µM, or 75 µg/l) in deionized water were each incubated with various concentrations of MT ($5-15$ µM) in phosphate buffered saline (PBS) for $1-2$ h. These solutions were subject to HPLC-ICPMS analysis of MT–As complexes. The MT-bound and free arsenic species were separated on a size exclusion column with phosphate buffer (pH 7.3) as a mobile phase, and were detected using the ICPMS. Both arsenic (m/z 75) and cadmium (m/z 114) were detected simultaneously to confirm the binding of arsenic with MT because cadmium was present in the native MT.

To study the binding stoichiometry between MT and As^{III}, MMA^{III}, and DMA^{III} species, a constant amount of MT was incubated with varying concentrations of the trivalent arsenic species to obtain MT to arsenic molar ratios of 20:1, 5:1, 1:1, 1:5, 1:20, and 1:100. The mixture solutions were incubated at room temperature for 2 h. They were diluted with 50% methanol immediately prior to electrospray ionization mass spectrometry (ESI-MS) analysis.

Initially, the MT/arsenic mixture solutions were acidified with 1% formic acid to remove Zn and Cd from the native MT and to simplify the mass spectra of MT–As complexes. Subsequent experiments were carried out without the acidification.

3. Results and discussion

3.1. Speciation of As^{III}, MMA^{III}, DMA^{III}, As^{V}, MMA^{V}, and DMA^{V}

The most commonly used arsenic speciation techniques involve a combination of chromatographic separation with spectrometric detection. Four methods have been reported for the speciation analysis of monomethylarsonous acid (MMA^{III}) and dimethylarsinous acid (DMA^{III}) (Table 2).

The first method used solvent extraction of trivalent arsenic species followed by hydride generation atomic absorption analysis (Hasegawa et al., 1994). As^{III}, MMA^{III}, and DMA^{III} were separated from pentavalent arsenic species by solvent extraction using diethyldithiocarbamate and carbon tetrachloride (CCl$_4$). The diethyldithiocarbamate–arsenic complexes were back-extracted with 0.1 M NaOH aqueous solution. The solution was subsequently analyzed for arsenic species using hydride generation atomic absorption after cold trapping and gas chromatographic separation. Because the pentavalent arsenic species were not extracted to the CCl$_4$ phase with diethyldithiocarbamate, the analysis of the extract accounted for the trivalent arsenic species. This method was used to analyze arsenic species in water. Minor amounts of MMA^{III} ($<0.012\ \mu g/l$) and DMA^{III} ($<0.014\ \mu g/l$) were found along with MMA^{V} ($<0.05\ \mu g/l$), DMA^{V} ($<0.76\ \mu g/l$), As^{III} ($\sim 0.2\ \mu g/l$), and As^{V} ($\sim 1.9\ \mu g/l$) in water samples from Lake Biwa, Japan (Hasegawa et al., 1994; Sohrin et al., 1997).

We combined HPLC separation with hydride generation atomic fluorescence spectrometry (HGAFS) detection for arsenic speciation analysis. HPLC separation of As^{III}, As^{V}, MMA^{V}, DMA^{V}, MMA^{III}, and DMA^{III} was performed on a reversed-phase column (ODS-3) using 5 mM tetrabutylammonium hydroxide as an ion-pairing agent. The separation was complete within 7 min and the elution order was As^{III}, MMA^{III}, DMA^{V}, MMA^{V}, DMA^{III}, and As^{V}. A post-column hydride generation-atomic fluorescence system provided high sensitivity detection for trace levels of arsenic species. The HPLC-HGAFS method resulted in the discovery of MMA^{III} and DMA^{III} in human urine (Aposhian et al., 2000a,b; Le et al., 2000a,b). It continues to be used for toxicological and epidemiological studies of arsenic (Cohen et al., 2002; Gong et al., 2002).

Table 2. Methods used for speciation analysis of MMA^{III} and DMA^{III}.

Method	Samples	References
Solvent extraction and HG-GC-AAS	Water, Lake Biwa, Japan	Hasegawa et al. (1994) and Sohrin et al. (1997)
HPLC-HGAFS	Human urine, Inner Mongolia	Aposhian et al. (2000a) and Le et al. (2000a,b)
HPLC-HGAFS	Human urine, Romania	Aposhian et al. (2000b) and Lu et al. (2001)
HPLC-ICPMS	Human urine, India	Mandal et al. (2001)
HG-GC-AAS	Human urine, Mexico	Del Razo et al. (2001)
HPLC-HGAFS	Rat urine	Cohen et al. (2002)

Other detectors, such as ICPMS, can also be used for detection of arsenic species separated using the same ion-pairing chromatography. Figure 1 shows a typical chromatogram from the HPLC-ICPMS analysis of As[III], MMA[III], DMA[V], MMA[V], DMA[III], and As[V]. Each analysis of the six arsenic species was complete within 5 min.

Anion exchange separation followed by ICPMS detection was used to determine MMA[III] and DMA[III] in human urine samples collected from India where the subjects were exposed to high levels of arsenic from drinking water (Mandal et al., 2001). A polymer-based anion exchange column (10 cm × 7.5 mm) was used, with a mobile phase containing 15 mM citric acid. The pH of the mobile phase was adjusted to 2.0 with nitric acid, and this mobile phase did not have any buffering effect. A subsequently modified method involved cation exchange separation with ICPMS detection (Shiobara et al., 2001). With a strong cation exchange column and a mobile phase containing 36 mM formic acid and 2 mM ammonium formate (pH 2.8), DMA[III] (~5 min) was separated from DMA[V] (~6.5 min). The method was used to determine DMA[III] uptake by red blood cells (Shiobara et al., 2001).

Another method used for the determination of MMA[III] and DMA[III] in urine samples involved a combination of selective hydride generation, cold trapping, gas chromatography, and atomic absorption detection. Several arsenic species could form arsines upon treatment with sodium borohydride in an acid medium: As[III] and As[V] gave AsH_3, MMA[V] and MMA[III] gave CH_3AsH_2, and DMA[V] and DMA[III] gave $(CH_3)_2AsH$. Because the boiling points of these arsines were different (°C): AsH_3, −55; CH_3AsH_2, 2; and $(CH_3)_2AsH$, 35.6, the arsenic species could be differentiated by using hydride generation with cryogenic trapping and gas chromatography. Arsines produced in a reaction vessel were swept to and trapped in a U-shaped tube immersed in

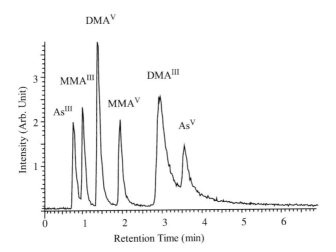

Figure 1. Speciation analysis of As[III], MMA[III], DMA[V], MMA[V], DMA[III], and As[V] using HPLC-ICPMS. A reversed-phase column (ODS-3, 150 mm × 4.6 mm, 3 μm particle size) was used for separation. A mobile phase solution (pH 5.9) contained 5 mM tetrabutylammonium hydroxide, 3 mM malonic acid, and 5% methanol, and its flow rate was 1.2 ml/min. An ICPMS was used for detection of arsenic at m/z 75. The concentrations of the arsenic species were As[III] (10 μg/l), MMA[III] (10 μg/l), DMA[V] (15 μg/l), MMA[V] (10 μg/l), DMA[III] (25 μg/l), and As[V] (10 μg/l). The injection volume was 20 μl.

liquid nitrogen. After complete trapping of arsines (usually took 20 min), liquid nitrogen was removed, and the U-tube was warmed up. The arsines evaporated upon heating, and were introduced to a detector (such as an atomic absorption spectrometer), where AsH_3, CH_3AsH_2, and $(CH_3)_2AsH$ were detected sequentially. Although both trivalent and pentavalent arsenic species could form arsines under acidic conditions (pH < 1), the generation of arsines from the pentavalent arsenic species were not efficient under high pH conditions. Therefore, by carefully controlling the pH of the reaction medium, e.g. pH higher than 6, the trivalent arsenic species (As^{III}, MMA^{III}, and DMA^{III}) could be selectively determined (Del Razo et al., 2001). The sum of trivalent and pentavalent arsenic species could be determined using a lower pH reaction medium (pH < 1). Thus, two separate analyses of each sample, carried out at different pH conditions, were required to selectively determine the trivalent arsenic species at the high pH and the total arsenic concentration at the low pH. It is important to note that the pH needs to be carefully selected to minimize the interference from the pentavalent arsenic species (As^V, MMA^V, and DMA^V) in the determination of the trivalent arsenic species (As^{III}, MMA^{III}, and DMA^{III}).

3.2. Stability and preservation of MMA^{III} and DMA^{III}

A main problem facing the quantitative analysis of MMA^{III} and DMA^{III} is their oxidative instability. MMA^{III} and DMA^{III} can be readily oxidized to MMA^V and DMA^V during sample collection, handling, and storage. It is conceivable that the MMA^{III} and DMA^{III} concentrations reported from the analyses of urine samples may represent only the residual amounts of these labile arsenic species (Aposhian et al., 2000a,b; Le et al., 2000a,b; Del Razo et al., 2001; Mandal et al., 2001).

We have compared the effects of the storage temperature (25, 4, and $-20°C$) and storage duration (up to 4 months) on the stability of MMA^{III} and DMA^{III} in deionized water and in human urine. MMA^{III} and DMA^{III} were spiked to deionized water and human urine samples, and the samples were stored under various conditions. The concentrations of arsenic were determined periodically. We found that MMA^{III} and DMA^{III} were much less stable than other arsenic species, and their stability depended on sample matrix and temperature (Gong et al., 2001).

Low temperature conditions (4 and $-20°C$) improved the stability of arsenic species over the room temperature storage condition. MMA^{III} in deionized water was relatively stable for almost 4 months, when stored at 4 or $-20°C$ with less than 10% of MMA^{III} oxidized to MMA^V. In contrast, most of MMA^{III} (>90%) in urine was oxidized to MMA^V over a 4-month period under the 4 or $-20°C$ storage conditions. At 25°C, MMA^{III} in urine was completely oxidized to MMA^V within a week (Fig. 2).

DMA^{III} in deionized water was stable for only 2–3 days, being rapidly oxidized to DMA^V. DMA^{III} in urine was completely oxidized to DMA^V within a day at 4 or $-20°C$. The conversion of DMA^{III} to DMA^V in urine at 25°C was complete in 17 h.

It is important to emphasize that the stability of arsenic species depends on the sample matrix, although what specific components of sample matrix affect arsenic stability is not clearly understood. We have shown previously that in some human urine samples, As^{III}, DMA^V, MMA^V, and As^V were stable for only 2 months when stored at $-20°C$, whereas in

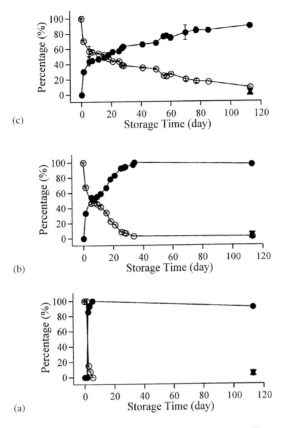

Figure 2. Effect of storage duration and temperature on the stability of MMA^{III} in urine. (a) 25°C, (b) 4°C, and (c) −20°C. (○) MMA^{III}, (●) MMA^{V}, (▲) As^{III}, and (▼) DMA^{V} (reprinted with permission from Gong et al., 2001).

most urine samples they could be stable for more than a year under the same storage conditions (Feldmann et al., 1999). Similarly, it has been demonstrated that the stability of MMA^{III} and DMA^{III} was quite different in urine samples collected from different volunteers (Del Razo et al., 2001).

We have made much effort to stabilize MMA^{III} and DMA^{III} using various potential preservatives. A number of reducing agents were initially tested because a reducing environment would minimize the oxidation of MMA^{III} and DMA^{III}. However, a main challenge of using the reducing agents was to avoid the reduction of As^{V}, MMA^{V}, and DMA^{V} species.

We found that the use of a complexing agent, DDDC, was able to improve the stability of MMA^{III} and DMA^{III} (Gong et al., 2003). Urine samples spiked with MMA^{III} and DMA^{III} were stored at −20°C for up to 4 months and were periodically analyzed for arsenic species. In the presence of DDDC (1–10 mM), MMA^{III} was found to be stable for 4 months (with a recovery of 85–95%). DMA^{III} was partially preserved. Approximately 80% of DMA^{III} remained after storage of 3 weeks and 10–40% remained after 4 months

(Gong et al., 2003). Therefore, the use of DDDC as a preservative is promising for stabilizing MMA^{III} and DMA^{III} for a period that is suitable for typical toxicological and epidemiological studies.

3.2.1. Binding of arsenic to MT

Studies of biochemical interactions between arsenic and proteins are crucial to a better understanding of arsenic health effects. Arsenite has been shown to interact with glutathione to produce 3:1 glutathione:arsenic complex (Serves et al., 1995). Cysteine was able to react with As^{III} to produce an As(Cys)$_3$ complex (Johnson and Voegtlin, 1930). Recently, Farrer et al. (2000) found that As^{III} was able to distort polypeptide structure in order to satisfy its preferred trigonal-pyramidal thiolate coordination. The peptides used were synthesized to ensure the presence of a thiolate for metal binding. Using model α-helical peptides containing two cysteine residues, Cline et al. (2003) further demonstrated that As^{III} binding could be a significant modulator of helical secondary structure. Making use of the high affinity of As^{III} for thiol groups, Tsien and coworkers (Griffin et al., 1998; Adams et al., 2002) have designed a fluorescent probe to label proteins that contained an introduced sequence rich in cysteine residues. The probe was a biarsenical fluorescein derivative. Previous studies have also recognized the roles of arsenic–protein interactions in the metabolism and possible health effects of arsenic (Chen et al., 1985; Ji and Silver, 1992; Stevens et al., 1999).

We demonstrate here the binding of trivalent arsenicals with MT. We first developed a technique that is based on size exclusion HPLC separation of the MT-bound arsenic from the unbound arsenic followed by specific detection of arsenic using ICPMS. Because of their size differences, the MT-bound arsenic species were separated from the unbound arsenicals. Experiments carried out using As^{III}, MMA^{III}, and DMA^{III} (1 μM) and MT of various concentrations (1–50 μM) confirmed that these trivalent arsenic species were able to bind with MT (Jiang et al., 2003).

To further understand the binding of arsenicals to MT and their binding stoichiometry, we used ESI-MS to examine various species resulting from the binding. Analyses of MT solutions by ESI-MS revealed that the binding of Zn and Cd in the native MT was pH dependent. At pH 6, both Zn and Cd remained intact on the MT. When the pH of the MT solution was decreased to 4.5, zinc was released, leaving only Cd on the MT. When the pH of the solution was further lowered to 2.0, Cd was also released from the protein, resulting in apo-MT. Figure 3 shows a typical spectrum from the ESI-MS analysis of apo-MT at pH 2.0. Characteristic ions (m/z) for the apo-MT included 1022.04 (6$^+$-charge state), 1226.26 (5$^+$-charge state), and 1532.64 (4$^+$-charge state). These correspond to a molecular mass of 6126 for the apo-MT, consistent with the expected value (6126.3).

Having established ESI-MS analysis for MT, we further studied complex species in reaction solutions containing 7 μM MT and varying concentrations of As^{III} (0.35–140 μM). The molar ratios of MT to As^{III} ranged from 5:1 to 1:20. Multiple charged ions (5$^+$ and 4$^+$) were predominant. With the increase of As^{III} concentration at a constant concentration of MT, the number of As^{III} atoms bound to the MT increased. At a low concentration of As^{III} relative to MT, the apo-MT ions were observed as the dominant species. When the As^{III} concentration was increased to equal the concentration of MT, MT–As$_2$ complex was the most abundant species, along with the MT–As, MT–As$_3$, and

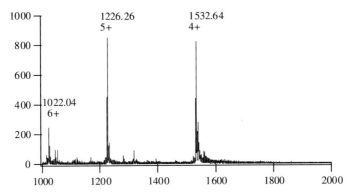

Figure 3. An ESI mass spectrum for apo-MT. The native MT was diluted in 0.5% formic acid (pH 2.0) and 50/50 (v/v) methanol/water. At this pH, Zn and Cd were removed from the MT. The solution was introduced at a flow rate of 10 μl/min and an electrospray voltage of 5500 V.

MT–As$_4$ complexes. In the presence of fivefold excess of As^{III} over MT, MT–As$_5$ and MT–As$_6$ complexes were detected, with MT–As$_5$ as the major species. Increasing As^{III} to an As:MT ratio of 20:1 resulted in the formation of the MT–As$_6$ complex as the dominant species. A further increase of As^{III} to an As:MT ratio of 100:1 did not show the formation of MT complex with more than 6 As^{III} moieties. It appeared that the maximum number of arsenic atoms bound to the MT was 6.

We carried out further experiments to confirm that the formation of the MT–As complexes occurred in solution, and not an artifact of the electrospray process. We examined the formation of MT–As complexes over different incubation periods in solution and the results indicated that the complex formation was dependent on the incubation in solution (Fig. 4). When As^{III} and MT (4:1 molar ratio) were mixed and immediately analyzed using ESI-MS, only MT (1226.26 at 5$^+$ and 1532.64 at 4$^+$ charges) and MT–As$_1$ (1240.66 at 5$^+$ and 1550.76 at 4$^+$ charges) species were observed. Repeat ESI-MS analysis of the same mixture after 1 h incubation demonstrated the presence of MT–As$_5$ and MT–As$_6$ species (Fig. 4b). These results suggest that the formation of MT–As complexes took place in the liquid phase and depended on the reaction time in solution.

The binding of each MT molecule with a maximum of 6 As^{III} in the presence of excess As^{III} is understandable as schematically illustrated in Scheme 1. MT contains 20 cysteine residues. Because each As^{III} is able to bind with three cysteines, the maximum number of As^{III} that can be bound on an MT molecule is 6.

Further studies using MMA^{III} and DMA^{III} binding with MT confirmed the arsenic binding stoichiometry. As shown in Scheme 1, MMA^{III} [CH$_3$As(OH)$_2$] would be able to bind with two thiols. DMA^{III} [CH$_3$AsOH] would be able to bind to only one thiol group. Therefore, up to 10 As(CH$_3$) and 20 As(CH$_3$)$_2$ could be bound to each MT when MMA^{III} and DMA^{III} are in excess. This was confirmed by analyzing reaction mixtures containing varying molar ratios of MMA^{III} over MT (1:5 to 50:1) and DMA^{III} over MT (1:5 to 200:1).

The expected and experimental masses of MT–[As(CH$_3$)]$_n$ are listed in Table 3. The mass difference between the MT–[As(CH$_3$)]$_n$ complex and the apo-MT is 87.9n,

G. Jiang et al.

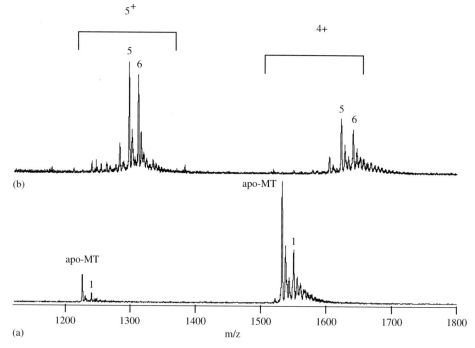

Figure 4. ESI mass spectra from the repeat analyses of a solution containing 1 μM MT and 4 μM As^{III}. The mixture of MT (1 μM) and As^{III} (4 μM) in deionized water was either analyzed immediately after mixing (a) or was incubated at room temperature for 1 h before analysis (b). The solution was diluted with 50% methanol and acidified with formic acid to pH 2.0 immediately prior to ESI-MS analysis. The peaks labeled with numbers were complexes of MT and As^{III}. The numbers on the peaks represent the number of As^{III} bound to the MT molecule. For example, peak 6 represents MT–As$_6$ (adapted from Jiang et al., 2003).

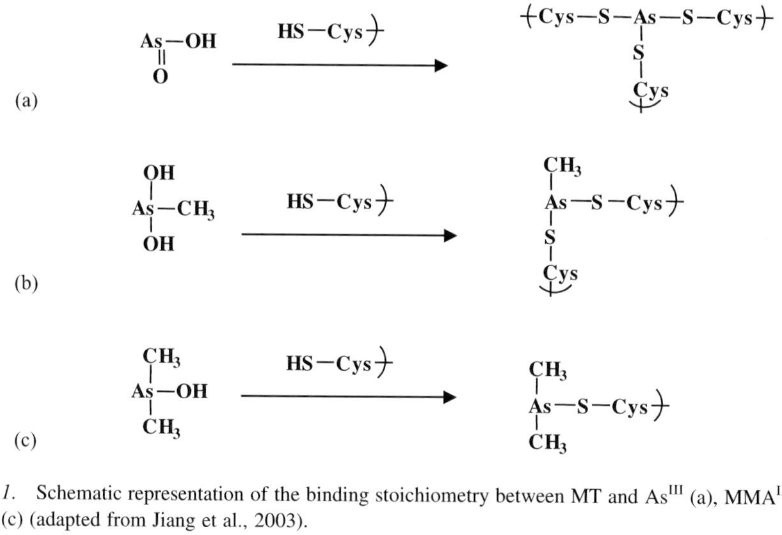

Scheme 1. Schematic representation of the binding stoichiometry between MT and As^{III} (a), MMA^{III} (b), and DMA^{III} (c) (adapted from Jiang et al., 2003).

Table 3. Expected and experimental molecular masses for unbound MT and MT bound with MMA^{III}.

Species	Theoretical values (Da)	Experimental values				
	Molecular mass	5$^+$ state (*m/z*)	4$^+$ state (*m/z*)	Molecular mass (Da)	Δm (Da)	Mass accuracy (ppm)
apo-MT	6126.30	1226.20	1532.65	6126.30	0	0
MT–As (CH$_3$)	6214.23	1243.97	1554.49	6214.41	0.18	29
MT–[As (CH$_3$)]$_2$	6302.16	1261.36	1576.50	6301.90	−0.26	−41
MT–[As (CH$_3$)]$_3$	6390.09	1279.30	1598.63	6390.99	0.9	142
MT–[As (CH$_3$)]$_4$	6478.02	1296.71	1620.60	6478.48	0.46	72
MT–[As (CH$_3$)]$_5$	6565.95	1314.09	1642.62	6565.97	0.02	4
MT–[As (CH$_3$)]$_6$	6653.88	1331.93	1664.52	6654.36	0.48	73
MT–[As (CH$_3$)]$_7$	6741.81	1349.52	1686.46	6742.22	0.41	62
MT–[As (CH$_3$)]$_8$	6829.74	1367.12	1708.62	6830.53	0.79	116
MT–[As (CH$_3$)]$_9$	6917.66	1384.56	1730.59	6918.09	0.43	62
MT–[As (CH$_3$)]$_{10}$	7005.59	1402.06	1752.39	7005.41	−0.18	−26

where n is the number of MMA^{III} moieties bound to MT and varies from 1 to 10. This mass difference is consistent with the expected mass of 87.9294 for AsCH$_3$ with a loss of two protons. This is due to the formation of AsCH$_3$ complex with MT through binding with two thiols (resulting in loss of the proton from each thiol group).

The observed and expected masses for MT–[As(CH$_3$)$_2$]$_n$ are listed in Table 4. The measured mass difference between the MT–[As(CH$_3$)$_2$]$_n$ complexes and the apo-MT is $103n$, where n is the number of As(CH$_3$)$_2$ moieties bound to MT and varies from 1 to 20. This is consistent with the expected mass of 103.9607 for As(CH$_3$)$_2$ with the loss of a proton.

To confirm the presence of MT–As complexes, we further analyzed the fragment ions of MT–As species by using tandem mass spectrometry. CID of the MT–As species resulted in arsenic-containing fragments. Figure 5 shows tandem mass spectra of MT–MMA^{III} and MT–DMA^{III} by selecting MT[As(CH$_3$)]$_7$ at *m/z* of 1686.46 (4$^+$) and MT[As(CH$_3$)$_2$]$_7$ at *m/z* of 1714.03 (4$^+$) as the parent ions, respectively. The peak observed at *m/z* of 163.9573 for MT[As(CH$_3$)]$_7$ corresponds to CH$_3$AsSC$_2$H$_4$N$^+$ (theoretical value 163.9515), resulting from AsCH$_3$ and cysteine binding. The As(CH$_3$)$_2$S$^+$ (136.9406) located at *m/z* of 136.9428 and As(CH$_3$)$_2$SC$_2$H$_5$N$^+$ (179.9828) at *m/z* of 179.9885 arise from the As(CH$_3$)$_2$–cysteine binding. Observation of these arsenic-containing fragments further supports the binding of trivalent arsenic with MT.

MMA^{III} and DMA^{III} are two crucial metabolites of arsenic biomethylation process. The use of As^{III}, MMA^{III}, and DMA^{III} is a unique system, allowing us to clearly illustrate the binding stoichiometry. The technique described here can be applied further to study interactions of arsenic with other proteins.

Table 4. Expected and experimental molecular masses for unbound MT and MT bound with DMA[III].

Species	Theoretical values (Da)	Experimental values			
	Molecular mass	4$^+$ charge stat (m/z)	Molecular mass (Da)	Δm (Da)	Mass accuracy (ppm)
apo-MT	6126.3	1532.51	6126.03	−0.27	−44
MT–As $(CH_3)_2$	6230.26	1558.57	6230.29	0.03	4
MT–[As $(CH_3)_2$]$_2$	6334.22	1584.86	6335.43	1.21	190
MT–[As $(CH_3)_2$]$_3$	6438.18	1610.89	6439.57	1.39	216
MT–[As $(CH_3)_2$]$_4$	6542.14	1636.75	6543.01	0.87	132
MT–[As $(CH_3)_2$]$_5$	6646.10	1662.27	6645.06	−1.04	−155
MT–[As $(CH_3)_2$]$_6$	6750.06	1688.27	6749.06	−1.0	−148
MT–[As $(CH_3)_2$]$_7$	6854.02	1714.03	6852.11	−1.91	−279
MT–[As $(CH_3)_2$]$_8$	6957.99	1740.20	6956.81	−1.18	−168
MT–[As $(CH_3)_2$]$_9$	7061.91	1765.77	7059.07	−2.84	−407
MT–[As $(CH_3)_2$]$_{10}$	7165.91	1792.69	7166.74	0.83	117
MT–[As $(CH_3)_2$]$_{11}$	7269.87	1818.60	7270.40	0.53	73
MT–[As $(CH_3)_2$]$_{12}$	7373.83	1844.70	7374.80	0.97	132
MT–[As $(CH_3)_2$]$_{13}$	7477.79	1870.93	7479.73	1.96	260
MT–[As $(CH_3)_2$]$_{14}$	7581.75	1896.72	7582.87	1.12	148
MT–[As $(CH_3)_2$]$_{15}$	7685.71	1922.78	7687.13	1.42	185
MT–[As $(CH_3)_2$]$_{16}$	7789.67	1948.80	7791.20	1.53	196
MT–[As $(CH_3)_2$]$_{17}$	7893.63	1974.71	7894.85	1.23	154
MT–[As $(CH_3)_2$]$_{18}$	7997.59	2000.98	7999.93	2.34	292
MT–[As $(CH_3)_2$]$_{19}$	8101.55	2026.71	8102.83	1.28	158
MT–[As $(CH_3)_2$]$_{20}$	8205.51	2052.85	8207.44	1.87	231

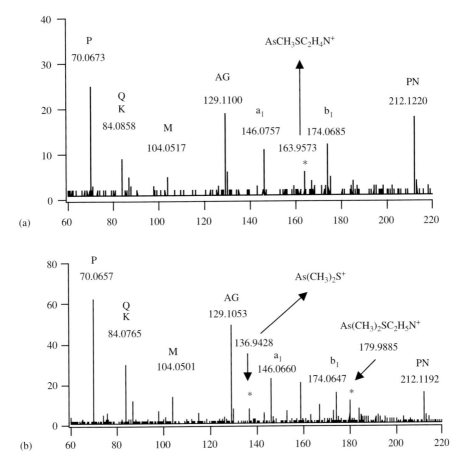

Figure 5. Partial ESI-MS/MS spectra showing the low mass region for MT–[As(CH₃)]₇ (a) and MT–[As(CH₃)₂]₇ (b). The ion at m/z of 1686.46 and 1714.03 with 4^+-charge state was fragmented at the collision cell. The peaks at m/z of 163.9573 in (a) and 136.9428, 179.9885 in (b) are the arsenic-related fragment ions with possible formula shown.

Acknowledgements

We thank the Natural Sciences and Engineering Research Council of Canada (NSERC), the Canadian Water Network NCE, American Water Works Association Research Foundation, and Alberta Health and Wellness for financial support.

References

Adams, S.R., Campbell, R.E., Gross, L.A., Martin, B.R., Walkup, G.K., Yao, Y., Llopis, J., Tsien, R.Y., 2002. New biarsenical ligands and tetracysteine motifs for protein labeling in vitro and in vivo: synthesis and biological applications. J. Am. Chem. Soc., 124, 6063–6076.

Aposhian, H.V., 1997. Enzymatic methylation of arsenic species and other new approaches to arsenic toxicity. Annu. Rev. Pharmacol. Toxicol., 37, 397–419.

Aposhian, H.V., Zakharyan, R.A., Wildfang, E.K., Healy, S.M., Gailer, J., Radabaugh, T.R., Bogdan, G.M., Powell, L.A., Aposhian, M.M., 1999. How is inorganic arsenic detoxified? In: Chappell, W.R., Abernathy, C.O., Calderon, R.L. (Eds), Arsenic Exposure and Health Effects. Elsevier, Amsterdam, pp. 289–297.

Aposhian, H.V., Gurzau, E.S., Le, X.C., Gurzau, A., Healy, S.M., Lu, X., Ma, M., Yip, L., Zakharyan, R.A., Maiorino, R.M., Dart, R.C., Tirus, M.G., Gonzalez-Ramirez, D., Morgan, D.L., Avram, D., Aposhian, M.M., 2000a. Occurrence of monomethylarsonous acid (MMA^{III}) in urine of humans exposed to inorganic arsenic. Chem. Res. Toxicol., 13, 693–697.

Aposhian, H.V., Zheng, B., Aposhian, M.M., Le, X.C., Cebrian, M.E., Cullen, W.R., Zakharyan, R.A., Ma, M., Dart, R.C., Cheng, Z., Andrews, P., Yip, L., O'Malley, G.F., Maiorino, R.M., Van Voorhies, W., Healy, S.M., Titcomb, A., 2000b. DMPS-arsenic challenge test: modulation of arsenic species, including monomethylarsonous acid, excreted in human urine. Toxicol. Appl. Pharmacol., 165, 74–83.

Burrows, G.J., Turner, E.E., 1920. A new type of compound containing arsenic. J. Chem. Soc. Transact., 117, 1373–1383.

Chen, C.M., Mobley, H.L., Rosen, B.P., 1985. Separate resistances to arsenate and arsenite (antimonate) encoded by the arsenical resistance operon of R factor R773. J. Bacteriol., 161, 758–763.

Cline, D.J., Thorpe, C., Schneider, J.P., 2003. Effects of As(III) binding on alpha-helical structure. J. Am. Chem. Soc., 125, 2923–2929.

Cohen, S.M., Arnold, L.L., Uzvolgyi, E., Cano, M., St. John, M., Yamamoto, S., Lu, X., Le, X.C., 2002. Possible role of dimethylarsinous acid in dimethylarsinic acid-induced urothelial toxicity and regeneration in the rat. Chem. Res. Toxicol., 15, 1150–1157.

Cullen, W.R., Reimer, K.J., 1989. Arsenic speciation in the environment. Chem. Rev., 89, 713–764.

Cullen, W.R., McBride, B.C., Reglinski, J., 1984. The reduction of trimethylarsine oxide to trimethylarsine by thiols: a mechanistic model for the biological reduction of arsenicals. J. Inorg. Biochem., 21, 45–60 (see also pp. 179–194).

Cullen, W.R., McBride, B.C., Manji, H., Pickett, A.W., Reglinski, J., 1989. The metabolism of methylarsine oxide and sulfide. Appl. Organomet. Chem., 3, 71–78.

Del Razo, L.M., Styblo, M., Cullen, W.R., Thomas, D.J., 2001. Determination of trivalent methylated arsenicals in biological matrices. Toxicol. Appl. Pharmacol., 174, 282–293.

Farrer, B.T., McClure, C.P., Penner-Hahn, J.E., Pecoraro, V.L., 2000. Arsenic(III)–cysteine interactions stabilize three-helix bundles in aqueous solution. Inorg. Chem., 39, 5422–5423.

Feldmann, J., Lai, V.W.M., Cullen, W.R., Ma, M., Lu, X., Le, X.C., 1999. Sample preparation and storage can change arsenic speciation in human urine. Clin. Chem., 45, 1988–1997.

Gong, Z., Lu, X., Cullen, W.R., Le, X.C., 2001. Unstable trivalent arsenic metabolites, monomethylarsonous acid and dimethylarsinous acid. J. Anal. At. Spectrom., 16, 1409–1413.

Gong, Z., Jiang, G., Cullen, W.R., Aposhian, H.V., Le, X.C., 2002. Determination of arsenic metabolic complex excreted in human urine after administration of sodium 2,3-dimercapto-1-propane sulfonate. Chem. Res. Toxicol., 15, 1318–1323.

Gong, Z., Lu, X., Wang, Z., Cullen, W.R., Le, X.C., 2003. Preservation of monomethylarsonous acid and dimethylarsinic acid using diethyldithiocarbamate for arsenic speciation analysis, submitted for publication.

Goyer, R.A., 1996. Toxic effects of metals. In: Klaassen, C.D. (Ed.), Casarett and Doull's Toxicology: The Basic Science of Poisons. McGraw-Hill, New York, pp. 696–698.

Griffin, B.A., Adams, S.R., Tsien, R.Y., 1998. Specific covalent labeling of recombinant protein molecules inside live cells. Science, 281, 269–272.

Hasegawa, H., Sohrin, Y., Matsui, M., Hojo, M., Kawashima, M., 1994. Speciation of arsenic in natural waters by solvent extraction and hydride generation atomic absorption spectrometry. Anal. Chem., 66, 3247–3252.

Ji, G., Silver, S., 1992. Reduction of arsenate to arsenite by the ArsC protein of the arsenic resistance operon of *Staphylococcus aureus* plasmid pI258. Proc. Natl Acad. Sci. USA, 89, 9474–9478.

Jiang, G., Gong, Z., Li, X-F., Cullen, W.R., Le, X.C., 2003. Interaction of trivalent arsenicals with metallothionein. Chem. Res. Toxicol., 16, 873–880.

Johnson, J.M., Voegtlin, C., 1930. Arsenic derivatives of cysteine. J. Biol. Chem., 89, 27–31.

Le, X.C., Ma, M., 1998. Short-column liquid chromatography with hydride generation atomic fluorescence detection for the speciation of arsenic. Anal. Chem., 70, 1926–1933.

Le, X.C., Ma, M., Wong, N.A., 1996. Speciation of arsenic compounds using high performance liquid chromatography at elevated temperature and selective hydride generation atomic fluorescence detection. Anal. Chem., 68, 4501–4506.

Le, X.C., Lu, X., Ma, M., Cullen, W.R., Aposhian, H.V., Zheng, B., 2000a. Speciation of key arsenic metabolic intermediates in human urine. Anal. Chem., 72, 5172–5177.

Le, X.C., Ma, M., Lu, X., Cullen, W.R., Aposhian, H.V., Zheng, B., 2000b. Determination of monomethylarsonous acid, a key arsenic methylation intermediate, in human urine. Environ. Health Perspect., 108, 1015–1018.

Lin, S., Cullen, W.R., Thomas, D.J., 1999. Methylarsenicals and arsinothiols are potent inhibitors of mouse liver thioredoxin reductase. Chem. Res. Toxicol., 12, 924–930.

Lin, S., Shi, Q., Nix, F.B., Styblo, M., Beck, M.A., Herbin-Davis, K.M., Hall, L.L., Simeonsson, J.B., Thomas, D.J., 2002. A novel S-adenosyl-L-methionine:arsenic(III) methyltransferase from rat liver cytosol. J. Biol. Chem., 277, 10795–10803.

Lu, X., Gong, Z., Ma, M., Cullen, W.R., Aposhian, H.V., Zheng, B., Gurzau, E.S., Le, X.C., 2001. Speciation of human urinary arsenic including the trivalent methylation metabolites. In: Chappell, W.R., Abernathy, C.O., Calderon, R.L. (Eds), Arsenic Exposure and Health Effects. Elsevier, Oxford, pp. 339–352.

Ma, M., Le, X.C., 1998. Effect of arsenosugar ingestion on urinary arsenic speciation. Clin. Chem., 44, 539–550.

Mandal, B.K., Ogra, Y., Suzuki, K.T., 2001. Identification of dimethylarsinous and monomethylarsonous acids in human urine of the arsenic-affected areas in West Bengal, India. Chem. Res. Toxicol., 14, 371–378.

Mass, M.J., Tennant, A., Roop, R.C., Cullen, W.R., Styblo, M., Thomas, D.J., Kligerman, A.D., 2001. Methylated trivalent arsenic species are genotoxic. Chem. Res. Toxicol., 14, 355–361.

Merijanian, A., Zingaro, R., 1966. Arsine Oxides. Inorg. Chem., 5, 187–191.

Nesnow, S., Roop, B.C., Lambert, G., Kadiiska, M., Mason, R.P., Cullen, W.R., Mass, M.J., 2002. DNA damage induced by methylated trivalent arsenicals is mediated by reactive oxygen species. Chem. Res. Toxicol., 15, 1627–1634.

NRC, 1999. Arsenic in Drinking Water. National Research Council, Washington, DC.

Petrick, J.S., Ayala-Fierro, F., Cullen, W.R., Carter, D.E., Aposhian, H.V., 2000. Monomethylarsonous acid (MMA^{III}) is more toxic than arsenite in Chang human hepatocytes. Toxicol. Appl. Pharmacol., 163, 203–207.

Petrick, J.S., Bhumasamudram, J., Mash, E.A., Aposhian, H.V., 2001. Monomethylarsonous acid (MMA^{III}) and arsenite: LD$_{50}$ in hamsters and in vitro inhibition of pyruvate dehydrogenase. Chem. Res. Toxicol., 14, 651–656.

Sampayo-Reyes, A., Zakharyan, R.A., Healy, S.M., Aposhian, H.V., 2000. Monomethylarsonic acid reductase and monomethylarsonous acid in hamster tissue. Chem. Res. Toxicol., 13, 1181–1186.

Serves, S.V., Charalambidis, Y.C., Sotiropoulos, D.N., Ioannou, P.V., 1995. Reaction of arsenic(III) oxide, arsenous and arsenic acids with thiols. Phosphorus, Sulfur, and Silicon and the Related Elements, 105, 109–116.

Shiobara, Y., Ogra, Y., Suzuki, K.T., 2001. Animal species difference in the uptake of dimethylarsinous acid (DMA^{III}) by red blood cells. Chem. Res. Toxicol., 14, 1446–1452.

Sohrin, Y., Matsui, M., Kawashima, M., Hojo, M., Hasegawa, H., 1997. Arsenic biogeochemistry affected by eutrophication in Lake Biwa, Japan. Environ. Sci. Technol., 31, 2712–2720.

Stevens, S.Y., Hu, W., Gladysheva, T., Rosen, B.P., Zuiderweg, E.R.P., Lee, L., 1999. Secondary structure and fold homology of the ArsC protein from the *Escherichia coli* arsenic resistance plasmid R773. Biochemistry, 38, 10178–10186.

Styblo, M., Thomas, D.J., 1995. In vitro inhibition of glutathione reductase by arsenotriglutathione. Biochem. Pharmacol., 49, 971–974.

Styblo, M., Delnomdedieu, M., Thomas, D.J., 1995. Biological mechanisms and toxicological consequences of the methylation of arsenic. In: Goyer, R.A., Cherian, G. (Eds), Toxicology of Metals – Biochemical Aspects, Handbook of Experimental Pharmacology. Springer, Berlin, pp. 407–433.

Styblo, M., Serves, S.V., Cullen, W.R., Thomas, D.J., 1997. Comparative inhibition of yeast glutathione reductase by arsenicals and arsenothiols. Chem. Res. Toxicol., 10, 27–33.

Styblo, M., Vega, L., Germolec, D.R., Luster, M.I., Del Razo, L.M., Wang, C., Cullen, W.R., Thomas, D.J., 1999. Metabolism and toxicity of arsenicals in cultured cells. In: Chappell, W.R., Abernathy, C.O., Calderon, R.L. (Eds), Arsenic Exposure and Health Effects. Elsevier, Amsterdam, pp. 311–323.

Vahter, M., 1999. Variation in human metabolism of arsenic. In: Chappell, W.R., Abernathy, C.O., Calderon, R.L. (Eds), Arsenic Exposure and Health Effects. Elsevier, Amsterdam, pp. 267–279.

Yamauchi, H., Fowler, B.A., 1994. Toxicity and metabolism of inorganic and methylated arsenicals. In: Nriagu, J.O. (Ed.), Arsenic in the Environment, Part II: Human Health and Ecosystem Effects. Wiley, New York, pp. 35–43.

Zakharyan, R.A., Wu, Y., Bogdan, G.M., Aposhian, H.V., 1995. Enzymatic methylation of arsenic compounds: assay, partial purification, and properties of arsenite methyltransferase and monomethylarsonic acid methyltransferase of rabbit liver. Chem. Res. Toxicol., 8, 1029–1038.

Zakharyan, R.A., Ayala-Fierro, F., Cullen, W.R., Carter, D.M., Aposhian, H.V., 1999. Enzymatic methylation of arsenic compounds. VII. Monomethylarsonous acid (MMA^{III}) is the substrate for MMA methyltransferase of rabbit liver and human hepatocytes. Toxicol. Appl. Pharmacol., 158, 9–15.

Zakharyan, R.A., Sampayo-Reyes, A., Healy, S.M., Tsaprailis, G., Board, P.G., Liebler, D.C., Aposhian, H.V., 2001. Human monomethylarsonic acid (MMA^{V}) reductase is a member of the glutathione-S-transferase superfamily. Chem. Res. Toxicol., 14, 1051–1057.

Arsenic Exposure and Health Effects V
W.R. Chappell, C.O. Abernathy, R.L. Calderon and D.J. Thomas, editors

Chapter 5

Arsenic in Yellowknife, North West Territories, Canada

William R. Cullen, Elena Polishchuk, Kenneth J. Reimer, Yongmei Sun, Lixia Wang and Vivian W.-M. Lai

Abstract

In the Giant mine, Yellowknife, NWT, Canada, about 260,000 tons of arsenic trioxide (ca. 80% pure) are stored underground in old mine workings and in specially constructed chambers. The options for safe management of this material are discussed. Some promising results, in terms of long-term leachability of material containing up to 50% mine dust, are being obtained from mine dust/bitumen materials. Many fungi such as Aleurodiscus farlowii *grow underground: some of these have been identified by using 28S ribosomal RNA gene (28S rRNA gene) analysis and their biochemistry with respect to arsenic metabolism is described. A white solid that is deposited in some regions of the mine is a mixture of calcium sulfate (Gypsum) and magnesium arsenate (Roesslerite).*

Keywords: arsenic trioxide, solidification/stabilization, bitumen, fungal identification, rRNA gene analysis, fungal metabolites

The first mineral claim in Yellowknife area was staked in 1935 and production began in 1948. The gold is associated with arsenopyrite and pyrite so a roaster (smelter) was commissioned in 1948. The stack gas, mainly arsenic trioxide and sulfur dioxide, was initially released into the atmosphere. However, this undesirable practice was mitigated in 1951 with the installation of an electrostatic precipitator to remove some of the arsenic trioxide. The arsenic recovery system was upgraded over the next decade so that the quantity released in 1949, 7300 kg/day (estimated), had dropped to 75 kg/day (estimated) in 1960. In 1991, two separate measurements of the daily emission gave 59 and 26 kg/day and the numbers were fairly constant until the mine ceased operation in 1999. In the late 1990s the daily recovery of the mine dust was 8–10 tons (Dillon, 1995). The practice of storing the recovered mine dust underground began in 1951 in the belief that the storage areas, initially abandoned mined-out sections, were either located in permafrost or in areas that would revert to permafrost once human activity had ceased in the vicinity. Unfortunately although permafrost conditions were never reestablished, the underground storage was continued and eventually special chambers were constructed to hold the mine dust. At the present time there are 15 storage areas, five of these are abandoned mine workings, and their size ranges from a chamber of 2294 m^3 containing about 3000 tons of dust to a mined-out stope of 25,740 m^3 containing about 30,000 tons of dust. The total amount of mine dust now in storage amounts to 260,000 tons and comprises about 200,000 tons of arsenic trioxide.

It is worth noting that by the end of 1998, the mine had processed 17,425,000 tons of ore to yield 6,954,250 ounces of gold (worth about $2 billion US).

Many options for the management of the mine dust have been considered including two essentially noninvasive ones that have considerable backing from an economic view, although they ultimately dodge the problem. These are: (1) leave the mine dust in place and pump the mine in perpetuity and (2) leave in place and induce permafrost conditions by means of thermoprobes inserted into the ground around the storage chambers.

Other options involve removal of the material for treatment on the surface and in the process circumvent in some way the obvious difficulties in recovering the dust because of the toxicity and because access is restricted to some of the storage chambers.

Because the dust contains gold, on average 18.1 g/ton, there has always been the "mantra" that the dust should be processed in such a way as to recover the gold to augment the costs. One such process that received considerable backing and remained on the table for many years was named WAROX (White arsenic oxide: note the direct avoidance of the word arsenic). In this process, the dust was to be sublimed to recover pure arsenic trioxide, free of antimony and iron, so that it could be used for wood preservative (CCA) manufacture. Gold would be recovered from the residue. Another process that was considered was to transport the material to another near-by gold mine, the Con Mine, for processing in their pressure oxidation plant. The gold would be recovered and the arsenic converted to Scorodite (ferric arsenate) and related minerals.

From the chemical point of view there is not much that can be easily and economically done to produce a safely disposable product from the dust. Reduction to arsenic metal would be possible (electrochemically with carbon) but the metal would have to be carefully stored and the process could be expensive. There is no obvious way of converting the dust directly to a stable "mineral" form of arsenic such as Scorodite.

Some chemical options become available if the dust is "mined" as a slurry. The most obvious of these would be to first oxidize the arsenic in solution to arsenate and then co-precipitate the product with ferric hydroxide, the so-called "ferric arsenate" of the water purification industry. The product from the water treatment plant is best thought of as a ferric oxyhydroxy phase (known as ferrihydrite), to which is adsorbed anions such as arsenate and arsenite. The co-precipitation is more efficient if the arsenic is in the (V) oxidation state, i.e. arsenate. There is controversy about the long-term stability of the product (Riveros et al., 2001). Dehydration can lead to instability. Recrystallization to geothite, $FeOOH$, with release of arsenate could occur, and there is the possibility of microbial reaction that again releases the arsenic (Ahmann et al., 1997). In spite of this, the US EPA (2001) considers that the product, which normally passes the TCLP test, can be disposed of in a nonhazardous landfill, even though it is well recognized that the TCLP is not of much value in evaluating the mobility of arsenic species (Hooper et al., 1998). The arsenic would certainly be removed from any solution by using this precipitation technology, but it is our opinion that special storage would be necessary, and in addition the process is not economically attractive (Kyle and Lunt, 1991). Conversion to the insoluble sulfide would require a source of sulfide ion, preferably not the toxic hydrogen sulfide, and the product is thermodynamically unstable with respect to the oxide in an oxic environment. Conversion to calcium arsenate has similar problems with respect to conversion to calcium carbonate and release of arsenic.

A review of the recent literature covering treatment technologies for arsenic contaminated media shows that seven methods have been used for solid matrices. Solidification/stabilization (S/S) and vitrification were the most commonly used.

These were followed by soil washing/acid extraction, prometallurgical treatment; *in situ* soil flushing; electrokinetics; and phytoremediation. All these technologies are available commercially.

In the Yellowknife situation vitrification would probably be too expensive and would probably generate off-gasses containing arsenic (Bauer, 1983), other than S/S, the rest are of limited applicability. The encapsulation of arsenic compounds in cement has been described in a number of publications (e.g. Dutre and Vandecasteele, 1998; Cullen et al., 2002) but usually at low loadings. It is believed that there is a chemical interaction between the arsenic species and the cement that enhances the binding of the arsenic species particularly arsenate. At U.B.C. we have found that pure arsenic trioxide can be incorporated into Portland cement at 1% loading, and the product sets easily. When this product is powdered and subjected to the standard acetic acid leaching procedure (TCLP) that is used to mimic the landfill environment, the leachate contains arsenic in the low ppb range. When the cement product is fabricated into small cylinders, the same leachate test gives very encouraging results: we found leachate concentrations of less than 50 ppm for loadings of up to 20% arsenic trioxide in the cement (Cullen et al., 2002).

Some other early studies indicated that bitumen (asphalt) could be considered as a candidate for stabilizing the mine dust if low concentrations of mine dust were used in the S/S mixture. In order to minimize the consumption of cement or bitumen, and hence reduce the waste volume, as well as the financial cost, we carried out a more detailed and systematic investigation on the encapsulating capacities of cement and bitumen.

1. Leaching tests from S/S material containing mine dust mixed with cement, or bitumen, or bitumen plus zeolite

Cement samples: Samples with different concentrations of mine dust (15, 17.5, 20, 22.5, and 25%) in commercial Portland cement were mixed in sealed plastic bottles and shaken for 1 h before the addition of water. Manually stirring the mixture produced a uniform blend that was transferred into a Ziploc plastic bag and allowed to cure for 28 days for later leaching tests, TCLP. The arsenic contents of the leaching solutions were well above of the cut-off value of 5 ppm: 15%, 275 ppm; 17.5%, 217 ppm; 20%, 232 ppm; 25%, 1120 ppm. These results, which are the average of two or three duplicate experiments, are sufficiently unsatisfactory that we abandoned further work with mine dust/cement S/S.

Bitumen samples: The bitumen, provided by McTar petroleum Co Ltd., Coquitlam, BC, Canada, was carefully heated to a semi-fluid and the appropriate amount of mine dust was added to the stirred bitumen. The stirring was continued until the mixture became homogeneous. The resulting mixture was allowed to cool to room temperature and stored to harden at 4°C. The mine dust/bitumen mixture was cracked into small pieces (freezing in liquid nitrogen is a help). An acetate buffer (pH 5.2) was used as a leaching solution. The arsenic contents of the leachates (Table 1) are satisfactory up to a loading of around 50%.

Bitumen plus zeolite samples: Mine dust, zeolite, and bitumen were mixed 1:1:1 ratio, in a similar manner, in the hope that some advantage might be gained by adding zeolites to the mixture. The leachate under the TCLP conditions gave an arsenic content of 3.95 ppm which is no great improvement over the "bitumen alone" samples. The zeolite was obtained from C_2C Mining Corporation, Calgary.

Table 1.   Arsenic content in the leachate of mine dust/bitumen.

Mine dust/bitumen (%)	15	17.5	20	22.5	25
Arsenic content of leachate (ppm)[a]	0.040	0.114	0.186	1.422	0.048
Mine dust/bitumen (%)	20	30	40	50	60
Arsenic content of leachate (ppm)[a]	2.45	4.25	3.30	5.50	26.90

[a] Average of two or three duplicate experiments.

2. Possible volatilization of arsenic trioxide and polycyclic aromatic hydrocarbons (PAHs)

Although it is not difficult to mix the mine dust with bitumen by heating the mixture, it is desirable that the process should be carried out at as low a temperature as possible to reduce energy consumption and the loss of toxic volatile arsenic trioxide into the atmosphere. The bitumen/mine dust mixing process (for both type A and C bitumen) was investigated by mounting a cold finger condenser inside the heated round bottom flask that contained the bitumen/mine dust mixture. The production of volatile arsenic trioxide could be easily monitored by visual inspection of the cold finger condenser. In some cases, the cold finger was rinsed and the washings were analyzed for arsenic. Some results are summarized in Table 2.

The cold fingers were also examined to see if they trapped any volatile PAHs. The results obtained by ASL Analytical Service Laboratories, Vancouver, BC, Canada, indicated that the concentrations of PAHs of concern were below detection limits.

Table 2.   Volatilization of arsenic trioxide.

Sample	Heating time (h)	Temperature (°C)	Volatile loss
Bitumen alone	24	200	None
	24	140	None
	24	130	None
Mine dust alone	24	200	Much
	24	140	Less
	24	130	Slight
40% mine dust in bitumen	24	200	Much
	8	140	Some
30% mine dust in bitumen	24	200	Much
	8	140	Some
	2	130	Trace[a]

[a] Quantitative analysis indicates that the volatile loss of arsenic trioxide is less than 0.01% of the original arsenic content of the mine dust/bitumen mixture.

3. Investigation of long-term leaching of bitumen/mine dust material

Once it had been established that the bitumen/mine dust S/S material passed the TCLP test, it was necessary to establish that the bitumen samples are durable and safe in an appropriate storage environment over time. There are no standard procedures for testing for long-term stability, so it was necessary to devise a model that would reflect a possible storage scenario. The following experiment is based on the possibility that a bitumen/mine dust monolith would be stored in an open or closed mine shaft that would receive water from the surface. The experiment covers the worst-case scenario in which the shaft is flooded and the monolith becomes immersed in water. In practice fragments of the S/S material, rather than a monolith, were used to provide a greater surface area and thus, test the arsenic retention more rigorously. The slow flow of water was modeled by draining aliquots from the bottom of the vessel periodically for analysis, and replacing them with equal volumes of fresh water. The experiment was carried out at 4°C to mimic the mine temperature.

Granules of mine dust/bitumen mixture (40%, 32.5 g) were loaded into a stainless steel metal net cylinder, to a height of 5 cm. The cylinder was 35 cm long with a diameter of 3.5 cm. The loaded cylinder was suspended inside a glass column that was fitted with a stopcock at the bottom. Deionized water, 650 ml, was added to completely immerse the mine dust/bitumen sample. A nitrogen flow was bubbled through the water periodically, particularly prior to sampling, to facilitate mixing and leaching. When required, 12 ml of the aqueous sample was drained through the stopcock at the bottom and an equal amount of fresh water was added to the top of leaching solution. The experiment has been running for nearly 3 years and the results are shown in Figure 1.

4. Leaching studies on mine dust/bitumen monoliths

Some additional experiments have been conducted with mine dust/bitumen cylinders and some results are shown in Figures 2 and 3. For these experiments, done in duplicate, a weighed amount of bitumen was carefully heated to a semi-fluid (*T* less than 140°C) and the appropriate amount of mine dust was added. The mixture was stirred until it became homogeneous, poured into 150 ml beakers lined with aluminum foil, and cooled to room

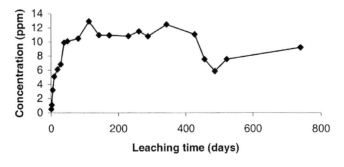

Figure 1. Long-term leaching test of mine dust/bitumen (40%).

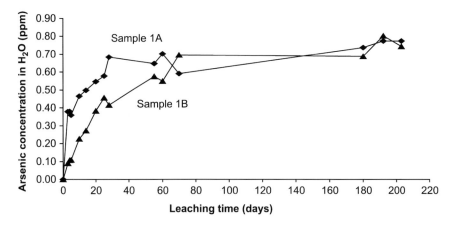

Figure 2.   Leaching of 47% mine dust/bitumen cylinders into deionized water.

temperature. The samples (*ca.* 75 ml) were then stored in a cold room at 5°C for several days. The aluminum foil was removed from the cylindrical forms which were then immersed in 600 ml beakers containing deionized water as the leaching solution (*ca.* 3 times the weight of the hardened mixture) and then returned to the cold room. Nitrogen gas was bubbled through the leaching solution for about 30 min before sampling the leachate on each sampling day. A sample of 10 ml was taken which was replaced with 10 ml of deionized water to keep the volume constant in the beaker. Figures 2 and 3 show the arsenic content of the leachate, determined by using ICP/MS or HG/AAS at different intervals of time and different mine dust concentrations.

A similar mixture was prepared that contained an equal weight of mine dust, zeolite, and bitumen. The leachate data from the cylindrical forms are shown in Figure 4.

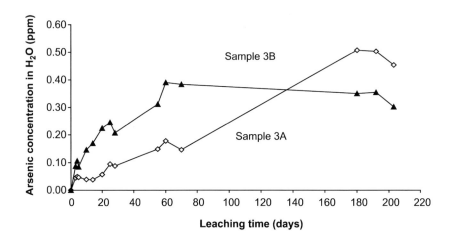

Figure 3.   Leaching of 29% mine dust/bitumen cylinders into deionized water.

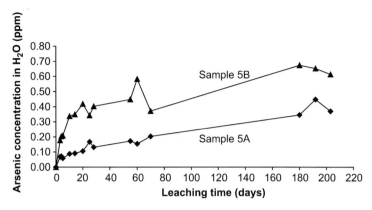

Figure 4. Leaching of 34% mine dust/bitumen/zeolite cylinders into deionized water.

A sample of bitumen was used as a blank. The concentration of the arsenic that leached out of the bitumen is below the detection limit, as was the PAH content.

5. Leaching from bitumen coated mine dust/bitumen cylinders

In order to model the situation where the S/S material of mine dust/bitumen would be stored in an enclosure that had been previously coated with bitumen, arsenic leaching from bitumen coated solid cylinders prepared from mine dust/bitumen (30–50% mine dust) was examined. Cylindrical forms were first prepared from the mixtures as described above. These were left in the cold room for several days, the foil wrapping was removed, and each cylinder was dipped into hot bitumen. The cooled samples were then placed in the cold room for several more days.

Leaching experiments were carried out as described above, and some results are shown in Figure 5.

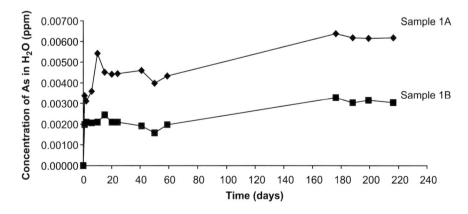

Figure 5. Leaching of bitumen coated 47% mine dust/bitumen cylinders into deionized water.

6. Discussion

The bitumen (asphalt) used in these experiments is the last cut in the petroleum refinery. These materials are complex mixtures and not well characterized. Most of the compounds are high molecular weight hydrocarbons that contain alcohol, acid, phenolic and thiol groups all of which can interact with the arsenic trioxide. Thus, it is not surprising that the oxide is well retained in the S/S mixture. The trend in the long-term experiment (Fig. 1) is flat. It should be noted that the leaching solution has not been changed and that fresh water is used only to replace the sample taken for the analysis. The initial concentrations in the leachate (Fig. 1) are higher than the concentrations observed in the more recent experiments (Figs. 2 and 3) in spite of the fact that the leachate to solid ratio is much higher (approximately 20:1 vs. 3:1). This may be a consequence of the use of fragmented material with higher surface area (Fig. 1) rather than solid cylinders (Figs. 2 and 3). In addition, there are probably some problems with mixing the materials on a small scale (stirring rods rather than mechanical mixers).

The leaching data from solid cylinders (Figs. 2 and 3) indicate that for a particular sample there is an initial gradual increase in the arsenic content of the leachate with time, as is seen in Figure 1. There appears to be a rough correlation with the amount of arsenic in the leachate and the amount in the sample. However, the data also indicate that there can be large variations in supposedly identical mixtures. This again may be the result of uneven mixing. Nevertheless, the arsenic concentration in all the leachates is satisfyingly low, even with mine dust concentrations approaching 50%.

The addition of zeolite does not seem to decrease the leachability of the arsenic as can be seen by comparing Figures 3 and 4. The higher arsenic concentrations in the leachate may be the result of the reduced bitumen content. Although the zeolite appears to increase the mechanical strength of the mixture there is probably no interaction between the zeolite and the arsenic compounds.

In order to reduce the surface exposure of a mine dust/bitumen mixture to water, it could be desirable to pour the mixture into an enclosure that had been pre-treated with bitumen in order to provide a sealed surface. To model this situation, pre-formed mine dust/bitumen cylinders were dipped into molten bitumen so that they became coated with fresh bitumen. These coated cylinders were then immersed in water at 5°C for leaching studies.

In practice the dipping proved to be difficult because the cylinder melted quite easily. The data shown in Figure 5 indicate the success of the experiment in that very little arsenic is leaching through the protective skin of bitumen (compare with Figure 2).

6.1. Underground deposits and growths

White solids: Samples of a white solid that accumulates on the rock and wooden surfaces in one area of the Giant mine (near B shaft) were collected and subjected to X-ray diffraction analysis. Two main minerals were present: Gypsum $CaSO_4 \cdot 2H_2O$ and Roesslerite $MgHAsO_4 \cdot 7H_2O$. The analytical results reflect this conclusion: As, 167,000 ppm; Mg, 47,900 ppm; Ca, 171,000 ppm and there is reasonable agreement between samples although there is some spatial and temporal variation. The atomic ratios suggest that other arsenic compounds are present. The iron concentration amounted to 13,000 ppm in one sample but generally the concentration was below 500 ppm.

The origin of this deposit is not obvious. The deposits appear on rock, wood, and metal and are absent from surfaces that are not in a direct line of an airflow down the tunnel. There is no deposit in the "shadow" of the airflow; thus only one side of a flat wooden surface or metal pipe is coated. Furthermore, the deposit is in a limited area. However, the chemical make up of the deposit seems to point to its formation *in situ* rather than being deposited from an air stream. Roesslerite is a mineral that is not commonly found in Canada (Roberts et al., 1990). Careful examination of the growth of the deposit over time should help unravel this mystery.

Fungal growths: A number of rock walls and roofs underground are covered with a rich fungal growth. Some of these growths have been collected for examination because of the possibility that they may be involved in processes that result in mobilization of the arsenic in the underground environment. The miners speak of "slimesicles" that emit gasses. The growths are also of interest because they are thriving in such an apparently hostile environment.

Some of the identifications that have been made, based on partial 28S rRNA gene analysis, are listed in Table 3. The procedures used in-house are outlined in a recent publication (Granchinho et al., 2002).

Very little is known about the biology or chemistry of these organisms and further studies could be rewarding. For example, we find that the pink fungus *A. farlowii*, which we have identified in two separate samples, grows on rock walls where the ambient water has an arsenic concentration of about 2000 ppm. The freeze-dried fungus contains 4650 ppm extractable arsenic (1:1, methanol:water); however, when grown in potato dextrose broth containing 100 ppm arsenate, only traces of arsenite

Table 3. Partial 28S rRNA gene analysis: matches from GenBank database.

Isolation method	Closest GenBank match	% Identity	Number of base pairs
B	Salal root associated fungus	96	398/412
B	*Phialophora gregata*	96	525/545
B	*Euascomycete* RFLP-type	95	528/552
A	*Phialophora gregata* and/or	86	243/281
	salal root associated fungus	85	224/287
A	*Aleurodiscus farlowii*	95	302/317
A	*Pseudeurotium zonatum*	96	405/520
C	*Nectria vilior*	98	373/377
A	*Sclerotinia veratri*	88	256/268
A	*Geomyces pannorum* var *asperulatus*	97	444/453

A: Direct sampling from the specimen as collected. B: A sample was grown on potato dextrose agar (PDA) and then the isolate was grown in potato dextrose broth (PDB). Frequent washing ensured axenic growth. C: A sample was grown in PDA to afford many colonies. The fungus was isolated and subsequently identified.

were observed and no arsenic methylation. Extensive reduction and limited methylation is seen in cultures of other isolates.

Epilogue. The material that was presented in San Diego contained some information about the use of principle component analysis to estimate background concentrations of arsenic in an industrially influenced region, a necessity for establishing site-specific criteria for cleanup. This material is now available elsewhere (Reimer et al., 2002). Two other related projects were described in poster format at the meeting: Reimer et al. "Arsenic in a deer mice food chain" and Ollson et al. "Influence of soil type on arsenic bioavailability and its consequences for human health risk assessment".

Acknowledgements

The project was supported by the Department of Indian and Northern Affairs, Canada, and by NSERC Canada. We are grateful for assistance and advice from Neill Thompson, Paula Spenser, and Ron Breadmore of the Royal Oak Project Team, DIAND, Yellowknife. The technical help of Dr C. Wang, S. Granchinho, C. Franz, A. Yuen, and M.-C. Delisle is gratefully acknowledged.

References

Ahmann, D., Krumholz, L.R., Hemond, H.F., Lovley, D.R., Morel, F.M.M., 1997. Microbial mobilization of arsenic from sediments of the Aberjona watershed. Environ. Sci., 31, 2923–2930.

Bauer, R.J., 1983. Arsenic: glass industry requirements. In: Lederer, W.H., Fensterheim, R.J. (Eds), Arsenic: Industrial, Biomedical, Environmental Perspectives. Van Nostrand Reinhold Company, New York, pp. 45–62.

Cullen, W.R., Andrewes, P., Fyfe, C., Grondey, H., Liao, T., Polishchuk, E., Wang, L., Wang, C., 2002. Solidification/stabilization of Adamsite and Lewisite in cement and the stability of arylarsenicals in soils. In: McGuire, R. (Ed.), Environmental Aspects of Converting Chemical Warfare Facilities to Peaceful Purpose. Kluwer Academic Publishers, Dordrecht, The Netherlands.

Dillon, M.M., 1995. Report, Government NWT: Air dispersion modeling of roaster slack emissions Royal Oak Giant Mine Yellowknife, NWT, #94-2491-01-01, revised May 1995.

Dutre, V., Vandecasteele, C., 1998. Immobilization mechanism of arsenic in waste solidified using cement and lime. Environ. Sci. Technol., 32, 2782–2787.

Granchinho, S.C.R., Franz, C.M., Polishchuk, E., Cullen, W.R., Reimer, K.J., 2002. Transformation of arsenic (V) by the fungus *Fusarium oxysporum melonis* isolated from the alga *Fucus gardneri*. Appl. Organomet. Chem., 16, 721–726.

Hooper, K., Iskander, M., Sivia, G., Hussein, F., Hsu, J., DeGuzman, M., Odion, Z., Llejay, Z., Sy, F., Petreas, M., Simmons, B., 1998. Toxicity characteristic leaching procedure fails to extract oxo-anion-forming elements that are extracted by municipal solid waste leachate. Environ. Sci. Technol., 32, 3825–3830.

Kyle, J.H., Lunt, D., 1991. An investigation of disposal options for arsenic trioxide produced from roasting operations. Proceedings of Extractive Metallurgy Conference, Perth, October 1991, pp. 347–353.

Reimer, K.J., Ollson, C.A., Koch, I., 2002. An approach for characterizing arsenic sources and risk at contaminated sites: application to gold mining sites in Yellowknife, NWT, Canada. In: Cai, Y., Braids, O. (Eds), Biogeochemistry of Environmentally Important Trace Elements. Oxford University Press, (accepted Feb. 02).

Riveros, P.A., Dutrizac, J.E., Spencer, P., 2001. Arsenic disposal practice in the metallurgical industry. Can. Metall. Quart., 40, 395–420.

Roberts, W.L., Campbell, T.J.; Rapp, G.R. (Eds), 1990. Encyclopedia of Minerals, 2nd edn. Van Nostrand Reinhold Company, New York.

US EPA, 2001. Fed. Regist. January 22, 2001.

Arsenic Exposure and Health Effects V
W.R. Chappell, C.O. Abernathy, R.L. Calderon and D.J. Thomas, editors
© 2003 Published by Elsevier B.V.

Chapter 6

Occurrence of public health and environmental hazards and potential remediation of arsenic-containing soils, sediments, surface water and groundwater at the Lava Cap Mine NPL Superfund site in Nevada County, California

G. Fred Lee and Anne Jones-Lee

Abstract

In 1999 the US EPA added the Lava Cap Mine area to the National Priority List (NPL) Superfund sites. This site is a former gold and silver mine located in Nevada County, California, near Nevada City. It is a Sierra-Nevada foothill wooded area, with low-density residential development. A risk assessment shows that there are significant potential human health and ecological risks associated primarily with arsenic in the mine site area tailings, the creeks that have received tailings discharges and the Lost Lake area. The US EPA has proposed several potential remediation approaches, which include containment with capping of existing contaminated areas or excavation of tailings and tailings-contaminated areas and disposal in a new landfill in the area or transported off-site for disposal.

An issue that will need to be resolved is the cleanup goal for groundwaters contaminated by waste-derived arsenic. While the contaminated soil and water cleanup objectives have not been established, the US EPA has indicated that the recently adopted 10 μg/l drinking water maximum contaminant level (MCL) is a potential cleanup objective for water contaminated by arsenic. However, there are questions as to whether a risk-based drinking water cleanup objective should be used which does not consider economic and political factors that were incorporated into the 10 μg/l MCL adopted by the US EPA. Another important issue is the ability of the standard RCRA landfill to prevent further groundwater pollution by landfilled arsenic-containing soils and tailings for as long as the landfilled tailings represent a threat.

Keywords: arsenic, Lava Cap Mine, Superfund, gold mine tailings, remediation, groundwater, landfills, Nevada County, CA

1. Introduction

The Lava Cap gold and silver mine was a shaft mine that started operating in 1861, and operated periodically until 1943. It was one of the largest gold mines in California. This mine is located about 8 km from Nevada City, California, in the foothills of the Sierra-Nevada mountains. The elevation of the mine site is about 818 m. Figure 1 shows a general map of the Lava Cap Mine Superfund site area.

The Lava Cap Mine site occupies about 12 ha. It is in a forested, low-density residential area, with a small stream and a log tailings dam located at the mine site, and another tailings storage area (Lost Lake) located several kilometers downstream. Lost Lake is a private lake surrounded by homes located about 2 km downstream from the Lava Cap Mine site. According to the US EPA (2002), in 1994 an estimated 1776 people lived within 1.6 km of the contaminated area and 24,091 lived within 6.4 km of the area. The ore

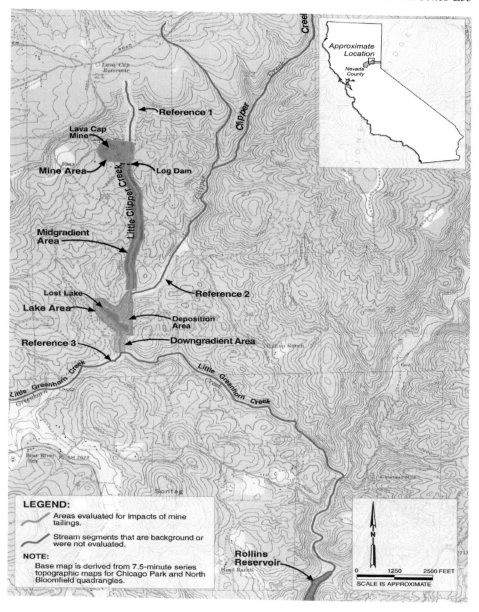

Figure 1. Map of Lava Cap Mine Superfund site area. Source: CH2M Hill (2001a).

processed for gold and silver recovery contained high levels of pyrite, arsenopyrite and galena. The ore was processed by crushing and grinding, followed by either flotation or cyanide treatment. Some of the ore was processed by amalgamation with mercury.

Following the partial collapse of a 9-m-high log-based tailings dam in January 1997, which released about 7650 m^3 of tailings to Little Clipper Creek (a small stream draining the area), Clipper Creek, Little Greenhorn Creek and Lost Lake (a tailings reservoir), in October 1997, the US EPA initiated a "removal action" to prevent further tailings release from the log-based dam. The tailings dam stores about 38,000 m^3 of tailings. Lost Lake was developed as a tailings storage area, with a 15-m-high earthen dam. It is estimated to contain about 115,000 m^3 of tailings. The Deposition Area just upstream of Lost Lake is estimated to contain 268,000 m^3 of tailings (see Figures 1 and 2).

CH2M Hill, Inc., of Sacramento, CA, with several subcontractors, is the US EPA contractor for the Remedial Investigation Feasibility Study (RI/FS). David Seter is the US

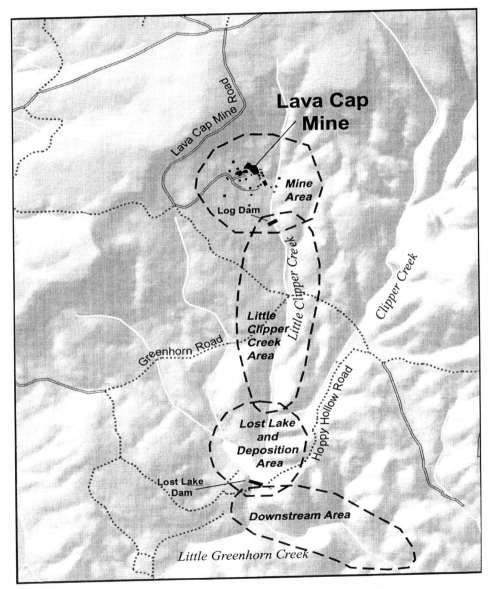

Figure 2. Lava Cap Mine Superfund site remediation areas. Source: US EPA (2001b).

EPA Region 9 project manager, with Don Hodge as the community involvement coordinator. The California Department of Toxic Substances Control, under the leadership of Steve Ross, is involved in reviewing the US EPA site investigation and remediation studies. Further, the California Department of Health Services has conducted a review (ATSDR, 2001) of the public health hazards of the site. The California Department of Fish and Game (Salocks, 1997) has conducted a review of the potential impacts of the tailings dam failure and the associated release of tailings on wildlife. The South Yuba River Citizens League (SYRCL) is the public representative for a US EPA Technical Assistance Grant. Janet Cohen is the Executive Director of SYRCL. The US EPA Technical Assistance Grant advisor is Dr G. Fred Lee, of G. Fred Lee & Associates, El Macero, California. The RI/FS is being conducted by the US EPA Region 9 since, at this time, there is no identified responsible party that can fund the investigation and cleanup. Funding is, therefore, being provided by the US EPA.

The US EPA Region 9 is the lead agency conducting the CERCLA remedial investigation and feasibility study for the purpose of:

- assessing the contamination associated with the Lava Cap Mine Superfund site and
- development of remedial alternatives for arsenic-contaminated soil, sediment, surface water and groundwater.

The US EPA has provided a summary of the Lava Cap Mine Superfund site issues on the Internet (US EPA, 2002). The Agency has released a series of contractor reports (CH2M Hill, 2001a,b,c, 2002) which serve as the background for the information provided in this paper on the characteristics of the site. The materials presented in this paper are largely derived from the investigations of the US EPA and its contractor, CH2M Hill.

2. Public health and environmental problems

After extensive monitoring of the mine area and downstream where tailings have been carried through deliberate discharges and the failure of the log-based tailings dam, it has been found that there is extensive surface soil, water and sediment contamination by arsenic derived from the tailings. Table 1 presents a summary of the data obtained in the studies that have been conducted thus far. Some of the soils have been found to contain as much as 34,000 mg/kg arsenic. As shown in Table 1, the reference areas selected by the US EPA typically contained on the average of 10–20 mg/kg arsenic, while the surface soils in the mine area contained on the average of about 2000 mg/kg. The surface soil area along Little Clipper Creek below the log dam had around 600 mg/kg arsenic. The tailings Deposition Area near Lost Lake contained about 500 mg/kg, and Lost Lake sediments had about 700 mg/kg arsenic. There is significant contamination by arsenic of surface areas that are readily accessible to the public, by the tailings releases that have taken place in the past.

A standard CERCLA RI/FS risk assessment has shown that the mine area, which includes the tailings dam, the soils and sediments along Little Clipper Creek, Clipper Creek, Little Greenhorn Creek and near Lost Lake, and downstream of Lost Lake to

Table 1. Summary of arsenic results in selected media – Lava Cap Mine RI, Nevada County, California.

Description	Media	Sample locations	Average	Minimum	Maximum	Units
Reference area 1 – above mine	Surface soil	14	21	5.2	95.3	mg/kg
	Sediment	7	25.8	17.9	44.3	mg/kg
	Surface water	6	0.3	ND	0.5	μg/l
	Groundwater	4	23.6	7	36.7	μg/l
Reference area 2 – Clipper Creek	Surface soil	3	13	7.6	20.0	mg/kg
	Sediment	6	13	10.9	16.0	mg/kg
	Surface water	10	0.3	ND	1.1	μg/l
Reference area 3 – Little Greenhorn Creek	Sediment	1	10.1	10.1	10.1	mg/kg
	Surface water	1	ND	ND	ND	μg/l
Source areas	Surface soil – mine buildings	9	10,000	848	31,200	mg/kg
	Surface soil – waste rock/tailings pile	6	1340	63.6	2070	mg/kg
	Subsurface soil	22	603	15.5	5360	mg/kg
	Sediment	1	9201	459	34,000	mg/kg
	Surface water – mine buildings	3	4022	68.8	14,300	μg/l
	Surface water – adit/seep/log dam	3	233	23.6	668	μg/l
	Groundwater	6	256	12.5	567	μg/l
	Ambient air	2	N/A	ND	0.021J	μg/l^3
Mine area	Surface soil – around mine buildings	13	2170	79.4	5570	mg/kg
	Surface soil – away from waste rock/tailings pile	25	370	4.7	1750	mg/kg
	Surface water	2	10.2	1.7	31.9	μg/l
	Groundwater – residential wells	4	154	11.2	528	μg/l
	Ambient air	1	N/A	ND	0.067	μg/l^3
Little Clipper Creek below the log dam	Surface soil	7	599	53.9	908	mg/kg
	Sediment	5	669	328	1150	mg/kg
	Groundwater – residential wells	11	9.7	ND	46.3	μg/l
	Surface water	5	132	19	285	μg/l
Deposition Area	Surface soil	34	459	10.2	913	mg/kg
	Sediment	8	615	398	892	mg/kg
	Ambient air	2	N/A	ND	ND	μg/l^3
	Subsurface soil	18	1430	719	2480	mg/kg
	Groundwater	4	1180	235	2410	μg/l
	Surface water	4	245	24.2	1160	μg/l

(*continued on next page*)

Table 1. (*continued*)

Description	Media	Sample locations	Average	Minimum	Maximum	Units
Lost Lake	Sediment	6	697	304	1140	mg/kg
	Surface soil	68	288	ND	848	mg/kg
	Ambient air	1	N/A	ND	ND	$\mu g/l^3$
	Groundwater – residential wells	7	0.2	ND	0.64	μg/l
	Surface water	3	28.4	5.8	70.6	μg/l
Downgradient of Lost Lake	Surface soil	2	403	261	673	mg/kg
	Sediment	5	753	38.5	2110	mg/kg
	Surface water	4	33	0.3	72.3	μg/l

Source: CH2M Hill (2002).
ND, non-detect; *J*, estimated concentration; N/A, not applicable because arsenic was either not detected or was only detected once.

some yet-undefined extent, are contaminated with sufficient arsenic to be a threat to human health through body contact (US EPA, 2000, 2001a). The risk assessment is based on the Guidelines for Conducting Remedial Investigation and Feasibility Studies under CERCLA (US EPA, 1989). The goal of the remedial investigation is to reduce the excess lifetime cancer risk associated with arsenic to the range of $10^{-4}-10^{-6}$ additional cancers. The remedial investigation non-cancer risk goal is to achieve background levels. Arsenic is the primary human health and ecological risk driver for the Lava Cap Mine site (CH2M Hill, 2001a; US EPA, 2001a). In addition, for ecological risk, antimony, cadmium, copper, cyanide, lead, mercury, silver and zinc have been found thus far to contribute to risk to aquatic and terrestrial wildlife (CH2M Hill, 2001b). Further, there is a potential for airborne tailings (dust) to be a problem for the residents and wildlife of the area, which has not been adequately evaluated at this time.

According to CH2M Hill (2001a), the total estimated lifetime cancer risk for residents and recreational users around Lost Lake is about 5×10^{-5}. Along Clipper Creek, it is about 1×10^{-3}. In the Deposition Area (see Figure 2) near Lost Lake, it is also about 1×10^{-3}. The non-cancer hazard risk for residents in the Lost Lake vicinity had an exposure quotient of 6.3. For residents along Clipper Creek, it was 6.3–10, and for recreational users of the Deposition Area above Lost Lake, it was 28.

There are seeps from abandoned mine shafts that are discharging high arsenic concentrations and other contaminants. Further, the interstitial waters within the tailings and below the tailings pile located near the mine contain elevated arsenic, with some groundwaters near the mine site having arsenic in excess of 100–500 μg/l. The groundwater in the mine area has a complex hydrogeology, consisting of a fractured rock aquifer system. With this aquifer system it will be difficult to trace the movement of mine-tailings-derived arsenic that has polluted the groundwaters in that site to down-groundwater-gradient areas residential wells.

The residents in the area use groundwater as a domestic water supply source. At a US EPA public information meeting held in November 2001, the public participants indicated that their greatest concern was the pollution of their domestic wells by arsenic derived from the mine and/or tailings. This situation is complicated by the fact that some of the groundwaters in the area have been found to contain naturally elevated arsenic. At this time, studies have not been done in sufficient detail to determine the origin of the arsenic present in a number of the domestic water supply wells. Further RI studies are underway to better define the extent of groundwater pollution at the site.

3. Remediation options/objectives

At this time, the cleanup objectives for the Lava Cap Mine Superfund site have not been established (CH2M Hill, 2001c). ARARs (Applicable, Relevant and Appropriate Requirements) are being developed for the site. These will be based on the regulatory requirements of the US EPA, the Central Valley Regional Water Quality Control Board, the California Department of Health Services and the California Department of Toxic Substances Control. Important issues that need to be defined are the appropriate degree of public health and environmental protection that should be achieved at the Lava Cap Mine site, and whether funds will be available to pay for this level of protection. With the Bush administration reducing Superfund support, there is a potential that funds for further investigation and, especially, remediation at the Lava Cap Mine site may be significantly reduced, and possibly eliminated.

3.1. Soil remediation

Davis et al. (2001) have recently published a summary of residential and industrial arsenic soil cleanup goals by target risk levels for Superfund sites. This information is presented in Table 2. Based on information in their paper, typically residential areas are cleaned up to about 25 mg/kg total arsenic, except in California, where the human health cleanup objectives have been about 3 mg/kg. The US EPA has yet to propose the cleanup goals for contaminated soils and sediments for the Lava Cap Mine Superfund site.

Table 2. Summary of residential and industrial cleanup goals by target risk level.

Target risk level	Residential cleanup goals (mg/kg)			Industrial cleanup goals (mg/kg)		
	Range	Mean[a]	n	Range	Mean[a]	n
1×10^{-6}	0.37–305	17	18	8–219	44	9
1×10^{-5}	30–250	68	5	21–500	65	11
1×10^{-4}	100–230	152	2	200–336	272	3

Source: Davis et al. (2001).
[a] The geometric mean was used as it best represented the central tendency of the data sets.

3.2. Groundwater remediation

There are several important issues relating to the establishment of the appropriate drinking water cleanup objective for those groundwater supplies that are polluted by arsenic wastes (mine tailings). The US EPA, after considerable technical and political debate, finally adopted a revised arsenic drinking water MCL of 10 μg/l. This MCL is to be implemented by 2006. This MCL is not necessarily based on risk. According to the NRC (2001), to achieve a one in a million lifetime additional cancer risk, the drinking water MCL should be set at about 0.1 μg/l. A risk-based drinking water MCL is not used, because many domestic water supplies contain arsenic in excess of 0.1 μg/l (Focazio et al., 1999; USGS, 2002).

One of the factors that played a major role in determining the 10 μg/l MCL was the cost of treatment of domestic water supplies to achieve a lower arsenic concentration (Frost et al., 2002). Even at 10 μg/l, there are about 4000 domestic water supplies in the US that will need to reduce arsenic in their treated water, at an average household cost of about \$32 per year, which translates to about 3 cents per person per day for large municipal systems. For small systems, the cost can be as high as 30 cents per person per day. One of the important issues that needs to be resolved at the Lava Cap Mine Superfund site is whether the 10 μg/l MCL is an appropriate groundwater cleanup objective, where it is clear that the arsenic derived from the mine/tailings is the primary source of the arsenic in the groundwater that is used or could be used for domestic purposes.

The Central Valley Regional Water Quality Control Board (CVRWQCB) and the US EPA regulate some constituents when derived from wastes, based on a risk-based approach, which requires greater cleanup than when based on a politically/economically based MCL. Precedent for the risk-based approach stems from the CVRWQCB's requirements for cleanup of waste-derived chloroform at the UCD/DOE LEHR National Superfund site, located on the University of California, Davis (UCD), campus. While UCD attempted to establish the US EPA trihalomethane (chloroform) MCL of 100 μg/l as the chloroform concentration goal for groundwater cleanup, the CVRWQCB determined that goal was inappropriate, because it was based on a variety of non-risk-related factors, and established a risk-based goal of about 1 μg/l. For a discussion of these issues, see the Davis South Campus Superfund Oversight Committee's (DSCSOC) website, http://members.aol.com/dscsoc.

3.3. Contaminated soil/sediment remediation options

At a US EPA public meeting held in November 2001, US EPA representatives discussed potential remediation options. Figure 2 defines the areas of concern as the Lava Cap Mine, Little Clipper Creek, Clipper Creek, Little Greenhorn Creek, Lost Lake and Deposition Area and downstream of Lost Lake. Table 3 presents a summary developed by the US EPA (US EPA, 2001b; CH2M Hill, 2002) on the potential impacts of the various remedial alternatives.

The remedial alternatives being considered by the US EPA include:

- "no action," required by CERCLA;
- institutional controls, where access to the contaminated areas is restricted;

- containment, with capping of existing areas, such as tailings stored near the mine site, excavation of tailings and tailings-contaminated soils along the streams, in the streambeds and in the Lost Lake area, with disposal either in a new landfill that would be constructed to contain the excavated tailings or after being trucked out of the area for off-site disposal.

"No action" is not an acceptable approach for the tailings-contaminated areas because of the widespread human health and ecological risk associated with the dispersion of tailings containing high levels of arsenic and some other constituents. The institutional controls approach will be difficult to implement reliably because of the widespread contamination that has occurred and the fact that the public has ready access to arsenic tailings-polluted areas.

3.4. Preliminary cost estimates

Table 3 lists the general aspects of the potential impacts of these various alternatives. At this time, the US EPA has not provided detailed information on any of the alternatives,

Table 3. Possible impacts to the community of different cleanup options.

Options	Impacts					
	Short term				Long term	
	Earth moving or quarrying	Truck traffic	Construction noise	Road building	Land use restriction	Surface water use restriction
Soil and sediment						
Excavation/off-site disposal	High	High	High	High	Least	Low
Excavation/on-site disposal	High	Locally high	High	Locally high	Landfill areas	Low
Capping and flood control	Medium	Medium	Medium	Locally high	Medium	Medium
Revegetation	Low–medium	Low–medium	Low–medium	Low	High	
Water management						
Upgrading Lost Lake dam, local rock	High	Locally high	Locally high	Locally high		Low or medium
Upgrading Lost Lake dam, imported rock	High	High	Locally high	High		Low or medium
Restoring Clipper Creek without Lost Lake	Medium	Medium	Medium	Medium		High (no lake)

Source: US EPA (2001b).

and while some preliminary costs have been developed for some of them, these costs are not comprehensive and do not consider some of the important long-term management issues that will have to be addressed as part of remediation of the site using certain of the approaches, such as landfilling. The preliminary cost estimates (Table 4) indicate that excavation with disposal in a new area landfill will cost on the order of $34 million, while excavation, trucking and off-site disposal are estimated to cost $100 million. Consideration is also given to treatment of the arsenic-containing soils and sediments to immobilize the arsenic. However, none of the potential treatment options appears to be economically feasible.

3.5. Remediation issues

There are a number of important long-term public health and environmental protection issues that need to be considered in developing an appropriate remediation approach, especially those associated with attempting to use capping of existing tailings/arsenic-polluted areas and/or the construction of an RCRA (Resource Conservation and Recovery Act) landfill cover or landfill in the area to store the contaminated soils and tailings. Typically today in municipal solid waste, industrial non-hazardous waste and hazardous waste management by landfills, the US EPA regulations do not provide high degrees of protection associated with the inevitable failure of the plastic sheeting and compacted clay

Table 4. Preliminary cost estimates by CH2M Hill (2002).

Area	Initial cost + 50 years limited monitoring and maintenance (million US $)
Mine buildings/tailings and waste rock pile	12
Excavation and on-site disposal	13
Excavation and off-site disposal	15
Little Clipper Creek, Clipper Creek	
Capping and channelization	1.9
Excavation and on-site disposal	0.65
Excavation and off-site disposal	0.8
Lost Lake and Deposition Area	
Drain Lake and cap sediment	8.5
Excavation and on-site disposal	19
Excavation and off-site disposal with removal of dam	83
Downstream of Lost Lake, Clipper Creek and Little Greenhorn Creek	
Excavation and on-site disposal	0.38
Excavation and off-site disposal	0.44
Total excavation and on-site disposal	34
Total excavation and off-site landfill	100

Source: US EPA (2001b).

landfill liners and associated covers that are used in Subtitle D (municipal and non-hazardous industrial waste) and Subtitle C (hazardous waste) landfills.

The Agency's approach focuses on limited-term containment of the waste material for a 30-year post-closure period. It is well known (see Lee and Jones-Lee, 1998a,b) that the liner systems that are being allowed in Subtitle C and D landfills will eventually fail to prevent leachate from migrating through the liners into the underlying groundwaters and, where groundwater surfaces through springs, to the surface waters of the area. Further, as discussed by Lee and Jones-Lee (1998a,b), the groundwater monitoring systems that are being allowed by the US EPA and the state regulatory agencies to be developed at many landfills are significantly deficient in detecting groundwater pollution when it first reaches the underlying groundwaters, before widespread off-site groundwater pollution occurs.

The state of California has significantly more explicit protection of public health and the environment in landfilling than the US EPA and many other states. California requires that the Subtitle C or D landfill be able to prevent impairment of the beneficial uses of groundwaters underlying the landfill for as long as the wastes in the landfill will be a threat. The issue that must be resolved is whether this requirement will be achieved in the remediation of the Lava Cap Mine Superfund site for all wastes (tailings) left in the area.

The basic problem with RCRA landfills is that the US EPA has failed to acknowledge and meaningfully address the protection of groundwaters from pollution by landfill leachate (or, in the case of the Lava Cap Mine site, tailings and arsenic-containing water) for as long as the wastes in the landfill will be a threat. The wastes in an RCRA Subtitle C or D landfill, which would contain tailings from the Lava Cap Mine site, will be a threat to pollute groundwaters, effectively, forever. The liner systems that are allowed, including a double composite liner system, eventually will fail to prevent significant leachate migration through the liner system. This will lead to off-site pollution, due to the fact that the groundwater monitoring systems that are typically used, under ideal conditions, have a low probability of detecting groundwater pollution (see Lee and Jones-Lee, 1998a,b). This situation is even more complex in a geological setting like that of the Lava Cap Mine area, where there is a fractured rock aquifer system, which is virtually impossible to reliably monitor for groundwater pollution.

While landfills can be developed for containment of solid wastes that will protect public health and the environment effectively forever, the initial cost of developing such landfills is about twice to three times that of conventional landfills. Further, and most importantly, there is an *ad infinitum* monitoring, operation and maintenance cost that must be borne by the responsible parties for managing the wastes in the landfill. The US EPA, as part of RCRA and in accord with the federal Congress' requirements, only requires minimal 30-year post-closure funding for limited monitoring and maintenance of the landfill containment system. There is no assured funding after the landfill has been closed for 30 years. This approach is obviously deficient in providing long-term public health protection from the waste contained in the landfill. Lee and Jones-Lee (1998a) have recommended that, at the time of establishing a landfill for waste management, a dedicated trust fund of sufficient magnitude to address all plausible worst-case failure scenarios for the landfill containment system and monitoring system be developed, which would be expected to generate sufficient funds to operate, monitor and maintain the landfill system forever.

Lee and Jones-Lee (1998a) have discussed approaches for developing landfills that could be used at the Lava Cap Mine site to contain, effectively forever, the tailings and arsenic-contaminated soils. This will involve a double composite lined system, with a leak detection system between the two liners. When leachate derived from the wastes that contain arsenic that is a threat to cause groundwater pollution passes through the upper composite liner, then either the wastes must be removed from the landfill, or a leak-detectable cover should be placed over the landfill that would be operated and maintained *ad infinitum* to prevent moisture from entering the landfill for as long as the wastes represent a threat.

4. Conclusions

The Lava Cap Mine Superfund site area is highly contaminated with tailings that contain, in some cases, greatly elevated arsenic. This arsenic is a threat to human health through contact recreation and potentially contaminated groundwater. Further work needs to be done to define the existing and potential groundwater pollution that has occurred due to the Lava Cap Mine and its tailings. In addition to the human health threat, the tailings are also a threat to aquatic and terrestrial life by arsenic and a variety of other heavy metals derived from the former mining activities.

While the US EPA has indicated that the 10 µg/l drinking water MCL could be used as the remediation goal for arsenic-polluted groundwaters, this goal is inappropriate, because it is not based on a risk-based approach, but considers economic and political factors. A risk-based groundwater cleanup objective should be adopted for those situations where either mine- or mine tailings-derived arsenic is the source of the arsenic that is polluting the groundwater above background.

The US EPA has presented some potential remediation approaches which will likely involve capping of some of the contaminated areas with an RCRA cap to reduce moisture entering into the area and to immobilize the tailings. There are important issues that must be addressed regarding the monitoring and maintenance of the cap to ensure that its integrity in preventing moisture from entering the tailings is maintained for as long as the capped tailings represent a threat. For many of the contaminated areas, it will likely be necessary to excavate with disposal in a local new landfill or with trucking to an off-site landfill. The preliminary estimated cost for excavation and onsite landfilling is on the order of $34 million, with off-site disposal costing on the order of $100 million. These costs do not include the true long-term costs associated with *ad infinitum* monitoring and maintenance of the landfill.

There are concerns about the adequacy of minimum RCRA landfills to be able to prevent further environmental pollution by the landfilled tailings and soils due to arsenic-containing leachate that will be generated within the landfill that will ultimately pass through the landfill liner system into the underlying groundwater system. Provisions should be made in remediation of the Lava Cap Mine site to ensure that the remediation approach adopted provides for a high degree of protection of public health and the environment for as long as the waste tailings represent a threat.

References

ATSDR, 2001. Public Health Assessment for Lava Cap Mine, Nevada City, Nevada County, California, EPA Facility ID: CAD983618893. Agency for Toxic Substances and Disease Registry, US Department of Health and Human Services, Athens, GA.

CH2M Hill, 2001a. Draft Human Health Risk Assessment Report for the Lava Cap Mine Superfund Site, Nevada County, California. Report to the US Environmental Protection Agency, CH2M Hill, Inc., Sacramento, CA, April.

CH2M Hill, 2001b. Final Draft Ecological Risk Assessment for Lava Cap Mine Superfund Site, Nevada City, Nevada County, California. Report to the US Environmental Protection Agency, CH2M Hill, Inc., Sacramento, CA, July.

CH2M Hill, 2001c. Public Release Draft Remedial Investigation Report; Lava Cap Mine Superfund Site, Nevada County, California. Report to the US Environmental Protection Agency, CH2M Hill, Inc., Sacramento, CA, November.

CH2M Hill, 2002. Preliminary Draft Feasibility Study for the Lava Cap Mine Superfund Site, Nevada County, California (not including alternative evaluation). Report to the US Environmental Protection Agency, CH2M Hill, Inc., Sacramento, CA, January.

Davis, A., Sherwin, D., Ditmars, R., Hoenke, K.A., 2001. An analysis of soil arsenic records of decision. Environ. Sci. Technol., 35, 4396.

Focazio, M.J., Welch, A.H., Watkins, S.A., Helsel, D.R., Horn, M.A., 1999. A Retrospective Analysis on the Occurrence of Arsenic in Ground-Water Resources of the United States and Limitations in Drinking-Water-Supply Characterizations. US Geological Survey Water-Resources Investigations Report 99-4279, http://co. water.usgs.gov/trace/pubs/wrir-99-4279/

Frost, F.J., Tollestrup, K., Craun, G.F., Raucher, R., Stomp, J., Chwirka, J., 2002. Evaluation of costs and benefits of a lower arsenic MCL. J. AWWA, 94 (3), 71–80.

Lee, G.F., Jones-Lee, A., 1998a. Assessing the potential of minimum Subtitle D lined landfills to pollute: alternative landfilling approaches. Proc. of Air and Waste Management Association 91st Annual Meeting, San Diego, CA, available on CD ROM as paper 98-WA71.04(A46), 40 pp., June. Also available at http://www.gfredlee.com

Lee, G.F., Jones-Lee, A., 1998b. Deficiencies in Subtitle D landfill liner failure and groundwater pollution monitoring. Presented at the NWQMC National Conference. Monitoring: Critical Foundations to Protect Our Waters, US Environmental Protection Agency, Washington, DC, July.

NRC, 2001. Arsenic in Drinking Water: 2001 Update. National Research Council, National Academy Press, Washington, DC, http://www.nap.edu

Salocks, C.B., 1997. Screening Level Evaluation of Potential Health Risks at Lava Cap Mine, Nevada City, California. California Department of Fish and Game Memorandum to Dan Ziarkowski, August 25.

US EPA, 1989. Risk Assessment Guidance for Superfund, Volume I: Human Health Evaluation Manual, Part A. EPA/540/1-89/002, US Environmental Protection Agency, December.

US EPA, 2000. Studies Indicate Arsenic is Primary Concern at Lava Cap Mine Superfund Site. US Environmental Protection Agency, Region IX, San Francisco, CA, September.

US EPA, 2001a. Lava Cap Mine Superfund Site: Arsenic Poses Unacceptable Risk to Human Health; Arsenic and Metals Pose Potential Risks to the Environment. US Environmental Protection Agency, Region IX, San Francisco, CA, November.

US EPA, 2001b. Lava Cap Mine Superfund Site: EPA Studies Cleanup Options, Seeks Public Input. US Environmental Protection Agency, Region IX, San Francisco, CA, November.

US EPA, 2002. Lava Cap Mine California. US Environmental Protection Agency website, June 11, http://yosemite.epa.gov/r9/sfund/overview.nsf/507c94f730e0ebf488256958005cda5f/0910a5ef115843c988256802 0076955?OpenDocument

USGS, 2002. Arsenic in ground water of the United States. US Geological Survey website, May 20, http://co. water.usgs.gov/trace/arsenic/

Arsenic Exposure and Health Effects V
W.R. Chappell, C.O. Abernathy, R.L. Calderon and D.J. Thomas, editors
© 2003 Published by Elsevier B.V.

Chapter 7

Investigation of arsenic release from sediment minerals to water phases

Tran Hong Con, Nguyen Thi Hanh, Michael Berg
and Pham Hung Viet

Abstract

Severe and widespread contamination by arsenic in groundwater and drinking water has been recently revealed in rural and suburban area of Vietnamese capital of Hanoi with similar magnitude observed in Bangladesh and West Bengal, India. This fact has prompted the need to investigate the possible mechanisms for such widespread contamination and develop suitable techniques for lowering arsenic concentrations in supplied water. In the present study, laboratory-scale experiments were performed to assess the possible release of arsenic from solid phase into water phase under both anaerobic and aerobic conditions. Various chemical equilibrations governing the speciation of different ions in water phase and alluvial sediment are discussed. Under anaerobic conditions, the release of arsenic seemed to be closely related to the content of MnO_2 in sediments, the reduction of $Fe(III)$ to $Fe(II)$ and the formed sulfur content. Elevated MnO_2 content may inhibit the release of arsenic to groundwater. The reduction of arsenic concentrations in water phases could be due to the formation of FeAsS or AsH_3. In aerobic conditions, the hydrothermal oxidation process was proposed as a plausible mechanism for the release of arsenic from arsenic-rich mineral surfaces to water phases. In acidic conditions, arsenic concentrations increased due to the more effective release from mineral surfaces. Under neutral medium ($pH \approx 7$), the arsenic release was less efficacious, which could be due to the co-precipitation onto ferric hydroxide. The possible mechanisms suggested in this study may be useful to explain the elevated contamination by arsenic in surface waters of upstream rivers of mountain areas in northern Vietnam and in underground water of rivers delta, and critical for moving to the next stage in developing suitable techniques for lowering arsenic concentrations in groundwater and drinking water in Vietnam.

Keywords: arsenic release aerobic, anaerobic, arsenic-rich mineral surfaces, sorption, groundwater, surface water

1. Introduction

In recent years, the naturally occurring contamination by arsenic in groundwater in Asian countries has received particular attention. Serious health effects caused by arsenic poisoning have been observed in Bangladesh and West Bengal, India (Chowdhury et al., 2000). In Vietnam, due to the similar composition of groundwater As in Bangladesh, elevated contamination by arsenic has been anticipated. Public media have also voiced concern that arsenic contamination in groundwater may become a key environmental problem in Asian developing countries in the 21st century (Christen, 2001).

    To address this issue, our laboratory has recently conducted a comprehensive monitoring survey of the status of arsenic contamination in groundwater and drinking water in Hanoi, the capital of Vietnam and its surrounding areas. Our results clearly demonstrated widespread and elevated contamination by arsenic in groundwater and supplied water in the suburban and rural areas of Hanoi (Berg et al., 2001). The magnitude of pollution was similar to that observed in Bangladesh, with a large number of well waters contained the arsenic concentrations exceeding the Vietnam standard of 0.05 mg/l. This severe situation has led us to continue to conduct further research towards the understanding of possible mechanisms of widespread arsenic contamination in groundwater and the development of suitable techniques for lowering arsenic concentrations. In 1992, Can (2001) observed extraordinarily high arsenic concentrations in some streams during his field investigation in the highlands area upstream of the Ma River, North Vietnam. Subsequent surveys showed that various arsenic-rich minerals, such as arsenopyrite, occurred widely in this area. The possible mechanism of arsenic contamination in stream water of a mountain area may be due to the weathering of arsenic-rich minerals (McArthur et al., 2001). In Vietnam, there is a relatively large pyrite mine located about 60 km far from Hanoi city. In addition, a number of gold mining sites with arsenopyrite are located over a large area of northern Vietnam. The gold mining activities in northern Vietnam have been extensive in recent years and arsenic-rich minerals have been distributed to the land surfaces during gold mining. If the hypothesis that weathering process may be a source of arsenic in surface waters, the contamination by arsenic in surface water would be a serious concern in the near future for a large area of northern Vietnam. In fact, we have tested surface water from the Read River, and found that arsenic concentrations in some locations reached a level of 0.09 mg/l, which exceeded the Vietnam standard level (0.05 mg/l). This fact has prompted us to examine the mechanism of arsenic release to surface water from mineral surfaces.

   In this study, we proposed the possible mechanisms of arsenic contamination in groundwater, based on our observations in well waters in upper and lower aquifers of Read River delta (Berg et al., 2001). The mechanisms of release of arsenic from various solid phases were investigated under anaerobic and aerobic conditions. These experimental investigations may be useful for understanding the mechanism of arsenic contamination in water sources and may be critical steps for developing suitable techniques for lowering arsenic concentrations in supplied water and drinking water in Vietnam.

2. Methodology

2.1. Experimental setup for the investigation of the mechanism of arsenic release to groundwater under anaerobic conditions

A batch experiment was performed to investigate the release of arsenic in groundwater. The experimental setup is described in Figure 1. The primary apparatus consisted of an anaerobic column, pump and the deoxygenation apparatus. The anaerobic column with a height of 700 mm and a diameter 45 mm was constructed in three layers. The first layer consists of humus collected from the surface layer. The second layer was coarse gravel (2–5 mm diameter) and the third layer contained clean sand spiked with 0.001% As (relative to the mass of the dry sand layers) in the form of AsO_4^{3-} and 0.005% MnO_2 co-precipitated with 0.1% Fe(III) in the form of $Fe(OH)_3$. The water phase was prepared to

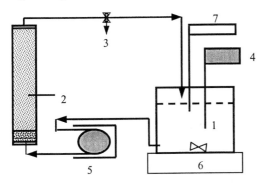

1: Water tank
2: Anaerobic column
3: Sampling valve
4: Deoxygenating apparatus
5: Circulative pump
6: Magnetic stirrer
7: Additional nutrient and traced
   elements pump

Figure 1. Schematic diagram of the experimental setup for the investigation of arsenic release from alluvial sediment under anaerobic conditions.

be similar to natural water with a composition as given in Table 1. To maintain the anaerobic condition throughout the system, water was pumped from tank (1) to the anaerobic column (2) from the lower end of the column through the layers and finally the water was pumped back to the tank (1). The experiment was run for 56 days continuously. Water samples were collected every 24 h at the upper end of an anaerobic column by a three-way valve and were analyzed for various parameters such as total arsenic, iron, manganese, nitrate, phosphate, ammonium, sulfate, sulfur and dissolved oxygen. Chemical analysis of these ions in water samples followed the methods reported in Standard Methods for Examination of Water and Waste Water (Table 2).

2.2. Experimental setup for the investigation of arsenic releases mechanism to surface water under aerobic conditions

The experimental setup is described in Figure 2. The main component is an aerobic column. The column has a 45 mm diameter and 700 mm length and was packed in two layers: the lower layer was the minerals derived from the weathering process. This mineral was taken from the Soc Son district, about 40 km north of Hanoi. The upper layer was the sand phase and spiked with arsenopyrite approximately 0.2% of the weight of the sand. Rainwater was

Table 1. Composition of water phase in an anaerobic experiment.

Composition	Concentration (M)
Ca^{2+}	1.0×10^{-3}
HCO_3^{-}	2.4×10^{-3}
NO_3^{-}	3.0×10^{-4}
SO_4^{2-}	5.2×10^{-3}
Na-glutamate	1.2×10^{-3}
PO_4^{3-}	6.0×10^{-7}
Mg^{2+}	6.0×10^{-5}

Table 2. Analytical methods for the determination of various parameters in anaerobic and aerobic experiments.

Composition	Analytical method
Arsenic	HVG-AAS
Fe, Mn	AAS
NO_3^-, PO_4^{3-}, NH_4^+	Colorimetry
SO_4^{2-}	Conductimetry
S^{2-}	Colorimetry (methylene blue)
Oxygen	Iodometry

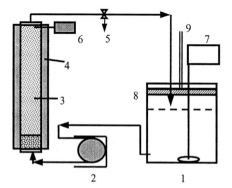

1: Water tank
2: Circulative pump
3: Sand and mineral packed column
4: Heating cover
5: Sampling valve
6: Thermometer
7: Air bubble generator/ oxygen absorber
8: Sol remover
9: Air outlet/ supplementary nutrient inlet

Figure 2. Schematic diagram of the experimental setup for the investigation of arsenic release from mineral surfaces under aerobic conditions.

used as a water phase. The water was continuously saturated with oxygen throughout the experiment. The temperature was kept constantly at 25 or 38°C by a temperature-regulation apparatus (4). The water phase was pumped into the aerobic column from the lower end and then back to the reservoir tank (1). The composition of the rainwater is given in Table 3.

3. Results and discussion

3.1. Release of arsenic from mineral surfaces to groundwater under anaerobic conditions

The anaerobic experiment was run for 56 days. The various parameters of the water phase in the outlet of the system were measured and results are expressed in Figures 3 and 4.

Table 3. Composition of water phase (rainwater) in an aerobic experiment.

Parameters	As (ppb)	SO_4 (ppm)	NO_3 (ppm)	Fe (ppm)	NH_4 (ppm)	HCO_3^- (ppm)
Concentration	0.8	11.6	8.2	0.087	2.4	13.6

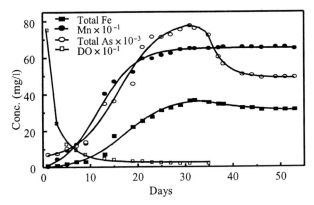

Figure 3. Chronological evolution of DO, Fe, Mn and As concentrations in the water phase of an anaerobic experiment.

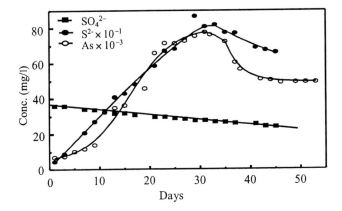

Figure 4. Chronological evolution of sulfate, sulfide and arsenic concentrations in the water phase of an anaerobic experiment.

Chronological variations of various ions in the water phase indicate that the reduction of MnO_2 to Mn(II) started under oxygen-depleted conditions (dissolved oxygen <2.4 mg/l). The Mn(II) concentrations increased and then remained relatively constant after 30 days. Fe concentrations initially increased slowly and the reduction of Fe(III) significantly increased after 15 days. At that time, more than 80% of MnO_2 was reduced to Mn^{2+}. This result indicates that under anoxic conditions, the reduction of MnO_2 to Mn(II) was faster than the reduction of Fe. This phenomenon can be explained by the higher redox potential of MnO_2/Mn^{2+} as compared to that of $Fe(OH)_3/Fe^{2+}$ in the relatively neutral medium (pH ≈ 7) (Nickson et al., 1998). From day 15 to 30, concentrations of Fe(II) increased rapidly and then declined and remained relatively constant until the end of the experiment. The suspended precipitation of FeS was formed in the system when Fe(II) concentrations decreased. Overall, the mechanisms of various processes involved in the anoxic system can be expressed by the following reactions:

$$\text{Organic matter and } SO_4{}^{2-} \xrightarrow{\text{Bacteria}} CH_4 + CO_2 + NH_4{}^+ + S_2{}^- + \cdots$$

$$MnO_2 + 2e + 4H^+ = \mathbf{Mn^{2+}} + 2H_2O$$

$$Fe(OH)_3 + e + H^+ = Fe(OH)_2 + H_2O$$

$$Fe(OH)_2 + 2CO_2 = \mathbf{Fe^{2+}} + 2HCO_3{}^-$$

$$Fe^{2+} + S^{2-} = \mathbf{FeS}$$

$$2Fe^{2+} + MnO_2 + 4H^+ = 2Fe^{3+} + Mn^{2+} + 2H_2O$$

$$Fe^{3+} + 3H_2O = Fe(OH)_3 + 3H^+$$

$$S^{2-} + H_2O \Leftrightarrow HS^- + OH^-$$

The dynamics of As in the anoxic water phase indicate that the release of As and Fe(II) to the water took place at the same time (Fig. 3). We hypothesize that arsenic in alluvial sediments is mainly present in the form of arsenate [As(V)] and predominantly adsorbed onto Fe(III) hydroxide. In the anaerobic experiment, when Fe(III) is reduced to Fe(II) in the presence of dissolved bicarbonate, the reduction of As(V) to As(III) takes place simultaneously (Nickson et al., 2000). In addition, when the MnO_2 content in the solid phase was still high, the re-oxidation of $AsO_3{}^{3-}$ to $AsO_4{}^{3-}$ and the re-precipitation of arsenate are plausible. The process can be depicted as follows:

$$FeAsO_4 + 3e + 2H^+ \Leftrightarrow Fe^{2+} + AsO_3{}^{3-} + H_2O$$

$$M_3(AsO_4)_2 + 4e + 4H^+ \Leftrightarrow 3M^{2+} + 2AsO_3{}^{3-} + 2H_2O$$

$$AsO_3{}^{3-} + MnO_2 + 2H^+ = AsO_4{}^{3-} + Mn^{2+} + H_2O$$

$$3M^{2+} + 2AsO_4{}^{3-} = M_3(AsO_4)_2$$

where M is the metal ion.

After 30 days, As and Fe(II) concentrations decreased, but not as a result of the lower reduction of Fe(III) to Fe(II). The decreased Fe(II) concentration is related to the formation of precipitated FeS. For the reduction of As concentration after 30 days, we hypothesize that there might be a formation of FeAsS and/or AsH_3. The reasons for the concomitant decrease of both As and Fe concentrations need further investigation.

The chemical processes described above suggest that the release of As under anaerobic reduction of organic matters in aquifers is closely related to the MnO_2 contents in alluvial sediments, the reduction of Fe(III) to Fe(II) and the amount of formed FeS. When MnO_2 levels in sediments are high, the reduction of Fe(III) and the release of As are inhibited. We also further tested the groundwater quality at various locations in Hanoi and revealed that elevated arsenic concentrations were found in iron-rich water. In contrast, in manganese-rich water, concentrations of Fe and As were low (for an example, see Table 4). These field observations are consistent to the results we found in our anaerobic experiment as discussed earlier. However, a number of chemical equilibrium as well as interactions among various phases may complicate the fate and behavior of arsenic in aquifers. Further studies are required for this topic.

Table 4. Mean concentrations of Fe, Mn and total As in groundwater collected from various locations in Hanoi city. MD, NH, NSL, LY, YP, PV, HD and TM refer to the following locations: Mai Dich, Ngoc Ha, Ngo Sy Lien, Luong Yen, Yen Phu, Phap Van, Ha Dinh and Tuong Mai, respectively.

Metals level (mg/l)	Location							
	MD	NH	NSL	LY	YP	PV	HD	TM
Total Fe	0.31	0.62	1.20	2.15	4.78	4.50	8.33	5.64
Mn^{2+}	0.95	1.63	0.86	0.37	0.38	0.11	0.13	0.26
Total arsenic	0.03	0.04	0.40	0.06	0.39	0.34	0.26	0.06

3.2. Release of arsenic from mineral surfaces to surface water under aerobic conditions

The aerobic experiment was run for 50 days at atmospheric conditions (25°C and 1 atm.) with oxygen-rich water (dissolved oxygen >7 mg/l). The results are shown in Figure 5. Sulfate concentrations in the water phase increased with a rate of about 0.25 mg/l per day. The oxidation of pyrite took place according to the following reactions:

$$2FeS_2 + 7O_2 + 2H_2O = 2Fe^{2+} + 4SO_4^{2-} + 4H^+$$

$$2Fe^{2+} + 1/2O_2 + 2H^+ = 2Fe^{3+} + H_2O$$

Arsenic concentrations also increased with time at a rate of 0.2 μg/l per day, about three orders of magnitude lower than the sulfate concentrations. However, the oxidation process that releases sulfate to the water phase took place in parallel with the release of

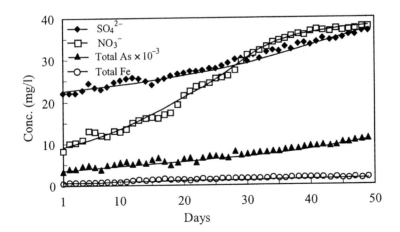

Figure 5. Chronological evolution of Fe, sulfate, nitrate and As concentrations in the water phase of an aerobic experiment.

arsenate from arsenopyrite:

$$4FeAsS + 11O_2 + 6H_2O = 4Fe^{2+} + 4AsO_3{}^{3-} + 4SO_4{}^{2-} + 12H^+$$

$$2AsO_3{}^{3-} + O_2 = 2AsO_4{}^{3-}$$

$$4Fe^{2+} + O_2 + 4H^+ = 4Fe^{3+} + 2H_2O$$

In addition, a certain amount of Fe was also released to the water. Nevertheless, Fe concentrations were actually found to be low and their chronological increase was negligible. Under oxic conditions of our experiment and pH range of 6.3–6.8, most Fe^{3+} ions released from the mineral surfaces were hydrolyzed and re-precipitated in the form of $Fe(OH)_3$ and remained in the solid phase. Only small amounts of Fe were dissolved in the water phase in the form of $Fe(OH)_2{}^+$ or bicarbonate (Barnes, 1997). This phenomenon led to the sorption of arsenate anions onto the solid phase and subsequently inhibited the release of arsenic to water phase. Sulfate ions were less well adsorbed and their concentrations were substantially greater than those of arsenate, as observed in Figure 5. During the experiment, the pH of the solutions slightly increased. When the pH was rapidly lowered through continuous CO_2 purging of the water solutions, Fe and As concentrations apparently increased. In this case, the re-dissolution of the precipitate $Fe(OH)_3$ may be a plausible explanation.

We also performed the same aerobic experiment at a higher temperature of 38°C. The oxidation process was faster as was evidenced by a more rapid elevation of sulfate concentrations. However, the chronological dissolution of Fe and As in the water phase remained relatively similar to that observed in experiment at 25°C. Nitrate concentrations increased at both temperatures due to the nitrification of ammonium that was present in the inlet solutions and the release of nitrate from the sediment material packed in the lower part of the aerobic column.

Considering these observations, we suggest that hydrothermal oxidation is a main factor for releasing arsenic from arsenic-rich mineral surfaces to surface water. The mechanism of this process was similar to the oxidation of sulfide to sulfate. At acidic pH values, As concentrations increased substantially. Under neutral pH conditions, the release of arsenic is less efficient due to the co-precipitation to iron hydroxide phases. This phenomenon may be a possible explanation for the elevated arsenic contamination in some streams observed in upstream of Ma River area, northern Vietnam. This river flows through a mountain-rock area and some gold mining sites with arsenic-rich minerals.

4. Conclusions

Weathering processes are a potential source for the release of arsenic from arsenic-rich minerals to water phases. Subsequently, arsenic is trapped in alluvial soil layers and again released to water phases during anaerobic bio-disintegration. These are naturally occurring processes that undergo a series of complicated mechanisms. Our preliminary experiments under both oxic and anoxic conditions provide insight into the mechanisms of widespread arsenic contamination in water resources of Vietnam. Such work is critical for us to move to the next stage in developing suitable techniques for removing and lowering arsenic concentration in groundwater and drinking water in Vietnam.

Acknowledgements

The authors are grateful to the Albert Kunstadter Family Foundation (New York) for valuable financial support. The authors also express their sincere thanks to students Nguyen Thi Hong Hanh and Pham Van Dien of CETASD as well as colleagues of EAWAG for the effective cooperation and useful discussion, respectively.

References

Barnes, H.L., 1997. Geochemistry of Hydrothermal Ore Deposits. Wiley, New York.

Berg, M., Tran, H.C., Nguyen, T.C., Pham, H.V., Schertenleib, R., Giger, W., 2001. Arsenic contamination of groundwater and drinking water in Vietnam: a human health threat. Environ. Sci. Technol., 35, 2621–2626.

Can, D.V., 2001. Preliminary assessment of the distribution, removal and accumulation of arsenic in hydrothermal depot bearing high arsenic content. Scientific Technical Communication on Geology, Department of Geology and Minerals of Vietnam, Hanoi, pp. 53–57.

Chowdhury, U.K., Biswas, B.K., Chowdhury, T.R., Samanta, G., Mandal, B., Basu, G.C., Chanda, C.R., Lodh, D., Saha, K.C., Mukherje, S.K., Roy, S., Kabir, S., Quamruzzaman, Q., Chakraborti, D., 2000. Groundwater arsenic contamination in Bangladesh and West Bengal, India. Environ. Health Perspect., 108, 393–397.

Christen, K., 2001. The arsenic threat worsens. Environ. Sci. Technol., 35, 286A–290A.

McArthur, J.M., Ravenscoft, P., Safinllah, S., Thirlwall, M.F., 2001. Arsenic in groundwater: testing pollution mechanisms for sedimentary aquifers in Bangladesh. Wat. Resour. Res., 37, 109–117.

Nickson, R.T., McArthur, J.M., Burgess, W.G., Ravenscroft, P., Ahmed, K.Z., Rahman, M., 1998. Arsenic poisoning of Bangladesh groundwater. Nature, 395, 338.

Nickson, R.T., McArthur, J.M., Burgess, W.G., Ravenscroft, P., Ahmed, K.Z., 2000. Mechanism of arsenic poisoning of groundwater in Bangladesh and West Bengal. Appl. Geochem., 15, 403–413.

Arsenic Exposure and Health Effects V
W.R. Chappell, C.O. Abernathy, R.L. Calderon and D.J. Thomas, editors
© 2003 Published by Elsevier B.V.

Chapter 8

Arsenic and heavy metal contamination of rice, pulses and vegetables grown in Samta village, Bangladesh

M.G.M. Alam, E.T. Snow and A. Tanaka

Abstract

Arsenic contaminated water from tube wells has become the major health problem threatening millions of people in Bangladesh. However, the arsenic (As) contaminated water is not just used for drinking, it is used to irrigate crops, and to wash and prepare food. Contamination of agricultural soils by long-term irrigation with As contaminated water can lead to contamination and phyto-accumulation of the food crops with As and other toxic metals. As a consequence, dietary exposure to As and other toxic metals may contribute substantially to the adverse health effects caused by the contaminated tube wells in Bangladesh. Various vegetables, rice, pulses and the grass pea were sampled in Samta village in the Jessore district of Bangladesh and screened for As, Cd, Cu, Pb and Zn by inductively coupled plasma atomic emission spectrometry and inductively coupled plasma mass spectrometry. These local food crops provide the majority of the nutritional intake of the people in this area and are of great importance to their overall health. In general, our data show the potential for some vegetables to accumulate heavy metals with concentrations of Pb greater than Cd. The concentrations of As and Cd were higher in vegetables than in rice and pulses. The concentration of Pb was generally higher in rice than in pulses and vegetables. However, some vegetables such as bottle ground leaf, ghotkol, taro, eddoe and elephant foot had much higher concentrations of Pb. Other leafy and root vegetables contained higher concentrations of Zn and Cu. Rice grown at Samta had increased Pb and As, but, considering an average daily intake of only 260 g rice per person per day, only the Pb is at concentrations which would be a health hazard for human consumption.

Keywords: arsenic, heavy metals, rice, pulses, vegetables, Samta village, Bangladesh

1. Introduction

The arsenic contamination of groundwater has caused a massive public health crisis in Bangladesh. Out of the 64 districts, 61 have As concentrations above the maximum permissible limit recommended by the World Health Organization (WHO) of $0.05 \ \text{mg} \ l^{-1}$, and therefore, at least 21 million people are now exposed to As poisoning (Fazal et al., 2001; Alam et al., 2002). In Bangladesh, groundwater is the main source for drinking, cooking and other household purposes. Even the agricultural system is mostly groundwater dependent as the large-diameter tube wells containing high concentrations of As (Alam et al., 2002) are used extensively for agricultural irrigation. As a consequence, As can be expected to be incorporated into the food chain from crops cultivated in these areas (Huq et al., 2001). It is also important to know whether other metal pollutants are present in high concentrations that could contribute to the conditions caused by the arsenicosis. Ingestion of As through routes such as contaminated food has not been adequately studied.

Bangladesh is predominantly a rural country with agriculture being the mainstay of the economy. The majority of the population is either directly or indirectly connected with agriculture. Rice (*Oryza sativa* indica) has been the main source of energy for the daily life of people in Bangladesh and it is by far the most important cereal grown in Bangladesh, accounting for nearly two-thirds of total domestic cereal production. With a per capita consumption of about $150 \, kg \, year^{-1}$, rice is the major staple food of Bangladesh, accounting for roughly 73% of caloric intake. Depending on local climate and soil conditions, typically three rice crops are grown: *aus* (planted in March/April and harvested in June/July); *aman* (planted in June/July and harvested in November/December); and *boro* (planted in December/January and harvested in April). Both *aus* and *aman* are largely rain-fed, while *boro* is mostly irrigated. Other important food crops include pulses such as beans, lentils and peas (known locally as *kheshari*, *masur* and *mung*). Pulses are the "meat" of Bangladesh, and the main source of protein for the most of its poor. In rural areas, pulses are consumed as *dal* or *dahl* (a soup type preparation) with rice. The grass pea (*Lathyrus sativus* L.; *khesari*) is cultivated over 33% of Bangladesh, accounting for 34% of pulse production (Malek et al., 1996), even though it contains a neurotoxic compound, β-*N*-oxalyl-diaminopropionic acid, which causes lathyrism in human beings. This crop is still cultivated because of its adaptability, low inputs, and its provision of cheap and quality cattle fodder.

Bangladesh has many horticultural crops with local production of more than 90 vegetables and 60 fruits. Major vegetable crops include potatoes, brinjal, bottle gourd, and taro. In an average Bengali home, the main meal would consist of boiled rice served with some sort of vegetables. Vegetables constitute an essential component of the diet, by contributing protein, vitamins, iron, calcium and other nutrients, which are usually in short supply. They also act as buffering agents for acid substances produced during the digestion process. However, these plants contain both essential and toxic elements over a wide range of concentrations. In this study, cereal, pulses and vegetables were collected from the As affected area of Samta village in the Jessore district, and the concentrations of As and other metals such as Cd, Cu, Pb and Zn were determined by inductively coupled plasma atomic emission spectrometry (ICP-AES) and inductively coupled plasma mass spectrometry (ICP-MS). Different kinds of vegetables, rice and pulses are grown during the year in tropical Bangladesh, but very little is known about the metal contents of local vegetables, rice and pulses. The purpose of this study is to analyze the metals in foods and to estimate the risks to health from certain elements in food. This has been assessed by comparing estimates of dietary exposures with the Provisional Tolerable Weekly Intakes (PTWIs) and Provisional Maximum Tolerable Daily Intakes (PMTDIs) recommended by the Joint Expert Committee on the Food Additives (JECFA).

2. Materials and methods

2.1. Study area

The study was conducted in Samta village of Jessore district, approximately 12 km from the border with West Bengal, India (Fig. 1). The total area of the village is 3.2 km^2, with

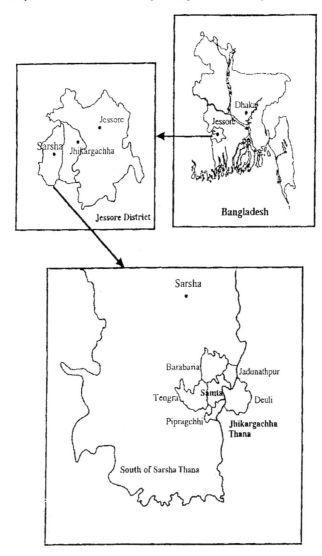

Figure 1. Location of Samta village, Bangladesh.

3606 people living in residential Samta village (1.5 km^2), with the remaining area being cultivated land. The average annual income of the villagers is Taka 32,000 (US $667). The small Betna river bounds the village to the east, but the main source of water for the villagers are 284, mainly shallow tube wells. There are also six deep tube wells that are used mainly for irrigation purposes. The average dissolved arsenic concentration of the contaminated water in Samta village was 0.24 mg l^{-1} and the highest arsenic concentration detected was 1.80 mg l^{-1}.

2.2. Sampling and pretreatment

Cereals, pulses and good quality fresh samples of vegetables commonly cooked by Bangladeshi's were collected in 1999. Cereals and pulses were taken from stocks to be consumed locally and were very probably also harvested locally, but in most cases the sites of production could not be identified and vegetables were sampled in home gardens with a random sampling procedure (replicate) and those were identified by a horticulturist (Table 1). The cleaning (removal of soil) of samples was performed by shaking and in the case of potatoes, taro, eddoe and elephant foot by means of a dry pre-cleaned vinyl brush. The samples were weighed, washed with distilled water to eliminate air-borne pollutants, dried in an oven at 65°C for 24 h, and then re-weighed to determine water content and ground to powder and used for metal analysis. Visible contaminants were hand-picked from the food samples. The leaves from amaranth plants, bottle gourd, drumstick leaf and Indian spinach were hand-picked, and the non-edible parts of lady's finger, raw plantain and brinjal were removed using a plastic knife. The edible portions of vegetables were washed three times with distilled water and finally rinsed with deionized water and dried in an oven at 65°C, ground using a ceramic-coated grinder and used for metal analysis.

Table 1. Cereal, pulses and vegetables giving the local, English, scientific and family name.

Local name (Bangla)	English name	Scientific name	Family
Cereal			
Aus	Rice	*Oryza sativa*	
Pulses			
Dal (Masur)	Red split lentil	*Lentil esculenta*	Fabaceae
Dal (Kheshari)	Grass pea	*Lathyrus sativus*	Fabaceae
Dal (Mung)	Pea	*Vigna radiata*	Fabaceae
Vegetables			
Alu	Potato	*Solanum tuberosum*	Solanaceae
Begoon	Brinjal	*Solanum melongena*	Solanaceae
Chal Kumra/Jali	Dhudi/Ash gourd	*Benincasa Hispida*	Cucurbitaceae
Chichinga	Snake gourd/Chichora	*Trichosanthes anguina*	Cucurbitaceae
Lau	Squash/Bottle ground	*Lagenaria siceraria*	Cucurbitaceae
Lau Sak (leaf)	Bottle ground	*Lagenaria siceraria*	Cucurbitaceae
Data Sak	Stem amaranth	*Amaranthus lividus*	Amaranthaceae
Dherosh	Okra/Ladies finger	*Abelmoschus esculentus*	Malvaceae
Kacha Pepe	Green Papaya	*Carica papaya*	Caricaceae
Kacha Kala	Mattock/Plantain	*Musa paradisiaca*	Musaceae
Kachu Lati	Taro	*Colocasia esculenta*	Araceae
Man Kachu	Taro	*Xanthasoma atrovirens*	Araceae
Mukhi Kachu	Eddoe	*Colocasia Schott*	Araceae
Ol Kachu	Elephant foot	*Amorphophallus campanulatus*	Araceae
Ghotkol	Ghotkol	*Typhonium trilobatum*	Araceae
Sajina Sak	Drumstick leaf	*Moringa oleifera*	Moringaceae
Pui Sakk (Sabuj)	Callaloo/Indian spinach	*Basella alba*	Basellaceace

2.3. Digestion and determination

After homogenization of the bulk sample, a small portion, typically 50 mg of dried and powdered samples, was transferred to a 7 ml Teflon PFA (perfluoroalkoxy) vial. An aliquot of ultrapure HNO_3 (1 ml) was added and then the cap of the vial closed tightly. The sample was kept at room temperature for 2–3 h. Thereafter, the caps were briefly opened to vent gases, resealed, and inserted into a stainless steel jacket. After sealing, the bomb was heated at 140°C for 4 h. After cooling the bomb overnight, the vial was opened slowly and carefully to allow gases to vent. 100 µl of HF was added to dissolve siliceous materials, then the sample heated at 120°C to evaporate the HF. After cooling, 100 µl of HNO_3 was added, and the sample heated to near dryness. This process was repeated twice more, then the residue was dissolved with 500 µl HNO_3, then made up to the final volume (50 ml) with deionized water. This digestion method is suitable for the determination of volatile elements because the digestion is completed in a doubly closed system. All samples were analyzed three times for Ca, Fe, Mg, Na, P, and Si and the trace elements As, B, Ba, Be, Cd, Co, Cr, Cu, Mn, Mo, Pb, and Zn by ICP-AES (Model ICAP-750, Nippon Jarrel Ash, Tokyo, Japan) and ICP-MS (Model HP 4500, Yokogawa Analytical Systems Inc., Tokyo, Japan). Standard solutions were prepared from SPEX certi Prep stock solutions. Sample blanks were analyzed after 7–10 samples. Detection limits were set at three times the standard deviation of the blanks. Three standard reference materials (SRMs) ("Rice Flour – unpolished", certified reference material (CRM) 10-a, National Institute for Environmental studies (NIES), Japan; "apple leaves1515" NIST SRM; and "Olea Europaea" BCR 62) were analyzed for each trace element, one per set. For matrix correction and the development of a refined metal correction, two internal standards, In and Bi were chosen to produce accurate ICP-MS results. The concentrations were calculated on a dry weight basis.

3. Results

The results concerning the SRM showed good agreement with the certified and reference values for all elements and found all certified and reference elements to be within 5–15% of expected values (85–110% recovery; Table 2). Based on the analysis of NIES CRM 10-a, NIST SRM No. 1515 and CRM No. 62 quantitative results were expected for 17 elements (Al, As, Ca, Cd, Cr, Cu, Co, Fe, K, Mg, Mn, Mo, Na, Se, Sr, Rb and Zn), all of which were detected in rice, pulses and vegetables but only five essential and/or toxic metals As, Cd, Cu, Pb and Zn are the focus of this paper. Food is not only the essential source of energy for daily life, but one of the unavoidable sources of pollutant element intake.

Metal concentrations in the Samta Village rice, pulses and vegetables are summarized in Table 3. Significant amounts of As were found in vegetables at Samta village compared with rice and pulses. As and Pb concentrations were higher in bottle gourd leaf, stem amaranth, Taro lati, elephant foot and ghotkol than in potato, lady's finger and Indian spinach but Cu and Zn concentrations were higher in brinjal, snake gourd, bottle gourd leaf, lady's finger, taro leaf and eddoe than in potato and ash gourd. Values quoted have not been corrected for analyte recoveries from certified reference materials. The minimum value for As was found in lady's finger (0.019 µg g^{-1}) and the maximum in ghotkol,

Table 2. Summary of certified and reference (*) metal concentrations in NIES CRMs 10-a, NIST 1515 and BCR 62.

Element	CRM 10-a		NIST 1515		BCR 62	
	Certified	Observed	Certified	Observed	Certified	Observed
			$\mu g\ g^{-1}$			
Al	3.0*	2.89 ± 0.006	286 ± 9	289.06 ± 3.42	450 ± 20	452.42 ± 5.34
As	0.17*	0.17 ± 0.004	0.038 ± 0.007	0.037 ± 0.004	0.2*	0.24 ± 0.002
Ca	93 ± 3	92.54 ± 0.28	–	–	–	–
Cd	0.023 ± 0.003	0.025 ± 0.002	0.014*	0.014 ± 0.002	0.10 ± 0.02	0.11 ± 0.01
Cr	0.07*	0.08 ± 0.001	0.3*	0.37 ± 0.003	2*	0.19 ± 0.003
Cu	3.5 ± 0.3	3.14 ± 0.4	5.64 ± 0.24	5.88 ± 0.13	46.6 ± 1.8	45.9 ± 0.9
Co	0.02*	0.02 ± 0.002	0.09*	0.089 ± 0.001	0.2*	0.25 ± 0.001
Fe	12.7 ± 0.7	12.56 ± 0.12	80*	82.53 ± 1.31	–	–
Mn	34.7 ± 0.008	31.73 ± 0.006	54 ± 3	53.41 ± 2.02	57 ± 2.4	58.13 ± 1.76
Mo	0.35 ± 0.05	0.43 ± 0.003	0.094 ± 0.013	0.096 ± 0.015	0.2	0.19 ± 0.002
Na	10.2 ± 0.3	10.32 ± 0.12	24.4 ± 1.2	24.20 ± 1.06	–	–
Se	0.06*	0.07 ± 0.007	0.050 ± 0.009	0.052 ± 0.004	0.1*	0.18 ± 0.002
Sr	0.3*	0.42 ± 0.005	25 ± 2	25.59 ± 1.95	–	–
Rb	4.5 ± 0.3	4.14 ± 0.002	9*	9.23 ± 0.06	0.2	0.22 ± 0.001
Zn	25.2 ± 0.8	22.38 ± 0.6	12.5 ± 0.3	12.72 ± 0.47	16 ± 0.7	17.02 ± 0.03
			wt%			
P	0.340 ± 0.007	0.339 ± 0.008	–	–	–	–
K	0.280 ± 0.008	0.280 ± 0.004	1.61 ± 0.02	1.60 ± 0.006	–	–
Mg	0.134 ± 0.008	0.132 ± 0.003	0.271 ± 0.008	0.270 ± 0.008	–	–

Values quoted on a dry weight basis. The minimum determinable limits (MDL) indicated are based on mean ± 10 standard deviations for the digested blanks (American Public Health Association, 1995).

taro (lati), and snake gourd (0.446, 0.440 and 0.489 μg g^{-1}, respectively). The Cd content in these leafy and non-leafy vegetables varied between 0.012 and 0.216 μg g^{-1}. The highest Cd levels were found in Ghotkol samples from Samta (0.216 μg g^{-1}). The concentration of Cd in rice and pulses varied between 0.02 and 0.03 μg g^{-1} and lower than all the vegetables studies in Samta village. Table 3 shows that rice in Samta village has the highest Pb concentration (7.71 μg g^{-1}) compared with other pulses and vegetables. The lowest Pb levels were found in lady's finger (0.143 μg g^{-1}) and the highest in ghotkol, elephant foot, bottle gourd leaf and stem amaranth (1.689, 0.967, 0.987 and 0.831 μg g^{-1}, respectively). The pulses and vegetables such as lady's finger, stem amaranth, bottle gourd leaf and ghotkol are better sources of Zn than other vegetables and rice in Samta village.

4. Discussion

Samta village was chosen as the model village in our study because we knew 90% of the tube wells in this village had arsenic concentrations above the Bangladesh standard of 0.05 mg l^{-1} (Yokota et al., 2001) and it is highly affected by As contamination in the groundwater. Tube wells with As concentrations of over 0.50 mg l^{-1} were distributed across a belt-like east–west zone in the southern part of the village. The high concentrations of As in the surface water in the Samta village is presumably derived from the agricultural use of contaminated irrigation water. In addition, chemical fertilizers and various pesticides are also used in the fields (Islam et al., 2000). The relatively neutral soil pH in the study area (4.9–8.8) suggests that As will be immobile in the local soil profile. Indeed, Matsumoto and Hosoda (2000) found that As content in sediments in Samta village was highest in the first upper muddy layer (3.0–261.5 mg kg^{-1}), decreasing with depth. Our study indicate that As concentrations in rice in Samta village were higher than the American brown easy cook rice (0.12 mg kg^{-1}), Supreme basmati Rice (<0.04 mg kg^{-1}), Minnesota quick wild rice (<0.04 μg g^{-1}) and Arborio risotto rice (<0.04 μg g^{-1}) (Ministry of Agriculture, Fisheries and Food, MAFF, 1999). The average per capita consumption of polished rice in Bangladesh is about 260 g d^{-1} (Masironi et al., 1977), resulting in an average intake of As from rice which was estimated to be 91 μg d^{-1}. This is below the Joint Expert Committee on Food Additives (JECFA) PTWI of 15 μg kg^{-1} body weight (equivalent to 130 μg d^{-1} for a 60 kg adult) for inorganic arsenic (World Health Organization, 1989). Vegetables make up about 16% of the total diet. The average per capita consumption of leafy and non-leafy vegetables in Bangladesh is 130 g per person per day for males and females of all ages as against the requirement of 200 g per person per day from the nutritional point of view (Hassan and Ahmad, 2000). Our study also indicates that As concentrations in leafy and non-leafy vegetables in Samta village were higher than the As concentrations in the UK where green vegetables, potatoes and other vegetables exhibited As concentrations of 0.003, 0.005 and 0.007 μg g^{-1} fresh weight, respectively (Ministry of Agriculture, Fisheries and Food, MAFF 1999). The As levels were also higher than in vegetables from the Republic of Croatia where the average As levels were 0.0004 μg g^{-1}) (Postruznik et al., 1996). The average intake of total As from leafy and non-leafy vegetables was estimated to be 27.78 μg d^{-1}. This is below the Joint Expert Committee on Food Additives (JECFA)

Table 3. Mean concentration (range) of As, Cd, Cu, Pb and Zn in rice, pulses and vegetables from Santa village, Jessore District, Bangladesh.

Vegetables	Elemental concentration (mg kg^{-1} dry weight)				
	As	Cd	Cu	Pb	Zn
Rice					
Aus	0.353 (0.164–0.584)	0.024 (0.008–0.042)	2.942 (2.271–3.394)	7.711 (2.604–15.893)	12.741 (10.392–15.354)
Pulses					
Red split lentil	0.042 (0.008–0.073)	0.015 (0.003–0.032)	12.823 (11.221–13.901)	0.594 (0.582–0.863)	56.701 (47.453–65.043)
Grasspea	0.031 (0.007–0.061)	0.003 (0.001–0.008)	9.791 (8.323–10.465)	1.483 (1.064–2.654)	45.564 (39.632–54.432)
Pea	0.033 (0.008–0.064)	0.034 (0.022–0.041)	14.083 (12.165–16.934)	0.644 (0.387–0.763)	34.052 (29.851–44.921)
Leafy vegetables					
Bottle ground leaf	0.306 (0.212–0.429)	0.089 (0.056–0.097)	16.6 (15.3–18.3)	0.987 (0.784–1.111)	43.3 (39.6–60.9)
Taro leaf	0.205 (0.154–0.297)	0.026 (0.19–0.028)	23.5 (21.2–25.4)	0.361 (0.298–0.383)	23.2 (21.4–25.2)
Drumstik leaf	0.185 (0.143–0.199)	0.038 (0.025–0.046)	6.35 (4.56–8.4)	0.207 (0.118–0.266)	21.7 (18.7–25.3)
Non-leafy vegetables					
Brinjal	0.187 (0.061–0.337)	0.128 (0.062–1.67)	12.49 (9.90–14.99)	0.288 (0.198–0.361)	22.3 (21.3–26.9)
Dhudi/Ash gourd	0.077 (0.069–0.087)	0.047 (0.039–0.058)	5.61 (5.01–7.52)	0.261 (0.232–0.493)	10.5 (9.3–13.5)
Snake gourd/Chichora	0.489 (0.344–0.631)	0.044 (0.035–0.064)	14.2 (13.8–16.7)	0.439 (0.377–0.696)	22.4 (21.6–25.5)
Potato	0.072 (0.047–0.096)	0.137 (0.109–0.184)	5.77 (5.44–7.04)	0.484 (0.272–0.851)	13.4 (11.7–18.8)
Stem amaranth	0.145 (0.105–0.287)	0.176 (0.110–0.203)	10.7 (9.6–12.6)	0.831 (0.520–0.990)	54.0 (51.2–69.6)
Okra/Ladies finger	0.019 (0.009–0.034)	0.097 (0.064–0.188)	13.0 (12.5–17.6)	0.143 (0.121–0.265)	48.7 (44.2–53.6)
Green Papaya	0.389 (0.288–0.530)	0.012 (0.009–0.026)	3.17 (2.16–4.05)	0.239 (0.187–0.288)	11.0 (7.14–13.54)
Plantain	0.083 (0.065–0.098)	0.024 (0.019–0.036)	3.83 (1.65–4.03)	0.303 (0.188–0.432)	30.6 (15.3–45.3)
Taro (lati)	0.440 (0.239–0.466)	0.030 (0.021–0.037)	12.2 (10.1–14.3)	0.276 (0.210–0.305)	19.7 (17.3–22.2)
Eddoe	0.071 (0.064–0.088)	0.042 (0.038–0.043)	10.9 (9.20–13.0)	1.639 (1.201–1.828)	22.4 (19.8–23.0)
Elephant foot	0.338 (0.231–0.456)	0.036 (0.027–0.045)	4.58 (3.69–6.88)	0.967 (0.854–1.685)	32.5 (28.3–36.6)
Ghotkol	0.446 (0.231–0.654)	0.216 (0.198–0.342)	5.67 (3.76–7.65)	1.689 (0.989–2.870)	38.2 (31.9–42.8)

PTWI of 15 µg kg^{-1} body weight (equivalent to 130 µg d^{-1} for a 60 kg adult) for inorganic arsenic (World Health Organization, 1989).

The average Cd concentration in Samta rice is the same as that of rice worldwide (0.02 mg kg^{-1}; Watanabe et al., 1989). Not surprisingly, therefore, the average daily intake of Cd (5.2 µg), does not exceed the recommended 60 µg d^{-1} (60 kg adult; World Health Organization, 1993). Our study indicates, however, that Cd levels were higher in vegetables from Samta village than in comparable leafy and non-leafy vegetables in Bombay, India, which exhibited Cd levels of 0.0149 and 0.0032 µg g^{-1}, respectively (Tripathi et al., 1997). Ysart et al. (1999) reported the Cd content of green vegetables, potatoes and other vegetables in the UK was of 0.006, 0.03 and 0.008 µg g^{-1} (fresh weight) respectively. Onianwa et al. (2000) reported that Cd levels in Nigerian leafy and fruity vegetables averaged 0.22 µg g^{-1} (dry weight) with a range of 0.09–0.62 µg g^{-1} (dry weight). The average daily intake of Cd from food products grown in Samta village is 9.45 µg, which does not exceed the JECFA PTWI of 60 µg d^{-1} (60 kg adult (World Health Organization, 1993).

Lead (Pb) concentrations in rice from Samta village are higher than in rice from Australia, Italy, Japan and the US (2.07, 6.97, 5.06 and 3.41 µg kg^{-1}, respectively; Zhang et al., 1996). Pb concentrations in green vegetables, potatoes and other vegetables are also high when compared to similar food products from the UK, where levels of 0.01, 0.01, and 0.02 µg g^{-1}, respectively, were found (Ysart et al., 1999), and also higher than in leafy vegetables and other vegetables from India which had Pb levels of 0.10 and 0.0041 µg g^{-1}, respectively (Tripathi et al., 1997). The average weekly intake of Pb from Samta rice is estimated to be 14 mg, which greatly exceeds the PTWI for Pb of 1.5 mg wk^{-1} for a 60 kg adult (World Health Organization, 1993). The average weekly intake of Pb from the leafy and non-leafy vegetables from Samta is estimated to be 0.523 mg, which does not exceed the JECFA PTWI for Pb of 1.5 mg wk^{-1} (equivalent to 210 µg d^{-1}) for a 60 kg adult (World Health Organization, 1993).

Copper and zinc are essential elements, although excess amounts can be toxic. Copper (Cu) concentrations in rice from Samta village are comparable with Japanese values (2.81 mg kg^{-1}; Ohmomo and Sumiya, 1981), but lower than in rice from China, Philippines and Taiwan (4.21, 3.9, and 4.4 mg kg^{-1}; Masironi et al., 1977; Suzuki et al., 1980). The average copper concentrations in leafy and non-leafy vegetables from Samta, 15.5 and 8.51 µg g^{-1}, respectively, are higher for leafy and non-leafy vegetables from Bombay, India (1.3045 and 0.5256 µg g^{-1} (Tripathi et al., 1997)), vegetables from Tianjin, China (0.651 µg g^{-1} (Zhou et al., 2000) and vegetables from southern India (<0.04 µg g^{-1} (Srikumar, 1993)). However, for the Samta population the dietary exposure to copper from vegetables is below the Provisional Maximum Tolerable Daily Intake (PMTDI) of 30 mg d^{-1} for a 60 kg person (World Health Organization, 1982a).

Zinc (Zn) concentrations in the Samta rice are lower than in rice grown in Japan, Indonesia or China. The weekly intake of Zn from Samta rice is estimated to be 21 mg, which is less than the recommended a PMTDI of 60 mg d^{-1} (World Health Organization, 1982b). The highest concentrations of Zn were found in samples of dry pulses (red split lentil, grass pea and pea), possibly correlated to the higher protein content of pulses compared with rice. The average weekly intake of Zn from Samta vegetables is estimated to be 25 mg, which is less than the recommended PMTDI of 60 mg d^{-1}. However, these levels are higher than the dietary allowances (RDA) recommended for humans by various governmental agencies, such as 15 and 12 mg d^{-1} for healthy adult men and women, respectively (National Research Council, 1989) and a reference intake of 9.5 mg d^{-1} for men and 7.1 mg d^{-1} for women (Comision Europea, 1993). The Zn levels obtained for leafy and non-leafy vegetable items in this study were higher than the Zn levels in Indian leafy vegetables and other vegetables (4.811, and 1.938 µg g^{-1} (Tripathi et al., 1997)) and in the UK's green vegetables, potatoes and other vegetables (3.4, 4.5 and 2.6 µg g^{-1} (Ysart et al., 1999)). The concentration of Cu, Zn and Cd in pulses compare well with those reported by Tripathi et al. (1997), but the concentrations of all the metals measured in the pulses, especially red split lentil, were higher than the UK Total Diet studies (Ysart et al., 1999).

5. Conclusion

We believe that arsenic is not the only agent responsible for the devastating health effects seen in Bangladesh. Other elements or compounds may contribute to a synergistic effect with arsenic and help endanger the human and animal health. The relative concentrations of Cd and Pb found in the foods studied were generally higher than those reported for the vegetables and cereals consumed in other parts of the world. However, some foods at Samta are also highly contaminated with the metalloid arsenic, surpassing greatly the national standards and international recommendations. Thus, an urgent and systematic study of the arsenic in the vegetables consumed and traded in Bangladeshi villages in general is recommended since the vegetables could significantly increase the intake of this toxic element, and contribute substantially to the ill health of the people consuming them. Therefore, regular monitoring of these metals and metalloids should be considered to help minimize contamination during farming.

Acknowledgements

The authors would like to thank Mr M. Harun-ar-Rashid, Executive Director, Agricultural Advisory Society (AAS) of Bangladesh for his help in collecting the samples, and Professor Md Golam Robanni, Department of Horticulture, Bangladesh Agricultural University for providing the local and scientific nomenclature. The research by ETS and MGMA at Deakin University was supported in part by the US Environmental Protection Agency's Science to Achieve Results (STAR) program and by the Electric Power Research Institute contract #WOEP-P4898/C2396.

References

Alam, M.G.M., Allinson, G., Stagnitti, F., Tanaka, A., Westbrooke, M., 2002. Arsenic contamination in Bangladesh groundwater: a great environmental and social disaster. Int. J. Environ. Health Res., 12 (3), 235–253.
American Public Health Association (APHA), 1995. In: Eaton, A.D., Clesceri, L.S., Greenberg, A.E. (Eds), Standard methods for the examination of water and wastewater. American Public Health Association, Washington, DC.

Comision Europea, 1993. Dictamen emitidoo, 11/12/1992, oficina de Publicaciones Oficiales de la Comunidad Europea.

Fazal, A.M., Kawachi, T., Ichion, E., 2001. Extent and severity of the groundwater arsenic contamination in Bangladesh. Water Int., 26 (3), 370–379.

Hassan, N., Ahmad, K., 2000. Intra-familial distribution of food in rural Bangladesh, Institute of Nutrition and food science, University of Dhaka, Bangladesh. Internet pages (http://www.unu.edu/unpress/food/8F064e/8F064E05.htm) (11/9/01).

Huq, I., Smith, E., Correll, R., Smith, L, Smith, J., Ahmed, M., Roy, S., Barnes, M., Naidu, R., 2001. Arsenic transfer in water soil crop environments in Bangladesh I: assessing potential arsenic exposure pathways in Bangladesh. In: Naidu, R., (ed.), Arsenic in the Asia-Pacific Region workshop "Managing arsenic for our Future" 20–23 November 2001, Adelaide, Australia. pp. 50–51.

Islam, M.R., Salminen, R., Lahermo, P.W., 2000. Arsenic and other toxic elemental contamination of groundwater, surface water and soil in Bangladesh and its possible effects on human health. Environ. Geochem. Health, 22, 33–53.

Malek, M.A., Sarwar, C.D.M., Sarker, A., Hassan, M.S., 1996. Status of grass pea research and future strategy in Bangladesh. In: Arora, R.K., Mather, P.N., Riley, K.W., Adham, Y., (Eds), Lathyrus Genetic Resources in Asia: Proceedings of a Regional Workshop, 27–29 December 1995. Indira Gandhi Agricultural University, Raipur, India, IPGRI Office for South Asia, New Delhi, pp. 7–12.

Masironi, R., Koirtyohann, S.R., Pierce, J.O., 1977. Zinc, copper, cadmium and chromium in polished and unpolished rice. Sci. Total Environ., 7, 27–43.

Matsumoto, T., Hosoda, T., 2000. Arsenic contamination and hydrogeological background in Samta village, western Bangladesh. Abstract, Pre-Congress Workshop (Bwo 10), Arsenic in Groundwater of Sedimentary Aquifers. The 31st International Geological Congress, 3–5 August 2000, Rio de Janeiro, Brazil, pp. 2–4.

Ministry of Agriculture, Fisheries and Food (MAFF), 1999. 1997 Total diet study: aluminium, arsenic, cadmium, chromium, copper, lead, mercury, nickel, selenium, tin and zinc. Food Surveillance Information Sheet, No. 191. HMSO, London.

National Research Council, 1989. Food and Nutrition Board, Recommended Dietary Allowances, 10th edn. National Academy of Sciences, Washington, DC.

Ohmomo, Y., Sumiya, M., 1981. Estimation of heavy metal intake from agricultural products. In: Kitagishi, K., Yamane, I. (Eds), Heavy Metal Pollution in Soils of Japan. Japan Scientific Press, Tokyo, pp. 235–244.

Onianwa, P.C., Lawal, J.A., Ogunkeye, A.A., Orejimi, B.M., 2000. Cadmium and nickel composition of Nigerian foods. J. Food Comp. Anal., 13, 961–969.

Postruznik, J.S., Bazulic, D., Kubala, H., 1996. Estimation of dietary intake of arsenic in the general population of the Republic of Croatia. Sci. Total Environ., 191, 119–123.

Srikumar, T.S., 1993. The mineral and trace element composition of vegetables, pulses and cereals of southern India. Food Chem., 46, 163–167.

Suzuki, S., Djuangshi, N., Hyodo, K., Soemarwoto, O., 1980. Cadmium, copper, and zinc in rice produced in Java. Arch. Environ. Contam. Toxicol., 9, 437–449.

Tripathi, R.M., Raghunath, R., Krishnamoorthy, T.M., 1997. Dietary intake of heavy metals in Bombay city, India. Sci. Total Environ., 208, 149–159.

Watanabe, T., Nakatsuka, H., Ikeda, M., 1989. Cadmium and lead in rice available in various areas of Asia. Sci. Total Environ., 80, 175–184.

World Health Organization, WHO, 1982a. Toxicological evaluation of certain food additive, Joint FAO/WHO expert committee on food additives, WHO Food Additives Series, Number 17, World Health Organization, Geneva.

World Health Organization, WHO, 1982b. Evaluation of certain food additives and contaminants, Technical Report Series, Number 683, World Health Organization, Geneva.

World Health Organization, WHO, 1989. Toxicological evaluation of certain food additives and contaminants, Food additive Series, Number 24, World Health Organization, Geneva.

World Health Organization, WHO, 1993. Evaluation of certain food additives and contaminants, WHO Technical Report Series, Number 837, World Health Organization, Geneva.

Yokota, H., Tanabe, K., Sezaki, M., Akiyoshi, Y., Miyata, T., Kawahara, K., Tsushima, S., Hironaka, H., Takafuji, H., Rahman, M., Ahmad, S.A., Sayed, M.H.S.U., Faruquee, M.H., 2001. Arsenic contamination of groundwater and pond water and water purification system using pond water in Bangladesh. Engng Geol., 60 (1–4), 323–331.

Ysart, G.E., Miller, P.F., Crews, H., Robb, P., Baxter, M., Delargy, C., Lofthouse, S., Sargent, C., Harrison, N., 1999. Dietary exposure estimates of 30 metals and other elements from the UK Total Diet Study. Food Addit. Contam., 16, 391–403.

Zhang, Z.W., Moon, C.S., Watanaabe, T., Shimbo, S., Ikeda, M., 1996. Lead content of rice collected from various areas in the world. Sci. Total Environ., 191, 169–175.

Zhou, Z.E., Fan, Y.P., Wang, M.J., 2000. Heavy metal contamination in vegetables and their control in China. Food Rev. Int., 16 (2), 239–255.

PART II
EPIDEMIOLOGY

Arsenic Exposure and Health Effects V
W.R. Chappell, C.O. Abernathy, R.L. Calderon and D.J. Thomas, editors
© 2003 Published by Elsevier B.V.

Chapter 9

Criteria for case definition of arsenicosis

D.N. Guha Mazumder

Abstract

In spite of the availability of a large number of publications reporting on studies of health effects of chronic As toxicity, no standard definition is available to characterize fully the clinical effects of chronic As exposure in man. Only a small number of reports are based on valid As exposure data, blinding of cases and controls, total assessment of clinical effect and correlation of these effects with As level in water and various biomarkers. It was, therefore, necessary to propose a clinical case definition of chronic arsenicosis.

Thirty-three papers on As-related health effects were reviewed directly and numerous cross references consulted. Giving weight to the quality of these papers, diagnostic criteria of chronic arsenicosis has been developed.

On the basis of a literature review, a minimum period of six months of ingestion of As is required for the diagnosis of arsenicosis. Normal As values of urine, hair and nails determined by proper techniques, e.g. atomic absorption spectrophotometry (AAS) or neutron activation analysis (NAA), have also been reviewed. Mean + 2SD of the normal values of the acceptable published reports have been calculated. From these, the upper limit of normal As level in urine, hair and nail was considered to be 0.05 mg/l, 0.8 mg/kg and 1.3 mg/kg, respectively.

Skin lesions, simulating arsenical dermatosis but caused by other systemic or skin diseases, have been reviewed. These conditions need to be considered in the differential diagnosis of arsenicosis. Over and above skin lesions, the criteria related to systemic manifestation and various As-related cancers were also considered for the preparation of case definition of chronic arsenicosis. A simplified algorithm of case definition, using dermatological criteria as the major clinical diagnostic parameter, has been developed.

Before reviewing the literature for a clinical case definition, it is essential to define the term arsenicosis. This will help to identify the criteria that will be needed for proper case definition. The term arsenicosis is meant to characterize various clinical manifestations caused by chronic arsenic (As) toxicity due to prolonged drinking of As-contaminated water or chronic exposure of As through other sources. It does not denote the clinical effect of acute exposure of high dose of arsenic for a short period of time as these are referred to as acute arsenic poisoning.

In spite of the availability of a large number of publications reporting on studies of health effects of chronic As toxicity, no standard definition is available to characterize fully the clinical effects of chronic As exposure in man. Earlier reports mainly highlighted the effects of the dermatological manifestations like pigmentation, keratosis and skin cancer. Later on, various non-cutaneous manifestations were also reported based on various clinical and epidemiological studies. However, only a small number of reports are based on valid As exposure data, blinding of cases and controls, total assessment of clinical effect and correlation of these effects with As level in water, urine, hair or nails, by estimation of As using sensitive techniques like AAS and NAA. It was therefore felt necessary to develop a clinical case definition of arsenicosis on the basis of review of clinical data obtained by objective criteria and validated by testing of water and biological samples in cases of chronically As-exposed people.

Keywords: Arsenicosis definition, Clinical features, Bio markers, Literature review, Diagnostic criteria, Diagnostic algorithm.

1. Review on clinical manifestations

Of the thirty-three papers on As-related health effects reviewed directly and numerous cross references consulted, only seven reports were found to be well designed with blinding of cases and controls. Neutron activation analysis (NAA) or atomic absorption spectrophotometry (AAS) was used for measurement of As concentrations in hair, nail or water and urine in 13 case reports. Validation of clinical findings with As exposure data by determining As level in water and correlation with As level in urine or hair or nail were done in six case series. Several epidemiological studies only characterized a single outcome like, ischaemic heart disease, hypertension, diabetes mellitus, etc.

2. Dermatological manifestations

Although chronic As toxicity produces varied non-malignant manifestations as well as cancer of skin and different internal organs, dermal manifestations such as hyperpigmentation and hyperkeratosis are diagnostic of chronic arsenicosis. The pigmentation of chronic As poisoning commonly appears in a finely freckled, "raindrop" pattern of pigmentation or depigmentation that is particularly pronounced on the trunk and extremities and has a bilateral symmetrical distribution. Pigmentation may sometimes be blotchy and involve mucous membranes such as the undersurface of the tongue or buccal mucosa (Black, 1967; Yeh, 1973; Tay, 1974; Saha, 1984,1995; Guha Mazumder et al., 1988, 1998a). The raindrop appearance results from the presence of numerous rounded hyperpigmented or hypopigmented macules (typically 2–4 mm in diameter) widely dispersed against a tan-to-brown hyperpigmented background (Tay, 1974) (Fig. 1). Although less common, other patterns include diffuse hyperpigmentation (melanosis) (Tay, 1974; Saha, 1984) and localized or patchy pigmentation, particularly affecting skin folds (Tay, 1974; Szuler et al., 1979; Luchtrath, 1983). So-called leukodermia or leukomelanosis (Saha, 1984, 1995) in which the hypopigmented macules take a spotty, white appearance usually occurs in the early stages of toxicity.

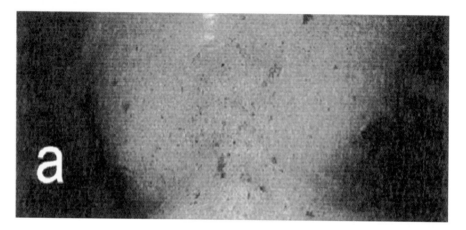

Figure 1. Pigmentation.

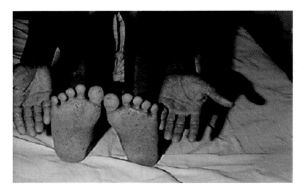

Figure 2. Keratosis.

Arsenical hyperkeratosis appears predominantly on the palms and the plantar aspect of the feet (Fig. 2), although involvement of the dorsum of the extremities and the trunk have also been described. In the early stages, the involved skin might have an indurated, grit-like character that can be best evaluated by palpation; however, the lesions usually advance to form raised, punctated, 2–4 mm wart-like keratosis that are readily visible (Tay, 1974). Occasional lesions might be larger (0.5–1 cm) and have a nodular or horny appearance occurring in the palm or dorsum of the feet. In severe cases, the hands and soles present with diffuse verrucous lesions. Cracks and fissures may be severe in the soles (Saha, 1984; Guha Mazumder et al., 1997). Histological examination of the lesions typically reveals hyperkeratosis with or without para keratosis, acanthosis, and enlargement of the rete ridges. In some cases, there might be evidence of cellular atypia, mitotic figures, in large vacuolated epidermal cells (Black, 1967; Tay, 1974; Ratnam et al., 1992; Alain et al., 1993; Guha Mazumder et al., 1998b). Yeh (1973) classified arsenical keratosis into two types: a benign type A, further subgrouped into those with no cell atypia and those with mild cellular atypia; and a malignant type B, consisting of lesions of Bowen's disease (intraepithelial carcinoma or carcinoma *in situ*), basal cell carcinoma or squamous cell carcinoma. The latter might arise in the hyperkeratotic areas or might appear on non-keratotic areas of the trunk, extremities or head (Sommers and McManus, 1953; Yeh, 1973).

A history of As exposure through inhalation or ingestion is helpful in corroborating a diagnosis of arsenicosis because skin manifestations such as diffuse melanosis cannot be differentiated from normal dark-complexioned farmers in the tropics who work in the field bare bodied under direct sunlight. However, spotty raindrop pigmentation of the skin distributed bilaterally and symmetrically over trunks and limbs is the best diagnostic feature of arsenical hyperpigmentation. Though spotty depigmented spots similarly distributed are also diagnostic for this condition, sometimes blotchy depigmented spots are seen and these need to be differentiated from other depigmented skin lesions like tinea versicolor, seborrheic dermatitis, etc. Diffuse hyperkeratitic lesions of the palms and soles are distinctive lesions of chronic arsenicosis. However, manual labourers, who work with bare hands, might have thickening of the palms. The thickening of palms in manual labourers are usually localised in the pressure points. Bare-footed farmers who work in the fields might have diffuse thickening of the soles. However, when the lesions become

Table 1.   Skin lesions to be differentiated from arsenicosis.

Pigmentation
   a) Pityriasis versicolor
   b) Idiopathic guttate hypomelanosis
   c) Actinic hypermelanosis
   d) Drug induced
   e) Melasma
   f) Pellagra
   g) Addison's disease
   h) Hereditary disorders of pigmentation, e.g. xeroderma pigmentosa
   i) Leprosy
   j) Leishmaniasis
Keratosis
   a) Tinea pedis
   b) Psoriasis
   c) Pityriasis rubra pilaris
   d) Verruca (warts)
   e) Eczema hyperkeratotic type
   f) Corns and calluses
   g) Lichon planus
   h) Discoid lupus erythematosus
   i) Hereditary palmar–plantar hyperkeratosis
   j) Reiter's syndrome

nodular the diagnosis becomes obvious. Presence of the following features will exclude the diagnosis of arsenical skin lesions: present since birth, scaling, elevation of the margin of the lesion and erythema surrounding the lesion. The various skin diseases which are to be differentiated from arsenical dermatosis are given in Table 1.

3. Systemic manifestations

Over and above arsenical skin lesions, other features of chronic arsenicosis are weakness, anaemia, peripheral neuropathy, hepatomegaly with portal zone fibrosis (with/without portal hypertension), chronic lung disease and peripheral vascular disease These features are manifested variably in different exposed populations and may also be caused by conditions unrelated to As exposure. Infrequent manifestations, which have been reported to occur by some investigators in people giving a history of chronic As exposure and which may also be unrelated are: conjunctivitis, keratitis, rhinitis, cardiovascular disease (ischaemic heart disease, hypertension), gastrointestinal disease, haematological abnormalities, cerebrovascular disease, dysosmia, perceptive hearing loss, cataract, nephropathy, solid oedema of the limbs and diabetes mellitus. Clinical features caused by chronic As toxicity as reported by various investigators are given in Table 2a and b.

   For the diagnosis of arsenic toxicity in humans, one needs to know the source and duration of exposure and the degree of contamination present in the source.

Table 2a. Literature review of clinical features of chronic arsenic toxicity.

Sl. no.	Reference	Arsenic level in water mg/l	Skin manifestation							Mee's line	GI symptoms					Liver and spleen						Cardiovascular system				
			Melanosis (diffuse)	Spotty pigmentation	Patchy pigmentation of skin/oral mucous membrane	Spotty depigmentation	Keratosis/diffuse	Nodules in palm/sole	Nodules in dorsum of hand, feet, legs, body		Nausea	Vomiting	Diarrhoea	Anorexia	Abd. pain	Liver enlargement	Liver fibrosis	Cirrhosis	Spleen enlargement	Ascites	Anaemia/pancytopenia	Arrhythmia	Isch. heart disease	Hypertension	Gangrene of feet/hand	Raynaud's phenomenon
1	Franklin et al. (1950)	Fowler's solu.	+	+			+	+										+	+	+	+					
2	Sommers and McManus (1953)	Fowler's solu.					+																			
3	Tseng et al. (1968)	0.01–1.82		+			+																			
4	Yeh (1973)	Fowler's solu.					+																			
5	Rosenberg (1974)	Up to 0.8		+			+						+			+	+	+					+		+	+
6	Zaldivar (1974), Zaldivar et al. (1981)	0.05–0.96	+	+		+	+			+			+	+		+	+	+	+	+			+		+	
7	Borgono et al. (1977)	0.6–0.8		+			+								+									+		+
8	Hindmarsh et al. (1977)	Fowler's solu.		+			+								+											
9	Cebrian et al. (1983)	0.41 (exp.) 0.0017 (cont)	+	+		+	+	+	+		+		+													
10	Saha (1984)	Up to 2.0	+	+		+	+	+		+																
11	Chakraborty and Saha (1987)	0.2–2.00	+	+		+	+	+	+							+			+	+						
12	Guha Mazumder et al. (1988)	0.2–2 (exp.) <0.05 (cont)	+	+			+	+		+	+	+	+	+	+	+		+	+	+						
13	Hotta (1989)	0.05–58	+	+		+							+			+										
14	Guha Mazumder et al. (1992)	Fowlers solu.		+			+									+	+		+	+	+	+	+		+	+
15	Cuzick et al. (1992)	0.7–0.93		+	+		+																			
16	Lai et al. (1994)	<0.1–0.6																								
17	Chen et al. (1994)	0.7–0.95																					+			
18	Chen et al. (1995)	0–3.59																						+		
19	Chiou et al. (1997)																									
20	Guha Mazumder et al. (1997, 1998a,b)	0.05–14.2	+	+			+	+			+		+		+	+	+		+	+	+				+	+

(continued on next page)

Table 2a. (continued)

Sl. no.	Reference	Nervous system								Respiratory system disease				Others			Cancer skin			Cancer internal organs							
		Peripheral neuritis	Paresthesia	Cerebrovascular dis.	Weakness	Mental change	Hearing loss	Dim vision	Headache	Chronic bronchitis/cough	Bronchiectasis	Pulmonary fibrosis	Broncho pneumonia/breathlessness	Diabetes mellitus	Conjunctivitis	Swelling hand/feet	Squamous cell	Basal cell	Bowen's disease	Bladder	Kidney	Lung	Liver	Prostate	Oesophagus	Colon	Breast
1	Franklin et al. (1950)											+				+											
2	Sommers and McManus (1953)																+	+	+	+	+	+		+	+	+	+
3	Tseng et al. (1968)																+	+	+								
4	Yeh (1973)																+	+	+								
5	Rosenberg (1974)									+																	
6	Zaldivar (1974), Zaldivar et al. (1981)				+					+			+				+						+				
7	Borgono et al. (1977)									+	+																
8	Hindmarsh et al. (1977)	+	+			+																					
9	Cebrian et al. (1983)								+							+											
10	Saha (1984)	+			+																						
11	Chakraborty and Saha (1987)									+							+										
12	Guha Mazumder et al. (1988)		+	+	+																						
13	Hotta (1989)	+	+	+	+	+	+	+		+					+												
14	Guha Mazumder et al. (1992)	+	+		+					+							+										
15	Cuzick et al. (1992)																			+							
16	Lai et al. (1994)													+													
17	Chen et al. (1994)																										
18	Chen et al. (1995)																+										
19	Chiou et al. (1997)			+																+	+				+		
20	Guha Mazumder et al. (1997, 1998a,b)	+	+		+				+	+		+			+	+		+	+			+	+	+			+

Table 2b. Literature review of clinical features of chronic arsenic toxicity.

Sl. no.	Reference	Arsenic Level in water mg/l	Skin manifestation								GI symptoms					Liver and spleen					Cardiovascular system					
			Melanosis (diffuse)	Spotty pigmentation	Patchy pigmentation of skin/oral mucous membrane	Spotty depigmentation	Keratosis/diffuse	Nodules in palm/sole	Nodules in dorsum of hand, feet, legs, body	Mee's line	Nausea	Vomiting	Diarrhoea	Anorexia	Abd. pain	Liver enlargement	Liver fibrosis	Cirrhosis	Spleen enlargement	Ascites	Anaemia	Arrhythmia	Isch. heart disease	Hypertension	Gangrene of feet/hand	Raynaud's phenomenon
21	Choprapawon and Rodcline (1997)	Up to 2.7		+			+	+																		
22	Luo et al. (1997)	0.05–0.95		+		+	+																			
23	Kiburn (1997)										+															
24	Chowdhury et al. (1997)	0–3.7	+	+		+	+																			
25	Ahmad et al. (1997)	>0.05	+	+		+	+						+		+			+							
26	Rahaman et al. (1998, 1999)			+		+	+	+															+		
27	Ahmad et al. (1999)	0.05–1.371	+	+		+	+	+							+			+		+					
28	Guha Mazumder et al. (1999)	Up to 3.4		+			+								+	+			+			+	+		+	
29	Ma et al. (1999)	>0.05–< 0.60		+		+	+	+					+			+						+	+			+
30	Chowdhury et al. (2000)	0.05–1.00	+	+	+	+	+	+								+			+	+						
31	Guha Mazumder et al. (2001)	> 0.05 (Exp) <0.05 (Cont.)		+			+				+					+										
32	Milton et al. (2001)	0.136–1.00	+	+		+	+	+											+							
33	Saha and Chakraborti (2001)		+	+	+	+	+	+	+							+	+		+						+	

(continued on next page)

Table 2b. (continued)

Sl. no.	Reference	Nervous system								Respiratory system disease				Others			Cancer skin			Cancer internal organs							
		Peripheral neuritis	Paresthesia	Cerebrovascular dis.	Weakness	Mental change	Hearing loss	Dim vision	Headache	Chronic bronchitis/cough	Bronchiectasis	Pulmonary fibrosis	Broncho pneumonia/breathlessness	Diabetes mellitus	Conjunctivitis	Swelling hand/feet	Squamous cell	Basal cell	Bowen's disease	Bladder	Kidney	Lung	Liver	Prostate	Oesophagus	Colon	Breast
21	Choprapawon and Rodcline (1997)	+														+	+	+									
22	Luo et al. (1997)	+	+		+	+											+	+	+								
23	Kiburn (1997)									+																	
24	Chowdhury et al. (1997)									+			+				+										
25	Ahmad et al. (1997)									+					+	+	+										
26	Rahaman et al. (1998, 1999)						+	+						+	+												
27	Ahmad et al. (1999)	+	+		+					+					+	+	+	+	+								
28	Guha Mazumder et al. (1999)	+	+		+					+					+	+	+	+	+		+						
29	Ma et al. (1999)	+	+	+						+					+	+	+	+	+								
30	Chowdhury et al. (2000)				+					+												+					
31	Guha Mazumder et al. (2001)	+								+					+	+	+	+	+	+		+					
32	Milton et al. (2001)									+							+										
33	Saha & Chakraborti 2001				+					+					+	+	+		+	+		+					

4. Arsenic exposure data

The natural concentration of total As in drinking water varies in different parts of the world. McCabe et al. (1970) investigated more than 18,000 community water supplies in the USA and found that less than 1% had As levels exceeding 0.01 mg/l. In a number of European countries and Japan, the WHO provisional guideline of 0.01 mg/l has been adopted as the standard for public drinking water supplies. The member states where the national standard for arsenic in drinking water remains at 0.05 mg/l include Bangladesh, China and India.

The duration of the patient's As exposure with the date of onset of symptoms does not follow a particular time frame. Arsenical skin lesions have been reported to occur in West Bengal after drinking As-contaminated water for one year or even less (Garai et al., 1984; Guha Mazumder et al., 1997). Rattner and Dorne (1943) reported the development of hyperpigmentation within 6–12 months of the start of treatment with As at a dose of 4.75 mg/day. Hyperkeratosis appeared after approximately three years. Hence exposure of As in highly contaminated water for a period of more than six months may be suggestive of producing clinical features of chronic As toxicity.

5. Biomarkers

On the basis of As metabolism data, three biomarkers of internal exposure are generally proposed for the diagnosis of As exposure, the urinary excretion of the element and its concentration in hair and nail (blood levels are generally too low and transient). Despite some encouraging reports, the use of As levels in hair and nail as indices of internal dose appears limited. Efforts are needed to develop a standardised procedure to solve the problem of external contamination of samples.

The concentration of total As in urine has often been used as an indicator of recent exposure because urine is the main route of excretion of most As species (Buchet et al., 1996; Vahter, 1994). A review of urinary As data after drinking safe water (As < 50 μg/l) is given in Table 3. The range of variation of mean urinary As value is quite wide being 5–39.9 μg/l. Any urine value more than 50 μg/l (mean ± 2SD being 47.5) (provided the subjects have not been taking sea food for the previous 4 days) could be taken as diagnostic of recent As exposure.

In people with no known exposure to As, the concentration of As in hair has been reported to vary from 0.02 to 0.81 mg/kg (range of mean data) (c.f. Table 4). Mean + 2SD of all the mean values of As level in hair of control cases was found to be 0.78. A value of 0.8 mg/kg has therefore been considered as upper limit of normal. The concentrations of As in hair are clearly elevated in people drinking water with high As concentration. For example, concentrations ranging from 3 to 10 mg/kg are commonly reported in people in areas of West Bengal that have high As concentrations in drinking water (Das et al., 1995). Normal As values in nails appear to range from 0.02 to 0.956 mg/kg (range of mean data c.f. Table 5. A value of 1.28 mg/kg was found to be the mean + 2SD of all the mean values of nail in the control cases reviewed. Hence 1.3 mg/kg has been taken as the upper limit of As level in nails in control cases. Although the As level in hair/nail has been found to be

Table 3. Literature data related to the relationship between low arsenic (As) in drinking water and the urinary excretion of the element (mean values).

Authors	n	As in water (µg As/l) mean	As in urine (µg As/l) mean	Range
Harrington et al. (1978)	67	10	38	
Kreiss et al. (1983)	95	9	35	
Vahter et al. (1995)	5	25	19	
	5	14	34	
Buchet et al. (1981, 1996))	135	<5	6	
	2			
Valentine et al. (1979)	32	<6	10.9	3.05–38.45
Das et al. (1995)	50	<50	32.2	13–55
Dhar et al. (1997)	62	4	31	
Mandal et al. (1996)	250	<10	24	
Gonsebatt et al. (1994)	34	30	34	
Del Razo et al. (1994)	22	10	39.3	12–104
Del Razo et al. (1997)	34	31	20	15–28
Garcia-Vargas et al. (1994)	31	20	19	
Wyatt et al. (1998)	10	9	16	
	10	15	15	
	10	30	28	
Kurttio et al. (1998)	9	<1 µg/l	5	4–44
Lin and Huang (1998)	25	32.75	39.9	

Mean, 24.79; SD, 11.35; 2SD, 22.70; Mean + 2SD, 47.49.

Table 4. Arsenic level in hair, mg/kg (control subjects) (water As < 0.05 mg/l).

	No. of samples	Method of analysis	Mean	SD	Range/median
1. Smith (1964)	1,000	NAA	0.81		0.51
Quoted by WHO, 1981	81%				
2. Leslie and Smith (1978)	52	NAA			0.01–0.4
3. Bencko and Sympon (1977)	Group		0.15		
4. Harrington et al. (1978)			0.4		
5. Valentine et al. (1979)	10		0.15	0.11	0.04–0.39
6. Das et al. (1995)	50	AAS	0.404	0.205	0175–0.694
7. Mandal et al. (1996)	250	AAS	0.204	0.105	0.175–0.494
8. Dhar et al. (1997)	62	AAS	0.41	0.18	0.12–0.85
9. Lin and Huang (1998)	25	AAS	0.22	0.08	
10. NRC (1999)					0.02–0.2

Mean, 0.348; SD, 0.216987; 2SD, 0.433974; Mean + 2SD, 0.781974.
NAA, neutron activation analysis; AAS, atomic absorption spectrophotometry.

Table 5. Arsenic level in nail, mg/kg (control subject) (water As < 0.05 mg/l).

	No. of samples	Method of analysis	Mean	SD	Range/median
1. Narang et al. (1987); Takagi et al. (1988)					0.02–0.5
2. Das et al. (1995)	50	AAS	0.956	0.119	0.80–1.12
3. Mandal et al. (1996)	250	AAS	0.756	0.109	0.50–1.10
4. Karagas et al. (1996)	11	NAA	0.14	0.02	
5. Dhar et al. (1997)	61	AAS	0.83	0.68	0.09–1.58
6. Lin and Huang (1998)	25	AAS	0.51	0.33	

Mean, 0.6384; SD, 0.322606; 2SD, 0.645213; Mean + 2SD, 1.283613.

elevated in people drinking As-contaminated water there is no correlation between As concentration in hair and nail and the degree of exposure (Guha Mazumder et al., 1997). Similarly, there is no correlation between the As level in hair and nail and clinical features of chronic As toxicity. In a village of West Bengal, all the 17 people drinking As-contaminated water had elevated hair and nail As, but only 8 had cutaneous lesions. Furthermore, out of 40 people with arsenical skin lesions in another village of West Bengal with a history of drinking As-contaminated water, normal hair and nail As values were found in 31 and 26 cases, respectively (Chowdhury et al., 1997).

Thus in the presence of typical spotty pigmentation (or/and depigmentation) with/without keratosis characteristic of chronic arsenicosis, elevated level of As in urine/hair/nail will confirm the diagnosis. However normal values of these biomarkers do not preclude the diagnosis of chronic arsenic toxicity if there is a definite history of intake of As-contaminated water for a prolonged period.

In the absence of typical skin manifestation of chronic arsenicosis, a major clinical feature, e.g. chronic lung disease, peripheral neuropathy, hepatomegaly with portal zone fibrosis and peripheral vascular disease need to be considered as probable manifestations of chronic As toxicity if there is both a history of chronic exposure of As and elevated level of As in any of those biomarkers.

6. Diagnostic criteria

It becomes evident that with the exception of cutaneous manifestations, other symptoms and signs of chronic arsenicosis are non-specific and can occur with other unrelated medical conditions. However, chronic lung disease, hepatomegaly, peripheral neuropathy and peripheral vascular disease are frequently reported by many investigators in chronic arsenic toxicity and hence may be considered as major systemic criteria. Other systemic features as described earlier that were also reported by a few investigators and frequently were not arsenic related may be considered as minor systemic criteria (*vide* Table 6). Further, cancer of skin (Bowen's disease, squamous cell and basal cell cancer) has been reported by a large number of investigators and hence is considered a major cancer

Table 6. Diagnostic criteria of chronic arsenicosis (clinical effect of chronic arsenic toxicity).

---

1. Prolonged (at least 6 months) intake of water with arsenic (As) levels greater than 0.05 mg/l or exposure of high As level from foods/air/medicine
2. Dermatological features characteristic of chronic arsenicosis
   a) Hyper/hypo: spotty/blotchy pigmentation of the body
   b) Diffuse/nodular keratosis of palms/sole (bilaterally symmetrical)
3. Non-cancer systemic manifestations
   a) Major: signs and symptoms of *chronic lung disease* (cough, breathlessness, chronic bronchitis, pulmonary fibrosis, chronic obstructive airway disease, bronchiectasis), *hepatomegaly* (non-cirrhotic portal fibrosis) with/without splenomegaly, cirrhosis of liver/ascites, peripheral *neuropathy*, peripheral *vascular disease* (Raynaud's phenomenon, gangrene of limbs, indolent ulcer)
   b) Minor: weakness, diabetes mellitus, hypertension, ischaemic heart disease, cardiac arrhythmia cerebrovascular accident, swelling of hands/feet, hearing defect, dim vision, headache, Mee's line in finger nails, conjunctivitis, chronic diarrhoea, anaemia, abdominal pain, anorexia, nausea
4. Cancers
   a) Major criteria: Bowen's disease, squamous cell skin cancer, basal cell skin cancer
   b) Minor criteria: cancer of urinary bladder, cancer of lung, cancer of liver, kidney cancer
5. Biomarkers of exposure: arsenic level in urine above 0.05 mg/l or in hair more than 0.8 mg/kg or in nail more than 1.3 mg/kg

---

criterion. Other internal cancers, e.g. bladder cancer, lung cancer and liver cancer, are reported by fewer investigators and are also infrequently associated with arsenicosis. Hence these are considered as minor cancer criteria. Thus, history of As exposure by drinking As-contaminated water and high level of As in urine and/or in hair and nails in association with those symptoms may help in the diagnosis of chronic arsenicosis. But its normal value in those materials does not exclude such diagnosis. Diagnostic criteria and algorithm of diagnosis and case definition of chronic As toxicity are summarized in Tables 7–8.

7. Outstanding questions on case definition of arsenicosis

Pigmentation and keratosis are considered diagnostic of chronic As toxicity. However, varied clinical manifestations have been reported to occur in As-exposed population. It needs to be ascertained whether arsenicosis could be diagnosed with specific clinical features in the absence of dermatological manifestations. Proper epidemiological studies comparing their incidence in As exposed and control population with similar age, sex and socioeconomic status need to be carried out. This will help in identifying a specific clinical feature, which could be considered diagnostic of chronic As toxicity. It needs to be emphasised that many people remain asymptomatic in spite of drinking As-contaminated water for many years. There is much variation in the incidence of As-related symptoms in an exposed population and only some of the As-exposed family show such features. Goldsmith and From (1980) evaluated the effects

Table 7. Simplified algorithm for case definition of arsenicosis.

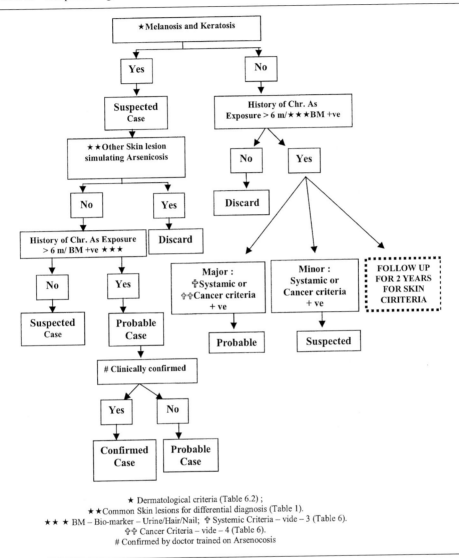

★ Dermatological criteria (Table 6.2) ;
★★Common Skin lesions for differential diagnosis (Table 1).
★ ★ ★ BM – Bio-marker – Urine/Hair/Nail; ✝ Systemic Criteria – vide – 3 (Table 6).
✝✝ Cancer Criteria – vide – 4 (Table 6).
Confirmed by doctor trained on Arsenocosis

of well water As (0.1–1.4 mg/l) on the health status of residents of Lassen county, California. No particular illness was found to have greater prevalence in groups exposed to elevated As level. Harrington et al. (1978) studied the exposure level and possible health effects of As in drinking water among residents of a 150 square miles area near Fairbanks, Alaska. The mean concentration of As in water was 0.22 mg/l with some values as high as 2.45 mg/l. No differences were found in signs, symptoms and physical examination findings in the various exposure categories. Valentine et al. (1985) surveyed groups of 20–57 residents in six US cities where drinking water

Table 8.   Case definition of arsenicosis.

Suspected arsenicosis – A:
  Typical As, skin lesion
  Hyper/hypo: spotty/blotchy pigmentation
  Diffuse/nodular/keratosis
  Bilaterally symmetrical
Suspected arsenicosis – B:
No skin lesion but coming from As endemic area
  a) H/O As exposure >6 m (water As more than 50 μg/l)/BM +ve
  b) Major systemic/cancer criteria +ve
Probable arsenicosis
I.
  a) Melanosis and keratosis (arsenical)
  b) H/O As expose >6 m/BM +ve
II.
  No skin lesion, coming from As endemic area
  a) H/O As exposure >6 m/BM +ve
  b) Major systemic/cancer criteria +ve
Confirmed arsenicosis
  a) Melanosis/keratosis (arsenical)
  b) H/O As exposure >6 m/BM +ve
  c) Clinically confirmed by excluding othe skin disease simulating arsenicosis
  With/without systemic/cancer criteria

---

concentrations of As ranged from 0.05 to 0.395 mg/l. No significant difference in the prevalence of gastrointestinal, dermal or neurological symptoms was detected between any of the groups studied. The reasons for non-expression of clinical manifestation of chronic As toxicity in many people exposed to prolonged intake of As-contaminated water needs further study.

References

Ahmad, S.A., Bandaranayake, D., Khan, A.W., Hadi, S.A., Uddein, G., Halim, M.A., 1997. Arsenic contamination in ground water and arsenicosis in Bangladesh. Int. J. Environ. Health Res., 7, 271–276.

Ahmad, S., Anderson, W.L., Kitichin, K.T., 1999. Dimethylarsinic acid effects on DNA damage and oxidative stress related biochemical parameters in B6C3F1 mice. Cancer Lett., 139, 129–135.

Alain, G., Tousignant, J., Rozenfarb, E., 1993. Chronic arsenic toxicity. Int. J. Dermatol., 32, 899–901.

Bencko, V., Sympon, K., 1977. Health aspects of burning coal with a high arsenic content. Environ. Res., 13, 378–385.

Black, M.D., 1967. Prolonged ingestion of arsenic. Pharm. J., Dec. 9, 593–597.

Borgono, J.M., Vicent, P., Venturino, H., Infante, A., 1977. Arsenic in the drinking water of the city of Antofagasta: epidemiological and clinical study before and after the installation of the treatment plant. Environ. Health Perspect., 19, 103–105.

Buchet, J.P., Lauwerys, R., Roels, H., 1981. Urinary excretion of inorganic arsenic and its metabolites after repeated ingestion of sodium metaarsenite by volunteers. Int. Arch. Occup. Environ. Health, 48, 111–118.

Buchet, J.P., Staessen, J., Roels, H., Lauwerys, R., Fagard, R., 1996. Geographical and temporal differences in the urinary excretion of inorganic arsenic: a Belgian population study. Occup. Environ. Med., 53, 320–327.

Cebrian, M.E., Albores, A., Aguilar, M., Blakely, E., 1983. Chronic arsenic poisoning in the north of Mexico. Hum. Toxicol., 2, 121–133.

Chakraborty, A.K., Saha, K.C., 1987. Arsenical dermatosis from tube well water in West Bengal. Indian J. Med. Res., 85, 326–334.

Chen, C.J., Lin, L.J., Hsueh, Y.M., Chiou, H.Y., Liaw, K.F., Horng, S.F., Chiang, M.H., Tseng, C.H., Tai, T.Y., 1994. Ischemic heart disease induced by ingested inorganic arsenic. In: Abernathy, C.O., Calderon, R.L., Chappell, W.R. (Eds), Arsenic Exposure and Health Effects. Science and Technology Letter, Northwood, pp. 83–90.

Chen, C.-J., Hsueh, Y.-M., Lai, M.-S., Shyu, M.-P., Chen, S.-Y., Wu, M.-M., Kuo, T.-L., Tai, T.-Y., 1995. Increased prevalence of hypertension and long-term arsenic exposure. Hypertension, 25, 53–60.

Chiou, H.-Y., Huang, W.-I., Su, C.-L., Chang, S.-F., Hsu, Y.-H., Chen, C.-J., 1997. Dose response relationship between prevalence of cardiovascular disease and ingested inorganic arsenic. Stroke, 28, 1717–1723.

Choprapawon, C., Rodcline, A., 1997. Chronic arsenic poisoning in Ronpibool Nakhon Sri Thammarat, the Southern Province of Thailand. In: Abernathy, C.O., Calderon, R.L., Chappell, W.R. (Eds), Arsenic Exposure and Health Effects, Vol. 6. Chapman and Hall, London, pp. 69–77.

Chowdhury, T.R., Mandal, B.K., Samanta, G., Basu, G.K., Chowdhury, P.P., Chanda, C.R., Karam, N.K., Lodh, D., Dhar, R.K., Das, D., Saha, K.C., Chakraborti, D., 1997. Arsenic in groundwater in six districts of West Bengal, India: the biggest arsenic calamity in the world: the status report up to August, 1995. In: Abernathy, C.O., Calderon, R.L., Chappell, W.R. (Eds), Arsenic Exposure and Health Effects, Chapman and Hall, London, pp. 93–111.

Chowdhury, U.K., Biswas, B.K., Chowdhury, T.R., Samanta, G., Mandal, B.K., Basu, G.C., Chanda, C.R., Lodh, D., Saha, K.C., Mukherjee, S.K., Roy, S., Kabir, S., Quamruzzaman, Q., Chakraborti, D., 2000. Ground water arsenic contamination in Bangladesh and West Bengal, India. Environ. Health Perspect., 108, 393–397.

Cuzick, J., Sasieni, P., Evans, S., 1992. Ingested arsenic, keratoses and bladder cancer. Am. J. Epidemiol., 1336 (4), 417–421.

Das, D., Chatterjee, A., Mandal, B.K., Samanta, G., Chakraborti, D., 1995. Arsenic in ground water in six districts of West Bengal, India: the biggest arsenic calamity in the world, part 2, arsenic concentration in drinking water, hair, nails, urine, skin-scale and liver tissue (biopsy) of the affected people. Analyst, 120, 917–924.

Del Razo, L.M., Hernandez, J.L., Garcia-Vargas, G., Ostiosky-Wegman, P., Cortinas De Nava, C., Cebrian, E., 1994. Urinary excretion of arsenic species in a human population chronically exposed to arsenic via drinking water. A pilot study. In: Abernathy, C.O., Calderon, R.L., Chappell, W.R. (Eds), Arsenic Exposure and Health Effects. Science and Technology Letter, Northwood, pp. 91–110.

Del Razo, L.M., Garcia-Vargas, G.G., Vargas, H., Albores, A., Gonsebatt, M.E., Montero, R., Ostrosky-Wegman, P., Kelsh, M., Cebrain, M.E., 1997. Altered profile of urinary arsenic metabolites in adults with chronic arsenicism. A pilot study. Arch. Toxicol., 71, 211–217.

Dhar, R.K., Biswas, B.K., Samanta, G., Mandal, B.K., Chakraborti, D., Roy, S., Jafar, A., Islam, A., Ara, G., Kabir, S., Khen, A.W., Akther Ahmed, S., Hadi, S.A., 1997. Groundwater arsenic calamity in Bangladesh. Curr. Sci., 73, 47–59.

Franklin, M., Bean, W.B., Hardin, R.C., 1950. Fowler's solution as an etiologic agent in cirrhosis. Am. J. Med. Sci., 518, 589–596.

Garai, R., Chakraborty, A.K., Dey, S.B., Saha, K.C., 1984. Chronic arsenic poisoning from tube well water. J. Indian Med. Assoc., 82, 34–35.

Garcia-Vargas, G.G., Del Razo, L.M., Cebrian, M.E., Albores, A., Ostrosky-Wegman, P., Montero, R., Gonsebatt, M.E., Lim, C.K., De Matteis, F., 1994. Altered urinary prophyrin excretion in a human population chronically exposed to arsenic in Mexico. Hum. Exp. Toxicol., 13, 839–847.

Goldsmith, S., From, A.H.L., 1980. Arsenic-induced atypical ventricular tachycardia. N. Engl. J. Med., 303, 1096–1098.

Gonsebatt, M.E., Vega, L., Montero, R., Garcia-Vargas, G., Del Razo, L.M., Albores, A., Cebrian, M.E., 1994. Lymphocyte replicating ability in individuals exposed to arsenic via drinking water. Mutat. Res., 313, 293–299.

Guha Mazumder, D.N., Chakraborty, A.K., Ghosh, A., Das Gupta, J.D., Chakraborty, D.P., Dey, S.B., Chattopadhya, N.C., 1988. Chronic arsenic toxicity from drinking tubewell water in rural West Bengal. Bull. WHO, 66, 499–506.

Guha Mazumder, D.N., Das Gupta, J., Chakraborty, A.K., Chatterjee, A., Das, D., Chakraborty, D., 1992. Environmental pollution and chronic arsenicosis in South Calcutta. Bull. WHO, 70 (4), 481–485.

Guha Mazumder, D.N., Das Gupta, J., Santra, A., Ghosh, A., Sarkar, S., Chattopadhaya, N., Charaboti, D., 1997. Non-cancer effects of chronic arsenicosis with special reference to liver damage. In: Abernathy, C.O., Calderon, R.L., Chappell, W.R. (Eds), Arsenic Exposure and Health Effects, Chapman and Hall, London, pp. 112–124.

Guha Mazumder, D.N., Das Gupta, J., Santra, A., Pal, A., Ghosh, A., Sarmar, S., 1998. Chronic arsenic toxicity in West Bengal – the worst calamity in the world. J. Indian Med. Assoc., 96 (1), 4–7 and 18.

Guha Mazumder, D.N., Ghoshal, U.C., Saha, J., Santra, A., De, B.K., Chatterjee, A., Dutta, S., Angel, C.R., Centeno, J.A., 1998. Randomized placebo-controlled trial of 2,3-dimercaptosuccinic acid in therapy of chronic arsenicosis due to drinking arsenic-contaminated subsoil water. Clin. Toxicol., 36 (7), 683–690.

Guha Mazumder, D.N., De, B.K., Santra, A., Dasgupta, J., Ghosh, N., Roy, B.K., Ghosal, U.C., Saha, J., Chatterjee, A., Dutta, S., Haque, R., Smith, A.H., Chakraborty, D., Angle, C.R. and Centeno, J.A., 1999. Chronic arsenic toxicity: epidemiology, natural history and treatment. In: Chappell, W.R., Abernathy, C.O. and Calderon, R.L. (Eds), Arsenic Exposure and Health Effects. Elsevier Science, London, 335–347.

Guha Mazumder, D.N., De, B.K., Santra, A., Ghosh, N., Das, S., Lahiri, S., Das, T., 2001. Randomized placebo-controlled trial of 2,3-dimercapto-1-propanesulfonate (DMPS) in therapy of chronic arsenicosis due to drinking arsenic contaminated water. Clin. Toxicol., 39, 665–674.

Harrington, J.M., Middaugh, J.P., Morse, D.L., Housworth, J., 1978. A survey of a population exposed to high concentrations of arsenic in well water in Fairbanks, Alaska. Am. J. Epidemiol., 108 (5), 377–385.

Hindmarsh, J.T., McLetchie, O.R., Leory, M.D., Heffernan, P.M., Hayne, O.A., Ellenberger, H.A., McCurdy, R.F., Thiebaux, H.J., 1977. Electrocardiagraphic abnormalities in chronic environmental arsenicalism. J. Anal. Toxicol., 1, 270–276.

Hotta, N., 1989. Clinical aspects of chronic arsenic poisoning due to environmental and occupational pollution in and around a small refining spot [in Japanese]. Nippon Taishitsugaku Zasshi [Jpn. J. Const. Med., 53 (1/2), 49–70.

Karagas, M.R., Steven Morris, J., Weiss, J.L., Connie, S.V.B., Robert Gremberg, E., 1996. Toenail samples as an indicator of drinking water arsenic exposure. Cancer Epidemiol., Biomarker Prev., 5, 849–852.

Kiburn, K.H., 1997. Neurobehavioral impairment from long-term residential arsenic exposure. In: Abernathy, C.O., Calderon, R.L., Chappell, W.R. (Eds), Arsenic Exposure and Health Effects, Vol. 14. Chapman and Hall, London, pp. 159–177.

Kreiss, K., Zack, M.M., Feldman, R.G., Niles, C.A., Chirico-Post, J., Sax, D.S., Landrigan, P.J., Boyd, M.H., Cox, D.H., 1983. Neurologic evaluation of a population exposed to arsenic in Alaskan well water. Arch. Environ. Health, 38 (2), 116–121.

Kurttio, P., Komulainen, H., Kakala, E., Rahelin, H., Pekkanen, J., 1998. Urinary excretion of arsenic species after exposure to arsenic present in drinking water. Arch. Environ. Contam. Toxicol., 34, 297–305.

Lai, M.-S., Hsueh, Y.-M., Chen, C.-J., Shyu, M.-P., Chen, S.-Y., Kuo, T.-L., Wu, M.-M., Tai, T.-Y., 1994. Ingested inorganic arsenic and prevalence of diabetes mellitus. Am. J. Epidemiol., 139, 484–492.

Leslie, A.C.D., Smith, H., 1978. Self-poisoning by the abuse of arsenic containing torices. Med. Sci. Law, 18, 159–162.

Lin, T.H., Huang, Y.L., 1998. Arsenic species in drinking water, hair, fingernails, and urine of patients with blackfoot disease. J. Toxicol. Environ. Health, 53, 85–93.

Luchtrath, H., 1983. The consequences of chronic arsenic poisoning among Moselle wine growers. Pathoanatomical investigations of post-mortem examinations performed between 1960 and 1977. J. Cancer Res. Clin. Oncol., 105, 173–182.

Luo, Z.D., Zhang, Y.M., Ma, L., Zhang, G.Y., He, X., Wilson, R., Byrd, D.M., Griffiths, J.G., Lai, S., He, L., Grumski, K., Lamm, S.H., 1997. Chronic arsenicism and cancer in Inner Mongolia – consequences of well-water arsenic levels greater than 50 µg/l. In: Abernathy, C.O., Calderon, R.L., Chappell, W.R. (Eds), Arsenic Exposure and Health Effects, Chapman and Hall, London, pp. 55–68.

Ma, H.Z., Xia, Y.J., Wu, K.G., Sun, T.Z., Mumford, J.L., 1999. Human exposure to arsenic and health effects in Bayingnormen, Inner Mongolia. In: Abernathy, C.O., Calderon, R.L., Chappell, W.R. (Eds), Arsenic Exposure and Health Effects. Elsevier Science, London, pp. 127–131.

Mandal, B.K., Chowdhury, T.R., Samanta, G., Basu, G.K., Chowdhury, P.P., Chanda, C.R., Lodh, D., Karan, N.K., Dhar, R.K., Tamili, D.K., Das, D., Saha, K.C., Chakraborti, D., 1996. Arsenic in groundwater in seven districts of West Bengal, India – the biggest arsenic calamity in the world. Curr. Sci., 70 (11), 976–986.

McCabe, L.J., Symons, J.M., Lee, R.D., Robeck, G.G., 1970. Survey of community water supply systems. J. Am. Water Works Assoc., 62, 670–687.

Milton, A.H., Hasan, Z., Rahman, A., Rahaman, M., 2001. Chronic arsenic poisoning and respiratory effects in Bangladesh. J. Occup. Health, 43, 136–140.

Narang, A.P.S., Chawla, L.S., Khurana, S.B., 1987. Levels of arsenic in Indian opium eaters. Drug Alcohol Depend., 20, 149–153.

NRC (National Research Council), 1999. Arsenic in Drinking Water. National Academic Press, Washington DC, 1177–1191.

Rahman, M., Tondel, M., Ahmad, S.A., Axelson, O., 1998. Diabetes mellitus associated with arsenic exposure in Bangladesh. Am. J. Epidemiol., 148, 198–203.

Rahman, M., Tondel, M., Chowdhury, I.A., Axelson, O., 1999. Relations between exposure to arsenic, skin lesions and glycosuria. Occup. Environ. Med., 56, 277–281.

Ratnam, K.V., Espy, M.J., Muller, S.A., Smith, T.F., Su, W.P., 1992. Clinicopathologic study of arsenic-induced skin lesions: no definite association with human papillomavirus. J. Am. Acad. Dermatol., 27, 120–122.

Rattner, H., Dorne, M., 1943. Arsenical pigmentation and keratoses. Arch. Dermatol. Syphilol., 48, 458–460.

Rosenberg, H.G., 1974. Systemic arterial disease and chronic arsenicism in infants. Arch. Pathol., 97 (6), 360–365.

Saha, K.C., 1984. Melanokeratosis from arsenic contaminated tube well water. Indian J. Dermatol., 29, 37–46.

Saha, K.C., 1995. Chronic arsenical dermatosed from tube-well water in West Bengal during 1983–87. Indian J. Dermatol., 40, 1–12.

Saha, K.C., Chakraborti, D., 2001. Seventeen years experience of arsenicosis in West Bengal, India. In: Chappell, W.R., Abernathy, C.O., Calderon, R.L. (Eds), Arsenic Exposure and Health Effect IV. Elsevier Science, Oxford, pp. 387–396.

Smith, H., 1964. The interpretation of the arsenic content of human hair. Forensic Sci. Soc. J., 4, 192–199.

Sommers, S.C., McManus, R.G., 1953. Multiple arsenical cancers of skin and internal organs. Cancer, 6, 347–359.

Szuler, I.M., Williams, C.N., Hindmarsh, J.T., Park-Dinesoy, H., 1979. Massive variceal hemorrhage secondary to presinusoidal portal hypertension due to arsenic poisoning. Can. Med. Assoc. J., 120, 168–171.

Takagi, Y.O., Matsuda, S., Imai, S., Ohmori, Y., Masuda, T., Vinson, J.A., Mehra, M.C., Puri, B.K., Kanie Woki, A., 1988. Survey of trace elements in human nails: an international comparison. Bull. Environ. Toxicol., 41, 690–695.

Tay, C.H., 1974. Cutaneous manifestations of arsenic poisoning due to certain Chinese herbal medicine. Aust. J. Dermatol., 15 (3), 121–131.

Tseng, W.P., Chu, H.M., How, S.W., Fong, J.M., Lin, C.S., Yeh, S., 1968. Prevalence of skin cancer in an endemic area of chronic arsenicism in Taiwan. J. Natl Cancer Inst., 40, 453–463.

Vahter, M., 1994. Species differences in the metabolism of arsenic. In: Abernathy, C.O., Calderon, R.L., Chappell, W.R. (Eds), Arsenic Exposure and Health Effects. Science and Technology Letter, Northwood, pp. 171–179.

Vahter, M., Concha, G., Nermell, B., Nilsson, R., Dulout, F., Natarajan, A.T., 1995. A unique metabolism of inorganic arsenic in native Andean women. Eur. J. Pharmacol., 293, 455–462.

Valentine, J.L., King, H.K., Spivey, G., 1979. Arsenic levels in human blood, urine and hair in response to exposure via drinking water. Environ. Res., 20, 24–32.

Valentine, J.L., Reisbord, L.S., Kang, H.K., Schluchter, M.D., 1985. Arsenic effects on population health histories. In: Mills, C.F., Bremner, I., Chester, J.K. (Eds), Trace Elements in Man and Animals – TEMA5. Proceedings of the Fifth International Symposium on Trace Elements in Man and Animals, Commonwealth Agricultural Bureau, Slough, UK.

Wyatt, C.J., Loper Quirogea, V., Olivas Acosta, R.T., Mendez, R.U., 1998. Excretion of arsenic in urine of children, 7–11 years exposed to elevated levels of As in the city water suppler in Hermosillo, Sonota, Mexico. Environ. Res., 78, 19–24.

Yeh, S., 1973. Skin cancer in chronic arsenicism. Hum. Pathol., 4, 469–485.

Zaldivar, R., 1974. Arsenic contamination of drinking water and food-stuffs causing endemic chronic poisoning. Beitr. Pathol., 151, 384–400.

Zaldivar, R., Prunes, L., Ghai, G., 1981. Arsenic dose in patients with cutaneous carcinomata and hepatic hemangio-enothelioma after environmental and occupational exposure. Arch. Toxicol., 47, 145–154.

Arsenic Exposure and Health Effects V
W.R. Chappell, C.O. Abernathy, R.L. Calderon and D.J. Thomas, editors

135

Chapter 10

Arsenic exposure alters purine metabolism in rats, mice, and humans

Luz María Del Razo, Eliud A. García-Montalvo
and Olga L. Valenzuela

Abstract

Inorganic arsenic (iAs) has long been known to be a human carcinogen. Additionally, its chronic exposure has been associated with a wide range of adverse health effects, including immune, cardiovascular, endocrine, and nervous system.

Purine metabolism is related to a variety of cell functions; energy conservation and transport, formation of coenzymes and of active intermediates of phospholipids and carbohydrate metabolism. Therefore, when a purine metabolic disorder exists any system can be affected. In purine metabolism, uric acid (UA) is the end product of biosynthesis de novo salvage and degradation. Consequently, measurement of UA in plasma and urine indicates a defect in purine metabolism.

In this study, the effect of iAs exposure in UA concentration was evaluated using three different models in vivo iAs[III]-exposure: (a) rats dosed subchronically at 1.2 mg/kg body weight daily for 6 weeks, (b) mice treated at 3, 6, or 10 mg/kg daily for 9 days, and (c) humans chronically exposed through contaminated drinking water from an endemic area, central Mexico.

The main finding of this work was the statistically significant association of hypouricemia and the hypouricosuria with iAs exposure. UA determination, in biological samples, is an easy and cheap analytical technique that could be used as a sensible marker of iAs exposure.

Knowledge of alteration of enzymes of purine metabolism is required to understand their possible relation with toxic effects associated with iAs exposure.

Keywords: arsenic, arsenic biomarker, purine metabolism, xanthine oxidase, uric acid

1. Introduction

Arsenic (As) is widely distributed in nature as a result of natural and anthropogenic contributions. Some of the As species are introduced into the environment as inorganic arsenic (iAs), through the consumption of contaminated water (Welch et al., 1999). iAs species undergo biomethylation in mammals, forming monomethylated and dimethylated arsenic species. The reduction of As from pentavalency to trivalency is a prerequisite for its oxidative methylation, thus both trivalent and pentavalent metabolites are intermediates or final products in the metabolic pathway, that are primarily excreted in the urine (Thomas et al., 2001).

The toxicity of As compounds highly depends on the oxidation state and chemical composition of the arsenical. Thus, the trivalent arsenic species, whether in the inorganic or organic form, are generally more toxic than the pentavalent forms of As. The toxicity of trivalent arsenicals lies in its ability to bind to the sulfur groups of essential cysteines in proteins. These groups can be important for the three-dimensional structure of proteins which affect the function of proteins and enzyme–substrate interactions. Because arsenicals in trivalency form bind to two sulfur groups, they can cross-link proteins, distorting their overall shape and impeding their function (Knowles and Benson, 1984).

Trivalent arsenicals are capable to inhibit many enzymes including xanthine oxidase (XO) (Coughlan et al., 1969), an enzyme involved in degradative-purine metabolism.

In human metabolism of purines, which are essential components of nucleic acids, uric acid (UA) is the end product of biosynthesis *de novo*, salvage and degradation of exogenous and endogenous purines. UA, an insoluble purine, is formed by sequential oxidation of hypoxanthine to xanthine, and this reaction is catalyzed by XO (Fig. 1).

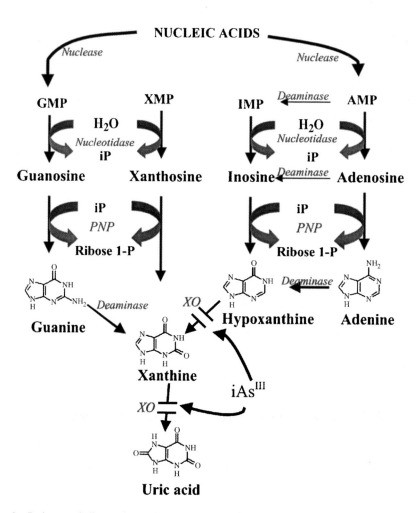

Figure 1. Purine metabolism pathway. Steps are catalyzed by catabolic enzymes. PNP: purine nucleotide phosphorilase and XO: xanthine oxidase.

Some animal species like birds and scaly reptiles enzymatically convert UA to a water-soluble substance *via* an enzyme called urate oxidase. Unfortunately, humans do not synthesize this enzyme and so cannot convert UA into a water-soluble compound. Consequently, any insoluble UA that is formed must be carried in the blood to the kidneys for elimination. The concentration of UA in the blood is related to the balance between UA production and excretion. Therefore, measurement of UA in plasma and urine will lead to an indication of a defect in purine metabolism.

In this work, we evaluate the effect of iAs exposure in UA concentration using three different models *in vivo* iAs^{III}-exposure: in rats, mice, and humans.

2. Methodology

2.1. Animals

The experimental animals were obtained from Cinvestav-IPN facility. They were maintained according to the Mexican Regulations of Animal Care and Maintenance (DOGM, 2001). Animals were housed in polycarbonate cages with hardwood chip bedding in groups of 5 and 10 animals for rats and mice, respectively. Rodent chow purine and tap water were available *ad libitum*. The animal room was on a 12-h dark/12-h light cycle (starting at 7.00 h) and the temperature and relative humidity were kept at $22 \pm 1°C$ and $50 \pm 10\%$, respectively.

2.1.1. Study 1: rats dosed subchronically with arsenite

Male Wistar rats (150–160 g) were divided into two groups ($n = 6$). One group of rats were given by oral gavage As as sodium arsenite at a dose of 1.2 mg/kg body weight daily for 6 weeks. The last one group was used as control. Each week, urine samples were obtained. Twenty hours previous urine collection, rats were housing individually in Nalgene metabolism cages (Nalge Co., Rochester, NY). These rats were sacrificed under anesthesia at the end of As treatment (6 weeks). Blood from rats was collected in EDTA tubes at the end of the As treatment. Plasma was separated by centrifugation at 2000g for 10 min.

2.1.2. Study 2: mice dosed with repeated dose of arsenite. Dose–response

Fifty-day-old female C57BL/6N mice were divided into four groups ($n = 10$) and given orally As as sodium arsenite at dose levels of 0, 3, 6, or 10 mg/kg daily for 9 days. Control mice were treated with water. Immediately after the last As dose, two mice were housed in each cage to accumulate enough urine during the 24 h postdosing collection time. During this time, urine samples were collected in plastic bottles on dry ice.

2.2. Human Study

2.2.1. Study 3: humans chronically exposed through contaminated drinking water

Participants who agreed to participate for this study were residents, between 15 and 51 years old, from an area located at the center of Mexico named Zimapan, Hidalgo. Ninety-seven (11 men and 86 women) participants had been exposed to very high levels of As in drinking water (150–1350 µg/l) since 1993. But, due to changes in the village water supply, the As content was reduced to average concentration of 130 µg As/l, 10 months previous to this study. The reference group consisted of 28 individuals (2 men and 26 women) selected from a village that had 10 µg As/l in its drinking water. The predominant form of As in the drinking water of the villages studied was pentavalent form ($<92\%$). All participants signed an informed consent form. After obtaining consent, each subject completed an exposure assessment questionnaire. This questionnaire requested demo-graphic and occupational information, medical history, and length of residence in their present home. Individuals who had been exposed to drugs that alter UA metabolism such as allopurinol, probenecid, sulphinpyrazone, colchicine, coumarin derivatives, and corticosteroids, in the previous 3 months before the study began, were excluded from the study.

A spot urine sample was collected from each individual. Urine was collected and stored in plastic bottles and immediately frozen in dry ice until analysis. Time weighted As exposure values (TWE) were estimated for each participant based on the history of well water consumption and their As concentrations (mg/years).

2.3. Uric acid assay

In the first study, UA in plasma and urine was analyzed by reversed-phase HPLC, using a Varian 8500 High Pressure Liquid Chromatograph with UV detection at 294 nm (Jauge and Del Razo, 1983). In the second and third studies, UA in urine was determined spectrophotometrically at 540 nm, using automated clinical analyzer Vitalab Eclipse Merck spectrophotometer by kit obtained from RANDOX (St. Fco., CA).

2.4. Arsenic analysis

Urine samples were assayed by hydride generation atomic absorption spectroscopy (HGAAS) according to Del Razo et al. (2001a). As species [iAs, monomethylated As (MAs) and dimethylated As (DMAs)] were reduced to their corresponding hydrides and then detected using a Perkin Elmer 3100 AAS.

Quality control for TAs (iAs + MAs + DMAs) included the analysis of freeze-dried urine standard reference material for toxic metals (SRM 2670) concurrently with urine samples from individuals and experimental animals. The certified concentration of the standard was 480 µg As/l and in our conditions we obtained 492 µg As/l (range 445–0.529). We attained an accuracy of 93–109% and 1–9% variation coefficient ($n = 6$).

2.5. Creatinine in human urine

Creatinine was measured by colorimetric automated method using a Vitalab Eclipse Merck spectrophotometer according to German Society of Clinical Chemistry. As concentrations in human urine were corrected for creatinine excretion.

2.6. Statistical procedures

Data analyses were performed using procedures available in Prism version 3.0. Associations among the urinary UA and arsenic concentration and treatment were measured using graphical methods, Pearson's correlation and linear correlation. In some cases, data were log transformed.

Comparisons between UA concentrations in exposed and control groups were made using ANOVA test, whereas to assess the differences of UA levels and the urinary levels of As species between the groups of experimental animals, the Duncan's multiple range test was used.

Student's *t*-test was used to test differences between exposure and respective control means. In all analyses, the level of significance chosen was $p < 0.05$.

3. Results

3.1. Study 1: rats dosed subchronically with arsenite

The results of subchronically arsenite administration on plasma UA levels in rats are given in Table 1. The data indicate that arsenite treatment produced a significant decrease ($p < 0.001$) in UA concentration after 6 weeks of daily treatment.

The time-course of urinary levels of UA in rats subchronically exposed to iAs^{III} are shown in Figure 2. We found that μg UA/min/kg body weight represents a better expression of UA in 24 h urine samples, due to differences in urine volume and body weight.

A significant decrease in urinary UA excretion was observed during iAs^{III} treatment (Fig. 2). The higher UA decrease was maintained during the first 3 weeks, while UA levels were increased after 3 weeks of daily treatment, but UA concentration was always lower than control group.

Table 1. Uric acid levels in plasma of rats subchronically exposed to arsenite (iAs^{III}) at dose of 1.2 mg/kg during 6 weeks.

Uric acid (mg UA/dl) median $\pm$ SD	
Control	iAs^{III}
3.28 $\pm$ 0.38	1.08 $\pm$ 0.07
($n = 6$)	($n = 6$)
	$p < 0.001$

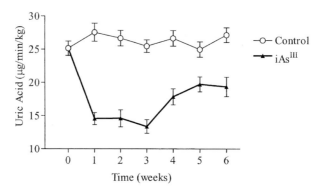

Figure 2. Urinary UA excretion in rats subchronically exposed to arsenite (iAs[III]) at a dose of 1.2 mg/kg during 6 weeks.

3.2. Study 2: mice dosed with repeated dose of arsenite. Dose–response

Figure 3 shows the dose–response urinary UA excretion observed in mice treated with iAs[III], where UA is expressed in µg UA/min/kg body weight. After 9 days of daily arsenite administration the urinary UA concentration was significantly decreased from 11.9 to 8.6, 7.2, and 4.4 µg UA/min/kg at doses of 0, 3, 6, or 10 mg/kg, respectively. All three arsenite-doses significantly reduced UA concentrations ($p < 0.005$ level).

The correlation of urinary arsenicals and UA are based on the results of individual mice (Fig. 4). A linear regression using 20 individual paired values for mouse urine UA and TAs gave a $R^2 = 0.640$ (Fig. 4). Much of the variability of this data set was contributed by different responses of individual mice, particularly in their UA responses to similar As concentrations.

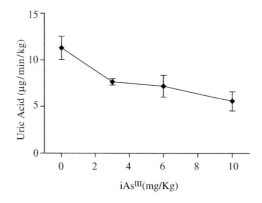

Figure 3. A plot of the mean of urinary UA vs. external administered dose of iAs[III] in mice.

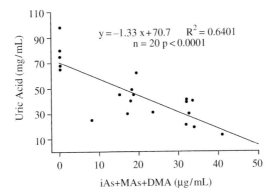

Figure 4. Linear correlation between the individual urinary As concentration in urine from mice. The straight line of the best fit has an intercept of 70.7 mg/l, a slope of 1.33, and an R^2 of 0.64.

Table 2. Urinary UA concentration, arsenic exposure values in urine and TWE in individuals chronically exposed to arsenic in their drinking water (Zimapan, Hidalgo, Mexico).

Parameters	Control ($n = 28$)	Exposed ($n = 97$)	p
	Median ± SD	Median ± SD	
(iAs + MAs + DMAs) in urine (μg/g creatinine)	39.6 ± 23.6	130.9 ± 177	0.007
TWE (mgAs × años)	0.017 ± 0.014	5.32 ± 2.86	<0.0001
Uric acid (mg/l)	26.5 ± 1.9	18.4 ± 2.0	0.015

3.3. Study 3: humans chronically exposed through contaminated drinking water

Mean As concentrations in urinary UA from exposure group were significantly lower than control individuals (26.5 vs. 18.4 mg/l, respectively; $p < 0.15$). The levels of urinary As in exposed group was significantly higher (131 μg/g creatinine) than in low exposure group (39.6 μg/g creatinine). Regarding the chronic As exposure value of TWE (Table 2), exposed group was 312-fold higher than control group.

4. Discussion

UA was formed primarily in the liver and was excreted by the kidney in the urine. The main finding in this work was the hypouricemia and the hypouricosuria associated with iAs exposure.

UA concentration found in plasma of rats represents 32.9% of control values when iAs^{III} was administered (Table 1). The inhibition of the enzyme XO by the interaction with arsenite may be the reason for this decrease. Arsenite ions are known to be inhibitory to XO and to the other molybdenum-containing hydroxylases, xanthine dehydrogenase, and aldehyde oxidase (Coughlan et al., 1969). XO is a complex metalloflavoprotein having three oxidation–reduction centers the molybdenum (Mo), flavin (FAD), and two distinct ferredoxin type Fe/S centers (Bray et al., 1975). The Fe/S centers are thought to facilitate electron transfer and to enhance the catalytic efficiency of the enzyme by maintaining the Mo and FAD in their reactive redox states Mo^{VI} and $FADH_2$ (Davis et al., 1982).

XO is a cytosolic enzyme that catalyzes an important physiologic reaction, namely the sequential oxidation of hypoxanthine to xanthine and UA as shown in Figure 1. There are several reports (Barber and Siegel, 1983; George and Bray, 1983; Hille et al., 1983) that have examined various aspects of the interaction of the Mo center of XO with arsenite. Arsenite is known to bind reversibly to the Mo center of the enzyme, preventing its reduction by purines, pteridines, and aldehydes (Coughlan et al., 1969). However, arsenite also binds to a reduced functional enzyme (George and Bray, 1983) and to oxidized desulfo (inactive) XO (Hille et al., 1983).

This study shows the consistently dramatic decrease of UA in urine from three different models of As exposure (Figs. 2 and 3; Table 2). In the case of rats subchronically dosed with arsenite at relatively low dose (0.05 of LD_{50}) during 6 weeks, the UA diminution of urinary UA levels could be correlated with the corresponding lowering of levels of plasma observed at the end of treatment. The coexistence of hypouricemia with hypouricosuria observed in rats does agree with XO inhibition. This alteration in purine metabolism was provoked by iAs^{III}. Nevertheless, *in vivo* iAs^V treatment in experimental rats also causes UA reduction but, this effect is lower than those caused by arsenite (Jauge and Del Razo, 1985). Arsenate (iAs^V) decreased UA formation, probably due to *in vivo* reduction of iAs^V to iAs^{III}. Evaluation of high levels of hypoxanthine in plasma and urine would confirm the XO inhibition caused by iAs^{III}.

The doses used in dose–response studies (Fig. 3) ranged from moderate to as high as mice can tolerate (a range of $3–10$ mg/kg that corresponds to $0.15–0.5$ of LD_{50}, respectively). In all mice dosed with iAs^{III}, the urinary UA concentration fell significantly ($p < 0.005$), in comparison with control mice. Because of differences in urine volume the UA concentration was expressed in the same way as the rat study (μg UA/min/kg body weight).

Table 2 shows the significant decrease of urinary UA observed in As exposed group, the median value represents 69% concentration from control group. As exposure was evaluated using the total concentration of the metalloid in water, the sum of iAs and its metabolites in urine, and the TWE. All these measurements were coherent with the historic knowledge of As exposure in each group.

Disorders of purine metabolism associated with the fall of UA, include deficiencies of another degradative-purine enzyme that cleaves purine nucleosides, such as purine nucleoside phosphorylase (PNP). The cytosolic and mitochondrial enzyme PNP catalyzes a slow hydrolysis of inosine in which the first catalytic site releases ribose, but binds tightly (1 pM) to the hypoxanthine in the absence of inorganic phosphate (iP) (Kline and Schramm, 1992). Hydrolysis of enzyme-bound inosine forms a stable enzyme-bound hypoxanthine complex and free ribose (Fig. 1). When iP or its analog, iAs^V are present in the cell, both are able to prevent formation of hypoxanthine complex (Kline and Schramm, 1992, 1993). PNP deficiency results in a failure to form hypoxanthine, xanthine, and UA in the purine salvage pathway (Fig. 1). Evaluation of low levels of hypoxanthine and high levels of deoxynucleotides could confirm the enzymatic alteration of PNP caused possibly by iAs^{III}.

Interestingly, it has recently been observed that PNP is responsible for the reduction of iAs^V to iAs^{III} when catalyzing the arsenolytic cleavage of inosine or guanosine in the presence of a suitable dithiol (Gregus and Németi, 2002; Radabaugh et al., 2002).

Inorganic and aromatic trivalent arsenicals have been shown to inhibit a number of enzymes due to their high affinity for thiols (Abernathy et al., 1999). Both XO and PNP are enzymes containing critical thiols groups. Inhibition of XO by iAs^{III} is recognized but the action of trivalent arsenicals on PNP needs to be studied. Recently Styblo et al. (2002) have shown that methylated trivalent arsenicals MAs^{III} and $DMAs^{III}$, intermediates of iAs metabolism, are more potent than iAs as enzyme inhibitors with catalytically active SH groups. Thus, it is probable that methylated arsenicals that contain As in $+3$ oxidation state can also interact with the molybdenum center [Mo (SH) (OH)] of XO or with some cysteines from PNP and contributes as a consequence to fall of UA. We have evidence of the urinary presence of MAs^{III} and $DMAs^{III}$ in mice and humans from study 2 and 3 of this chapter (data not shown).

A marker of imbalance in the purine metabolism enzyme pattern such as decrease of XO, is linked with breast (Alsabti, 1980; Vaidya et al., 1998), bladder (Alsabti, 1979; Durak et al., 1994), and liver (Stirpe et al., 2002) cancerous tissues. On the other hand, a decrease in PNP activity in humans would cause the accumulation of high levels of deoxynucleotides, which would be toxic for T lymphocytes. Therefore, deficiency of PNP is associated with specific T-cell immune suppression (Markert, 1991).

Other important aspects of purine metabolism is related to the antioxidant defense ability of UA. It has been shown to be a major antioxidant in human serum and was postulated to have a biological role in protecting tissues against the toxic effects of oxygen radicals (Nieto et al., 2000; Waring et al., 2001). Because iAs^{III} is an oxidative/nitrosative stress agent, able to generate reactive species and free radicals (Del Razo et al., 2001b), the reduction of UA caused by As exposure may contribute to the increase of the stress response elicited by arsenicals.

Purine catabolism may be an unappreciated component of the homeostatic response to cellular stress and may play a critical role in several disease states.

In conclusion, As exposure causes hypouricemia and hypouricosuria. UA determination, in biological samples, is an easy and inexpensive analytical technique that can be used as a sensible marker of iAs exposure. Knowledge of alteration of enzymes of purine metabolism is required to understand their possible relation with toxic effects associated with iAs exposure.

Acknowledgements

The authors thank Martha B. Cruz (SSA-Pachuca, Hidalgo) and Betzabe Calderón (SSA-Zimapán, Hidalgo) for their assistance in human study and Carolina Aguilar and Luz del Carmen Sanchez (Cinvestav-IPN) for their excellent technical support. This work was partially supported by Conacyt-Mexico (grant 38471-M).

References

Abernathy, C.O., Liu, Y.P., Longfellow, D., Aposhian, H.V., Beck, B., Fowler, B., Goyer, R., Menzer, R., Rossman, T., Thompson, C., Waalkes, M., 1999. Arsenic: health effects, mechanisms of actions, and research issues. Environ Health Perspect., 107, 593–597.

Alsabti, E., 1979. Serum xanthine oxidase in bladder carcinoma. Urol. Int., 34, 233–236.

Alsabti, E., 1980. Serum xanthine oxidase in breast carcinoma. Neoplasma, 27, 95–99.

Barber, M.J., Siegel, L.M., 1983. Electron paramagnetic resonance and potentiometric studies of arsenite interaction with the molybdenum centers of xanthine oxidase, xanthine dehydrogenase, and aldehyde oxidase: a specific stabilization of the molybdenum(V) oxidation state. Biochemistry, 22, 618–624.

Bray, R.C., Barber, M.J., Dalton, H., Lowe, D.J., Coughlan, M.P., 1975. Iron–sulphur systems in some isolated multi-component oxidative enzymes. Biochem. Soc. Trans., 3, 479–482.

Coughlan, M.P., Rajagopalan, K.V., Handler, P., 1969. The role of molybdenum in xanthine oxidase and related enzymes. Reactivity with cyanide, arsenite, and methanol. J. Biol. Chem., 244, 2658–2663.

Davis, M.D., Olson, J.S., Palmer, G., 1982. Charge transfer complexes between pteridine substrates and the active center molybdenum of xanthine oxidase. J. Biol. Chem., 257, 14730–14737.

Del Razo, L.M., Styblo, M., Cullen, W.R., Thomas, D.J., 2001a. Determination of trivalent methylated arsenicals in biological matrices. Toxicol. Appl. Pharmacol., 174, 282–293.

Del Razo, L.M., Quintanilla-Vega, B., Brambila-Colombres, E., Calderon-Aranda, E.S., Manno, M., Albores, A., 2001b. Stress proteins induced by arsenic. Toxicol. Appl. Pharmacol., 177, 132–148.

DOGM, 2001. Diario Oficial del Gobierno Mexicano. Norma oficial Mexicana NOM-062-ZOO-1999. "Especificaciones técnicas para la producción, cuidado y uso de los animales de laboratorio" In Spanish.

Durak, I., Perk, H., Kavutcu, M., Canbolat, O., Akyol, O., Beduk, Y., 1994. Adenosine deaminase, 5'nucleotidase, xanthine oxidase, superoxide dismutase, and catalase activities in cancerous and noncancerous human bladder tissues. Free Radic. Biol. Med., 16, 825–831.

George, G.N., Bray, R.C., 1983. Reaction of arsenite ions with the molybdenum center of milk xanthine oxidase. Biochemistry, 22, 1013–1021.

Gregus, Z., Nemeti, B., 2002. Purine nucleoside phosphorylase as a cytosolic arsenate reductase. Toxicol. Sci., 70, 13–19.

Hille, R., Stewart, R.C., Fee, J.A., Massey, V., 1983. The interaction of arsenite with xanthine oxidase. J. Biol. Chem., 258, 4849–4856.

Jauge, P., Del Razo, L.M., 1983. Urinary uric acid determination by reversed-phase high pressure liquid chromatography. J. Liquid Chromatogr., 6, 845–860.

Jauge, P., Del Razo, L.M., 1985. Uric acid levels in plasma and urine in rats chronically exposed to arsenic As(III) and As(V). Toxicol. Lett., 26, 31–35.

Kline, P.C., Schramm, V.L., 1992. Purine nucleoside phosphorylase. Inosine hydrolysis, tight binding of the hypoxanthine intermediate, and third-the-sites reactivity. Biochemistry, 31, 5964–5973.

Kline, P.C., Schramm, V.L., 1993. Purine nucleoside phosphorylase. Catalytic mechanism and transition-state analysis of the arsenolysis reaction. Biochemistry, 32, 13212–13219.

Knowles, F.C., Benson, A.A., 1984. The enzyme inhibitory form of inorganic arsenic. Z. Gesamte Hyg., 30, 625–626.

Markert, M.L., 1991. Purine nucleoside phosphorylase deficiency. Immunodefic. Rev., 3, 45–81.

Nieto, F.J., Iribarren, C., Gross, M.D., Comstock, G.W., Cutler, R.G., 2000. Uric acid and serum antioxidant capacity: a reaction to atherosclerosis? Atherosclerosis, 148, 131–139.

Radabaugh, T.R., Sampayo-Reyes, A., Zakharyan, R.A., Aposhian, H.V., 2002. Arsenate reductase II. Purine nucleoside phosphorylase in the presence of dihydrolipoic acid is a route for reduction of arsenate to arsenite in mammalian systems. Chem. Res. Toxicol., 15, 692–698.

Stirpe, F., Ravaioli, M., Battelli, M.G., Musiani, S., Grazi, G.L., 2002. Xanthine oxidoreductase activity in human liver disease. Am. J. Gastroenterol., 8, 2079–2085.

Styblo, M., Drobná, Z., Jaspers, L., Lin, S., Thomas, D.J., 2002. The role of biomethylation in toxicity and carcinogenicity of arsenic: a research update. Environ. Health Perspect., 110 (Suppl. 5), 767–771.

Thomas, D.J., Styblo, M., Lin, S., 2001. The cellular metabolism and systemic toxicity of arsenic. Toxicol. Appl. Pharmacol., 176, 127–144.

Vaidya, S.M., Kamlakar, P.L., Kamble, S.M., 1998. Molybdenum, xanthine oxidase and riboflavin levels in tamoxifen treated postmenopausal women with breast cancer. Indian J. Med. Sci., 52, 244–247.

Waring, W.S., Webb, D.J., Maxwell, S.R., 2001. Systemic uric acid administration increases serum antioxidant capacity in healthy volunteers. J. Cardiovasc. Pharmacol., 38, 365–371.

Welch, A.H., Helsel, D.R., Focazio, M.J., Watkins, S.A., 1999. Arsenic in the ground water supplies of the United States. In: Chappell, W.R., Abernathy, C.O., Calderon, R.L. (Eds), Arsenic Exposure and Health Effects. Elsevier, Amsterdam, pp. 9–17.

Arsenic Exposure and Health Effects V
W.R. Chappell, C.O. Abernathy, R.L. Calderon and D.J. Thomas, editors

Chapter 11

Risk analysis of non-melanoma skin cancer incidence in arsenic exposed population

Vladimír Bencko, Jiří Rameš, Miloslav Götzl, Petr Franěk
and Marián Jakubis

Abstract

We analyzed a database of 1503 non-melanoma skin cancer (NMSC) cases collected over 20 years (four 5-year intervals) in a region polluted by emissions from burning of coal with high arsenic content ranging between 900 and 1500 g per metric ton of dry coal.

Exposure assessment was based on biological monitoring (hair and urine). Determination of arsenic was done in groups of 10-year-old boys (in non-occupational settings) at different localities situated up to 30 km from the local power plant.

The age and gender standardized incidence of NMSC (each confirmed by biopsy histological examination) in a district with a population ~125,000 in non-occupational settings ranged from 45.9 to 93.9 in men and from 34.6 to 81.4 in women per 100,000 (study base 1325 thousand men/year and 1337 thousand women/year) while relevant data for occupational settings (male workers of power plant burning arsenic reach coal) ranged from 44.6 to 10 317 per 100,000 (study base 27 thousand men/year). Smoking was carefully registered in all cancer patients including NMSC cases and the potential contribution of both arsenic and smoking was analyzed. Our analysis found a positive correlation of human cumulative arsenic exposure and incidence of NMSC.

Keywords: cancer epidemiology, biological monitoring, arsenic toxicity, non-melanoma skin cancer

1. Introduction

The trace element content of coal is known to vary by the specific geological conditions of mines (Niu et al., 1997; Thornton and Farago, 1997).

Ecological and human health aspects including neurotoxicity and immuno-toxicity phenomena encountered in humans exposed in environmental and occupational settings of the excessive contamination of the environment by arsenic due to burning the local coal of high arsenic content were described and summarized in detail in the vicinity of power plant in Prievidza district, Central Slovakia (Bencko, 1997). This paper reports on an extension of our previous study of non-melanoma skin cancer (NMSC) incidence (Bencko et al., 1999) by a further 5 years to achieve a 20-year follow-up interval.

2. Material & methods

2.1. Study base

Our population-based epidemiological study of transitional type; beginning in the mid-1970s covers the entire population of the Prievidza district, Central Slovakia, with the primary goal of following up the incidence of all types of malignancies in this area. Our study attempted to obtain a complete register of malignant tumors within an administrative unit of about 125,000 people. This project was feasible due to our previous national health care system, which operated in this country. Each cancer patient or any person suspected of any malignancy was referred to the district oncologist who was responsible for the final diagnosis and therapy of the patient. Originally, our intent was to perform a 10-year study. However, the data collection efforts and the comprehensive nature of the health care system permitted extending this study to 20 years. The study was initiated in 1976. The results of the first year were eliminated as the system of data collection and questionnaire construction and implementation were fine-tuned.

The district was split in two areas marked off by a seven-km circle around the power-plant burning coal with high arsenic content. This circle was established using biological monitoring of human exposure within the particular locality. The exposure rates were established by the analysis of hair and urine samples for arsenic content.

2.2. Exposure assessment

To describe the human exposure in environmental settings arsenic determination was carried out on hair, urine, and blood samples taken from groups of 10-year-old boys, each group numbering 20–25 individuals, residing in the region polluted by arsenic. The samples were taken from the boys living at various residential distances up to approximately 30 km away from the source of emissions. In all the materials examined, elevated concentrations of arsenic were found. On the basis of the results obtained, the most advantageous material for the estimation of non-occupational exposure and especially to demonstrate environmental pollution seems to be hair, in spite of some problems with the decontamination procedure involved. The results corresponded to the theoretical ideas on spreading of emissions from elevated sources in the open air including a sophisticated modeling procedures developed later on (Nieuwenhuijsen et al., 2001) and tend to establish the applicability of arsenic determination in the hair as suitable means for monitoring exposure to arsenic. Considerable variability among individual arsenic values in the hair makes group examination a necessity (Bencko, 1995). The same applies to the blood and urine sampling, which is complicated by several technical difficulties concerning sampling and storage of the collected samples. Levels in urine reflect the quantities of arsenic inhaled or ingested after their absorption into the blood, and give a more realistic picture of possible total daily intake during the last 48–72 h.

The criterion for higher exposure was arsenic content exceeding, on the arithmetic mean, hair concentrations of 3 μg/g of arsenic. About two-tenths of the study population lives within a 7-km radius of the exposed region. Values up to 1 μg/g are considered normal. For example, the population in Prague showed approximately 0.2 μg/g, which is less than one-tenth of the mean value, which predominated in this previously heavily emission-loaded area near Prievidza (Fabiánová and Bencko, 1995).

3. Results & discussion

Preliminary analysis of our database of NMSC each confirmed histologically suggests a significant increase in skin basalioma cancer incidence in the most polluted part of the district compared with the data relevant for the rest of the district during the first 5-year period (Tables 1 and 2). The difference gradually disappeared and the last 5-year interval showed a reverse situation. The only difference is that neither is of statistical significance. Development of NMSC incidence during the first three 5-year intervals corresponds with a gradual decrease of arsenic emissions from power plant due to installation of the more effective electrostatic precipitators removing a solid phase of emissions and building of higher stacks-spreading emissions to more distant places of the district. A gradual decrease in NMSC incidence in the immediate vicinity of the power plant followed after decrease in emissions (see Figure 1) and so is dose-dependent. So it is as well the increase of NMSC incidence in originally less polluted rest of the district. In the less polluted part of the district, to collect a dose necessary to trigger blastic transformation of skin cells needed a longer exposure. The same pattern of development is in both sexes, men and women. The incidence of NMSC is markedly influenced by exposure to arsenic in occupational settings as can been seen from Table 3. Measurements, conducted quite recently, have revealed that the significantly increased arsenic concentrations exceeding the established hygienic limit (MAC) values for arsenic in occupational settings occur mainly during boiler-cleaning operations. Considering, however, the relatively long period of latency, so frequently described in arsenic-caused cancers, we may assume that the changed tumor mortality pattern was a result of arsenic exposures during the years characterized by the much less favorable hygienic conditions at the workplaces from the late 1950s to the mid-1970s.

As a result of radical reduction of emissions the main interests now are focused on the late effects of the previous occupational and environmental exposure to arsenic in the former heavily polluted region (Bencko et al., 1980; Fabiánová et al., 1994; Fabiánová and Bencko, 1995; Buchancová et al., 1998).

3.1. Confounding

Potential confounding was recently described in the paper by Pesch et al. (2002). Tobacco smoking is not an established NMSC risk factor, and there are conflicting results in the literature (Lear et al., 1998; Chuang and Brashear, 1999; van Dam et al., 1999; Karagas et al., 2001). A real potential confounding in our case can be UV radiation (Rossman, 1999; Seidl et al., 2001) due to quite exceptional climatic situation in the Upper Nitra valley open to southwest and representing one of the sunniest regions of Slovakia. But it should be stressed, that the hours of sunshine are the same in both the more exposed part of the district and in the remainder.

Table 1. NMSC incidence in population living in the vicinity of the power-plant burning the coal of high arsenic content and in the rest of the district (males only).

	1977–1981		1982–1986		1987–1991		1992–1996	
	Exp. cases (p-years)	Non-exp. cases (p-years)	Exp. cases (p-years)	Non-exp. cases (p-years)	Exp. cases (p-years)	Non-exp. cases (p-years)	Exp. cases (p-years)	Non-exp. cases (p-years)
Absolute number	44	125	32	134	30	142	27	222
Expected number	23.8		20.8		18.6		24.7	
Non-standardized rate	98.4	45.8	77.6	46.5	81.9	46.9	78.2	70.9
Age standardized rate	93.9	45.9	66.9	45.9	65.2	46.0	57.8	65.7
Person/year	(44,730)	(273,205)	(41,249)	(288,368)	(36,649)	(303,029)	(34,507)	(313,087)
Statistical parameters (confidence interval (p = 0.05))		Min–max		Min–max		Min–max		Min–max
SMR	2.05		1.46		1.42		0.88	
Mantel–Haenszel estimate	2.02	1.43–2.85	1.46	0.99–2.15	1.39	0.94–2.05	0.88	0.59–1.31
Chi-square	16.62		3.70		2.66		0.38	
Probability	<0.01	S	0.05	S	0.10	NS	0.54	NS

Table 2. NMSC incidence in population living in the vicinity of the power-plant burning the coal of high arsenic content and in the rest of the district (females only).

	1977–1981		1982–1986		1987–1991		1992–1996	
	Exp. cases (p-years)	Non-exp. cases (p-years)	Exp. cases (p-years)	Non-exp. cases (p-years)	Exp. cases (p-years)	Non-exp. cases (p-years)	Exp. cases (p-years)	Non-exp. cases (p-years)
Absolute number	46	118	32	134	22	165	25	205
Expected number	22.7		20.1		20.1		23.4	
Non-standardized rate	104.9	43.3	78.3	46.3	59.7	53.9	70.9	65.9
Age standardized rate	81.4	34.6	54.4	37.4	39.8	42.7	37.5	56.1
Person/year	(43,869)	(272,729)	(40,869)	(289,593)	(36,870)	(306,319)	(35,263)	(311,304)
Statistical parameters (confidence interval (p = 0.05))		Min–max		Min–max		Min–max		Min–max
SMR	2.35		1.45		0.93		0.67	
Mantel–Haenszel estimate	2.25	2.23–2.27	1.47	1.45–1.48	0.88	0.88–0.90	0.70	0.46–1.06
Chi-square	21.20		3.84		0.29		2.89	
Probability	<0.01	S	0.05	S	0.59	NS	0.09	NS

Table 3. NMSC incidence in workers of the power-plant burning the coal of high arsenic content (ENO) and in the rest of the district (males only).

	1977–1981		1982–1986		1987–1991		1992–1996	
	Exp. cases (p-years)	Non-exp. cases (p-years)	Exp. cases (p-years)	Non-exp. cases (p-years)	Exp. cases (p-years)	Non-exp. cases (p-years)	Exp. cases (p-years)	Non-exp. cases (p-years)
Absolute number	3	166	7	159	9	163	11	238
Expected number	3.5		3.7		3.7		4.5	
Non-standardized rate	45.0	53.3	95.0	49.3	123.6	49.0	173.3	69.7
Age standardized rate	44.6	68.4	355.5	61.0	10,317	59.7	7 357	63.8
Person/year	(6,667)	(311,268)	(7,371)	(322,246)	(7,284)	(332,394)	(6,302)	(341,209)

Statistical parameters (confidence interval (p = 0.05))

		Min–max		Min–max		Min–max		Min–max
SMR	0.65		5.83		172.00		4.38	
Mantel–Haenszel estimate	0.83	0.26–2.62	2.70	1.27–5.75	3.55	1.81–6.95	4.49	2.47–8.14
Chi-square	0.10		7.16		15.60		29.28	
Probability	0.75	NS	0.01	S	<0.01	S	<0.01	S

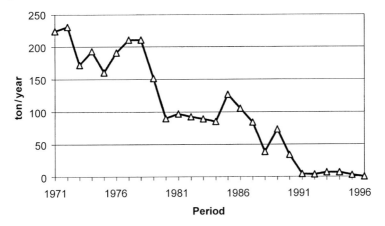

Figure 1. Arsenic in power plant emissions per year.

The potential difference in real exposure to UV radiation could arise from different lifestyles of villages and town dwellers (e.g. gardening), but both lifestyle groups are proportionally represented in both exposure groups.

4. Conclusion

Currently, we are performing a meta-analysis of the database on the malignant tumors obtained by the population-based epidemiological study within EXPASCAN, INCO-COPERNICUS project. Future analysis will include all types of malignancies, lung carcinoma which has already been associated with arsenic exposure (ATSDR, 1999) along with exposure assessment due to cigarette smoking carefully registered in our study. The main objective of our present activity is to obtain as precise as possible arsenic exposure in skin basalioma cases especially in occupationally exposed ones in collaboration with the Slovak National Cancer Registry database (Pleško et al., 2000) and the district and regional Institutes of Public Health in Prievidza and Banská Bystrica, respectively.

Acknowledgements

This study was supported by EC INCO COPERNICUS EXPASCAN grant ERBBIC 15 CT98-0325. Statistical analysis was made in collaboration with EuroMISE Center Cardio, supported by project LN00B107 Ministry of Education of the Czech Republic.

References

ATSDR, 1998. Toxicological profile for arsenic. US Department of Health and Human Services, p. 341.

Bencko, V., 1995. Use of human hair as a biomarker in the assessment of exposure to pollutants in occupational and environmental settings. Toxicology, 101, 29–39.

Bencko, V., 1997. Health aspects of burning coal with a high arsenic content: the Central Slovakia experience. In: Abernathy, C.O., Calderon, R.L., Chappell, W.R. (Eds), Arsenic, Exposure and Health Effects. Chapman & Hall, New York, pp. 84–92.

Bencko, V., Symon, K., Štálnik, L., Bátora, J., Vančo, E., Šrandová, E., 1980. Rate of malignant tumor mortality among coal burning power plant workers occupationally exposed to arsenic. J. Hyg. Epidemiol. (Praha), 24 (3), 278–284.

Bencko, V., Rameš, J., Götzl, M., 1999. Preliminary incidence analysis in skin basalioma patients exposed to arsenic in environmental and occupational settings. In: Abernathy, C.O., Calderon, R.L., Chappell, W.R. (Eds), Arsenic Exposure and Health Effects. Elsevier, pp. 201–206.

Buchancová, J., Klimentová, G., Kniková, M., Meško, D., Gáliková, E., Kubík, J., Fabiánová, E., Jakubis, M., 1998. A health status of workers of a thermal power station exposed for prolonged periods to arsenic and other elements from fuel. Cent. Eur. J. Public Health, 6, 29–36.

Chuang, T.Y., Brashear, R., 1999. Risk factors of non-melanoma skin cancer in United States veterans patients: a pilot study and review of literature. J. Eur. Acad. Dermatol. Venereol., 12, 126–132.

Fabiánová, E., Bencko, V., 1995. Central European study on health impact of environmental pollution. Final report, project PHARE EC/91/HEA/18. Brussels, Belgium, European Union, 1995.

Fabiánová, E., Hettychová, L'., Hrubá, F., 1994. Occupational exposure assessment and bioavailability of arsenic. Final report. EPRI Research Agreement RP 3370-12, pp. 106.

Karagas, M.R., Stukel, T.A., Morris, J.S., Tosteson, T.D., Weiss, J.E., Spencer, S.K., Greenberg, E.R., 2001. Skin cancer risk in relation to toenail arsenic concentrations in a US population-based case–control study. Am. J. Epidemiol., 153 (6), 559–565.

Lear, J.T., Tan, B.B., Smith, A.G., Jones, P.W., Heagerty, A.H., Strange, R.C., Fryer, A.A., 1998. A comparison of risk factors for malignant melanoma, squamous cell carcinoma and basal cell carcinoma in the UK. Int. J. Clin. Pract., 52, 145–149.

Nieuwenhuijsen, M.J., Rautiu, R., Ranft, U., Bencko, V. Cordos, E., 2001. Exposure to arsenic and cancer risk in central and east Europe. Final report, project EXPASCAN IC 15 CT98 0325. Brussels, Belgium, European Union, March 31.

Niu, S., Cao, S., Shen, E., 1997. The geochemistry of arsenic. In: Abernathy, C.O., Calderon, R.L., Chappell, W.R. (Eds), Arsenic, Exposure and Health Effects. Chapman & Hall, New York, pp. 78–83.

Pesch, B., Ranft, U., Jakubis, P., Nieuwenhuijsen, M.J., Hergemoller, A., Unfried, K., Jakubis, M., Miskovic, P., Keegan, T., 2002. Environmental arsenic exposure from a coal-burning power plant as a potential risk factor for non-melanoma skin carcinoma: results from a case–control study in the district of Prievidza, Slovakia. Am. J. Epidemiol., 155 (9), 798–809.

Pleško, I., Severi, G., Obšitníková, A., Boyle, P., 2000. Trends in the incidence of non-melanoma skin cancer in Slovakia, 1978–95. Neoplasma, 47 (3), 137–142.

Rossman, T.G., 1999. Arsenic genotoxicity may be mediated by interference with DNA damage-inducible signaling. In: Abernathy, C.O., Calderon, R.L., Chappell, W.R. (Eds), Arsenic Exposure and Health Effects. Elsevier, pp. 233–241.

Seidl, H., Kreimer-Erlacher, H., Back, B., Soyer, H.P., Hofler, G., Kerl, H., Wolf, P., 2001. Ultraviolet exposure as the main initiator of p53 mutations in basal cell carcinomas from psoralen and ultraviolet A-treated patients with psoriasis. J. Invest. Dermatol., 117 (2), 365–370.

Thornton, I., Farago, M., 1997. The geochemistry of arsenic. In: Abernathy, C.O., Calderon, R.L., Chappell, W.R. (Eds), Arsenic, Exposure and Health Effects. Chapman & Hall, New York, pp. 1–16.

van Dam, R.M., Huang, Z., Rimm, E.B., 1999. Risk factors for basal cell carcinoma of the skin in men: results from the Health Professionals Follow-up Study. Am. J. Epidemiol., 150, 459–468.

Arsenic Exposure and Health Effects V
W.R. Chappell, C.O. Abernathy, R.L. Calderon and D.J. Thomas, editors
© 2003 Elsevier B.V. All rights reserved.

Chapter 12

Effect of arsenic-contaminated drinking water on skin cancer prevalence in Wisconsin's Fox River Valley

Lynda Knobeloch and Henry Anderson

Abstract

Several townships in Wisconsin's Fox River Valley overlie an aquifer that is rich in arsenic-bearing sulfide minerals. More than 20,000 families in these townships obtain their drinking water from private wells that are at risk of arsenic contamination. In an effort to evaluate the scope of this problem and to assess the effect of chronic arsenic exposure on the development of non-melanoma skin cancer, 18 towns sponsored well water testing programs. As part of this effort, families were invited to submit a water sample for arsenic analysis and to complete a self-administered questionnaire that requested information about their residential history, water use habits and health. Arsenic exposure and health history information was provided by 2233 families comprising 6669 individuals between the ages of 0 and 100 years old. Analysis of data from this cohort indicates that long-term ingestion of water that contains more than 5 ug of arsenic per liter significantly increases the risk of skin cancer. A history of cigarette use was also associated with higher skin cancer rates and was additive to the carcinogenic effect of arsenic.

Keywords: arsenic, well, groundwater, skin cancer, Wisconsin, Outagamie, Winnebago, tobacco

1. Background

Inorganic arsenic is a well-known human poison that causes a wide array of adverse health effects. The World Health Organization (1996) and U.S. Environmental Protection Agency (1998) classify inorganic arsenic as a human carcinogen. Long-term ingestion of contaminated drinking water has been found to increase the risk of basal and squamous cell cancers of the skin. Arsenic exposure has also been linked to tumors of the bladder, kidney, liver and lung (Wu et al., 1989; Chen and Wang, 1990; Chiou et al., 1995; Tsuda et al., 1995). Exposure to inorganic arsenic may also cause thickening and discoloration of the skin (Tay, 1974; Saha, 1995; Mandal et al., 1996), nausea and diarrhea (Nevens et al., 1990), decreased production of blood cells (Eichner, 1984; Rezuke et al., 1991), abnormal heart rhythm (Glazener et al., 1968), blood vessel damage (Tseng et al., 1961) and numbness in the hands and feet (Murphy et al., 1981; Donofrio et al., 1987). Recent investigations have also linked arsenic exposure to the development of diabetes mellitus (Lai et al., 1994; Rahman and Axelson, 1995).

In 1987, a groundwater study conducted by the Department of Natural Resources identified a large vein of arsenic in a bedrock layer found at the interface of the St. Peter Sandstone and Sinnippee Dolomite. This formation stretches from southern Brown County into Outagamie and Winnebago Counties and lies beneath more than 20,000 private water supply wells (see map in Figure 1). Water samples collected from 1943 private wells between 1992 and 1993 contained arsenic concentrations that ranged from less than 2 to 12,000 ug/l. Arsenic levels in nearly 20% of these water supplies exceeded the new federal drinking water standard of 10 ug/l (Stoll et al.). Residents whose daily water intake provided a dose of 50 ug or more of arsenic were significantly more likely to report a diagnosis of skin cancer than others (Haupert et al., 1996).

Recent water quality testing seems to indicate that arsenic levels in this aquifer are increasing over time. Although the geochemistry is poorly understood, this trend appears to correlate with regional groundwater withdrawal and may involve the introduction of oxygen and chlorine disinfectants into the aquifer. Some experts have expressed concern that this trend could create a regional shortage of potable water.

One of the greatest challenges facing public health officials is convincing families to monitor arsenic levels in their private wells and to follow water use advisories. During 2000 and 2001, 19 townships sponsored water testing programs during which families were able to submit water samples for arsenic analysis at a reduced cost. This testing was tied to a health study conducted by the Wisconsin Department of Health and Family Services. Funding for this study was provided by the Wisconsin Groundwater Coordinating Council and the Wisconsin Department of Natural Resources. This chapter summarizes findings from that initiative.

2. Methods

Nineteen townships in the arsenic-impacted region sponsored water testing programs during which residents were encouraged to submit a well water sample for arsenic analysis and to complete a self-administered water use and family health questionnaire. Approximately one month after the water samples were collected, residents were invited to attend an informational meeting at their local town hall. Water test results were distributed immediately prior to the meeting. Each meeting included presentations on the hydrogeology of the region, well construction and abandonment, the health effects of inorganic arsenic, and household water treatment options.

2.1. Questionnaires

Information on residential history, drinking-water consumption, the use of a water treatment system and health was collected using a self-administered questionnaire that was distributed with the water sampling kits. Families were directed to complete the questionnaires and return them along with their well water samples.

2.2. Arsenic measurements

Well water samples were collected by homeowners and submitted for analysis between July 2000 and January 2002. All samples were acidified prior to analysis and were unfiltered due to concerns that iron-bound arsenic would be removed by filtration. Total arsenic was determined using graphite furnace atomic absorption. The limit of detection (LOD) and limit of quantitation (LOQ) were 1 and 2.4 ug/l, respectively.

2.3. Estimates of arsenic exposure

Arsenic exposures were estimated using the concentration of arsenic in the household water supply, the average daily water intake and the length of exposure in years. Daily intakes and cumulative exposures, expressed as ug/day and ug/l-yr, were developed using the water arsenic concentration and either the number of glasses of water consumed per day or the number of years that each individual had consumed water from their well.

3. Results

3.1. Cohort characteristics

2233 families comprising a total of 6669 individuals returned self-administered water use and health surveys. Nineteen townships participated in this study. The number of participants and the mean and maximum arsenic levels for each township are listed in Table 1.

Table 1. Summary of arsenic levels and participation by township.

Township	No. of participants	Mean arsenic level (ug/l)	Maximum arsenic level (ug/l)	No. of wells tested (% >10 ug/l)
Algoma	1384	29.9	3100	456 (34)
Black Creek	311	7.6	76	101 (24)
Black Wolf	28	1.4	3	12 (0)
Bovina	98	3.3	15	36 (6)
Center	697	12.9	233	237 (30)
Cicero	170	4.6	58	51 (14)
Clayton	657	11.7	266	219 (26)
Ellington	413	1.5	18	141 (4)
Freedom	584	5.0	240	167 (11)
Grand Chute	416	2.3	66	138 (6)
Greenville	450	4.4	79	142 (9)
Maple Creek	81	4.3	14	27 (15)
Omro	135	6.8	41	61 (20)
Osborn	346	5.6	142	108 (10)
Seymour	275	18.9	1300	93 (17)
Utica	110	8.7	72	41 (25)
Vinland	219	10.4	327	88 (27)
Winchester	66	2.8	34	23 (9)
Winneconne	224	3.6	46	92 (11)
Total	6669	10.9	3100	2233 (20)

Table 2. Distribution of drinking water levels among survey participants.

Arsenic level (ug/l)	No. of people (%)	No. of aged <6 yr (%)	No. of aged >64 yrs (%)	No. of households (%)
<1.0	2728 (40.9)	209 (39.6)	335 (44.5)	920 (41.2)
1–4.9	1565 (23.5)	115 (21.8)	187 (24.9)	544 (24.4)
5–9.9	988 (14.8)	81 (15.3)	97 (12.9)	322 (14.4)
10–19.9	645 (9.7)	42 (8.0)	82 (10.9)	208 (9.3)
>19.9	743 (11.1)	75 (14.2)	51 (6.8)	239 (10.7)
Total	6669 (100)	528 (100)	752 (100)	2233 (100)

3.2. Exposure assessment

Many study participants had very little exposure to arsenic-contaminated water and are at low risk for arsenic-related health problems (see Table 2). More than 40% of the wells that were sampled provided water that contained no detectable arsenic (<1.0 ug/l). An almost equal number had an arsenic level between 1 and 9.9 ug/l. Approximately 20% of the samples had an arsenic level of 10 ug/l or higher. Slightly more than 10% of families consumed water that had an arsenic level greater than 20 ug/l. This level is twice the current federal standard. Preschool-aged children living in these communities were twice as likely to consume water that had an arsenic concentration above 20 ug/l than people over the age of 65 years (14.2 vs. 6.8%). Because of their rapid growth and developing nervous and endocrine systems, young children are potentially more susceptible to the toxic effects of arsenic than adults.

Approximately half of the participants in this study (2940 of 6669) had consumed their well water for at least 10 years (Table 3). Only 408 residents over the age of 34 reported more than 9 years of exposure to well water that had an arsenic level of ≥10 ug/l.

4. Long-term arsenic exposure estimates

Lifetime arsenic exposure categories displayed in Table 4 were developed by multiplying the concentration of arsenic in the well water by the number of years an individual had

Table 3. Distribution of drinking water exposure times among respondents.

Years of water use	No. of people	(%)	No. with As ≥10 ug/l	No. with As ≥20 ug/l	No. with As ≥50 ug/l	No. with As ≥100 ug/l
<2	722	10.8	150	82	29	9
2–9	2999	45.0	655	293	108	30
10–19	1617	24.2	348	187	47	10
20+	1323	19.8	230	68	30	8
Total	6669	100	1398	650	264	157

Table 4. Distribution of lifetime arsenic exposure estimates.

Arsenic level (ug/l-yr)	No. of residents	No. of aged <18 yr	No. of adults
<10	3415	1029	2386
10–100	1949	572	1376
>100	1305	230	1060
>200	740	117	623
>300	469	74	395
>500	222	33	189
>1000	89	11	78

consumed the water. This method provides an exposure estimate with the unit ug/l-yr. It is not a quantitative dose estimate, which would be more appropriately expressed, in total milligrams of arsenic per kilogram of body weight. The ug/l-yr estimates are one method that can be used to place research subjects into low, medium and high exposure cohorts based on their long-term exposure potentials.

4.1. Skin cancer incidence

One hundred fifteen individuals who ranged in age from 35 to 96 years reported a previous diagnosis of non-melanoma skin cancer. The overall skin cancer prevalence rate for residents aged 35 years and over was 3.00%. Reporting rates varied within this cohort and were significantly higher among those aged 65 and older, men and cigarette smokers (see Table 5).

In an effort to evaluate the effect of chronic arsenic ingestion on skin cancer occurrence, rates were calculated for adults aged 35 years and over who had consumed their well water for 10 years or longer. Because cigarette use and age were strongly associated with the risk of skin cancer, the cohort was subdivided by age group and smoking status.

Table 5. Skin cancer rates among adults aged 35 years or more.

Group	Cases/population	Rate	OR (95% CI)	Mean age (years)
All adults	115/3828	3.00	NA	53.4
Ages 35–64	55/3076	1.79	1.00	48.3
Ages 65 and above	60/752	7.98	4.74 (3.21–7.01)	73.4
Women	42/1893	2.22	1.00	53.1
Men	73/1935	3.77	1.73 (1.16–2.59)	53.6
Non-smokers	64/2662	2.40	1.00	52.7
Cigarette smokers	51/1166	4.37	1.86 (1.26–2.74)	54.8

4.2. Ages 40–64 years

Non-smokers between the ages of 40 and 64 were 54% more likely to report a diagnosis of skin cancer if their water contained at least 5 ug arsenic per liter; however, this difference was not statistically significant. Smokers whose water was low in arsenic were not significantly more likely to report a diagnosis of skin cancer than non-smokers. However, smokers who consumed water that contained more than 5 ug arsenic per liter reported the highest rate of skin cancers among this age group. The prevalence among this subgroup was nearly twice the rate reported by non-smokers (2.88 vs. 1.66%) (see Table 6).

4.3. Ages 65 years and over

Among residents aged 65 years or more, cigarette use and arsenic-contaminated water were both associated with a higher prevalence of skin cancer. The rate among cigarette users was nearly twice as high as the rate among non-smokers (11.0 vs. 5.8%) (see Table 6).

Cigarette smoking appeared to synergize the effect of long-term exposure to arsenic-contaminated water. Nearly one in five smokers in this age group who had long-term exposure to arsenic-contaminated well water reported a diagnosis of skin cancer. This prevalence rate was more than five times higher than that observed among age-matched residents with neither risk factor.

In an effort to further elucidate the effects of arsenic-contaminated water on skin cancer prevalence, exposures were estimated using drinking water arsenic concentrations, daily arsenic doses and cumulative lifetime doses (see Table 7). For the purposes of this analysis, the cohort was limited to adults aged 50 and older who had consumed their well water for at least 10 years. Each method of estimation revealed a dose-related increase in skin cancer rates with the effect being strongest among smokers. Regardless of the method used to estimate dose, skin cancer prevalence was greatest among smokers who drank water that had an arsenic level of at least 5 ug/l. Using the drinking water arsenic level as

Table 6. Skin cancer rates vs. arsenic concentration, age and cigarette use.

Age group	Cigarette user	As concentration (ug/l)	Skin cancer prevalence	Rate per 100	OR (95% CI)
40–64 yr	No	<5	10/601	1.66	1.00
	No	≥5	9/354	2.54	1.54 (0.57–4.15)
	Yes	<5	5/284	1.76	1.06 (0.31–3.40)
	Yes	≥5	4/139	2.88	1.75 (0.46–6.18)
≥65 yr	No	<5	13/283	4.59	1.00
	No	≥5	10/115	8.69	1.98 (0.78–4.99)
	Yes	<5	10/135	7.41	1.66 (0.66–4.18)
	Yes	≥5	12/64	18.75	4.79 (1.92–11.96)

Analysis restricted to residents who had consumed their water for 10 years or longer.

Table 7. Risk ratios and 95% confidence intervals for skin cancer and arsenic exposure.

Arsenic exposure estimate	Cigarette user	No. (rate per 100)	OR (95% CI)
Concentration in water (ug/l)			
< 1.0	No	12/394 (3.04)	1.00
1–4.9	No	8/211 (3.79)	1.28 (0.53–3.09)
≥ 5	No	8/203 (3.94)	1.66 (0.80–3.46)
< 1.0	Yes	7/119 (5.88)	1.25 (0.46–3.36)
1–4.9	Yes	16/310 (5.16)	1.99 (0.69–5.59)
≥ 5	Yes	15/162 (9.26)	3.25 (1.40–7.60)
Cumulative dose (ug/l-yr)			
< 1.0	No	12/396 (3.03)	1.00
1–199	No	13/293 (4.44)	1.46 (0.43–3.05)
≥ 200	No	11/218 (5.04)	1.66 (1.12–6.44)
< 1.0	Yes	8/212 (3.77)	1.16 (0.68–3.16)
1–199	Yes	13/168 (7.74)	2.68 (0.75–3.71)
≥ 200	Yes	9/112 (8.03)	2.80 (1.05–7.35)

Analysis restricted to residents aged ≥ 50 years who had consumed their water for ≥ 10 years.

the exposure estimate, the crude odds ratio for this group was 3.25 which was higher than the combined odds ratios for arsenic and smoking of 1.91. This synergistic effect was seen, regardless of the method used to estimate long-term arsenic exposure.

5. Discussion

This report summarizes arsenic exposure and skin cancer incidence among nearly 7000 Wisconsin residents whose drinking water is supplied by private wells that are located in an area affected by arsenic-contaminated groundwater. Information about well construction, water use and skin cancer incidence was obtained from self-completed questionnaires. This research study is the largest epidemiological study that has ever been conducted in Wisconsin related to contaminated drinking water. The Winnebago and Outagamie county health departments and township officials coordinated the distribution and collection of water sample kits and questionnaires.

Major strengths of this study are the large cohort size, the availability of current arsenic measurements on each water supply and detailed information about exposure times. The use of a self-administered water use and health questionnaire has the advantage of collecting personal information that is not available on medical records. However, self-administered questionnaires are also subject to recall errors and are sometimes incomplete. In addition, residents who were concerned about their water supply or had health problems are sometimes more motivated to participate in the survey than others. It seems unlikely that these methodological weaknesses could explain the association that was observed between long-term arsenic exposure and cigarette smoking with skin cancer prevalence, however.

Our findings are consistent with previous reports that long-term exposure to inorganic arsenic increases the risk of non-melanoma skin cancer. The observation that this effect is more pronounced among current and past cigarette smokers is novel and merits additional investigation. The use of tobacco products has been shown to increase the risk of squamous cell carcinoma. In a recent study conducted by the Leiden University Medical Center, the relative risk of squamous cell carcinoma was 2.0 among smokers vs. non-smokers after adjustment for age, sex and sun exposure (De Hertog et al., 2001). Tobacco use was not associated with a higher risk of basal cell carcinoma, however. Our study did not involve a review of medical records and many residents who reported skin cancer diagnoses were unsure whether they were treated for basal or squamous cell carcinoma. Of 27 individuals who were able to provide information on the type of skin cancer they were treated for, 17 listed basal cell tumors, 7 listed basal and squamous cell tumors and 3 listed squamous cell tumors.

We hope to continue to monitor the incidence of skin cancer in the arsenic-impacted communities of Wisconsin's Fox River Valley. The families that participated in this study provide an ideal cohort for follow-up because their exposure to arsenic is already well documented. Future studies should include a detailed assessment of tobacco use in addition to a comprehensive assessment of exposure to inorganic arsenic exposure from water and other sources, such as arsenic-containing wood preservatives and seafood.

References

Chen, C.J., Wang, C.J., 1990. Ecological correlation between arsenic level in well water and age-adjusted mortality from malignant neoplasms. Cancer Res., 50, 5470–5474.

Chiou, H.Y., Hsueh, Y.M., et al., 1995. Incidence of internal cancers and ingested inorganic arsenic: a seven year follow-up study in Taiwan. Cancer Res., 55, 1296–1300.

De Hertog, S.A., Wensveen, C.A., Bastiaens, M.T., Kielich, C.J., Berkhout, M.J., Westendorp, R.G., Vermeer, B.J., Bouwes Bavinck, J.N., 2001. Relation between smoking and skin cancer. J. Clin. Oncol., 19 (1), 231–238.

Donofrio, P.D., Wilbourn, A.J., et al., 1987. Acute arsenic intoxication presenting a Gailain-Barre like syndrome. Muscle Nerve, 10, 114–120.

Eichner, E.R., 1984. Erythroid karyorrhexis in the peripheral blood smear in severe arsenic poisoning: a comparison with lead poisoning. Am. J. Clin. Pathol., 81, 533–537.

Glazener, F.S., Ellix, J.G., Johnson, P.K., 1968. Electrocardiographic findings with arsenic poisoning. Calif. Med., 109, 158–162.

Haupert, T., Wiersma, J.H., Goldring, J., 1996. Health effects of ingesting arsenic-contaminated groundwater. Wisc. Med. J.,.

Lai, M.S., Hsueh, Y.M., et al., 1994. Ingested inorganic arsenic and prevalence of diabetes mellitus. Am. J. Epidemiol., 139, 484–492.

Mandal, B.K., Chowdury, T.R., et al., 1996. Arsenic in groundwater in seven districts of West Bengal, India – the biggest arsenic calamity in the world. Curr. Sci., 70, 976–986.

Murphy, M.J., Lyon, L.W., Taylor, J.W., 1981. Subacute arsenic neuropathy: clinical and electrophysiological observations. J. Neurol. Neurosurg. Psychiatry, 44, 896–900.

Nevens, F., Fevery, J., et al., 1990. Arsenic and non-cirrhotic portal hypertension: a report of eight cases. J. Hepatol., 11, 80–85.

Rahman, M., Axelson, O., 1995. Diabetes mellitus and arsenic exposure: a second look at case-control data from a Swedish copper smelter. Occup. Environ. Med., 52, 773–774.

Rezuke, W.N., Anderson, C., et al., 1991. Arsenic intoxication presenting as a myelodysplastic syndrome: a case report. Am. J. Hematol., 36, 291–293.

Saha, K.C., 1995. Chronic arsenical dermatoses from tube-well water in West Bengal during 1983–1987. Indian J. Dermatol., 40, 1–12.

Stoll, R., Burkell, R., La Plant, N. Naturally-occurring arsenic in sandstone aquifer water supply wells of NE Wisconsin. WI Groundwater Research and Monitoring Project Summaries, WI DNR PUBL-WR-423-95.

Tay, C.H., 1974. Cutaneous manifestations of arsenic poisoning due to certain Chinese herbal medicines. Australas. J. Dermatol., 15 (3), 121–131.

Tseng, W.P., Chen, W.Y., et al., 1961. A clinical study of blackfoot disease in Taiwan: an epidemic of peripheral vascular disease. Mem. Coll. Med. Natl Taiwan Univ., 7, 1–18.

Tsuda, T., Nagira, T., et al., 1995. Ingested arsenic and internal cancers: a historical cohort study followed for 33 years. Am. J. Epidemiol., 141, 198–209.

U.S. Environmental Protection Agency, 1998. IRIS Document for Arsenic, updated 1998.

World Health Organization, 1996. WHO Guidelines for Drinking Water Quality. Vol. 2 – Health Criteria and Other Supporting Information (ISBN 92 4 154480 5).

Wu, M.M., Kuo, T.L., Hwang, Y.H., Chen, C.J., 1989. Dose–response relation between arsenic concentration in well water and mortality from cancers and vascular diseases. Am. J. Epidemiol., 130, 1123–1132.

PART III
BIOMARKERS AND ANIMAL MODELS

Arsenic Exposure and Health Effects V
W.R. Chappell, C.O. Abernathy, R.L. Calderon and D.J. Thomas, editors
© 2003 Elsevier B.V. All rights reserved.

Chapter 13

Alteration of GSH level, gene expression and cell transformation in NIH3T3 cells by chronic exposure to low dose of arsenic

Yu Hu, Ximei Jin, Guoquan Wang and Elizabeth T. Snow

Abstract

It is well established that arsenic toxicity is postulated to be primarily due to the binding of As(III) to sulfhydryl-containing enzymes. However, the mechanism of carcinogenesis induced by arsenic is still unclear. The interaction of arsenic with GSH and related enzymes seems a very important issue regarding mechanism of arsenical induced toxicity or carcinogenesis. The purpose of this work is to investigate the effect of chronic exposure to low dose of As(III) on GSH level, gene expression and cell transformation in NIH3T3 cells. The results showed that long-term, low dose arsenic treatment makes 3T3 cell more resistant to acute arsenic treatment. There were morphology changes after long-term arsenic treatment. First, partially immortalized 3T3 cell became immortalized. In addition, the cells were doubling more quickly than the control cells and attained higher density than the control cells at confluence. Second, cells treated with 0.1 μM As(III) exhibited anchorage-independent growth. Arsenic could enhance GSH level at 0.5–10 μM dose of arsenic in 24 h treatment and decrease it at 25 μM and above. In long-term treatment with low dose of arsenic, GSH levels were decreased. As(III) can increase both glutathione S-transferase (GST) and glutathione reductase (GR) activities at low dose (0.5–10 M), but decreased GST and GR activities at 25 M and higher dose of arsenic, while in long-term As(III) treatment, GST and GR activities are increased. Both long-term and short-term treatments with As(III) can induce GR gene expression. GPx mRNA levels were decreased both in acute and chronic arsenic-treated cells. Chronic treatment with As(III) also decreased the p53 mRNA level. Taken together, our results suggest that As(III) can alter GST, GR enzyme activities as well as GSH level and related gene expression both in long-term and short-term treatment but in a different manner in different doses. Alteration of cellular GSH level by As(III) might play an important role in gene expression and arsenic-induced cell transformation.

Keywords: trivalent arsenic [As(III)], glutathione (GSH), gene expression, cells transformation, carcinogenesis

1. Introduction

Arsenic is associated with the occurrence of adverse health effects, including skin alterations, peripheral vascular disease and cancer, in humans exposed by drinking water as well as by occupational exposure (Sommers and Mcmanus, 1953; Hu et al., 1988; Snow, 1992; Rossman, 1998; Abernathy et al., 1999). However, the mechanism of carcinogenesis induced by arsenic is still unclear. It appears that humans are particularly sensitive to arsenic-induced malignancies, at least when compared with rodent species. The lack of knowledge of carcinogenic mechanism of action together with the apparent sensitivity of human populations creates even more concern for the adverse potential of this important environmental pollutant. It is well established that arsenic toxicity is postulated to be primarily due to the binding of As(III) to sulfhydryl-containing enzymes (Leonard and Lauwerys, 1980). Glutathione (L-γ-glutamyl-cysteiny-glycine or GSH), which is present in most cells at millimolar concentration, is one of the most important intracellular antioxidants (Meister, 1988) as well as potentially a target for arsenic (Kalyanaraman, 1995). The role of GSH in the protection of cells from various stressors, including oxidants and metals, is well documented (Arrick, 1982). Numerous toxic or potentially toxic compounds are taken up or removed from the cell by GSH-mediated pathways (Ballatori, 1994). It has been suggested that GSH-dependent methylation and protein binding are the major mechanisms for arsenic detoxification by mammals and other organisms (Aposhian, 1997). GSH is also essential in the maintenance of introcellular redox balance and the essential thiol status of proteins (Deleve and Kaplowitz, 1991).

Glutathione related enzymes such as glutathione reductase (GR) and glutathione peroxidase (GPx) function either directly or indirectly as antioxidants, and glutathione S-transferase (GST) plays an important role in metabolic detoxification. Reduction of GSSG is catalyzed by GR which uses NAPDH to provide reducing potential (Kirkman et al., 1987; Meister, 1994). Endogenous hydrogen peroxide (H_2O_2) is removed by GSH in the presence of GPx and can also be removed by catalase in the peroxisomes. GPx prevents the formation of reactive species (ROS) by catalyzing the reduction of hydroperoxides, such as H_2O_2, to their metabolizable hydroxyl forms with the concomitant formation of oxidized glutathione (GSSG) (Wendel, 1980; Pickett and Lu, 1989). GST catalyzes the reaction between GSH and electrophilic substrates, which is an integral step in the detoxification and elimination of diverse classes of toxic chemical compounds and thus prevents toxic injury in several tissues (Keen et al., 1976; Habig and Jakoby, 1981; Pickett and Lu, 1989; Mari and Cederbaum, 2000). Either GPx or GST can reduce organic hydroperoxides in the presence of GSH. Glutathione-metal conjugates are redox active and can provide a very important pathway for directed free radical formation and DNA damaging potential (Klein et al., 1998). Alteration of GSH level was also reported to be a key signal for activating gene expression. Increasing evidence shows that arsenic could promote oxidative stress and in turn causes cellular injury and DNA damage (Snow, 1992; Guyton et al., 1996; Wang and Rossman, 1996; Klein et al., 1998). The interaction of arsenic with GSH and related enzymes seems a very important issue regarding mechanism of arsenical induced toxicity or carcinogenesis. Our previous study showed that GSH-related enzymes, such as GR, GST and GPx, seem not to be direct targets of arsenic (Chouchane and Snow, 2001). The purpose of this work is to investigate the effect of As(III) on GSH level and gene expression both in short-term and long-term treatment in 3T3 cells and its possible role in cell transformation.

2. Material and methods

2.1. Chemicals

Sodium arsenite ($NaAsO_2$), sodium phosphate (dibasic and monobasic), EDTA (sodium salt), sodium chloride, lauryl sulfate (SDS; sodium dodecylsulfate), agarose, sodium acetate, glycine, ammonium peroxydisulfate, trizma base, sodium citrate, mopes, hepes, neutral red, polyoxyethylene-sorbitan monolaurate (tween 20) were purchased from Sigma Chemical Co. (St. Louis, MO). 5,5′dithiol-bis (2-nitrobenzoic acid) (DTNB), NAPDH, glutathione (reduced form), glutathione (oxidized form) and GR were purchased from Boehringer Mannheim. GR and glutathione peroxidase primers were purchased from Sigma Chemical Co. (St. Louis, MO).

2.2. Cell culture and treatment

Partially immortalized NIH3T3 mice fibroblast cells were treated with sodium arsenite for short term (0.1–25 μM, 3–24 h) as well as long term (0.1 and 0.5 μM, 56–133 days) in DMEM medium containing 10% fetal bovine serum (FBS), 100 IU/ml penicillin at 37°C with 5% CO_2 and 95% O_2. After arsenic exposure, cells were challenged with or without TPA, H_2O_2 for 3 h. Cell extracts were obtained after each treatment time point for activity assay.

2.3. Cellular toxicity assay

Cellular toxicity was measured by using neutral red uptake assay. A minor modification of the neutral red absorption spectrophotometric test (Little et al., 1996) was used to measure cell viability. Briefly, neutral red dye was dissolved in complete tissue culture medium at a concentration of 50 μg/ml and 200 μl was added to each well. Following a 3 h incubation (37°C, 5% CO_2, 95% O_2) the medium was removed and cells were washed twice (200 μl/well) with 1% paraformadehyde–10% $CaCl_2$. Neutral red dye taken up by viable cells was extracted by addition of 1% ethanoic acid in 50% ethanol. Optical density (OD) was measured at 570 nm with a microplate reader (BioRad).

2.4. Assay for anchorage-independent growth

1×10^4 AS(III)-treated cells were plated in 4 ml of 0.3% agar in DMEM with 10% FBS overlaid onto a solid layer of 0.6% agar in DMEM with 10 FBS. The culture was maintained for 3–4 weeks and colonies were counted. The plating efficiency in soft agar was determined relative to the total number of cells plated.

2.5. Glutathione assay

Glutathione concentration was determined by using the method of Clark (Clark et al., 1996). Briefly, aliquots (10 μl) of glutathione standard were added to a 96-well microtiter plate in a duplicate column down the plate. Samples extracted in PBS were aliquoted in the same manner. A solution of 8.14 ml NAPDH (0.5 mg/ml) buffer [71.5 mM sodium phosphate (dibasic) 71.5 mM sodium phosphate (monobasic) and 6.3 mM EDTA], 1.969 ml DTNB (2.4 mg/ml) and 8.014 ml distilled water was prepared immediately before use, mixed and poured into a dispensing trough. 165 μl aliquot of the mixture was added to each well. The plate was then incubated at room temperature for 20 min prior to the addition of 40 μl GR per well. The final concentration of each component in the reaction was 0.22 mM NAPDH, 0.53 mM DTNB and 0.0027 U/μl GR, respectively, in a total volume of 215 μl per well. The plate was shaken for 10 s and read every 30 s at 405 nm for 3 min at room temperature. Raw data were expressed as the mean change in OD per minute. Protein concentration was measured by using a BioRad Kit and the final results were adjusted to *n* mole GSH per μg protein.

3. Enzyme activity assay

3.1. Glutathione S-transferase assay

GST activity was determined spectrophotometrically at 340 nm by assessing the formation of conjugated glutathione and 1-chloro-2,4-dinitrobenzene (CDNB) (Habig and Jakoby, 1981). 20 μg of cell extracts from arsenic-treated cells were added to 1 ml reaction mixture containing 73 mM phosphate buffer (pH 6.5), 0.73 mM EDTA, 2.5 mM GSH, 1 mM CDNB and 3.3% (v/v) ethanol at 25°C. The rate of increase in the absorbance at 340 nm was monitored over a 3 min period, and the enzyme activity in micromolar per minute was determined by the least-squares fit. Measuring the rate in the absence of cell extracts makes correction for spontaneous reaction.

3.2. Glutathione reductase assay

GR enzyme activity was determined using protocols described by Styblo and Thomas (1995) following the rate of decrease in NAPDH absorbance due to the reduction of GSSG. 20 μg of cell extracts from arsenic-treated cells were added to 1 ml of 0.15 M phosphate buffer (pH 7.0) containing 6 mM EDTA, and 0.1 mM GSSG at 37°C. After incubation for 2 min, the reaction was started by the addition of NAPDH (90.23 mM). The rate of change in the absorbance of NAPDH at 340 nm ($\varepsilon 340 = 6.22$ mM^{-1} cm^{-1}) was monitored with an UV/Vis Beckman spectrophotometer, model DU-530, using the kinetic program at 10 s intervals for 3 min. The rate of nonenzyme reaction was determined in a mixture without cell extract (blank). The net rate of enzymatic oxidation of NAPDH was used to calculate enzyme activity (micromoles NAPDH consumed per minute per μg protein).

4. Gene expression

4.1. Reverse transcription-polymerase chain reaction (RT-PCR)

GR and GPx probe were synthesized using 1st Strand cDNA Synthesis kit for RT-PCR (AMV) (Boehringer Mannheim Inc.). Primers were designed according to specific sequences as following, and purchased from Sigma. Mouse GR primers: forward: 5'-GGTGGTGGAGAGTCACAA-3'; reverse: 5'-CATCCCTTTTCTGCTTGA-3'. GPx primers: forward: 5'-ACCCTCTTCCCTGTTCCT-3'; reverse: 5'-GAGAAGGCATA CACGGTG. P53 primers: forward: 5'-CACTGCATGGACGATCTG-3'. reverse: 5'-CAG GAGCTATTACACATG-3'.

4.2. Isolation of RNA and Northern blot assay

Whole cell RNA was isolated using QIAGEN RNAeasy kit. 5 μg of total RNA was loaded in 1% agarose gel, electrophoresed, transferred to Hybond-N membrane (Amersham Pharmacia Biotech) and probed with ^{32}p dCTP-labeled GR, GPx and p53 PCR products. The membrane was hybridized at 65°C for overnight in 25% SSPE, 0.5% SDS, 25% Denhardt's solution and 100 μg/ml salmon sperm DNA. Then the membrane was washed twice in 2 × SSPE, 0.1% SDS for 20 min, and once in 1 × SSPE, 0.1% SDS for 20 min, and exposed to Kodak-X-OMAT film for auto radiography at − 80°C with an intensifying screen for 4–6 days. Signal for GR, GPx and p53were detected and quantified on a densitomiter (BioRad). Both rRNA and GAPDH were used as standard control for RNA sample loading.

5. Results

5.1. Cellular toxicity

In order to determine LC_{50} of arsenic in 3T3 cells and to select an appropriate concentration of arsenic for long-term treatment, neutral red uptake was measured in 3T3 cells after exposure to 0, 0.1, 0.5, 1, 5, 10, 25, 50, 100 μM As(III) for 24 h. The nontoxic concentrations of arsenic (0.1 and 0.5 μM As) were selected as long-term treatment doses. Medium was changed once a week and maintained at the same concentration of arsenic. Neutral red uptake was measured again after long-term arsenic treatment (up to 133 days) with arsenic. We found that treated 3T3 cells were more resistant to arsenic after long-term, lower dose arsenic treatment (Fig. 1).

5.2. Morphology changes and anchorage-independent growth

We also observed that there were morphological changes after long-term arsenic treatment. First, partially immortalized 3T3 cells became immortalized. Usually, mouse fibroblast cells grow for about 30–40 passages and then cells become moribund (i.e. easily detached from the substratum). However, after long-term arsenic treatment the cells grow vigorously for more than 55 passages; i.e. these arsenic-treated cells have been immortalized. In addition, arsenic-treated cells doubled more quickly than did control cells and were denser than controls at confluency. Cells treated with 0.1 μM As(III) for > 110 days exhibited anchorage-independent growth (Fig. 2). This result indicated that long-term treatment with low dose As(III) induced more aggressive cell growth and a transformed phenotype.

5.3. Alteration of GSH level in arsenic-treated cell

Previous study showed the cellular GSH level to be an important determinant of arsenic sensitivity (Yang et al., 1999). To investigate how arsenic affects GSH level, both in short-term and long-term treatment, we measured GSH levels in arsenic-treated 3T3 cells. The results showed that arsenic enhanced GSH level in cells exposed to 0.5–10 μM As for

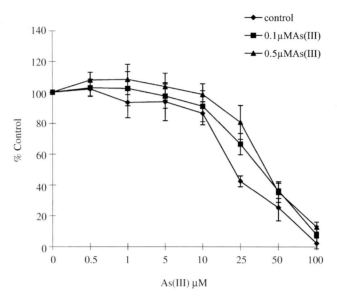

Figure 1. Effect of As(III) on Neutral red uptake in NIH3T3 cells. Cells were treated with 0.1 and 0.5 μM As(III) for more than 70 days. Arsenic toxicity was measured using neutral red uptake assay.

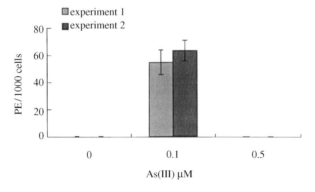

Figure 2. Anchorage-independent growth induced by chronic exposure to As(III). 1×10^4 As(III)-treated cells were plated in 4 ml of 0.3% agar in DMEM with 10% FBS overlaid onto a solid layer of 0.6% agar in DMEM with 10 FBS. The culture was maintained for 3–4 weeks and colonies were counted. The plating efficiency in soft agar was determined relative to the total number of cells plated.

24 h and decreased GSH levels in cells cultured with greater than 25 μM As (Fig. 3). In contrast, after long-term treatment with low concentration of arsenic, GSH level was decreased (Fig. 4). These results demonstrate that short or long-term exposure to arsenic exerts distinct effects on cellular GSH levels. Neutral red uptake results also indicated that 3T3 cells were more resistant to arsenic after long-term exposure to a nontoxic dose of arsenic.

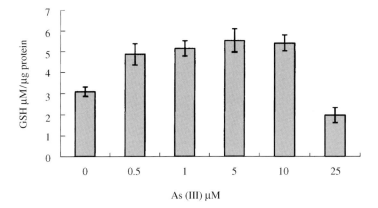

Figure 3. Effect of short-term arsenic treatment on GSH level in NIH3T3 cells. Cells were treated with 0.5, 1, 5, 10, 25 μM As(III) for 24 h. 20 μg of cell extracts from arsenic-treated cells were used to measure GSH. Measuring rate in the absence of cell extracts makes correction for spontaneous reaction. GSH levels were increased in cells exposed to 0.5 –10 μM As(III). GSU level was significantly decreased at 25 μM ($P < 0.01$).

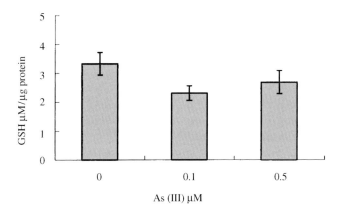

Figure 4. Effect of long-term arsenic treatment on GSH level in NIH3T3 cells. Cells were treated with 0.1 and 0.5 μM As(III) for 56–133 days and GSH levels were determined by Clarke's method. GSH level in 3T3 cells was decreased after chronic low dose arsenic treatment ($P < 0.01$).

5.4. GST and GR activity in arsenic-treated cell

Our previous study had shown that As(III) does not affect the activities of purified GST, GR and GPx activities even at near millimolar levels (Chouchane and Snow, 2001). In this work, we investigated whether or not arsenic can affect the activities of these enzymes in arsenic-treated 3T3 tibroblast cells. The results showed that As(III) can increase both GST and GR activity at low dose (0.5–10 μM), but decreased GST and GR activity at 25 μM and higher doses of arsenic (data not shown) in acute treatment. This is paralleled with

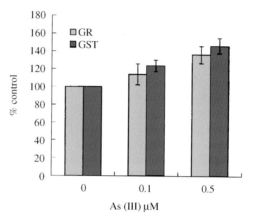

Figure 5. Effect of long-term arsenic treatment on GR and GST activity in NIH3T3 cells. Cells were treated with 0.1 and 0.5 μM As(III) for 56–133 days. GST activity was measured by the formation of a GSH–CDNB complex in 1 ml reaction buffer containing 2.5 mM GSH and 1 mM CDNB and 20 μg protein mixture incubated for 2 min at 25°C. GR activity was measured by NAPDH consumption in 1 ml reaction buffer containing 0.1 mM GSSG, 20 μg protein, and 0.23 mM NAPDH that was incubated for 2 min at 37°C. Data was presented as percentage control. GST and GR activities were increased in As-treated 3T3 cells.

GSH level in short-term arsenic treatment. With long-term As(III) treatment, GST and GR activity showed an increasing trend (Fig. 5).

5.5. GR, GPx and P53 gene expression in arsenic-treated 3T3 cell

We measured GR and GPx mRNA level in As(III)-treated 3T3 cells. We found that basal GR mRNA level is pretty low, but GR mRNA level increased both in time course and dose response manner in 3T3 cells after 3, 6 and 24 h acute treatment. The highest enhancement is about fourfold higher than that of the control at 5 μM As(III) for 24 h treatment (data not shown). After long-term treatment with 0.1 and 0.5 μM As(III), GR mRNA level was increased, while GPx mRNA level was decreased both in acute (data not shown) and chronic arsenic treatment. We measured p53 mRNA level in As(III)-treated 3T3 cells. We found that p53 mRNA level is increased both in time course and dose response manner in 3T3 cells after 3, 6 and 24 h acute treatment (data not show), while after long-term treatment with 0.1 and 0.5 μM As(III) p53 mRNA level is decreased (Fig. 6).

6. Discussion

In this study, we investigated the effect of As(III) on GSH levels and on the activities of relevant enzymes such as GST, GR as well as GR and GPx gene expression and As-induced cell transformation in NIH3T3 fibroblasts. The results showed that As(III) could alter GSH level both in short-term and long-term arsenic treatment. GSH level was increased after 24 h treatment with 0.5–10 μM As(III), and decreased at 25 μM As(III) and above. In contrast, cellular GSH level was decreased after long-term treatment with 0.1 and 0.5 μM As(III), which is paralleled by arsenic resistance after chronic treatment

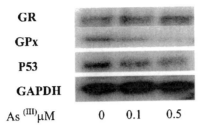

GR	
GPx	
P53	
GAPDH	

As $^{(III)}$μM   0   0.1   0.5

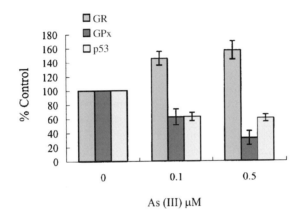

Figure. 6. Effect of As(III) on p53, GR and GPx gene expression. Cells were treated with 0.1 and 0.5 μM As(III) for 85–133 days, then the whole cell RNAs were extracted with RNAeasy kit (QIAGEN). 5 μg of total RNA sample was separated in 1% agarose gel and transferred into Hybond-N membrane and then probed with ^{32}p dCTP-labeled GR, GPx and p53 PCR products. Signals for GR, GPx and p53 were detected and quantified on a densitomiter.

with low concentrations of arsenic. Our results differ from a previous report of increased GSH levels in arsenic-resistant Chinese hamster ovary cells (SA7) (Lee et al., 1989). Cellular GSH system plays an important role in arsenic detoxification in mammalian cells.

GSH also helps maintain the normal reduced state of the cells and counteracts the deleterious effects of oxidative stress by reducing protein disulfides, and by scavenging free radicals and reactive oxygen intermediates (Meister, 1995; Salvemini et al., 1999). The physiological role of GSH as an antioxidant has been described and substantiated in numerous disorders reflecting increased oxidation as a result of an abnormal GSH metabolism. GSH is synthesized in most mammalian cells by the activity of two ATP-requiring GSH-synthesizing enzymes, γ-glutamycysteine synthetase (γ-GSC) and glutathione synthetase. γ-GSC catalyzes the rate-limiting step of GSH synthesis (Seelig and Meister, 1985). GSH synthesis is also thought to be an important factor in cellular defense against stress (Goldwin et al., 1992). Extensive evidence showed arsenic-induced oxidative stress. (Rossman et al., 1991; Snow, 1992; Guyton et al., 1996; Klein et al., 1998; Flora, 1999). It has been suggested that following cellular exposure to oxidants levels of the primary cellular sulfhydryl reductant GSH decrease and levels of its oxidation product (GSSG) increase (Li et al., 1994; Janero and Yarwood, 1995). This increase in the

GSSG:GSH ratio is associated with attendant modification and altered function of various structural protein, enzymes and transcription factors (Walters and Gilbert, 1986; Liu et al., 1996). Oxidatively stressed retinas were capable of regenerating GSH and re-establishing a normal GSSG:GSH ratio (Jahngen-Hodge et al., 1997). It has been reported that depletion of the cellular level of GSH enhances arsenic-induced proto-oncogene activation, which might contribute to subsequent cellular transformation (Shimizu et al., 1998). Our results indicate that cytotoxicity induced by As(III) is closely related to cellular GSH level. Increased GSH level after As(III) exposure may be due to induction of γ-GSC and glutathione synthetase which could increase GSH synthesis. Our results also show changes in cellular morphology in cells treated with 0.1 μM As(III) for up to133 days. In addition, cells chronically treated with 0.1 μM As(III) exhibit anchorage-independent growth. This suggests that long-term treatment with low dose As(III) might produce cells with a more aggressive growth phenotype. Our results also showed that arsenic exposure could alter p53 and c-jun expression (data not shown) both in acute and chronic treatment but in different manners. It is well established that both p53 and c-jun play an important role in cell signal transduction as well as proliferation (Dong, 2002; Spruill et al., 2002; Eferl et al., 2003). Decreasing of GSH level together with alteration of gene expression may play an important role in arsenic-induced cell transformation.

GST and GR play a very important part in metabolic detoxification and antioxidants system (Keen et al., 1976; Habig and Jakoby, 1981; Kirkman et al., 1987; Pickett and Lu, 1989; Meister, 1994). GSTs, which belong to a supergene family of phase II detoxification enzymes, are involved in the conjugation of wide range of electrophilic xenobiotics, including carcinogens and mutagens to the endogenous nucleophile GSH (Keen et al., 1976; Habig and Jakoby, 1981; Mannervik and Danielson, 1988; Pickett and Lu, 1989; Awasthi et al., 1994; Hayes and Pulford, 1995).

GR is the enzyme responsible for maintaining glutathione (GSH) in its reduced form by catalyzing the reaction of glutathione disulfide (GSSG): GSSG + NAPDH + H$^+$ $\rightarrow$ 2GSH + NAPD$^+$. Because most cellular functions of glutathione require the reduced form, the enzyme has a key role in the biochemistry of glutathione (Kirkman et al., 1987; Calberg and Mannervik, 1988; Meister, 1994). Previous studies showed that arsenic had no effect on purified GST and GR activity at micromolar level, which suggested that GST and GR are not direct targets of As(III) (Chouchane and Snow, 2001). When GR and GST are measured in extracts from human keratinocytes treated with micromolar concentration of As(III) they have specific activities different from those of the same enzymes in extracts from untreated cells (Snow et al., 1999). They may be activated or inhibited by treatment of cultured cells with micromolar concentration of As(III), depending on the dose and the time of treatment. In this study, both GST and GR activities in 3T3 cells treated with micromolar levels of As(III) for short term as well as long term were significantly enhanced. At high concentrations of As(III) (>25 μM), which is a cytotoxic level, GST and GR activities were decreased.

GR mRNA level also increased in both long-term and short-term As(III) treatment, which paralleled changes in enzyme activity. GPx mRNA levels were decreased with either acute or chronic arsenic treatment. Our results show, for the first time, the effects of As(III) on GST and GR activities after long-term treatment with sub-micromolar As(III). This is more important than short-term treatment because it is more representative of the human exposure to arsenic.

Taken together, our results suggest that As(III) can alter GST and GR enzyme activities as well as GSH levels and related gene expression both in long-term and short-term treatment. The pattern of effects depends on the intensity and duration of exposure to arsenic. Cytotoxicity induced by As(III) is closely related with GSH levels and the activities of related enzymes. Changes in GSH levels and the activities of these enzymes may play very important roles in arsenic-induced carcinogenesis.

Acknowledgements

This work was supported in part by the U.S. Environmental Protection Agency's Science to Achieve Results (STAR) program, USA, Chinese National Nature Science Foundation (CNNSF) and the Center for Cellular and Molecular Biology, School of Biological and Chemical Sciences, Deakin University, Australia.

References

Abernathy, C.O., Liu, Y.P., Longfellow, D., Aposhian, H.V., Beck, B., Fowler, B., Goyer, R., Menzer, R., Rossman, T., Thompson, C., Waalkes, M., 1999. Arsenic: health effects, mechanisms of actions, and research issues. Environ. Health Perspect., 107 (7), 593–597.

Aposhian, H.V., 1997. Enzymatic methylation of arsenic species and other new approaches to arsenic toxicity. Annu. Rev. Pharmacol. Toxicol., 37, 397–419.

Arrick, B.A., 1982. Glutathione depletion sensitize tumor cells to oxidative cytolysis. J. Biol. Chem., 257, 1231–1237.

Awasthi, Y.C., Sharma, R., Singhal, S.S., 1994. Human glutathione S-transferases. Int. J. Biochem., 26, 295–308.

Ballatori, N., 1994. Glutathione mercaptides as transport forms of metals. Adv. Pharmacol., 27, 271–298.

Carlberg, I., Mannervik, B., 1988. Reduction of 2,4,6-trinitrobenzenesulfonate by glutathione reductase and the effect of $NADP^+$ on the electron transfer. J. Biol. Chem., 261 (4), 1629–1635.

Chouchane, S., Snow, E.T., 2001. In vitro effect of arsenical compounds on glutathione-related enzymes. Chem. Res. Toxicol., 14 (5), 517–525.

Clark, J.B., Fortes, M.A., Giovanni, A., Brewater, D.W., 1996. Modification of an enzymatic glutathione assay for microtiter plate and the determination of glutathione in rat primary cortical cells. Toxicol. Method, 6 (4), 223–230.

Deleve, L., Kaplowitz, N., 1991. Glutathione metabolism and its role in hepatotoxicity. Pharmacol. Ther., 52, 287–305.

Dong, Z., 2002. The molecular mechanisms of arsenic-induced cell transformation and apoptosis. Environ. Health Perspect., 110 (5), 757–759.

Eferl, R., Ricci, R., Kenner, L., Zenz, R., David, J.-P., Rath, M., Wagner, E.F., 2003. Liver tumor development: c-Jun antagonizes the proapoptotic activity of p53. Cell, 112, 181–192.

Flora, S.J., 1999. Arsenic induced oxidative stress and its reversibility following combined administration of N-acetylcysteine and meso-2,3-dimercaptosuccinic acid in rats. Clin. Exp. Pharmacol. Physiol., 26 (11), 865–869.

Goldwin, A.K., Meister, A., O'Dwyer, P.J., Huang, C.S., Hamilton, T.C., Anderson, M.E., 1992. High resistance to cisplatin in human ovarian cancer cell lines is associated with marked increase of glutathione synthesis. Proc. Natl Acad. Sci. USA, 89, 3070–3074.

Guyton, K.Z., Xu, Q.B., Holbrook, N.J., 1996. Induction of the mammalian atress response gene GADD 153 by oxidative stress: role of AP-1 element. Biochem. J., 314, 547–554.

Habig, W.H., Jakoby, W.B., 1981. Glutathione S-transferases (rat and human). Methods Enzymol., 77, 218–231.

Hayes, J.D., Pulford, D.J., 1995. The glutathione-S-transferases supergene family: regulation of GST and contribution of the isoenzymes to cancer chemoprotection and drug resistance. Crit. Rev. Biochem. Mol. Biol., 30, 445–600.

Hu, Y., Wang, G.Q., Hong, X.F., Hua, Z.H., 1988. Effect of arsenic on micro-circulation-Raynaud's disease caused by drinking high arsenic well water. Chin. Prev. Med., 22, 149–151.

Jahngen-Hodge, J., Obin, M.S., Gong, X., Shang, F., Nowell, T.R., Jr., Gong, J., Abasi, H., Blumberg, J., Taylor, A., 1997. Regulation of ubiquitin-conjugating enzymes by glutathione following oxidative stress. J. Biol. Chem., 272 (45), 28218–28226.

Janero, D.R., Yarwood, C., 1995. Oxidative modulation and inactivation of rabbit cardiac adenylate deaminase. Biochem. J., 306, 421–427.

Kalyanaraman, B., 1995. Thiyl radical in biological systems: significant or trivial?. In: Rice-Evans, C., Halliwell, B., Lunt, G.G. (Eds), Free Radicals and Oxidative Stress: Environment, Drugs and Food Additives. Portland Press, London, pp. 55–63.

Keen, J.H., Habig, W.H., Jakoby, W.B., 1976. Mechanism for the several activities of the glutathione S-transferases. J. Biol. Chem., 251, 6183–6188.

Kirkman, H.N., Galiano, S., Gaetani, G.F., 1987. The function of catalase-bound NADPH. J. Biol. Chem., 262, 662–666.

Klein, C.B., Snow, E.T., Frenkel, K., 1998. Molecular mechanisms in metal carcinogenesis: role of oxidative stress. In: Aruoma, O.I., Halliwell, B. (Eds), Molecular Biology of Free Radicals in Human Disease. OICA International, pp. 79–137.

Lee, T.C., Wei, L., Cang, W.J., Ho, I.C., Lo, J.F., Jan, K.Y., Huang, H., 1989. Elevation of glutathione levels and glutathione S-transferase activity in arsenic-resistant Chinese hamster ovary cells. In Vitro Cell. Dev. Biol., 25 (5), 442–448.

Leonard, A., Lauwerys, R.R., 1980. Carcinogenicity, teraogenicity and mutagenicity of arsenic. Mutat. Res., 75 (1), 49–62.

Li, C.-K., Chai, Y.C., Zhao, W., Thomas, J.A., Hendrich, S., 1994. S-Thiolation and irreversible oxidation of sulfhydryls on carbonic anhydrase III during oxidative stress: a method for studying protein modification in intact cells and tissues. Arch. Biochem. Biophys., 308, 231–239.

Little, M.C., Gawkrodger, D.J., Macneil, S.C., 1996. Chromium- and nickel-induced cytotoxicity in normal and transformed human keratinocytes: an investigation of pharmacological approaches to prevention of Cr(VI)-induced cytotoxicity. Br. J. Dermatol., 134, 199–207.

Liu, H., Lightfoot, R., Stevens, J.L., 1996. Activation of heat shock factor by alkylating agents is triggered by glutathione depletion and oxidation of protein thiols. J. Biol. Chem., 271, 4805–4814.

Mannervik, B., Danielson, U.H., 1988. Glutathione transferases – structure and catalytic activity. CRC Crit. Rev. Biochem., 23, 283–337.

Mari, M., Cederbaum, A.I., 2000. CYP 2E1 over-expression in HepG2 cells induces glutathione synthesis by transcriptional activation of γ-glutamylcysteine synthetase. J. Biol. Chem., 275, 15563–15571.

Meister, A., 1988. Glutathione metabolism and its selective modification. J. Biol. Chem., 263 (33), 17205–17208.

Meister, A., 1994. Glutathione, ascorbate, and cellular protection. Cancer Res., 54 (Suppl.), 1969s–1975s.

Meister, A., 1995. Glutathione metabolism. Methods Enzymol., 251, 3–7.

Pickett, C.B., Lu, A.Y., 1989. Glutathione S-transferase: gene structure, regulation, and biological function. Annu. Rev. Biochem., 58, 743–764.

Rossman, T.G., 1998. Arsenic. In: Rom, W.N. (Ed.), Environmental and Occupational Medicine. Lippencott-Raven, Philadephia, pp. 1011–1019.

Rossman, T.G., Goncharova, E.I., Nadas, A., Dolzhanskaya, N., 1991. Chinese hamster cells expressing antisense to metallothionein become spontaneous mutators. Mutat. Res., 373, 75–85.

Salvemini, F., Franze, A., Lervolino, A., Filosa, S., Salzano, S., Urisini, M.V., 1999. Enhanced glutathione level and oxidoresistance mediated by increased glucose-6-phophate dehydrogenase expression. J. Biol. Chem., 274, 2750–2757.

Seelig, G.F., Meister, A., 1985. Gamma-glutamycysteine synthetase from erythrocytes. Methods Enzymol., 113, 390–392.

Shimizu, M., Hochadel, J.F., Fulmer, B.A., Waalkes, M.P., 1998. Effect of glutathione depletion and metallothionein gene expression on arsenic induced cytotoxicity and c-myc expression in vitro. Toxicol. Sci., 45 (2), 204–211.

Snow, E.T., 1992. Metal carcinogenesis: mechanistic implications. Pharma. Ther., 53, 31–65.

Snow, E., Hu, Y., Yan, C., Chouchane, S., 1999. Modulation of DNA repair and glutathione levels in human keratinocytes by micromolar arsenite. In: Chappell, W.R., Abernathy, C.O., Calderon, R.L. (Eds), Arsenic

Exposure and Health Effect: Proceedings of the Third International Conference on Arsenic Exposure and Health Effect, July 12–15, 1998. Elsevier Science, Oxford, UK, pp. 243–251.

Sommers, S.C., Mcmanus, R.G., 1953. Multiple arsenical cancer of skin and internal organs. Cancer, 6, 347–356.

Spruill, M.D., Song, B., Whong, W.Z., Ong, T., 2002. Proto-oncogene amplification and overexpression in cadmium-induced cell transformation, 5, 2131–2144.

Styblo, M., Thomas, D.J., 1995. In vitro inhibition of glutathione reductase by arsenotriglutathione. Biochem. Pharmacol., 49, 971–977.

Walters, D.W., Gilbert, H.F., 1986. Thiol/disulfide redox equilibrium and kinetic behavior of chicken liver fatty acid synthetase. J. Biol. Chem., 261, 13135–13143.

Wang, Z., Rossman, T.G., 1996. The carcinogenicity of arsenic. In: Chang, L.W. (Ed.), Toxicology of Metals. CRC Press, Boca Raton, FL, pp. 219–227.

Wendel, A., 1980. Glutathione peroxidase. Enzymatic Basis of Detoxication, Academic Press, San Diego, pp. 333–353.

Yang, C.H., Huo, M.L., Chen, J.C., Chen, Y.C., 1999. Arsenic trioxide sensitivity is associated with low level of glutathione in cancer cells. Br. J. Cancer, 81, 796–799.

Arsenic Exposure and Health Effects V
W.R. Chappell, C.O. Abernathy, R.L. Calderon and D.J. Thomas, editors

Chapter 14

Laboratory and field evaluation of potential arsenic exposure from mine tailings to grazing cattle

Jack C. Ng, Scott L. Bruce and Barry N. Noller

Abstract

In Australia, the main future use of rehabilitated mine land is to stock grazing animals. Criteria for environmental management and rehabilitation of mine sites have become more stringent with the increasing awareness of the potential of harmful elements including arsenic in tailings. Bioavailability data are lacking to provide realistic health risk assessment of arsenic from mine tailings in Australian conditions. For the evaluation of comparative bioavailability, groups of three cattle were fed 5 days a week, a diet spiked with mine tailings or sodium arsenate or arsenite for 8 months. Blood, biopsy of the muscle and liver were periodically collected to monitor arsenic accumulation. At necropsy, blood, muscle, liver, kidney and other saleable tissues were measured for arsenic concentrations. For field validation, cattle were allowed to graze on rehabilitated tailings facilities over 6.5–8 months. The field data confirm the relative low bioavailability of arsenic from mine tailings to grazing cattle. The rehabilitated mine tailings under the test conditions appear to be suitable for grazing animals with no foreseen adverse health effects. Results obtained from this animal model should be a useful tool for rehabilitation design of mined land in order to minimise adverse health effects on animals and humans. More data of this type will help regulatory agencies to develop guideline values and policy in relationship to mine closure.

Keywords: arsenic, bioavailability, mine tailings, mining waste, risk assessment, cattle

1. Introduction

The mine closure plan outlined in the United Nations Environment Programme, Berlin II (2000) – Guidelines for Mining and Sustainable Development (UNEP, 2002) highlights the main aim of site rehabilitation as: "the need to reduce the risk of pollution, to restore the land and landscape, to prevent further degradation and to provide for future economic use". These guidelines provide a model for various countries to use. Hence, in line with these criteria, rehabilitation of mine sites in Australia has become more stringent with the increasing awareness of the potential for detrimental effects on the environment and human health from exposure to harmful metals and metalloids liberated from mine waste. The strategic framework for mine closure of ANZMEC/Minerals Council of Australia (2000) identifies: "to establish a set of indicators which will demonstrate the successful completion of the closure process" and "the need for targeted research to assist both government and industry in making better and more informed decisions".

A key issue in Australia is the utilisation of mined land for future pastoral activity (Bruce et al., 2001). In particular, the state of Queensland faces an increasing need for such criteria, as it supports economic mining and livestock industries. There are a variety of important regions in Queensland where mining of ores incorporating both base and precious metals is carried out. Following mining of the ore, extraction or separation of the economic item is required. Usually crushing and grinding are necessary to improve surface area and to allow efficient extraction by reagents or separation. The resulting waste products on a mine site, based on open cut mining, are waste rock and tailing material, usually held in separately constructed landforms or structures which minimise their erosion and transport of contaminants. Key contaminants present in waste from base metal mining and other sulfidic deposits are metals (e.g. cadmium, copper, lead and zinc) and metalloids (arsenic, antimony, bismuth and selenium). Queensland's mineral production alone totalled over \$8 billion (Australian dollars) in the year 1999–2000 (QDNRM, 2002), providing significant amounts of government revenue and jobs annually. Equally, Queensland supports a lucrative beef industry, including important export trade to the USA and Japan, earning over \$2.5 billion in the year 2000–2001 (QDPI, 2002). Considering the relative size and scope of these two industries, it is important to ensure that transfer of contaminants such as arsenic and other harmful elements from rehabilitated mine land does not create a health risk to *via* consumption of meat, to both local and international markets.

One of the key aspects of risk assessment is the bioavailability of contaminants, including metals and metalloids from the mine waste material to grazing animals, where bioavailability refers to the ratio of the element absorbed compared to the amount ingested in the material in question. Traditionally, in the absence of specific bioavailability data, this is assumed to be 100%. In an effort to quantify metal and metalloid bioavailability, estimates for mine waste have been conducted using a variety of animal models including rats, rabbits, guinea pigs, dogs, monkeys and swine (Davis et al., 1992; Freeman et al., 1992, 1993, 1994, 1995; Groen et al., 1994; Ng and Moore, 1996; USEPA, 1997; Ng et al., 1998) and various *in vitro* models (Ruby et al., 1993, 1996, 1999; Rodriguez et al., 1999). However, to quantify the realistic bioavailability of metals and metalloids such as arsenic from mine waste to grazing cattle, experiments specifically utilising these animals, are required. Other key aspects, including *in situ* metal dose rate and exposure pathways, are necessary to complete a comprehensive risk assessment tool, required to evaluate the potential for metal and metalloid accumulation and associated contamination of grazing cattle.

Hence, it was the aim of this study to investigate the realistic bioavailability of arsenic and other metals and metalloids (data not shown) from mine tailing materials under controlled experimental conditions, and to validate this with grazing cattle under field conditions. Studies of this type will permit realistic health risk assessment of rehabilitated mine facilities, under Australian conditions.

2. Materials and methods

2.1. Cattle controlled-feeding trial design

The trial was conducted at the University of Queensland Veterinary Science Farm, Pinjarra Hills, Brisbane, following approval by the Queensland Health Scientific Services Animals Ethics Committee (QHSS-AEC) (NRC-1/00/19). A number of Angus cross and

Santa Gertrudis cross heifers, approximately 300 kg in weight, were purchased and allocated randomly to treatment groups. After appropriate health check and acclimatisation for 1 month, each animal was kept in a 1.5 m × 3 m pen from Monday to Friday morning, when they were released into an adjacent paddock until the following Monday. Water was supplied to each pen *via* a self-filling trough and all pens were cleaned daily.

2.2. Dosing protocol

Tailing materials were collected from various mines in Queensland, Australia, which included the site of an *in situ* grazing trial conducted directly on the rehabilitated tailing material. In the field grazing trial (see below) two separate paddocks were created with different metal and metalloid concentrations, whereas for the controlled-feeding trial, material taken from both paddocks was mixed together. Initially all materials were air-dried and homogenised in a cement mixer before being sieved (<1 mm^2) to remove rocks and large pieces of organic matter. The cement mixer was cleaned thoroughly before each new tailing material was homogenised. The tailing material was not sieved further as the aim of this trial was to replicate a field scenario whereby cattle would be ingesting a mixture of particle sizes.

For the controlled-feeding trial, cattle were dosed once daily 5 days a week by spiking a measured amount of either the tailing material or the positive control solution (sodium arsenate or sodium arsenite) into a ration of meal pellets (approximately 2 kg) at a dose rate of 0.5 mg As/kg of body weight. Once the meal pellets were consumed, the animals were fed with wheat or oaten chaff. The negative control animals received the same ration of meal pellets and chaff only. The dose of tailing material was recalculated monthly for each animal after they were weighed.

For the field validation grazing trial, cattle were allowed to graze *ad lib* within the allocated paddock (see detail later).

2.3. Blood and tissue sampling

Blood samples collected into lithium heparinised tubes from the tail vein, as well as muscle (internal abdominal oblique) and liver biopsies were taken from all experimental cattle over the experimental period (day 0, 63 days, 98 days and 179 days) by an experienced veterinary surgeon. Liver sampling was undertaken by inserting a specially made stainless steel tissue biopsy needle into the liver of the living animal, once locally anaesthetised and disinfected. None of the samples were exposed to instruments or containers which could have caused contamination with metals or metalloids.

2.4. Necropsy

At the completion of the trial (approximately 300 days) all animals were sacrificed, when samples of blood, muscle and other tissues including liver, kidney, tongue, spleen, lung, heart, gut and brain were collected.

2.5. Elemental analysis

Liver, muscle and blood from each of the four biopsy sampling events, along with all samples collected during necropsy, were digested in nitric acid before being analysed by inductively coupled plasma mass spectrometry (ICP-MS) (Ng et al., 1998). As part of quality control/quality assurance (QC/QA) program, QC samples were analysed after every 6–10 specimens in a run. The relative standard deviation (RSD) ($n = 20$) of the ICP-MS results using a certified reference standard solution (ICPMO 111-1; EM Science, Gibbstown, NJ, USA) was 5.8%. An in-house rock digest gave an RSD of 3.2% ($n = 6$).

The nine metals and metalloids included in the Australian and New Zealand Food Authorities Maximum Permitted Food Guidelines (ANZFA, 1994) (antimony, arsenic, cadmium, copper, lead, mercury, selenium, tin and zinc) were chosen for analysis, as well as nickel and chromium. These guidelines were subsequently updated by ANZFA (2000), when maximum permitted concentrations (MPC) were replaced with maximum levels (ML) for contaminants in food (Table 1). The new guidelines do not, however, include many of the existing metals and metalloids. Amongst those not specified for meat and other edible bovine products were metals and metalloids relevant to mine waste material, including arsenic and lead, and hence, the previous MPCs are referred to in this report.

2.6. Cattle field grazing trial design

Two mine sites (referred to mine Site 1 and mine Site 2) in North Queensland, Australia were chosen for field validation metal uptake studies in which grazing cattle were allowed to graze on rehabilitated mined land consisting various types of mining wastes including tailings, heap leach material and waste rock dump (WRD) material. In this 9-month trial, accumulation and potential contamination by metals and metalloids of known bioavailability (from controlled-feeding study) were validated under high intensity grazing. Management procedures to reduce potential risk, if required, would then be

Table 1. Maximum permitted concentrations and maximum levels as set by the Australian and New Zealand Food Authority in 1994 and 2000.

	Sb	As	Cd	Cr	Cu	Pb	Hg	Ni	Se	Sn	Zn
MPC (mg/kg) (ANZFA, 1994)											
Bovine muscle	1.5	1.0	0.05	na	10	0.5	0.03	na	1.0	150	150
Bovine tissue	1.5	1.0	2.5 (kidney)	na	100	1.0	0.03	na	2.0	150	150
Bovine liver	1.5	1.0	1.25	na	100	1.0	0.03	na	2.0	150	150
ML (mg/kg) (ANZFA, 2000)											
Bovine muscle	na	na	0.05	na	na	0.1	na	na	na	na	na
Bovine tissue	na	na	2.5 (kidney)	na	na	0.5	na	na	na	na	na
Bovine liver	na	na	na	na	na	0.5	na	na	na	na	na

proposed based on these outcomes, as opposed to exposure based on a lighter grazing pressure under normal farming practices. Therefore, if contamination did not occur during the trial under heavily stocked conditions, further investigations would be unwarranted.

2.7. Mine Site 1

The field grazing trial included a control paddock situated approximately 50 km away from the rehabilitated mine waste paddocks (hereafter referred to as the "background paddock"), not affected by mining activities, and two paddocks within the tailing dam rehabilitated area [tailing paddock 1 (TP1) and tailing paddock 2 (TP2)] for the intensive grazing trial (Fig. 1), with different average metal and metalloid concentrations, and acid-soil conditions (Bruce et al., 2002).

TP1 having a higher acid potential and hence regarded as a "hot spot". The grazed area was initially restricted to 0.8 ha for the first 2 months of the experimental period. Due to low consumption of the standing dry herbage over this period, it was decided to slash the existing 0.8 ha area to encourage fresh regrowth.

To ensure that sufficient pasture was available for stock, the restricted grazing area was enlarged. The restricted area tripled in "tailing paddock 1", to become 2.4 ha and doubled in "tailing paddock 2" to become 1.6 ha, based on the dry matter yield of herbage within each of the paddocks. This design was not intended to determine the potential for contamination under a lighter grazing intensity, as it was premised that the regulatory body usually prefers to accept proposed land use management based on the worst-case scenario. If unacceptable levels of contamination occurred under the proposed trial design,

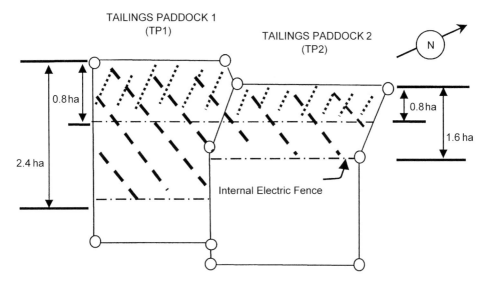

Figure 1. Schematic diagram showing the internal movable electric fence used to enlarge grazing area in the two tailing paddocks. The dotted-line area (···) represents the initial enclosed area and the dash-lined area (– – –) represents the grazing area after the internal electric fence was moved and the grazing area enlarged.

Table 2.   Average concentrations of metals (mg/kg) in the composite samples of soils obtained from the background and two treatment paddocks at mine Site 1.

	TP1	TP2	BP	Combined controlled-feeding material
As	280	370	<15	310
Cd	6.3	16	<1	11
Cr	35	40	11	36
Cu	150	140	8.5	170
Hg	<0.2	<0.2	<0.2	<0.2
Ni	22	25	4	24
Pb	170	150	<10	170
Sb	<10	<10	<10	<10
Se	<15	<15	<15	<15
Sn	<15	<15	<15	<15
Zn	750	1400	36	1090

recommendations for appropriate land management could be made based on a risk assessment.

Pasture condition, botanical composition, pasture nutrient concentrations and soil fertility parameters of the control and treatment sites were measured (data not shown). The concentrations of metals in the background and two treatment paddocks are given in Table 2.

Although grass was found to contain adherent soil, uptake from grass consumption was much less than from soil as cattle ingesting soil through grazing activity received 90% of the total dose.

A key experimental technique was *in situ* measurement of uptake by periodic muscle and liver biopsy, and blood sampling. Several organ systems were sampled at necropsy when the trial was terminated after 8 months. Examples of liver biopsy results are shown in Figure 2.

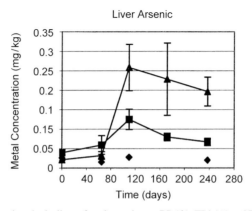

Figure 2.   Arsenic concentrations in the liver of cattle grazing on BP (◆), TP1 (▲) and TP2 (■) over 8 months at mine Site 1.

2.8. Mine Site 2

In this trial over 6.5 months, the accumulation of metals and the subsequent potential for contamination were tested under a high intensity of grazing. This generated results based on a worst-case scenario, whereby cattle would have the maximum exposure to the soil similar to the trial at mine Site 1.

Based on pasture biomass assessments conducted on site, and a daily stock consumption estimate of 2.5% of body weight, 5 ha was proposed as the paddock size for the trial including heap leach paddock (HLP), WRD and background paddock (BP) (undisturbed natural land).

Similar to mine Site 1 trial, all three paddocks were strip-grazed, using movable (electrified) internal fences to provide fresh feed on a controlled basis. This approach has the advantage of forcing exposure to the soil surface in a shorter period, thus maximising the potential for metal consumption to occur within the trial period. An area of 2 ha was provided for the initial 2 months within each of the three trial paddocks. Another 1 ha was then provided at the end of the 2 months, and the final 2 ha was provided after 4 months.

Periodic blood, muscle and liver biopsies were taken for multi-elemental analyses as for mine Site 1. The metal concentrations of the three paddocks at mine Site 2 are given in Table 3. Examples of liver biopsy results are shown in Figure 3.

3. Results

3.1. Cattle controlled-feeding trial – arsenic accumulation and bioavailability

The highly soluble sodium arsenate and sodium arsenite were used for the positive control groups. The concentrations of arsenic continued to accumulate in the liver

Table 3. Concentrations of metals in the composite soils of BP, HLP and WRD paddock at mine Site 2.

Element	BP (mg/kg)	HLP (mg/kg)	WRD (mg/kg)
As	110	620	250
Cd	1.9	4.5	3
Cr	26	16	28
Cu	73	2600	1440
Hg	<0.2	0.5	<0.2
Ni	15	35	34
Pb	180	790	470
Sb	<10	23	<10
Se	<15	19	<15
Sn	<15	71	<15
Zn	250	6800	3400

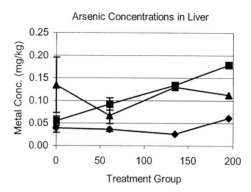

Figure 3.   Arsenic concentrations in the liver biopsy samples of cattle grazing on BP (◆), HLP (▲) and WRD (■) at mine Site 2.

before beginning to plateau off after approximately 150 days (Fig. 4). Consequently, the MPCs in the liver samples of the positive control animals exceeded the Australian and New Zealand Food Authority guidelines (ANZFA, 1994) for arsenic in the liver tissue (1.0 mg/kg). The accumulation of arsenic in the liver of the animals dosed with tailing materials varied markedly, however, were appreciably lower compared to the positive control groups. It was assumed that the oral dosing with sodium arsenate and sodium arsenite had a 100% bioavailability. The relative bioavailability (RBA) was then calculated by comparing, after the subtraction from the initial background values, the accumulation end-points of the positive controls and the tailings groups (Fig. 4). The highest RBA of arsenic in a historic tailing used in the controlled-feeding trial was found to be 29%. The RBA in the tailing material collected from the field grazing trial site, was only about 2%.

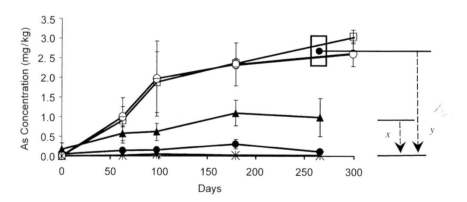

Figure 4.   Arsenic accumulation in the liver of cattle giving a daily oral dose of 0.5 mg As/kg body weight 5 days a week in the form of sodium arsenite (□), sodium arsenate (○), new mine tailing (●) or historic tailing (▲) compared to negative control group (*). Where x represents the relative uptake of arsenic at the end of the experiment by the liver of historic tailing group subtracting from that of control; y represents the average uptake of the two positive control groups (sodium arsenite or sodium arsenate); and RBA $= 100 * x/y$.

3.2. Cattle field grazing trail – tissue distribution metals and metalloids

Arsenic accumulation was significant during the first 100 days of the trial in the liver for animals in the two tailing paddocks compared to the background group, after which a steady state was reached below the MPC guidelines (Fig. 2).

Excluding the liver analysed during the trial period, the kidneys of animals in both TP1 and TP2 compared to the BP (Table 4) had significantly higher arsenic.

No appreciable accumulation occurred in the muscle or offal tissue (gut, tongue, spleen, lung, heart and brain) during the trial period for any of the 11 metals and metalloids analysed, and hence, none of the MPCs were exceeded (Table 4).

3.3. Cattle field grazing trail – plant analysis

Although the concentrations of arsenic in the grasses found on the two tailing paddocks were markedly higher than the grass found on the BP (Bruce et al., 2002), the percent contribution in the grass and adhered dust, to the estimated total dose rate varied (Table 5). That is, the percent contribution of arsenic from the plants and adhered dust to the estimated total dose rate of this element was relatively low. The total daily dose rate was calculated by adding the estimated daily intake of plant material (unwashed) and the dose of arsenic received from the ingestion 1 kg of tailing material.

4. Discussion

Excessive doses of both nutritionally-essential metals and metalloids, and those for which there are no nutritional requirements, may lead to interference with homeostatic mechanisms, accumulation and eventually toxic effects in organisms. A number of factors influence the toxicity of these elements, as follows: the dose, the toxicity of the element, the bioavailability to the organism, the ability of the element to transform in the body to become more or less toxic and the capacity of the organism to excrete the metal or metalloid.

The toxicity of metals and metalloids is generally associated with their aqueous solubility, with the most soluble being salts such as nitrates and chlorides. Having entered the body, toxic metals and metalloids compete with more benign and essential elements, thus inhibiting their function (Seawright, 1989). It is well understood that once absorbed, metals and metalloids such as arsenic and lead concentrate in the liver, after which they are distributed to other organs, accumulating in organs such as the kidneys. The liver can be repeatedly sampled with relative ease without risk to the animals health, and for this reason, the liver is an ideal target organ to test metal and metalloid concentrations and associated accumulation, especially when attempting to detect only small changes.

4.1. Relative bioavailability

Data on bioavailability of toxic elements in mammals from mine waste in Australian conditions are lacking. It is obvious that any risk assessment that assumes 100%

Table 4.   Metal and metalloid concentrations (mean, mg/kg) for all tissues sampled at necropsy for the TP1, TP2 and the BP. None of the concentrations exceeded the MPC in any of the tissues.

Element	Sb	As	Cd	Cr	Cu	Pb	Hg	Ni	Se	Sn	Zn
Kidney											
TP1	0.00	0.43*	1.02*	0.30	7.31	0.13*	0.00	0.02	2.00	0.00	19.29
TP2	0.00	0.21*	1.07*	0.30	4.86	0.06	0.00	0.02	1.75	0.00	29.26
BG	0.00	0.07	0.41	0.32	3.45	0.03	0.01	0.01	1.92	0.00	19.29
Liver											
TP1	0.00	0.20*	0.32*	0.43	11.20	0.10	0.00	0.01	0.34	0.01	31.33
TP2	0.00	0.07*	0.41*	0.42	12.54	0.05	0.00	0.02	0.22	0.00	44.24
BG	0.00	0.02	0.04	0.46	34.26	0.04	0.00	0.01	0.36	0.01	27.48
Gut											
TP1	0.00	0.09	0.00	0.17	0.44	0.02	0.00	0.15	0.09	0.00	14.46
TP2	0.00	0.02	0.00	0.18	0.54	0.04	0.00	0.15	0.04	0.00	13.02
BG	0.00	0.01	0.00	0.16	0.43	0.03	0.00	0.45	0.07	0.00	12.88
Tongue											
TP1	0.10	0.08	0.02	0.47	0.93	0.00	0.00	0.03	0.30	0.00	29.12
TP2	0.07	0.10	0.05	0.45	0.88	0.02	0.03	0.06	0.29	0.03	26.33
BG	0.02	0.08	0.02	0.51	1.04	0.00	0.00	0.04	0.28	0.00	29.47
Spleen											
TP1	0.00	0.09	0.02	0.35	0.82	0.03	0.00	0.03	0.30	0.00	16.48
TP2	0.00	0.04	0.02	0.29	0.63	0.02	0.00	0.02	0.25	0.00	18.41
BG	0.00	0.01	0.00	0.30	0.67	0.01	0.00	0.01	0.33	0.00	20.42
Lung											
TP1	0.00	0.06	0.02	0.31	1.05	0.03	0.00	0.01	0.19	0.01	14.28
TP2	0.00	0.03	0.03	0.31	0.91	0.01	0.00	0.02	0.15	0.00	13.65
BG	0.00	0.01	0.00	0.32	1.38	0.01	0.00	0.01	0.24	0.00	15.35
Heart											
TP1	0.00	0.07	0.00	0.33	3.82	0.02	0.00	0.01	0.21	0.01	16.39
TP2	0.00	0.03	0.00	0.32	3.84	0.01	0.00	0.01	0.16	0.01	16.54
BG	0.00	0.01	0.00	0.31	3.80	0.01	0.00	0.02	0.24	0.01	17.29
Muscle											
TP1	0.00	0.11	0.00	0.35	0.67	0.01	0.00	0.01	0.17	0.01	40.82
TP2	0.00	0.08	0.00	0.41	0.70	0.02	0.00	0.01	0.15	0.01	40.45
BG	0.00	0.01	0.00	0.33	0.67	0.01	0.00	0.00	0.15	0.02	39.38
Brain											
TP1	0.00	0.04	0.00	0.27	2.15	0.02	0.00	0.01	0.14	0.00	8.92
TP2	0.00	0.01	0.00	0.27	2.63	0.02	0.00	0.01	0.12	0.00	9.68
BG	0.00	0.00	0.00	0.26	3.03	0.01	0.00	0.01	0.14	0.00	9.87

The asterisks indicate concentrations that differ significantly ($P < 0.05$) from the BP animals.

bioavailability could be overly conservative and may result in unnecessary and expensive remediation. This is particularly pertinent to the mining industry, where site remediation can encompass time consuming and expensive procedures. It is equally important not to assume that the bioavailability of toxic metals in mine waste is insignificant to impact

Table 5. The percent contribution of arsenic from plants and adhered dust, and from direct ingestion, to the estimated total dose rate.

Location	Percent contribution to diet	
	Plants and adhered dust	Ingestion 1 kg soil
TP1		
Arsenic	7	93
TP2		
Arsenic	3	97
BP		
Arsenic	18	82

upon environmental health of plants, animals and humans. It is therefore, essential to obtain site-specific bioavailability data to evaluate the realistic risk under Australian conditions. It has been shown that the bioavailability of arsenic in soils of an anthropogenic source (Ng and Moore, 1996) is higher than that of natural source of mineralisation (Ng et al., 1998). The present study highlights the variation in bioavailability of arsenic for different tailing materials, compared to a highly soluble ("bioavailable") control solution. This is directly associated with the solubility of the tailing material in the digestive system of the animal model, which can be attributed to a number of physical and chemical properties of the mineral matrix. For example, the mineral composition, degree of encapsulation and the particle size all affect the solubility, and consequently, the bioavailability of metals and metalloids (Davis et al., 1992). The large variation in arsenic RBA observed in the controlled-feeding experiment may also be attributed to the age of the tailing materials used, as the tailing material with the highest arsenic RBA (29%) was collected from an historic gold mine in Central Queensland, where the material had been deposited above ground approximately 100 years ago. Comparatively, the tailing material collected from the rehabilitated gold mine in North Queensland (arsenic RBA: 2%) where the field grazing trial was conducted, was less than 5 years old, and has undergone more contemporary extraction processes.

4.2. Field validation – metal and metalloid accumulation

It was the nature of the field grazing trial to create an intensive grazing scenario, whereby the animals were restricted to a small area of the rehabilitated tailings dam over a 9-month period. Having maximised the tailing exposure, the bioavailability of metals and metalloids from the material would determine the extent of tissue accumulation and contamination. During the trial period, the liver biopsy samples indicated only minimal accumulation of arsenic and none of which exceeded the MPC of 1 mg/kg, or were predicted to, in 2 years had the trial continued. Arsenic concentrations in the kidneys were higher in the two treatment groups compared to the background group, confirming the liver biopsy results.

4.3. Field validation – pathways of metal and metalloid exposure

The potential pathways of metal and metalloid contamination include ingestion *via* plant material and the adhered tailing dust, and ingestion of tailing material directly. The relative contribution of these pathways can depend on the element involved. Thornton and Abrahams (1983) reported arsenic within plant tissue as being a minor contributor to metal ingestion in cattle, compared to zinc. Tailing material containing metals and metalloids adhered to the plant tissue is ingested during grazing, or may be ingested directly from the ground as a consequence of normal grazing behaviour, or as a deliberate behaviour known as "pica". This behaviour usually occurs in animals lacking essential dietary elements (Fraser, 1974; Beaver, 1994). Also, in areas of exposed tailing, evaporites containing metals and metalloids are able to form, and deliberate ingestion by cattle of these evaporites has been reported (Noller et al., 1997). Such "direct ingestion" of soil (not associated with plant material) may be up to 10% of the daily dry matter intake, equating to as much as 1 kg of soil per day (Healy, 1968; Thornton and Abrahams, 1983). The variation in percent contribution of the arsenic to the total dose rate estimate highlights the extent of plant uptake of this element. That is, the percent contribution of arsenic to the daily dose rate exceeded 80% for all treatment groups, including the background, suggesting the grass species found on these sites did not hyperaccumulate these elements and highlighting the importance of plant–arsenic uptake when considering strategies to minimise exposure.

4.4. On-site management considerations

It is possible to identify the ways in which exposure can be limited to reduce the potential for contamination by considering the exposure pathways outlined above. Firstly, it must be made clear that a proportion of the arsenic dose is received from the plant material. Even though the bioavailability of arsenic from plant material may be low, species chosen for remediation of tailing should be those that do not accumulate arsenic in excessive concentrations. The adaptability of the grass species to tailing material, their suitability to the environment and pasture needs are important considerations. However, their ability or inability to accumulate arsenic should also play a role in the design stage of revegetation of any mine waste material being exposed to grazing animals.

Secondly, the dust adhered to plant material potentially contributes a major percentage of arsenic to the diet. This further highlights the importance of maximising ground cover across the tailing facility to minimise the potential for tailing dust to accumulate on the plant material. This is especially important in the first few years after development, before accumulation of leaf litter on the surface of the tailing is significant in reducing splash effect from rain impacting on the tailing material in the wet season, or dust dispersion in the dry season. In order to reduce the risk associated with these exposed tailing areas, a number of management procedures could be adopted. For example, fertilising in the first few years during pasture establishment is essential to maximise percent ground cover, together with follow-up seeding to ensure consistent ground cover is maintained. Also, selection of acid or salt tolerant species of grasses and trees may be necessary.

5. Conclusions

As no quantitative risk-based guidelines currently exist for rehabilitated mined land in Australia, site specific information is necessary in order to better understand the environmental and health risks following mine closure. Controlled-feeding experiments provide such quantitative information that is able to demonstrate variations in the RBA of arsenic (and other metals and metalloids) from different tailing materials. This highlights the importance of better understanding the factors that influence bioavailability of these materials. Further understanding of the relationship between bioavailability, accumulation and source material requires specific measurement of metal species (Thornton, 1997), knowledge of geochemical factors (Davis et al., 1992) and effects of mixtures if a more comprehensive risk-based assessment procedure is required. *In situ* field grazing trials are also necessary to validate the bioavailability estimated from controlled-feeding experiments. The minimal accumulation and associated contamination observed in the field grazing trial, discussed above, supports the observed low RBA estimate determined for the same material dosed under controlled experimental conditions. Bioavailability, however, is one of many factors that needs to be considered when assessing the risks from metal and metalloid uptake associated with cattle grazing on rehabilitated mine waste. The exposure to the mine waste material is considered to be influenced by the following, namely: the percent ground cover, the potential for metal and metalloid hyperaccumulation by vegetation and the stocking duration and intensity. Furthermore, the interaction of mixed metals influencing the bioavailability is the topic of another research project. These are all important considerations when developing management strategies for rehabilitated mine waste and mine close-out.

Acknowledgements

National Research Centre for Environmental Toxicology (EnTox) is funded by Queensland Health, Griffith University, Queensland University of Technology and the University of Queensland. We would like to thank the Australian Research Council (ARC) for providing a SPIRT grant to B.N.N. and J.C.N. (No. 0980021006) including the provision of an APAI scholarship for S.L.B. Financial support from Kidston Gold Mine, Australian Centre for Mine and Energy Research, EPA (Queensland) and Department of Natural Resources and Mines is acknowledged. Expert veterinary assistance was provided by Professor Alan Seawright, Dr Helen Byrnes and Dr Alison Gunn.

References

ANZFA, 1994. Maximum Permitted Concentrations in Foods. Australian and New Zealand Food Authority, Australia.

ANZFA, 2000. Maximum Limits in Foods. Australian and New Zealand Food Authority, Australia.

ANZMEC/Minerals Council of Australia, 2000. Strategic Framework for Mine Closure. Australian and New Zealand Minerals and Energy Council: Minerals Council of Australia, Canberra, Australia.

Beaver, B., 1994. Soil eating. The Veterinarian's Encyclopedia of Animal Behaviour, Iowa State University Press, Ames, 254 pp.

Bruce, S.L., Noller, B.N., Grigg, A.H., Mullen, B.F., Mulligan, D.R., Marshall, I., Moore, M.R., Olszowy, H., O'Brien, G., Dreyer, M.L., Zhou, J.X., Ritchie, P.J., Currey, N.A., Eaglen, P.L., Bell, L.C., Ng, J.C., 2001. Targeted research on the impact of arsenic and lead from mine tailings to develop quantitative indicators for mine closure. Full Manuscript on CD-1611179 19030a. Proceedings of 26th Annual Minerals Council of Australia Environmental Workshop. Hotel Adelaide International, 14–17 October 2001. Minerals Council of Australia. pp. 1–18.

Bruce, S.L., Noller, B.N., Grigg, A.H., Mullen, B.F., Mulligan, D.R., Ritchie, P.J., Currey, N.A., Ng, J.C., 2003. A field study conducted at Kidston Gold Mine, to evaluate the impact of arsenic and zinc from mine tailing to grazing cattle. Toxicol. Lett., 137, 23–34.

Davis, A., Ruby, M.V., Bergstrom, P.D., 1992. Bioavailability of arsenic and lead in soils from the Butte, Montana, mining district. Environ. Sci. Technol., 26, 461–468.

Fraser, A.F., 1974. Anomalous ingestive behaviour. Farm Animal Behaviour, Bailliere Tindall, London, 177 pp.

Freeman, G.B., Johnson, J.D., Killinger, J.M., Liao, S.C., Feder, P.I., Davis, A.O., Ruby, M.V., Chaney, R.L., Lovre, S.C., Bergstrom, P.D., 1992. Relative bioavailability of lead from mining waste soil in rats. Fundam. Appl. Toxicol., 1, 388–398.

Freeman, G.B., Johnson, J.D., Killinger, J.M., Liao, S.C., Davis, A.O., Ruby, M.V., Chaney, R.L., Lovre, S.C., Bergstrom, P.D., 1993. Bioavailability of arsenic in soil impacted by smelter activities following oral administration in rabbits. Fundam. Appl. Toxicol., 21, 83–88.

Freeman, G.B., Johnson, J.D., Liao, S.C., Feder, P.I., Davis, A.O., Ruby, M.V., School, R.A., Chaney, R.L., Bergstrom, P.D., 1994. Absolute bioavailability of lead acetate and mining waste lead in rats. Toxicology, 91, 151–163.

Freeman, G.B., Johnson, J.D., Schoof, R.A., Ruby, M.V., Davis, A.O., Dill, J.A., Liao, S.C., Laplin, C.A., Bergstrom, P.D., 1995. Bioavailability of arsenic in soil and house dust impacted by smelter activities following oral administration in Cynomolgus monkeys. Fundam. Appl. Toxicol., 28, 215–222.

Groen, K., Vaessen, H.A.M.G., Kliest, J.J.G., deBoer, J.L.M., van Ooik, T., Timmerman, A., Vlug, R.G., 1994. Bioavailability of inorganic arsenic from bog ore-containing soil in the dog. Environ. Health Perspect., 102, 182–184.

Healy, W.B., 1968. Ingestion of soil by dairy cows. NZ J. Agric. Res., 11, 487–499.

Ng, J.C., Moore, M.R., 1996. Bioavailability of arsenic in soils from contaminated sites using a 96 hour rat blood model. In: Langley, A., Markey, B., Hill, H. (Eds), The Health Risk Assessment and Management of Contaminated Sites. Department of Health and Family Services and the Commonwealth Environment Protection Agency. Contaminated sites monograph series. No. 5:1996. pp. 355–363.

Ng, J.C., Kratzmann, S.M., Qi, L., Crawley, H., Chiswell, B., Moore, M.R., 1998. Speciation and bioavailability: risk assessment of arsenic contaminated sites in a residential suburb in Canberra. Analyst, 123, 889–892.

Noller, B.N., Eapaea, M.P., Parry, D.L., 1997. Transport of arsenic in water from tropical mines, through sequential extraction procedures. Proceedings Seventh Asian Chemical Congress 7ACC 1997, 16–20, Hiroshima, Japan, 81 pp.

QDNRM, 2002. The economic significance of mining to Queensland. Retrieved September 19, 2002 from Queensland Department of Natural Resources and Mines: http://www.nrm.qld.gov.au/mines/commodities/minerals_worth.html

QDPI, 2002. Queensland Beef Industry Profile. Retrieved September 19, 2002 from Queensland Department of Primary Industries: http://www.dpi.qld.gov.au/extra/pdf/bsu/Beef2002.pdf

Rodriguez, R.R., Basta, N.T., Casteel, S.W., Pace, L.W., 1999. An in vitro gastrointestinal method to estimate bioavailable arsenic in contaminated soils and solid media. Environ. Sci. Technol., 33 (4), 642–649.

Ruby, M.V., Davis, A., Link, T.E., Shoof, R., Chaney, R.L., Freeman, G.B., Bergstrom, P., 1993. Development of an in vitro screening test to evaluate the in vitro bioaccessibility of ingested mine-waste material. Sci. Total Environ., 27, 2870–2877.

Ruby, M.V., Davis, A., Link, T.E., Shoof, R., Eberle, S., Sellstone, C.M., 1996. Estimation of lead and arsenic bioavailability using a physiologically based extraction test. Environ. Sci. Technol., 30 (2), 422–430.

Ruby, M.V., Schoof, R., Brattin, W., Goldade, M., Post, C., Harnois, M., Moseby, D.E., Casteel, S.W., Berti, W., Carpenter, M., Edwards, D., Ceagin, D., Chappell, W., 1999. Advances in evaluating the oral bioavailability of inorganics in soil for use in human health risk assessment. Environ. Sci. Technol., 33 (21), 3697–3705.

Seawright, A.A., 1989. Metals, metalloids and other inorganic substances. Animal Health in Australia Volume 2 (Second Edition) Chemical and Plant Poisons, A.G.P.S., Canberra, pp. 187–231.

Thornton, I., 1997. Sources and pathways of arsenic exposure in South-West England: health implications. In: Chappell, W.R., Abernathy, D.O., Cothern, C.R. (Eds), Arsenic Exposure and Health Effects. Chapman & Hall, London, pp. 61–71.

Thornton, I., Abrahams, P., 1983. Soil ingestion – a major pathway of heavy metals into livestock grazing contaminated land. Sci. Total Environ., 28, 287–294.

UNEP, 2002. Guidelines for Mining and Sustainable Development. United Nations Environment Programme, Berlin II, 2002. Retrieved September 23, 2002, from: http://www.mineralresourcesforum.org/workshops/ Berlin/

USEPA, 1997. Relative bioavailability of arsenic in mining wastes. Superfund/Office of Environmental Assessment, Document Control Number: 4500-88-AORH.

Arsenic Exposure and Health Effects V
W.R. Chappell, C.O. Abernathy, R.L. Calderon and D.J. Thomas, editors
© 2003 Elsevier B.V. All rights reserved.

Chapter 15

Does arsenic require a carcinogenic partner?

Toby G. Rossman, Ahmed N. Uddin, Fredric J. Burns
and Maarten C. Bosland

Abstract

Epidemiological studies show an association between inorganic arsenic in drinking water and increased risk of cancers, yet animal models for arsenic carcinogenesis have not been successful. This lack hinders mechanistic studies of arsenic carcinogenesis. Previously, we found that low concentrations of arsenite are not mutagenic in short-term assays, but can enhance the mutagenicity of other agents. This comutagenic effect appears to result from inhibition of DNA repair by arsenite, but not via inhibition of DNA repair enzymes. Rather, arsenite disrupts p53 function and enhances proliferative signaling [Mutat. Res. 478 (2001) 159]. Failure to find animal models for arsenic carcinogenesis might indicate that arsenite is not a complete carcinogen, but rather acts as an enhancing agent (not a promoter, but a cocarcinogen). To test this hypothesis, Skh1 mice were given 10 mg/l sodium arsenite in drinking water (or not) and irradiated with 1.7 kJ/m² solar UVR three times weekly. After 26 weeks, no tumors appeared in any organs in control mice or in mice given arsenite alone, but irradiated mice given arsenite had a 2.4-fold higher skin tumor yield than did mice given UVR alone. The tumors were mostly squamous cell carcinomas, and appeared earlier, were larger, and more invasive in mice given UVR plus arsenite. These results support the hypothesis that arsenic acts as a cocarcinogen with a second (genotoxic) agent by inhibiting DNA repair and/or enhancing positive growth signaling. Possible partners for arsenic carcinogenesis are discussed.

Keywords: arsenic, mice, cells, comutagen, cocarcinogen, proliferation

1. Epidemiological evidence for a cocarcinogenic role for arsenic

Arsenic contamination of drinking water is a worldwide problem. It is especially severe in the Bengal region of India and in parts of Bangladesh, Taiwan, Chile, Argentina, Mexico and Mongolia, but is also of concern in the United States, including some Western states, parts of the Midwest, and in small areas of New England. Chronic arsenic exposure is of concern mainly because of its carcinogenic effects. The increase in cancer risk observed in epidemiological studies is attributed mainly to the presence of inorganic trivalent arsenic (IARC, 1980; Tinwell et al., 1991). The association between skin cancer and arsenic ingestion in drinking water was seen in studies in Taiwan, Chile, Argentina and Mexico (Tseng et al., 1968; Borgono et al., 1977; Tseng, 1977; Cebrian et al., 1983; Hopenhayn-Rich et al., 1998). Squamous cell carcinoma and basal cell carcinoma, but not malignant melanoma, are associated with arsenic in drinking water (Guo et al., 2001). In addition, significantly elevated standard mortality ratios for internal cancers, particularly of the bladder and lung, are also associated with arsenic in drinking water (Tseng, 1977; Chen et al., 1992; Smith et al., 1992).

However, a number of studies have shown synergy between arsenic and other environmental contaminants in the induction of cancers. The earliest evidence of this synergy was seen in an occupational setting. For example, Swedish copper smelter workers who were exposed to arsenic and who smoked tobacco showed much higher mortality from lung cancer compared with either single exposure (Pershagen et al., 1981). In reviewing this subject, Hertz-Piccioto (2001) calculates that the synergistic excess fraction of lung cancer (i.e. the proportion of cases among those with two exposures that would not have occurred had only one of the exposures been present) ranges from 30 to 54% for smoking and industrial exposure to arsenic. Synergy also occurs in lung cancer mortality in miners exposed to radon (ionizing radiation) and arsenic-containing dusts (Xuan et al., 1993). A Japanese population exposed to arsenic in drinking water showed increased lung cancer risk, but in addition, there was strong synergy between smoking and arsenic ingestion (Tsuda et al., 1995). Similar results were seen in a Taiwan population (Chiou et al., 1995) where there was no increased risk in non-smokers, but a relative risk of 2.45 in smokers in the arsenic endemic area. A recent report suggests that factors other than arsenic in drinking water contribute to the high incidence of bladder cancers in Taiwan, which occur even in the absence of arsenic exposure (Yang et al., 2002).

2. At low concentrations, arsenite is a comutagen

Unlike many carcinogens, arsenite is not a mutagen in bacteria and is only weakly mutagenic in mammalian cells at high (toxic) concentration at single gene loci (Rossman et al., 1980; Rossman, 1998). Because arsenite is such a poor mutagen at endogenous loci such as the X-linked *hprt*, attempts have been made to find genetic markers more likely to be able to detect large deletions such as those induced by ionizing radiation and oxygen radicals. Large deletion mutations are often lethal when they extend past the *hprt* locus on the X chromosome because nearby essential genes are also deleted and there is only one functional X chromosome in mammalian cells. G12 cells, a transgenic derivative of Chinese hamster V79 cells (Klein and Rossman, 1990) can detect clastogens such as ionizing radiation that induce deletions because the *E. coli gpt* target gene is inserted into Chinese hamster chromosome 1 (Klein et al., 1994). Even if the deletion is very large and extends beyond the *gpt* locus into essential genes, there is a second chromosome 1 in the cell that will have a copy of those genes. Figure 1 shows the mutagenic and cytotoxic effects of arsenite in these cells. The mutagenicity is weak and only reaches significance at toxic doses. Similar results are seen in a mouse lymphoma cell line, which can tolerate deletions at the TK locus due to its autosomal location (Moore et al., 1997), in a Chinese hamster CHO transgenic cell transfected with *E. coli gpt* (Meng and Hsie, 1996) and in A_L cells, which are CHO-K1 cells containing a single copy of human chromosome 11, which can suffer deletions when exposed to 7 μM arsenite (Hei et al., 1998).

Arsenite alone does, however, induce many effects at the chromosomal level. Chromosome aberrations, aneuploidy, sister chromatid exchanges and micronucleus induction (a marker of chromosome damage or aneuploidy) have all been demonstrated in

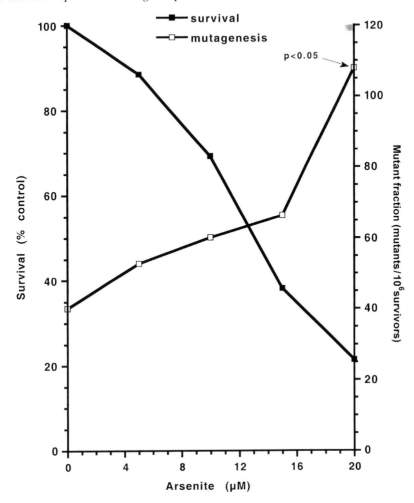

Figure 1. Cytotoxicity and mutagenicity of arsenite at the transgenic *gpt* locus in Chinese hamster G12 cells. The cells are exposed to arsenite after attachment for 24 h. The mutagenesis assay is described in Klein and Rossman (1990).

arsenite-treated cells (Rossman, 1998; Basu et al., 2001). Micronuclei are found in the bone marrow of mice treated with arsenite (Tinwell et al., 1991; Tice et al., 1997) and in buccal cells, lymphocytes and exfoliated bladder cells from exposed humans (Warner et al., 1994; Basu et al., 2002).

Low concentrations of arsenite, which are not mutagenic, are able to enhance the mutagenicity of other carcinogens such as ultraviolet (UV) A, B or C radiation (Rossman, 1981; Lee et al., 1985; Li and Rossman, 1991), ethyl methanesulfonate (Jan et al., 1986), crosslinking agents (Lee et al., 1986), and methylnitrosourea (MNU) (Li and Rossman, 1989a). Arsenite also enhances the clastogenic effects of UV, X-rays and diepoxybutane (Jha et al., 1992; Wiencke and Yager, 1992; Wang et al., 1994). Of particular interest is a recent report that arsenite potentiates the mutagenicity and the formation of DNA adducts by benzo(a)pyrene (BaP) in mouse hepatoma cells that are capable of metabolizing BaP (Maier et al., 2002). BaP, a polycyclic aromatic hydrocarbon (PAH), is formed by incomplete combustion of organic matter such as tobacco smoke, and is often present as a cocontaminant with arsenic.

There is evidence that low concentrations of arsenite can block both base and nucleotide excision repair (Li and Rossman, 1989b; Hartwig et al., 1997; Lynn et al., 1997; reviewed in Rossman, 1998). In the case of MNU, whose lesions are repaired by base excision repair, we demonstrated that arsenite inhibits a late step in DNA repair, probably the ligation step (Li and Rossman, 1989b). Results using the single cell comet assay have confirmed that arsenite prevents the completion of DNA repair in human white blood cells and in human SV40-transformed fibroblasts, consistent with an inhibition of DNA ligase and/or repair synthesis (Hartmann and Speit, 1996). However, neither DNA ligase I nor III (the latter is implicated in base excision repair) can be inhibited by arsenite concentrations many fold higher than those that can inhibit DNA repair in cells (Li, 1989; Li and Rossman, 1989b; Hu et al., 1998). It thus appears that at least for base excision repair, the effects of arsenite are indirect, perhaps resulting from cell signaling changes.

In human and other eukaryotic cells, the cell cycle is regulated by activating or deactivating cyclins and cyclin-dependent kinases (CDKs). Specific inhibitors such as p21WAF1/CIP1 (here called p21) negatively regulate CDKs. When p21 binds to cyclin/CDK2, 4 or 6 complexes, cell cycle arrest takes place (Sherr, 2000). p21 is itself regulated by both p53-dependent and p53-independent pathways (Gartel and Tyner, 1999). In a study on the effects of arsenite on DNA damage-inducible signaling, we showed that in cells treated with arsenite and ionizing radiation, the p53-dependent increase in p21 expression, normally a block to cell cycle progression after DNA damage, is deficient (Vogt and Rossman, 2001). This would allow DNA replication to take place on a damaged template, which would increase mutagenesis. In addition, we and others have found that low (non-toxic) exposure to arsenite enhances signaling that leads to cell proliferation (Germolec et al., 1997, 1998; Barchowsky et al., 1999; Trouba et al., 2000; Chen et al., 2001; Vogt and Rossman, 2001). We suggest that the absence of normal p53 functioning along with increased growth signaling in the presence of DNA damage both contribute to defective DNA repair and account for the comutagenic effects of arsenite. This model is summarized in Figure 2. If damaged DNA is replicated, it may be mutated or lost due to chromosome breaks.

In addition to inhibiting repair of some DNA lesions, arsenite can apparently also increase the DNA adduct level in cells treated with BaP (Maier et al., 2002). There is also evidence that treatment of cells with arsenite results in changes in cytosine methylation in DNA. In human adenocarcinoma A549 cells, arsenite treatment increased methylation in the p53 promoter (Mass and Wang, 1997). In contrast, when rat liver TRL1215 cells are transformed by arsenite, the DNA is globally hypomethylated (Zhao et al., 1997), a result not inconsistent with specific gene hypermethylation. It has been shown that methylation of CpG islands in the p53 promoter increases the level of adduction of BaP metabolites (Tang et al., 1999), but it is not known whether this is the mechanism by which BaP adducts are increased in arsenite-treated cells.

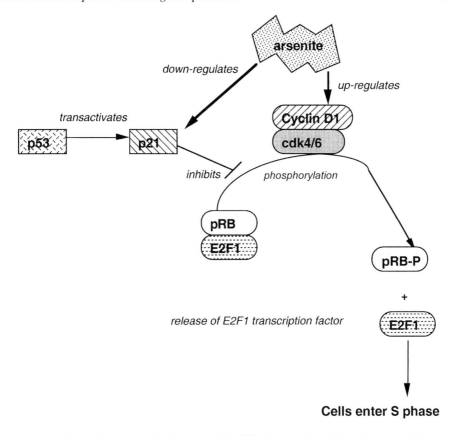

Figure 2. Model showing how arsenite increases cell proliferation (entering S phase) by up-regulating cyclin D1 and down-regulating p21.

3. Arsenic as a cocarcinogen

Many attempts to find an animal model for arsenic carcinogenesis over the past 30 years have failed (National Research Council, 2000). Because of this, arsenic compounds are the only compounds that IARC considers to have sufficient evidence for human carcinogenicity, but inadequate evidence for animal carcinogenicity (Wilbourne et al., 1986). A few reports of arsenic-induced carcinogenesis in animals using intratracheal installation under extreme conditions exist (Ivankovic et al., 1979; Ishinishi et al., 1983; Pershagen et al., 1984; Yamamoto et al., 1987). Despite these positive results, it must be kept in mind that very toxic doses of arsenic compounds were required for tumor induction, and the exposure route was not comparable to the drinking water exposure in humans.

Unlike laboratory animals, humans are always exposed to a mixture of toxicants for long periods of time. Because arsenite is not significantly mutagenic at endogenous loci in bacterial or mammalian cells at concentrations giving high levels of survival (see above), it is sometimes called a tumor promoter. There is little evidence for this view, as negative results have been obtained in bioassays testing arsenite as a promoter (Milner, 1969; Germolec et al., 1997). Arsenic compounds also failed to induce tumors in animals when tested as initiators in two-stage carcinogenesis assays (IARC, 1980; National Research Council, 2000).

Based on arsenite's effects on DNA repair and its comutagenic activity, we have previously suggested that arsenite might act as a cocarcinogen with sunlight to cause skin cancer (Li and Rossman, 1991). We recently developed an experimental protocol intended to simulate lifetime human exposure to arsenic in drinking water along with solar radiation (Rossman et al., 2001). Weanling strain Skh1 hairless (*hr/hr*) mice were given drinking water containing 10 mg/l sodium arsenite ($\sim$5770 μg/l As) starting at 21 days after birth. This concentration of arsenite had no significant effect on their growth or health, compared to control mice. Starting 3 weeks after arsenite exposure began, the mice were irradiated three times per week with 1.7 kJ/m^2 solar spectrum ultraviolet radiation (UVR).

No tumors were seen in unexposed mice, nor in mice exposed only to arsenite, confirming the lack of carcinogenicity by oral arsenite alone found in other studies. Arsenite in the drinking water increased the skin tumor yield in mice exposed to UVR 5 3-fold by day 126 after the start of UVR and 2.4-fold at the end of the experiment (6 months), compared with mice exposed to UVR alone. The tumors in mice exposed to arsenite + UVR appear earlier than with UVR alone (Fig. 3) and are larger and more highly invasive squamous cell carcinomas (Rossman et al., 2001).

This is the first demonstration that arsenite can enhance the onset and growth of malignant skin tumors induced by a genotoxic carcinogen in mice. For the first time, we now have the means to study and evaluate cancer risk scenarios in a *bona fide* animal model. Because arsenite exposure is also associated with lung and bladder cancers, it is likely that a cocarcinogenic approach may lead to animal models for these cancers as well. An experiment is underway to test whether arsenite will enhance tobacco smoke-induced lung cancer.

The development of an animal model for arsenic-related carcinogenesis now makes it possible to study the molecular mechanism of arsenic's cocarcinogenicity in depth and to develop chemopreventive strategies. Our *in vitro* studies have led to the hypothesis that arsenite acts as a cocarcinogen by blocking DNA repair and/or enhancing cell proliferation through unique signaling mechanisms. The blocking of DNA repair may occur *via* a signaling event that controls DNA repair (e.g. by a p53-dependent or independent process) or may result from the enhanced cell proliferation (e.g. by up-regulating cyclin D1), thus forcing the cell to replicate damaged DNA (Fig. 1).

4. Cocarcinogenic partners of arsenic

The population in Taiwan is the best-studied population regarding the association between arsenic in drinking water and skin cancer. The exposures there were of greater duration than in other parts of the world and the skin cancers were already extensive by the late 1960s. UVR from sunlight is the most prominent carcinogen in our natural environment and the most important cause of skin cancers (de Gruijl et al., 2001). There is a common misconception that arsenic-related skin cancers occur on parts of the body that are not

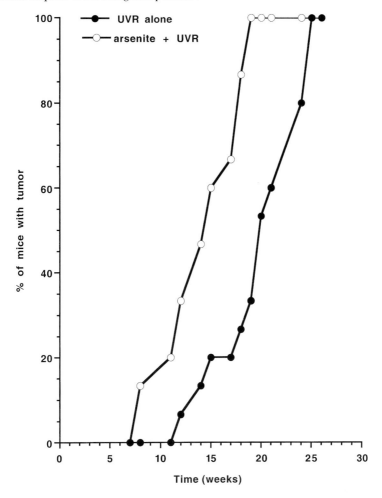

Figure 3. Percent of mice with skin tumors after UVR alone or UVR + arsenite in drinking water. Details of this experiment appear in Rossman et al. (2001).

exposed to sun. The data from Taiwan clearly show that this is not the case (Table 1). If we consider as "unexposed" parts of the body the axilla, groin, genitals, buttock, palm and sole (the latter two because the thick keratin layer prevents sunlight penetration), then only 5.2% of skin cancers appeared in "unexposed" parts of the body in men.

Another thing to notice in Table 1 is that skin cancers are more frequent in males. In populations not exposed to arsenic, the incidence of non-melanoma skin cancer is generally higher in males (Marks, 1995), an effect attributed to both recreational and occupational exposure to sunlight. Although the risk of squamous cell carcinoma shows a dose/response relationship, the risk of basal cell carcinoma appears to increase with increasing exposure to sunlight only at low levels of exposure, but then levels off or falls at high levels of exposure (English et al., 1997). The sites of tumors differ between males and females. Males in this region of Taiwan work outdoors as farmers, fishermen

Table 1.   Sites of skin cancers in an arsenic endemic area of Taiwan.

Site	Male	Female	Total
Exposed			
Face	1	4	5
Scalp	3	1	4
Neck	6		6
Chest wall	21	3	24
Abdominal wall	23	1	24
Back	36	24	60
Flank	6	3	9
Shoulder	3	1	4
Upper arm	4	1	5
Forearm, wrist	5	1	6
Dorsal hand	5	1	6
Finger	5	4	9
Knee		1	1
Back of knee	2	4	6
Leg, ankle	12	7	19
Dorsal foot	1	1	2
Toe		2	2
Thigh	12	10	22
Non-exposed			
Buttock	2		2
Axilla	1		1
Groin	2	2	4
Genitals		1	1
Palm	2	4	6
Sole	1	7	8
Total	153	94	247

Data from Yeh et al. (1968).

and in salt flats. Furthermore, because Taiwan has a subtropical climate, the men usually work wearing little more than shorts; hence, the amount of sun exposure on chest, abdomen and back is high. The women spend more time indoors, and even when they work in agriculture, they tend to cover up because of modesty and the desire for light skin (Lung Chi Chen, personal communication). The type of clothing worn in Taiwan is not a complete block to UV from the sun (Wang et al., 2001).

Besides sunlight, non-melanoma skin cancer risk has been found to increase after exposure to ionizing radiation (Karagas et al., 1996). Some examples of populations at increased risk from radiation include uranium (and some other) miners, radiologists, atomic bomb survivors and individuals treated for tinea capitis or thymic enlargement by X-rays. An association between tobacco smoking and squamous cell carcinoma of the skin was found (relative risk 2.3) (De Hertog et al., 2001). Tobacco smoking is also a major cause of lung and bladder cancers (Peto et al., 1992) and therefore might be expected to synergize with arsenic exposure in causing these cancers. Papillomavirus infection contributes to the etiology of some human cancers, including skin (along with UVR) (zur Hausen, 1996). Heavy exposure to PAHs occurs in some occupations such as aluminum production, coke production, iron and steel foundries, tar distillation, shale oil extraction, wood impregnation, roofing, road paving, carbon black production, chimney sweeping and occupations with exposure to diesel exhaust, which also contain carcinogenic nitro-PAHs (Boffetta et al., 1997). Dermal exposure to PAHs increases risk of skin cancer, while inhalation leads to increased lung and bladder cancers. These and some other possible partners in arsenic carcinogenesis are listed in Table 2.

Finally, mention must be made of malnutrition as a possible "partner" in arsenic carcinogenesis. DNA oxidation, DNA strand breaks and chromosome breaks have been detected in cells or animals with various deficiencies, including folate, selenium, niacin, vitamin C, vitamin E, vitamin B6, zinc and iron (Ames, 2001). The partnership between folate deficiency and arsenite has recently been demonstrated in mice. When mice are fed a folate-deficient diet and then given arsenite, about a twofold enhancement of micronuclei in peripheral blood lymphocytes was seen compared with arsenite alone

Table 2.   Possible partners for arsenic-induced cancers of the skin, lung and bladder.

Agent or industry	Cancer(s)
Ultraviolet light (sunlight)	Skin
Papillomavirus (with UV)	Skin
Kerosine[a]	Skin
Shale oils	Skin
Mineral oils (metal machining, printing, cotton/jute spinning)	Skin
Ionizing radiation	Skin, lung
Polycyclic aromatic hydrocarbons (coal tar, coke, coal burning)	Skin, lung, bladder
Tobacco smoke	Skin, lung, bladder
Asbestos	Lung
Aflotoxin B1[a] (moldy peanuts or grain)	Lung
DDT (possibly)	Lung
Petroleum refining	Lung
Sugar and confectionary industry	Lung
Non-metallic mineral and stone products	Lung
Newspaper publishing	Lung, bladder
Plastics, synthetics, resins industry	Lung, bladder
Benzidine (dyes)	Bladder
Aromatic amines	Bladder
4-Aminobiphenyl	Bladder
Phenacetin	Bladder
Schistosome infection	Bladder
Tires and inner tubes industry	Bladder

Information in this table was based on the following references: Boffetta et al. (1997), English et al. (1997), Tolbert (1997), Ward et al. (1997), Nessel (1999) and Tomatis (2000).
[a] Sufficient evidence in animals.

(McDorman et al., 2002). Because few of the micronuclei exhibited kinetochore staining, the micronuclei were derived from broken chromosomes. Thus, folate deficiency enhances arsenite clastogenesis.

Acknowledgements

This work was supported by NIEHS grant ES09252 and is part of The Nelson Institute of Environmental Medicine and the Kaplan Cancer Center programs supported by grant ES00260 from the National Institute of Environmental Health Sciences and grant CA16087 from the National Cancer Institute. We thank Eleanor Cordisco for help with the manuscript preparation.

References

Ames, B.N., 2001. DNA damage from micronutrient deficiencies is likely to be a major cause of cancer. Mutat. Res., 475, 7–20.

Barchowsky, A., Roussel, R.R., Klei, L.R., James, P.E., Ganju, N., Smith, K.R., Dudek, E.J., 1999. Low levels of arsenic trioxide stimulate proliferative signals in primary vascular cells without activating stress effector pathways. Toxicol. Appl. Pharmacol., 159, 65–75.

Basu, A., Mahata, J., Gupta, S., Giri, A.K., 2001. Genetic toxicology of a paradoxical human carcinogen, arsenic: a review. Mutat. Res., 488, 171–194.

Basu, A., Mahata, J., Roy, A.K., Sarlar, J.N., Poddar, G., Nandi, A., Sarkar, P.K., Dutta, P.K., Banerjee, A., Das, M., Ray, K., Roychaudhury, S., Natarajan, A.T., Nilsson, R., Giri, A.K., 2002. Enhanced frequency of micronuclei in individuals exposed to arsenic through drinking water in West Bengal, India. Mutat. Res., 516, 29–40.

Boffetta, P., Jourenkova, N., Gustavsson, P., 1997. Cancer risk from occupational and environmental exposure to polycyclic aromatic hydrocarbons. Cancer Causes Control, 8, 444–472.

Borgono, J.M., Vicent, P., Venturino, H., Infante, A., 1977. Arsenic in the drinking water of the city of Antofagasta: epidemiological and clinical study before and after the installation of a treatment plant. Environ. Health Perspect., 19, 103–105.

Cebrian, M.E., Albores, A., Aguilar, M., Blakely, E., 1983. Chronic arsenic poisoning in the north of Mexico. Human Toxicol., 2, 121–133.

Chen, C.-J., Chen, C.W., Wu, M.-M., Kuo, T.-L., 1992. Cancer potential in liver, lung, bladder and kidney due to ingested inorganic arsenic in drinking water. Br. J. Cancer, 66, 888–892.

Chen, H., Liu, J., Merrick, B.A., Waalkes, M.P., 2001. Genetic events associated with arsenic-induced malignant transformation: applications of cDNA microarray technology. Mol. Carcinog., 30, 79–87.

Chiou, H.-Y., Hsueh, Y.-M., Liaw, K.-F., Horng, S.-F., Chiang, M.-H., Pu, Y.-S., Lin, J.S.-N., Huang, C.-H., Chen, C.-J., 1995. Incidence of internal cancers and ingested inorganic arsenic: a seven-year follow-up study in Taiwan. Cancer Res., 55, 1296–1300.

de Gruijl, F.R., van Kranen, H.J., Mullenders, L.H.F., 2001. UV-induced DNA damage, repair, mutations and oncogenic pathways in skin cancer. J. Photochem. Photobiol., 63, 19–27.

De Hertog, S.A.E., Wensveen, C.A.H., Bastiaens, M.T., Kielich, C.J., Berkhout, M.J.P., Westendorp, R.G.J., Vermeer, B.J., Bavinck, J.N.B., 2001. Relation between smoking and skin cancer. J. Clin. Oncol., 19, 231–238.

English, D.R., Armstrong, B.K., Kricker, A., Fleming, C., 1997. Sunlight and cancer. Cancer Causes Control, 8, 271–283.

Gartel, A.L., Tyner, A.L., 1999. Transcriptional regulation of the p21 (WAF1/CIP1) gene. Exp. Cell Res., 246, 280–289.

Germolec, D.R., Spalding, J., Boorman, G.A., Wilmer, J.L., Yoshida, T., Simeonova, P.P., Bruccoleri, A., Kayama, F., Gaido, K., Tennant, R., Burleson, F., Dong, W., Lang, R.W., Luster, M.I., 1997. Arsenic can mediate skin neoplasia by chronic stimulation of keratinocyte-derived growth factors. Mutat. Res., 386, 209–218.

Germolec, D., Spalding, J., Yu, H.-S., Chen, G.S., Simeonova, P.O., Humble, M.C., Bruccoleri, A., Boorman, G.A., Foley, J.F., Yoshida, T., Luster, M.I., 1998. Arsenic enhancement of skin neoplasia by chronic stimulation of growth factors. Am. J. Pathol., 153, 1775–1785.

Guo, H.R., Yu, H.S., Hu, H., Monson, R.R., 2001. Arsenic in drinking water and skin cancers: cell-type specificity. Cancer Causes Control, 12, 909–916.

Hartmann, A., Speit, G., 1996. Effect of arsenic and cadmium on the persistence of mutagen-induced DNA lesions in human cells. Environ. Mol. Mutagen., 27, 98–104.

Hartwig, A., Groblinghoff, U.D., Beyersman, D., Natarajan, A.T., Filon, R., Mullenders, L.H.F., 1997. Interaction of arsenic(III) with nucleotide excision repair in UV-irradiated human fibroblasts. Carcinogenesis, 18, 399–405.

Hei, T.K., Liu, S., Waldren, C., 1998. Mutagenicity of arsenic in mammalian cells: role of reactive oxygen species. Proc. Natl Acad. Sci. USA, 95, 8103–8107.

Hertz-Piccioto, I., 2001. Interactions between arsenic and other exogenous exposures in relation to health outcomes. In: Chappell, W.R., Abernathy, C.O., Calderon, R.L. (Eds), 4th International Conferences on Arsenic Exposure and Health Effects. Elsevier Sciences, B.V., Amsterdam, pp. 173–180.

Hopenhayn-Rich, C., Biggs, M.L., Smith, A.H., 1998. Lung and kidney cancer mortality associated with arsenic in drinking water in Cordoba, Argentina. Int. Epidemiol. Assoc., 27, 561–569.

Hu, Y., Su, L., Snow, E.T., 1998. Arsenic toxicity is enzyme specific and its affects on ligation are not caused by the direct inhibition of DNA repair enzymes. Mutat. Res., 408, 203–218.

IARC (International Agency for Research on Cancer), 1980. IARC Monographs on the Evaluation of Carcinogenic Risk of Chemicals to Man. Some Metals and Metallic Compounds, Vol. 23. World Health Organization, Lyon, France.

Ishinishi, N., Yamamoto, A., Hisanaga, A., Inamasu, T., 1983. Tumorigenicity of arsenic trioxide to the lung in Syrian golden hamsters by intermittent instillations. Cancer Lett., 21, 141–147.

Ivankovic, S., Eisenbrand, G., Preussmann, R., 1979. Lung carcinoma induction in BD rats after single intratracheal instillation of an arsenic-containing pesticide mixture formerly used in vineyards. Int. J. Cancer, 34, 786–788.

Jan, K.Y., Huang, R.Y., Lee, T.C., 1986. Different modes of action of sodium arsenite, 3-aminobenzamide, and caffeine on the enhancement of ethyl methanesulfonate. Cytogenet. Cell Genet., 41, 202–208.

Jha, A.N., Noditi, M., Nilsson, R., Natarajan, A.T., 1992. Genotoxic effects of sodium arsenite on human cells. Mutat. Res., 284, 215–221.

Karagas, M.R., Mcdonald, J.A., Greenberg, E.R., Stukel, T.A., Weiss, J.E., Baron, J.A., Stevens, M.M., 1996. Risk of basal cell and squamous cell skin cancers after ionizing radiation therapy. For the skin cancer prevention study group. J. Natl Cancer Inst., 88, 1848–1853.

Klein, C.B., Rossman, T.G., 1990. Transgenic Chinese hamster V79 cell lines which exhibit variable levels of gpt mutagenesis. Environ. Mol. Mutagen., 16, 1–12.

Klein, C.B., Su, L., Rossman, T.G., Snow, E.T., 1994. Transgenic gpt V79 cell lines differ in their mutagenic response to clastogens. Mutat. Res., 304, 217–228.

Lee, T.-C., Huang, R.Y., Jan, K.Y., 1985. Sodium arsenite enhances the cytotoxicity clastogenicity and 6-thioguanine-resistant mutagenicity of ultraviolet light in Chinese hamster ovary cells. Mutat. Res., 148, 83–89.

Lee, T.-C., Lee, K.C., Tseng, Y.J., Huang, R.Y., Jan, K.Y., 1986. Sodium arsenite potentiates the clastogenicity and mutagenicity of DNA crosslinking agents. Environ. Mutagen., 8, 119–128.

Li, J.-H., 1989. Ph.D. Thesis, New York University.

Li, J.-H., Rossman, T.G., 1989. Mechanism of comutagenesis of sodium arsenite with N-methyl-N-nitrosourea. Biol. Trace Element Res., 21, 373–381.

Li, J.-H., Rossman, T.G., 1989. Inhibition of DNA ligase activity by arsenite: a possible mechanism of its comutagenesis. Mol. Toxicol., 2, 1–9.

Li, J.-H., Rossman, T.G., 1991. Comutagenesis of sodium arsenite with ultraviolet radiation in Chinese hamster V79 cells. Biol. Metals, 4, 197–200.

Lynn, S., Lai, H.T., Gurr, J.-R., Jan, K.Y., 1997. Arsenite retards DNA strand break rejoining by inhibiting DNA ligation. Mutagenesis, 12, 353–358.

Maier, A., Schumann, B.L., Chang, X., Talaska, G., Puga, A., 2002. Arsenic co-exposure potentiates benzo[a]pyrene genotoxicity. Mutat. Res., 517, 101–111.

Marks, R., 1995. The epidemiology of non-melanoma skin cancer: who, why and what can we do about it. J. Dermatol., 22, 853–857.

Mass, M.J., Wang, L., 1997. Arsenic alters cytosine methylation patterns of the promoter of the tumor suppressor gene p53 in human lung cells: a model for a mechanism of carcinogenesis. Mutat. Res., 386, 263–277.

McDorman, E.W., Collins, B.W., Allen, J.W., 2002. Dietary folate deficiency enhances induction of micronuclei by arsenic in mice. Environ. Mol. Mutagen., 40, 71–77.

Meng, Z., Hsie, A.W., 1996. Polymerase chain reaction-based deletion analysis of spontaneous and arsenite-enhanced gpt mutants in CHO-AS52 cells. Mutat. Res., 356, 255–259.

Milner, J.E., 1969. The effects of ingested arsenic exposure on methyl cholanthrene induced skin tumours in mice. Arch. Environ. Health, 18, 7–11.

Moore, M., Harrington-Brock, K., Doerr, C., 1997. Relative genotoxic potency of arsenic and its methylated metabolites. Mutat. Res., 386, 279–290.

National Research Council, 2000. Arsenic in Drinking Water. National Academy Press, Washington, DC.

Nessel, C.S., 1999. A comprehensive evaluation of the carcinogenic potential of middle distillate fuels. Drug Chem. Toxicol., 22, 165–180.

Pershagen, G., Wall, S., Taube, A., Linnman, L., 1981. On the interaction between occupational arsenic exposure and smoking and its relationship to lung cancer. Scand. J. Work Environ. Health, 7, 302–309.

Pershagen, G., Nordberg, G., Bjorklund, N.E., 1984. Carcinomas of the respiratory tract in hamsters given arsenic trioxide and/or benzo(a)pyrene by the pulmonary route. Environ. Res., 34, 227–241.

Peto, R., Lopez, A.D., Boreham, J., Thun, M., Heath, C., Jr., 1992. Mortality from tobacco in developed countries: indirect estimation from national vital statistics. Lancet, 339, 1268–1278.

Rossman, T.G., 1981. Enhancement of UV-mutagenesis by low concentrations of arsenite in *E. coli*. Mutat. Res., 91, 207–211.

Rossman, T.G., 1998. Molecular and genetic toxicology of arsenic. In: Rose, J. (Ed.), Environmental Toxicology: Current Developments. Gordon and Breach, Amsterdam, pp. 171–187.

Rossman, T.G., Stone, D., Molina, M., Troll, W., 1980. Absence of arsenite mutagenicity in *E. coli* and Chinese hamster cells. Environ. Mutagen., 2, 371–379.

Rossman, T.G., Uddin, A.N., Burns, F.J., Bosland, M., 2001. Arsenic is a cocarcinogen with solar ultraviolet radiation for mouse skin: an animal model for arsenic carcinogenesis. Toxicol. Appl. Pharmacol., 176, 64–71.

Sherr, C.J., 2000. The Pezcoller lecture: cancer cell cycles revisited. Cancer Res., 60, 3689–3695.

Smith, A.H., Hopenhayn-Rich, C., Bates, M.N., Goeden, H.M., Hertz-Picciotto, I.H., Duggan, I., Wood, R., Kosnett, M., Smith, M.T., 1992. Cancer risk from arsenic in drinking water. Environ. Health Perspect., 97, 259–267.

Tang, M.S., Zheng, J.B., Denissenko, M.F., Pfeifer, G.P., Zheng, Y., 1999. Use of UvrABC nuclease to quantify benzo[a]pyrene diol epoxide-DNA adduct formation at methylated versus unmethylated CpG sites in the p53 gene. Carcinogenesis, 20, 1085–1089.

Tice, R.R., Yager, J.W., Andrews, P., Crecelius, E., 1997. Effect of hepatic methyl donor status on urinary excretion and DNA damage in B6C3F1 mice treated with sodium arsenite. Mutat. Res., 386, 315–334.

Tinwell, H., Stephens, S.C., Ashby, J., 1991. Arsenite as the probable active species in the human carcinogenicity of arsenic: mouse micronucleus assays on Na and K arsenite, orpiment, and Fowler's solution. Environ. Health Perspect., 95, 205–210.

Tolbert, P.E., 1997. Oils and cancer. Cancer Causes Control, 8, 386–405.

Tomatis, L., 2000. The identification of human carcinogens and primary prevention of cancer. Mutat. Res., 462, 407–421.

Trouba, K.J., Wauson, E.M., Vorce, R.L., 2000. Sodium arsenite-induced dysregulation of proteins involved in proliferative signaling. Toxicol. Appl. Pharmacol., 164, 161–170.

Tseng, W.P., 1977. Effects and dose response relationships of skin cancer and blackfoot disease with arsenic. Environ. Health Perspect., 19, 109–119.

Tseng, W.P., Chu, H.M., How, S.W., Fong, J.M., Lin, C.S., Yeh, S., 1968. Prevalence of skin cancer in an endemic area of chronic arsenicism in Taiwan. J. Natl Cancer Inst., 40, 453–4635.

Tsuda, T., Babazono, A., Yamamoto, E., Kurumatani, N., Mino, Y., Ogara, T., Kishi, Y., Aoyama, H., 1995. Ingested arsenic and internal cancer: a historical cohort study followed for 33 years. Am. J. Epidemiol., 141, 198–209.

Vogt, B., Rossman, T.G., 2001. Effects of arsenite on p53, p21 and cyclin D expression in normal human fibroblasts – a possible mechanism for arsenite's comutagenicity. Mutat. Res., 478, 159–168.

Wang, T.C., Huang, J.S., Yang, V.C., Lan, H.J., Lin, C.J., Jan, K.Y., 1994. Delay of the excision of UV light-induced DNA adducts is involved in the coclastogenicity of UV light plus arsenite. Int. J. Radiat. Biol., 66, 367–372.

Wang, S.Q., Kopf, A.W., Marx, J., Bogdan, A., Polsky, D., Bart, R.S., 2001. Reduction of ultraviolet transmission through cotton T-shirt fabrics with low ultraviolet protection by various laundering methods and dyeing: clinical implications. J. Am. Acad. Dermatol., 44, 767–774.

Ward, E.M., Burnett, C.A., Ruder, A., Davis-King, K., 1997. Industrties and cancer. Cancer Causes Control, 8, 356–370.

Warner, M.L., Moore, L.E., Smith, M.T., Kalman, D.A., Fanning, E., Smith, A.H., 1994. Increased micronuclei in exfoliated bladder cells of individuals who chronically ingest arsenic-contaminated water in Nevada. Cancer Epidemiol. Biomarkers Prevent., 3, 583–590.

Wiencke, J.K., Yager, J.W., 1992. Specificity of arsenite in potentiating cytogenetic damage induced by the DNA crosslinking agent diepoxybutane. Environ. Mol. Mutagen., 19, 195–200.

Wilbourne, J., Haroun, L., Heseltine, E., Kaldor, J., Partensky, D., Vainio, V., 1986. Response of experimental animals to human carcinogens: an analysis based upon the IARC Monographs programme. Carcinogenesis, 7, 1853–1863.

Xuan, X.Z., Lubin, J.H., Li, J.Y., Yang, L.F., Luo, A.S., Lan, Y., Wang, J.Z., Blot, W.J., 1993. A cohort study in southern China of tin miners exposed to radon and radon decay products. Health Physics, 64L, 120–131.

Yamamoto, A., Hisanaga, A., Ishinishi, N., 1987. Tumorigenicity of inorganic arsenic compounds following intratracheal instillations to the lungs of hamsters. Int. J. Cancer, 40, 220–223.

Yang, M.H., Chen, K.K., Yen, C.C., Wang, W.S., Chang, Y.H., Huang, W.J., Fan, F.S., Chiou, T.J., Liu, J.H., Chen, P.M., 2002. Unusually high incidence of upper urinary tract urothelial carcinoma in Taiwan. Urology, 59, 681–687.

Yeh, S., How, S.W., Lin, C.S., 1968. Arsenical cancer of skin histological study with special reference to Bowen's disease. Cancer, 21, 312–339.

Zhao, C.Q., Youngh, M.R., Diwan, B.A., Coogan, T.P., Waalkes, M.P., 1997. Association of arsenic-induced malignant transformation with DNA hypomethylation and aberrant gene expression. Proc. Natl Acad. Sci. USA, 94, 10907–10912.

zur Hausen, J., 1996. Papillomavirus infections – a major cause of human cancer. Biochim. Biophys. Acta, F55–78.

Chapter 16

Carcinogenicity of dimethylarsinic acid and relevant mechanisms

Min Wei, Hideki Wanibuchi, Keiichirou Morimura
and Shoji Fukushima

Abstract

In the present experiments, we examined carcinogenicity of a major organic metabolite of arsenic, dimethylarsinic acid (DMA), in male F344 rats in a 2-year carcinogenicity test, in addition to assessing genetic alteration patterns in the induced tumors. Furthermore, to test the hypothesis that reactive oxygen species (ROS) may play a role in DMA carcinogenesis, 8-hydroxy-2′-deoxyguanosine (8-OHdG) formation in urinary bladder was examined. From weeks 97–104, urinary bladder tumors were observed in rats treated with DMA at dietary doses of 50 ppm (tumor incidence, 26%) and 200 ppm (39%), but not at doses of 0 and 12.5 ppm. Mutation analysis showed these tumors to have a low rate of H-ras mutations. No alterations of the p53, K-H-ras or β-catenin genes were found, whereas aberrant protein expression of p27^{kip1}, cyclin D1 and COX-2 could be demonstrated in the urinary bladder lesions by immunohistochemistry. 8-OHdG formation level in urinary bladder DNA was significantly increased after treatment with 200 ppm DMA in the drinking water for 2 weeks, as compared with the controls. The present work provides evidence that DMA is a complete carcinogen for the rat urinary bladder. DMA-induced urinary bladder tumors may occur as the result of accumulations of diverse genetic alterations and induction of cell proliferation at the two highest doses. Relevance of these results to humans is discussed.

Keywords: dimethylarsinic acid, carcinogenicity, urinary bladder, genetic alteration, rat

1. Introduction

Arsenic is a known human carcinogen with the development of neoplasms in humans exposed to arsenic in the general environment and in industry. Unfortunately, acute and chronic arsenic exposure still remains a major public health problem in many countries. In particular, drinking water contamination results in increased incidences of cancers at multiple organ sites, especially the skin, lung, and urinary bladder, but possibly also in the liver, kidney, nasal cavity, and prostate, in highly exposed populations. However, precise mechanisms by which arsenic induces cancer are unknown, in large part due to the lack of an appropriate animal model.

We have focused on dimethylarsinic acid (DMA), a major metabolite of arsenic in most mammals, including humans. Humans have continuous exposure to DMA from arsenic in their drinking water, production or use of arsenic-containing herbicides and ingestion of foods contaminated with these herbicides and naturally occurring arsenic from the soil. We have previously demonstrated that DMA promotes carcinogenesis in the urinary bladder, kidney, liver, and thyroid gland of F344 rats in an *in vivo* multi-organ carcinogenesis bioassay (Yamamoto et al., 1995). Wanibuchi et al. (1996, 1997) indicated that DMA exerts promoting potential in a dose dependent manner with regard to urinary bladder and liver carcinogenesis in F344 rats, possibly via a mechanism involving stimulation of cell proliferation and DNA damage caused by oxygen radicals. Of particular importance, the findings from our animal studies revealed effects at sites where human arsenic-associated cancers develop, such as the urinary bladder, liver and kidney, suggesting that DMA exposure may be relevant to the carcinogenic risk of arsenic to humans. This led us to conduct a 2-year bioassay to determine the carcinogenicity of DMA in F344 rats. Previously, we reported that administration of DMA induced urinary bladder tumors in male F344 rats at doses of 50 and 200 ppm in drinking water for 2 years (Wei et al., 1999). A brief report of a 2-year carcinogenicity study in F344 rats fed DMA in the diet showed that it produced bladder tumors at doses of 40 or 100 ppm, and the effect was greater in female rats compared to males (van Gemert and Eldan, 1998). In addition, no urinary bladder tumors were observed in mice treated with DMA in long-term carcinogenicity tests (van Gemert and Eldan, 1998; Salim et al., 2002).

This paper summarized the results of carcinogenicity of DMA from our recent rat studies (Wei et al., 1999; 2002). The first objective in the studies was to determine the carcinogenicity of DMA in F344 rats when orally administered in the drinking water for a 2-year period. The second objective was to elucidate possible mechanisms involved in DMA bladder carcinogenesis by analyzing DMA-induced urinary bladder tumors for mutations of *p53*, *ras*, and β-catenin genes. Protein expression of cell cycle regulatory factors p53, p27[kip1], and cyclin D1 in normal epithelium, preneoplastic and neoplastic lesions of rat urinary bladder was assessed by immunohistochemistry. To test the hypothesis that reactive oxygen species (ROS) may play a role in DMA bladder carcinogenesis, we evaluated formation of 8-hydroxy-2′-deoxyguanosine (8-OHdG) in urinary bladder after a short-term DMA treatment, this lesion being most commonly used as a marker for evaluation of oxidative DNA damage (Floyd, 1990; Nakae et al., 1997). Moreover, cyclooxygenase-2 (COX-2) expression was examined in DMA-induced rat bladder tumors as it has been shown to play a role in the development of preneoplastic and neoplastic lesions in the human and rat urinary bladder (Mohammed et al., 1999; Kitayama et al., 2000). We also examined the possible role of defective DNA mismatch repair activity in DMA-induced urinary bladder tumors by analyzing microsatellite instability, reported in various human malignant tumors including human urinary bladder tumors (Mao et al., 1996). Finally, we discuss the possible implications of the results in DMA urinary bladder carcinogenesis.

2. Carcinogenicity of DMA in male F344 rats

To determine the carcinogenicity of DMA, we administered DMA in the drinking water to male, 10-week-old, F344 rats at concentrations 0, 12.5, 50 and 200 ppm (36 rats in each group), respectively, for 104 weeks (Wei et al., 1999, 2002). Rats that had died or were killed under ether anesthesia when becoming moribund during the study or killed at the end of the study at week 104 were autopsied for macroscopic and histopathological examinations. Ten rats from each group received an i.p. injection of 100 mg/kg body weight of BrdU (Sigma Chemical Co., St Louis, MO) 1 h before autopsy. All major organs were excised and fixed in 10% buffered formalin. After adequate fixation, they were cut and processed for paraffin embedding and routinely stained with hematoxylin and eosin for histological examination.

Urinary samples from the rats at weeks 30, 60, and 100 were used for assessment of urine chemistry, including sodium, potassium, chloride, and calcium (Hitachi-710 Electrolyte Analyzer, Tokyo, Japan). Combined ion chromatography (IC, model 7000, Yokogawa Analytical Systems, Tokyo, Japan) with inductively coupled plasma mass spectrometry (ICP-MS, model 4500; Hewlett–Packard Co., DE) were used for the determination of arsenic species in the urine samples.

There was no evidence of increased morbidity or mortality after DMA treatment. There were no statistically significant differences for the overall survival rates between groups during the experiment. The most common cause of death in all groups was leukemia, with grossly enlarged spleen and usually diffuse infiltrates of leukemic cells in the liver and spleen. DMA did not have adverse effects on food consumption, but caused an increase in water consumption at 50 and 200 ppm. The final body weights were not significantly different between the groups. No treatment-related adverse effects were apparent in the hematological and serum biochemical data among groups.

Data for the numbers and incidences of rats with urinary bladder lesions are summarized in Table 1. The reported rates are for animals surviving until at least week 97, when the first urinary bladder tumor was found. Urinary bladder tumors were observed in 8 of 31 and 12 of 31 animals in 50 and 200 ppm DMA groups, respectively. Incidences of carcinomas were 6/31 (19%) and 12/31 (39%) in 50 and 200 ppm DMA groups. Two rats in 200 ppm DMA group with bladder papillomas also had carcinomas. Papillary or nodular (PN) hyperplasias, preneoplastic lesions in the urinary bladder, were observed in 12 and 14 rats in 50 and 200 ppm DMA groups (31 rats each), respectively. No such lesions were observed in groups treated with 0 or 12.5 ppm DMA. Carcinomas were histopathologically transitional cell carcinomas (TCCs).

In all groups, the incidences of tumors except for those in the urinary bladder were typical for male F344 rats and did not exceed the historical control incidences/range of

Table 1. Incidences of urinary bladder lesions in rats treated with DMA.

DMA (ppm)	Effective no. of rat[a]	PN hyperplasia (%)	Papilloma (%)	TCC (%)	No. of tumor-bearing rats (%)
0	28	0	0	0	0
12.5	33	0	0	0	0
50	31	12 (39)[b]	2 (6)	6 (19)[c]	8 (26)[d]
200	31	14 (45)[b]	2 (6)	12 (39)[b]	12[e](39)[b]

PN, papillary or nodular; TCC, transitional cell carcinoma.
[a] The effective number of rats was indicated as the number alive at week 97, when the first bladder tumor was found.
[b] $p < 0.01$ (vs. 0 ppm group).
[c] $p < 0.05$ (vs. 0 ppm group).
[d] $p < 0.001$ (vs. 0 ppm group).
[e] Two rats with bladder papillomas also had carcinomas.

the NTP. There were no significant differences in the incidences of these tumors at any dose level.

The urinary pH did not differ significantly among groups. The concentrations of sodium, potassium, chloride and calcium were decreased in the rats treated with DMA in a dose-dependent manner, with statistical significance being reached in the 50 and 200 ppm DMA groups. The decrease in urine electrolytes was presumably due to increased urinary volume. These results confirmed previous observations showing the carcinogenic action of DMA in rat urinary bladder is not correlated with urine pH (Wanibuchi et al., 1996).

The urinary concentrations of arsenic metabolites at week 100 are indicated in Table 2. Arsenic compound levels increased in a dose-dependent manner except for AsBe. Major compounds were DMA itself and trimethylarsine oxide (TMAO), with small amounts of monomethylarsonic acid (MMA) and tetramethylarsonium (TeMA). As described in previous reports, two unidentified metabolites, peak 1 and peak 2 (Wanibuchi et al., 1996; Yoshida et al., 1998), were also found in DMA-treated groups but not the controls.

3. Genetic alteration patterns in the induced rat urinary bladder lesions

To elucidate the possible mechanisms involved in DMA bladder carcinogenesis, we also elucidate the genetic alteration patterns in the induced rat urinary bladder lesions (Wei et al., 2002).

Twenty paraffin-embedded DMA-induced rat urinary tumors (18 TCCs and 2 papillomas) were examined for mutations in exons 1 and 2 of the H- and K-*ras* oncogenes, in exons 5–8 of the *p53* tumor suppressor gene and the β-catenin gene using PCR-SSCP and direct sequencing techniques. Mutations in exon 1 of H-*ras* were demonstrated in two TCCs. No mutations were found in *p53*, K-*ras*, and β-catenin genes.

We also examined the possible role of defective DNA mismatch repair activity in DMA-induced urinary bladder tumors by analyzing microsatellite instability. Eighteen microsatellite loci interspersed throughout the rat genome were analyzed in 16 DMA-induced urinary bladder carcinomas, but no alterations were detected.

Urinary bladder epithelial cell proliferation following treatment with DMA was determined by BrdU incorporation. The data are presented as BrdU labeling indices for morphologically normal bladder epithelium. A significant increase was noted for the groups treated with 50 and 200 ppm DMA when compared with controls (data not shown). Urinary bladders from a total of 40 rats from the groups in which bladder tumors were observed (50, 200 ppm DMA-treated groups), as well as 10 rats from the 12.5 ppm DMA and control groups, respectively, were examined for p53, p27^{kip1}, cyclinD1 and COX-2 by immunohistochemistry. Results of immunohistochemical assessment in the rat urinary bladder lesions are shown in Figure 1.

Positive p27^{kip1} staining was noted within the nuclei of the urothelial cells in the control group and in morphologically normal appearing urothelium and simple hyperplasia in DMA-treated groups. The majority of PN hyperplasias also stained intensely; in marked contrast, almost all TCC and papillomas demonstrated a heterogeneous pattern of significantly reduced p27^{kip1} immunoreactivity. Thus 16 of 18 (89%) TCCs, and 3 of 4 (75%) papillomas demonstrated decreased p27^{kip1} expression. It is worthy to note that

Table 2. Urinary concentrations of arsenic compounds at week 100 in male F344 rats treated with DMA.

DMA (ppm)	No. of samples	Concentration (μg/ml)						
		MMA	DMA	TMAO	TeMA	AsBe	Peak 1	Peak 2
0	10	<0.01	0.08 ± 0.03	0.04 ± 0.01	<0.01	0.37 ± 0.11	0	0
12.5	10	0.01 ± 0.01	3.1 ± 1.7	2.7 ± 1.6	0.01 ± 0.01	0.32 ± 0.16	0.08 ± 0.07	0.23 ± 0.13
50	10	0.02 ± 0.02	20.3 ± 8.5	9.5 ± 2.5	0.04 ± 0.02	0.34 ± 0.11	2.01 ± 0.68	2.14 ± 0.76
200	10	0.07 ± 0.01	44.1 ± 5.4	36.6 ± 9.1	0.27 ± 0.04	0.45 ± 0.06	5.45 ± 1.81	6.99 ± 0.13

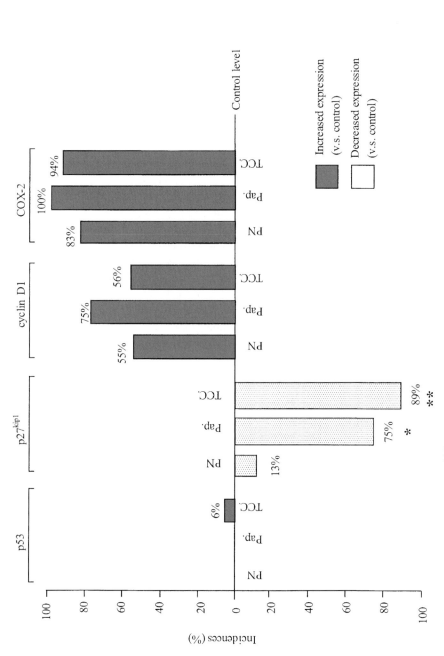

Figure 1. Immunohistochemical assessment of p53, p27^{kip1}, cyclin D1 and COX-2 expression in rat urinary bladder lesions induced by DMA. Significantly different from PN hyperplasia at * $p < 0.05$; ** $p < 0.001$. PN, papilliary or nodular hyperplasia; Pap, papilloma; TCC, transitional cell carcinoma.

$p27^{kip1}$ is significantly downregulated in papillomas and TCCs when compared with PN hyperplasias. In fact, no staining was observed in 1 papilloma and 10 TCCs.

All 10 control specimens of the control group did not show positive cyclin D1 nuclear staining in the urothelium. In contrast, all 40 specimens from the DMA-treated group displayed occasional nuclear cyclin D1 staining in small stretches of normal appearing epithelium, including all 10 bladders examined from rats given 12.5 ppm DMA in which no bladder lesion was observed. The cyclin D1 overexpression phenotype, defined as positive immunoreactivity in the nuclei of >5% of neoplastic cells, was found in 26 of 47 (55%) PN hyperplasias, 3 of 4 papillomas (75%), and 10 of 18 TCCs (56%). There was no significant difference in incidences of cyclin D1 overexpression between the bladder lesion types.

COX-2 staining was localized to the cytoplasm of tumor or preneoplastic cells but was not present in normal urothelial cells. Increased expression was noted in 17 of 18 (94%) TCCs, 4 of 4 (100%) papillomas, and 39 of 47 (83%) PN hyperplasias. Normal epithelial cells in the control group and morphologically normal urothelium of the DMA-treated groups generally did not show immunoreactivity for COX-2, providing a negative internal control for each specimen. However, positive COX-2 staining was also occasionally noted in the simple hyperplasia and morphologically normal epithelium adjacent to tumors.

Only one TCC (6%) demonstrated nuclear accumulation of p53 protein in more than 5% of the nuclei. We did not find any statistically significant association between $p27^{kip1}$, cyclin D1 and COX-2 expression in the DMA-induced urinary bladder lesions although most of bladder lesions showed decreased $p27^{kip1}$ expression and increased cyclin D1 and COX-2 expression simultaneously. However, no mutually exclusive distribution pattern was observed.

4. Increased 8-OHdG formation in urinary bladder after short-term DMA treatment

To test the hypothesis that ROS may play a role in DMA carcinogenesis, 8-OHdG formation in urinary bladder was examined (Wei et al., 2002). Forty, 10-week-old, male F344 rats were divided into two equal groups, and given DMA at concentrations of 0 and 200 ppm, respectively, in the drinking water for 2 weeks. The rats were killed under ether anesthesia and their urinary bladder removed, immediately frozen in liquid nitrogen and stored at $-80°C$ until used for analysis. Tissues from pairs of urinary bladders were pooled as samples for detection of 8-OHdG formation in nuclear DNA. 8-OHdG formation level was significantly increased in DMA-treated rats ($1.76-0.59/10^5$dG) compared to the controls ($1.21-0.13/10^5$dG).

5. Discussion

The present study demonstrated that DMA is carcinogenic for the urinary bladder of F344 male rats, but lacks other organ-specific carcinogenic effects, particularly regarding the liver and kidney in which promoting effects on rat carcinogenesis have been shown and arsenic-associated tumors have been reported in humans. We also found that DMA-induced rat urinary bladder tumors had a low rate of H-*ras* mutations and no mutations in *p53*, K-*ras* or β-catenin genes. Furthermore, we demonstrated that induction of cell proliferation might contribute to the carcinogenicity of DMA via mechanisms involving oxidative stress and/or alterations in cell cycle regulatory protein, p27[kip1] and cyclin D1. DMA is a major metabolite of arsenic in most mammals. Thus, the results from the present study would appear directly relevant to the carcinogenic risk of arsenic.

The bladder-specific carcinogenic effect of DMA in rat may indicate (1) longer exposure to DMA in the urinary bladder than in other organs due to urinary retention; (2) promoting activities of DMA on multiple organs need to be considered for assessing carcinogenic effects of arsenic in humans; (3) humans may be more sensitive than experimental animals to cancer induction by arsenic (Goering et al., 1999; Huff et al., 2000) or (4) DMA may not be the only carcinogenic compound involved in arsenic carcinogenesis.

p53 is the most frequently mutated tumor suppressor gene described so far in human and experimental animal cancers including urinary bladder cancer. Examination of the molecular changes in the *p53* tumor suppressor gene can contribute to our understanding the nature of carcinogenic activity. In contrast to our previous finding of frequent *p53* mutations in *N*-butyl-*N*-(4-hydroxybutyl)nitousamine-induced rat urinary bladder cancer (Masui et al., 1996), no such lesions were found in DMA-induced rat urinary bladder tumors examined in the present study. Our results indicated that DMA differs from BBN, which is genotoxic, and pathways other than the p53 pathway must be involved in the etiology of the DMA-induced rat urinary bladder tumors. Two previous studies reported high frequency of *p53* mutations in arsenic-related bladder and skin tumors from the endemic area of blackfoot disease in Taiwan (Shibata et al., 1994; Hsu et al., 1999). However, the discrepancy might be partly explicable by the fact that development of malignancy in humans is a complex multistep process, and many factors may affect the likelihood that cancer will develop. Therefore, it is reasonable to hypothesize that genetic alterations found in human cancers may be results of factors such as smoking and exposure to other arsenicals. DMA is considered to be a clastogenic agent (ATSDR, 1999) and negative in most mutagenicity studies (EPA, 1997). In light of these actions, mutations of H-*ras* in two TCCs could occur indirectly by oxidative damage or cytotoxicity of DMA (Hei et al., 1998; Arnold, 1999).

It is well established that disruption of the normal cell cycle is a critical step in cancer development. Because alterations in the *p53* gene were lacking, we focused our attention on the expression of the tumor suppressor gene p27[kip1]. p27[kip1] functions as a p53 independent negative cell cycle regulator involved in G1 arrest, and the reduction in the protein level of p27[kip1] have been reported in a variety of human cancers and is likely to provide a selective growth advantage (Steeg and Abrams, 1997). The strong p27[kip1] staining in the majority of preneoplastic lesions noted here is consistent with the established role of p27[kip1] as a tumor suppressor gene counteracting proliferative signals generated by DMA exposure. The reduced expression of p27[kip1] protein in almost all TCCs and papillomas suggests a role in malignant progression in DMA-induced rat bladder tumors.

Overexpression of cyclin D1 in human urinary bladder tumors could be a key regulatory event leading to cell proliferation and tumorigenesis. In the present study, we also found that positive cyclin D1 nuclear staining appeared in small stretches of histologically normal appearing epithelium following DMA treatment, even in animals in which no bladder lesions were observed histologically, as well as in most local lesions. The present data strongly suggested that cyclin D1 induction is one of the early events in DMA-induced rat bladder carcinogenesis. However, the fact that failure to induce bladder tumors at a dose of 12.5 ppm DMA suggests that increased cyclin D1 associated with such a low dose may be insufficient for DMA bladder carcinogenesis under the present conditions. In addition, the observed existence of tumors without increased cyclin D1 expression but featuring down-regulation of p27^{kip1} expression may mean that it is no longer necessary for at least a subset of tumors in the later stages. The actions of cyclin D1 are regulated by CDK inhibitors such as p27^{kip1}, which control its ability to activate CDK4 and CDK6. Therefore, those tumors are likely a result, at least in part, of either increased degradation or transcription of p27^{kip1}. The lack of any exclusive distribution pattern indicates that alterations may occur independently and there might exist other mechanisms by which DMA affects cell cycle regulation. Defects in a cell cycle check point may be responsible for the genomic instability (Hartwell, 1992). We can conclude that such abnormalities are frequent in DMA bladder carcinogenesis and might induce genomic instability despite the rarity of mutations in the present study.

Increasing evidence supports the hypothesis that ROS may play a role in DMA carcinogenesis. It is reported that metabolism of substances by the P-450 enzyme system can generate oxygen-free radicals (Parke and Ioannides, 1990; Klaunig et al., 1998). Our recent finding that an increase in hepatic P450 levels, especially in CYP2B1 protein in rat livers after treatment with 100 ppm DMA suggests that DMA is metabolized by P450 in rat liver and could represent a mechanism by which DMA generate ROS (Nishikawa et al., 2002). Alternatively, the possibility also exits that cytotoxicity of DMA may involve the generation of ROS since xenobiotic chemicals can produce ROS by either direct or indirect means (Trush and Kensler, 1991). Among ROS-induced forms of DNA damage, 8-OHdG is typical and most commonly used as a marker for quantitative analysis (Floyd, 1990). The finding of a significant increase in DMA-treated rats in the present study suggests that DMA treatment causes DNA damage via ROS generation, as shown earlier for the rat liver (Wanibuchi et al., 1997).

COX-2 expression, shown to be involved with the development of preneoplastic and neoplastic lesions in the human and rat bladder (Mohammed et al., 1999; Kitayama et al., 2000), was also diffused in the majority of TCCs, papillomas and PN hyperplasias in the present study. The occasional positive COX-2 staining noted in the morphologically normal epithelium adjacent to tumor, observed also in human invasive TCC of the bladder, may indicate that neoplastic cells can exert paracrine effects through the release of cytokines and/or growth factors (Mohammed et al., 1999). ROS are known to play a crucial role in the expression of COX-2 (Sen and Packer, 1996), so that this may also be a molecular marker of oxidative stress (Romanenko et al., 2000).

Based on the observations in the present experiment and the results from the literature, potential modes of action for DMA with regard to rat urinary bladder carcinogenesis are given in Figure 2. We propose two possible mechanisms by which

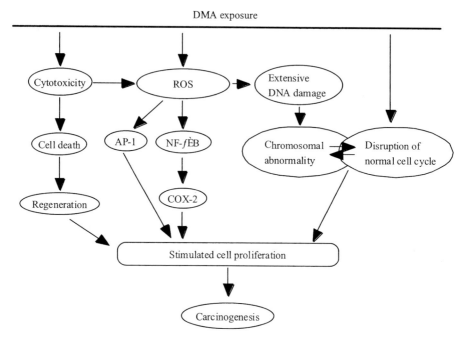

Figure 2. Potential modes of action underlying DMA carcinogenesis in rat urinary bladder.

ROS could be involved in DMA carcinogenesis in rats: (1) DMA-initiated ROS may cause specific molecular changes resulting in the activation of transcription factors such as AP-1 and NF-B. Although AP-1 and NF-B were not investigated here, ROS has been shown to cause synthesis of AP-1, and activation of AP-1 or NF-B promotes carcinogenesis (Schenk et al., 1994). DMA induces cell proliferation and gene expression in the bladder epithelium associated with AP-1 in mice (Simeonova et al., 2000). It should be remembered that ROS are known to play a crucial role in the expression of COX-2 through activating nuclear factor-B (Sen and Packer, 1996). (2) DMA could cause chromosomal abnormalities by generation of ROS and resultant DNA single strand breaks and DNA-protein crosslinks (Dong and Luo, 1993). It is generally accepted that tumor development occurs as the result of accumulation of genetic alterations. The various genetic alterations induced by DMA may not be the result of independent mechanisms. Some modes may be operating concurrently or sequentially. DMA-initiated defects in cell cycle checkpoints could give rise to genomic instability, while DMA-induced DNA damage would be expected to affect the expression of cell cycle regulators. Moreover, the fact that oxygen radicals may participate in the carcinogenic process, including the stages of initiation, promotion, and progression, suggests a reasonable mechanism by which DMA may act in all three phases. In addition, an alternative or complementary mechanism suggested by Cohen et al. for rat bladder carcinogenesis, is cytoxicity and regeneration (Arnold et al., 1999; Cohen et al., 2001). The ultimate relationships between oxidative stress, toxic stress and genetic alterations in arsenic carcinogenesis remain to be determined.

References

Arnold, L.L., Cano, M., St John, M., Eldan, M., van Gemert, M., Cohen, S.M., 1999. Effects of dietary dimethylarsinic acid on the urine and urothelium of rats. Carcinogenesis, 20 (11), 2171–2179.

ATSDR, 1999. Toxicological Profile for Arsenic (Update). Agency for Toxic Substances and Disease Registry, Atlanta, GA.

Cohen, S.M., Yamamoto, S., Cano, M., Arnold, L.L., 2001. Urothelial cytotoxicity and regeneration induced by dimethylarsinic acid in rats. Toxicol. Sci., 59, 68–74.

Dong, J.T., Luo, X.M., 1993. Arsenic-induced DNA-strand breaks associated with DNA-protein crosslinks in human fetal lung fibroblasts. Mutat. Res., 302, 97–102.

Environmental Protection Agency, 1997. Report on the Expert Panel on Arsenic Carcinogenicity. National Center for Environmental Assessment, US Environmental Protection Agency, Washington, DC.

Floyd, R.A., 1990. The role of 8-hydroxyguanine in carcinogenesis. Carcinogenesis, 11, 1447–1450.

Goering, P.L., Aposhian, H.V., Mass, M.J., Cebrian, M., Beck, B.D., Waalkes, M.P., 1999. The enigma of arsenic carcinogenesis: role of metabolism. Toxicol. Sci., 49, 5–14.

Hartwell, L., 1992. Defects in a cell cycle checkpoint may be responsible for the genomic instability of cancer cells. Cell, 71, 543–546.

Hei, T.K., Liu, S.X., Waldren, C., 1998. Mutagenicity of arsenic in mammalian cells: role of reactive oxygen species. Proc. Natl Acad. Sci. USA, 95, 8103–8107.

Hsu, C.H., Yang, S.A., Wang, J.Y., Yu, H.S., Lin, S.R., 1999. Mutational spectrum of p53 gene in arsenic-related skin cancers from the blackfoot disease endemic area of Taiwan. Br. J. Cancer, 80, 1080–1086.

Huff, J., Chan, P., Nyska, A., 2000. Is the human carcinogen arsenic carcinogenic to laboratory animals? Toxicol. Sci., 55, 17–23.

Kitayama, W., Denda, A., Yoshida, J., Sasaki, Y., Takahama, M., Murakawa, K., Tsujiuchi, T., Tsutsumi, M., Konishi, Y., 2000. Increased expression of cyclooxygenase-2 protein in rat lung tumors induced by *N*-nitrosobis(2-hydroxypropyl)amine. Cancer Lett., 148, 145–152.

Klaunig, J.E., Xu, Y., Isenberg, J.S., Bachowski, S., Kolaja, K.L., Jiang, J., Stevenson, D.E., Walborg, E.F., Jr., 1998. The role of oxidative stress in chemical carcinogenesis. Environ. Health Perspect., 106 (Suppl. 1), 289–295.

Mao, L., Schoenberg, M.P., Scicchitano, M., Erozan, Y.S., Merlo, A., Schwab, D., Sidransky, D., 1996. Molecular detection of primary bladder cancer by microsatellite analysis. Science, 271, 659–662.

Masui, T., Dong, Y., Yamamoto, S., Takada, N., Nakanishi, H., Inada, K., Fukushima, S., Tatematsu, M., 1996. p53 mutations in transitional cell carcinomas of the urinary bladder in rats treated with *N*-butyl-*N*-(4-hydroxybutyl)-nitrosamine. Cancer Lett., 105, 105–112.

Mohammed, S.I., Knapp, D.W., Bostwick, D.G., Foster, R.S., Khan, K.N., Masferrer, J.L., Woerner, B.M., Snyder, P.W., Koki, A.T., 1999. Expression of cyclooxygenase-2 (COX-2) in human invasive transitional cell carcinoma (TCC) of the urinary bladder. Cancer Res., 59, 5647–5650.

Nakae, D., Kobayashi, Y., Akai, H., Andoh, N., Satoh, H., Ohashi, K., Tsutsumi, M., Konishi, Y., 1997. Involvement of 8-hydroxyguanine formation in the initiation of rat liver carcinogenesis by low dose levels of *N*-nitrosodiethylamine. Cancer Res., 57, 1281–1287.

Nishikawa, T., Wanibuchi, H., Ogawa, M., Kinoshita, A., Morimura, K., Hiroi, T., Funae, Y., Kishida, H., Nakae, D., Fukushima, S., 2002. Promoting effects of monomethylarsonic acid, dimethylarsinic acid and trimethylarsine oxide on induction of rat liver preneoplastic glutathione S-transferase placental form positive foci: a possible reactive oxygen species mechanism. Int. J. Cancer, 100, 136–139.

Parke, D.V., Ioannides, C., 1990. Role of cytochromes P-450 in mouse liver tumor production. Prog. Clin. Biol. Res., 331, 215–230.

Romanenko, A., Morimura, K., Wanibuchi, H., Salim, E.I., Kinoshita, A., Kaneko, M., Vozianov, A., Fukushima, S., 2000. Increased oxidative stress with gene alteration in urinary bladder urothelium after the Chernobyl accident. Int. J. Cancer, 86, 790–798.

Salim, E.I., Wanibuchi, H., Morimura, K., Wei, M., Mitsuhashi, M., Yoshida, K., Endo, G., Fukushima, S., 2003. Carcinogenicity of dimethylarsinic acid in p53 heterozygous knockout and wild-type C57BL/6J mice. Carcinogenesis, 24 (2), 335–342.

Schenk, H., Klein, M., Erdbrugger, W., Droge, W., Schulze, O.K., 1994. Distinct effects of thioredoxin and antioxidants on the activation of transcription factors NF-kappa B and AP-1. Proc. Natl Acad. Sci. USA, 91, 1672–1676.

Sen, C.K., Packer, L., 1996. Antioxidant and redox regulation of gene transcription. FASEB J., 10, 709–720 (see comments).

Shibata, A., Ohneseit, P.F., Tsai, Y.C., Spruck, C.H., 3rd, Nichols, P.W., Chiang, H.S., Lai, M.K., Jones, P.A., 1994. Mutational spectrum in the p53 gene in bladder tumors from the endemic area of blackfoot disease in Taiwan. Carcinogenesis, 15, 1085–1087.

Simeonova, P.P., Wang, S.Y., Toriuma, W., Kommineni, V., Matheson, J., Unimye, N., Kayama, F., Harki, D., Ding, M., Vallyathan, V., Luster, M.I., 2000. Arsenic mediates cell proliferation and gene expression in the bladder epithelium: association with activating protein-1 transactivation. Cancer Res., 60, 3445–3453.

Steeg, P.S., Abrams, J.S., 1997. Cancer prognostics: past, present and p27. Nat. Med., 3, 152–154 (news; comment).

Trush, M.A., Kensler, T.W., 1991. An overview of the relationship between oxidative stress and chemical carcinogenesis. Free Radic. Biol. Med., 10, 201–209.

van Gemert, M., Eldan, M., 1998. Chronic Carcinogenicity Assessment of Cacodylic Acid. The Third International Conference on Arsenic Exposure and Health Effects, Book of Abstracts, pp. 113.

Wanibuchi, H., Yamamoto, S., Chen, H., Yoshida, K., Endo, G., Hori, T., Fukushima, S., 1996. Promoting effects of dimethylarsinic acid on *N*-butyl-*N*-(4-hydroxybutyl)nitrosamine-induced urinary bladder carcinogenesis in rats. Carcinogenesis, 17, 2435–2439.

Wanibuchi, H., Hori, T., Meenakshi, V., Ichihara, T., Yamamoto, S., Yano, Y., Otani, S., Nakae, D., Konishi, Y., Fukushima, S., 1997. Promotion of rat hepatocarcinogenesis by dimethylarsinic acid: association with elevated ornithine decarboxylase activity and formation of 8-hydroxydeoxyguanosine in the liver. Jpn. J. Cancer Res., 88, 1149–1154.

Wei, M., Wanibuchi, H., Yamamoto, S., Li, W., Fukushima, S., 1999. Urinary bladder carcinogenicity of dimethylarsinic acid in male F344 rats. Carcinogenesis, 20, 1873–1876.

Wei, M., Wanibuchi, H., Morimura, K., Iwai, S., Yoshida, K., Endo, G., Nakae, D., Fukushima, S., 2002. Carcinogenicity of dimethylarsinic acid in male F344 rats and genetic alterations in induced urinary bladder tumors. Carcinogenesis, 23 (8), 1387–1397.

Yamamoto, S., Konishi, Y., Matsuda, T., Murai, T., Shibata, M.A., Matsui, Y.I., Otani, S., Kuroda, K., Endo, G., Fukushima, S., 1995. Cancer induction by an organic arsenic compound, dimethylarsinic acid (cacodylic acid), in F344/DuCrj rats after pretreatment with five carcinogens. Cancer Res., 55, 1271–1276.

Yoshida, K., Inoue, Y., Kuroda, K., Chen, H., Wanibuchi, H., Fukushima, S., Endo, G., 1998. Urinary excretion of arsenic metabolites after long-term oral administration of various arsenic compounds to rats. J. Toxicol. Environ. Health, Pt. A, 54, 179–192.

PART IV
MODE OF ACTION

Arsenic Exposure and Health Effects V
W.R. Chappell, C.O. Abernathy, R.L. Calderon and D.J. Thomas, editors

Chapter 17

Enzymology and toxicity of inorganic arsenic

H. Vasken Aposhian, Robert A. Zakharyan, Sheila M. Healy,
Eric Wildfang, Jay S. Petrick, Adriana Sampayo-Reyes,
Philip G. Board, Dean E. Carter, D.N. Guha Mazumder
and Mary M. Aposhian

Abstract

The biotransformation of inorganic arsenate to DMA involves a series of enzymatic steps. We have previously reported that the enzyme responsible for the reduction of arsenate to arsenite is human liver arsenate reductase. Our recent studies based on amino acid homology and other properties demonstrate that human liver arsenate reductase and human purine nucleoside phosphorylase (PNP) are identical proteins. The reaction requires inosine and dihydrolipoic acid. The latter is the most potent naturally occurring dithiol in this reaction. GSH is relatively inactive. PNP is an essential enzyme involved in purine and thus nucleic acid metabolism. Arsenate reductase/PNP will not reduce MMA^V.

The reduction of MMA^V to the very toxic and reactive MMA^{III} is catalyzed by human liver MMA^V reductase, which our lab has demonstrated to be identical to the new omega member of the glutathione-S-transferase superfamily. MMA^V reductase has an absolute requirement for GSH. Most of the other members of the glutathione-S-transferase superfamily will not reduce MMA^V.

Although at the time of this writing, the sequence of the human arsenic methyltransferase is not known, in vitro and in vivo studies will be correlated to compile a scaffold for the interactions of inorganic arsenic and its methylated metabolites, including MMA^{III} and DMA^{III}, with body constituents to attempt an explanation for the toxicity of ingested inorganic arsenic.

Keywords: arsenic enzymes of humans, MMA^{III} toxicity, GSH transferase $\emptyset$, biotransformation of arsenic, arsenate reductase/PNP, fate of arsenic

Abbreviations:

MMA, a generic term including MMA^{III} and MMA^V; MMA^{III}, monomethylarsonous acid; MMA^V, monomethylarsonic acid; DMA, a generic term including DMA^{III} and DMA^V; DMA^V, dimethylarsinic acid; DMA^{III}, dimethylarsinous acid; inorg As, inorganic arsenic; DMPS, sodium 2,3-dimercapto-1-propane sulfonate; PAGE, polyacrylamide gel electrophoresis; CySH, cysteine; LC, liquid chromatography; DHLP/α-DHLP, dihydrolipoic acid; MS, mass spectrometry; PIR, protein information resource; SDS, sodium dodecyl sulfate; CYT19, a protein.

1. Introduction

The approach of our group for studying the arsenic problem has been to identify and purify the enzymes of humans involved in the metabolism of inorganic arsenic (Aposhian, 1997;

Aposhian et al., 1999; Sampayo-Reyes et al., 2000; Zakharyan et al., 2001; Radabaugh et al., 2002). We basically are enzymologists and believe a great deal can be learned in an unambiguous manner if enzymes are purified and then used to study and characterize the metabolic step they catalyze. In addition, we have studied the arsenic species in urine of humans in a number of countries where people are chronically exposed to inorganic arsenic in their drinking water, resulting in the identification of MMA[III] in human urine (Aposhian et al., 2000a,b). Our overall goal has been to understand how the human body processes inorganic arsenic.

The purpose of this chapter is to step back and reflect on what we have found, how the work of others fits into our thinking, and what remains to be accomplished. It needs to be stated clearly that we do not study arsenic carcinogenicity. Rather, we study its biotransformation and its molecular mechanisms of toxicity with the expectation that the knowledge coming from our laboratory and others will be of help to those trying to unravel the mechanisms of arsenic carcinogenicity (Cohen et al., 2001; Rossman et al., 2001; Waalkes et al., 2001).

A summary of documented groundwater arsenic incidents throughout the world has been reported by Chakraborti et al. (2002). Their paper also pointed out the extent of the problem in West Bengal and Bangladesh along with the inept national and international bureaucracies that have been involved. A reading of the Chakraborti paper and others (Aposhian, 2001; Liu et al., 2002a) will again emphasize the need for environmental justice especially for the poor.

Inorganic arsenic metabolism in humans can now be summarized as shown in Figure 1. Beginning with arsenate or arsenite, the important metabolites in humans appear to be MMA[V], MMA[III], DMA[V], and DMA[III] (Aposhian et al., 2000b; Le et al., 2000a,b; Del Razo et al., 2001).

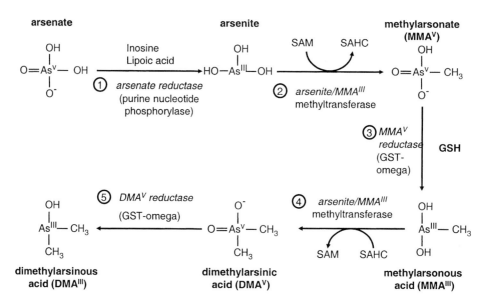

Figure 1. Enzyme pathways for inorganic arsenic biotransformation in the human.

2. Human liver arsenate reductase

This enzyme is the first step in the bioactivation of inorganic arsenate as a human carcinogen (Fig. 1). We have purified the enzyme and, after SDS-PAGE, two discrete bands were found. One of these bands was a 34 kDa protein (Radabaugh and Aposhian, 2000; Radabaugh et al., 2002). Each band was excised from the gel and sequenced by LC–MS/MS. Sequest analyses were performed against the OWL database SWISS-PROT with PIR. Mass spectra analysis matched the 34 kDa human protein of interest with human purine nucleoside phosphorylase (PNP). The peptide fragments analyzed were equal to 40.1% of the total protein and were 100% identical to the corresponding regions of the human PNP. This enzyme does not require glutathione. It prefers reduced lipoic acid (Radabaugh et al., 2002), which is the most active of the naturally occurring thiols and dithiols (Fig. 2).

Because we were not able to obtain the DNA recombinant human PNP for our studies, we used calf spleen PNP. The overall homology of calf spleen PNP and human PNP is 87% and it is 100% within the active site (Williams et al., 1984; Bzowska et al., 1995). MMA^V is not reduced by PNP.

The arsenate-reducing activity of calf thymidine phosphorylase was 5% and of rabbit muscle phosphorylase B <0.3% as compared to calf PNP (Radabaugh et al., 2002). Phosphate can compete with arsenate for arsenate reductase. The PNP mutation in knockout mice, however, is lethal so only cultures of cells can be obtained from them. The knockouts cannot be used to any great extent *in vivo*.

The reduction of arsenate by arsenate reductase/PNP is summarized in Scheme 1.

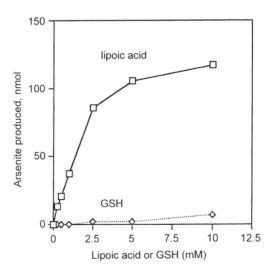

Figure 2. Dihydrolipoic acid and the reduction of arsenate by PNP. The assay (100 μl) was performed using inosine 10 mM, arsenate 6 mM, EDTA 1 mM, and PNP 1 μg in 0.1 M MOPS buffer, pH 7.5, 30 min at 37°C.

Scheme 1: Reduction of arsenate is catalyzed by PNP arsenolysis reaction in the presence of α-DHLP

1. Purine nucleoside + H_2PO_4 $\xrightarrow{\text{PNP}}$ purine base + ribose-1-phosphate.

2. Purine nucleoside + arsenate $\xrightarrow{\text{PNP}}$ purine base + ribose-1-arsenate $\xrightarrow{H_2O}$ ribose + arsenate.

3. Purine nucleoside + arsenate $\xrightarrow{\text{PNP}}$ purine base + ribose-1-arsenate $\xrightarrow[\text{DHLP}]{H_2O}$ arsenite + ribose + lipoic acid (oxidized form).

Thus, it appears that arsenate reductase/PNP is the major route for the reduction of arsenate to the more toxic arsenite (Radabaugh et al., 2002). The Gregus group in Hungary also has shown the importance of PNP for arsenate reduction (Gregus and Németi, 2002; Németi and Gregus, 2002).

PNP is an essential enzyme of purine metabolism. Many sophisticated biochemical studies of it are available (for a review, see Bzowska et al., 2000). It is specific for purine and purine analogues, ribonucleosides, and deoxyribonucleosides. Pyrimidines or pyridines are not active as substrates. PNP catalyzes the cleavage of the glycoside bond of ribo- and deoxyribonucleosides in the presence of inorganic orthophosphate to yield the purine base and ribose (deoxyribose) 1-phosphate (Parks et al., 1981; Stoeckler, 1984; Markert, 1991; Montgomery et al., 1993; Erion et al., 1997). The arsenolysis reaction catalyzed by PNP is similar to phosphorolysis, but the product ribose-1-arsenate is unstable and rapidly hydrolyzed to ribose and arsenate (Parks and Agarwall, 1972; Kline and Schramm, 1993).

The identification of human liver arsenate reductase and PNP as the same protein has implications for the present public health calamity involving the millions of people drinking water containing carcinogenic amounts of arsenate. Arsenate is believed to mimic phosphate in many metabolic processes in the human. There is extensive literature on inhibition of PNP and its potential role in the treatment of disease (Bzowska et al., 2000). The inhibition of arsenate reduction by therapeutic agents might decrease the production of arsenite and its resulting carcinogenesis.

3. Arsenic methyltransferases

The next biotransformation step is catalyzed by arsenic methyltransferase (Fig. 1). Our lab has been studying this enzyme for many years (Zakharyan et al., 1995, 1996, 1999; Healy et al., 1997). These enzymes catalyze reactions in which the methyl group of S-adenosyl methionine (SAM) is transferred to either arsenite or MMA[III]. The specific activities of arsenite methyltransferase in mouse tissue have been determined (Healy et al., 1998). The specific activity in the testes > kidney > liver > lung. Of all the enzymes involved in inorganic arsenic metabolism, this has been the most elusive in humans. Our laboratory reported the purification and properties of rabbit liver arsenic methyltransferase in 1995 (Zakharyan et al., 1995). The arsenite methyltransferase and MMA[III] methyltransferase were shown to be present on one protein (Zakharyan et al., 1995). The arsenic methyltransferase of Chang human hepatocytes was also purified (Zakharyan et al., 1999b).

No one has been successful as yet in demonstrating that surgically removed human tissue has arsenic methyltransferase activity. Our laboratory has analyzed biopsied samples of human liver from patients chronically exposed to inorganic arsenic in their drinking water as well as liver biopsies from control subjects without being able to detect such activity. Such samples were sent to us by Dr. Guha Mazumder of Calcutta. In addition, we have tested surgically removed liver samples from our university's medical center as well as samples purchased throughout the world. Again no arsenic methyltransferase activity could be detected. Why activity cannot be measured in surgically removed human tissue is at present unknown.

We do not know what this means. For the human enzyme, either we have incorrect assay conditions or an incorrect methyl donor. Although we cannot detect these enzyme activities in surgically removed tissue, we can find them in the Chang human hepatocyte cell line (Zakharyan et al., 1999). Others have found activity in human primary cells (Styblo et al., 2000), but not surgically removed tissues.

Most of our work defining this enzyme has used rabbit liver as the enzyme source. A number of different purification protocols have been used. The most recent one is summarized in Table 1. Three different investigators in our laboratory have tried to isolate and sequence the rabbit liver arsenic methyltransferase. Each has identified a protein different from that identified by the other two investigators. Activity in an enzyme solution that gives a single band on SDS-PAGE is not sufficient evidence. Very few purified protein bands isolated from tissues contain a single pure protein. If the amount of protein placed in the gel were increased, it would not be surprising to find additional proteins. Lin et al. (2002) have proposed that arsenic methyltransferase is a protein called CYT19 because of SDS-PAGE and sequence analysis of a rat liver protein. It will be very important for them to detect this protein and its activity in surgically removed human tissue.

4. Is methylation of inorganic arsenic, a detoxication mechanism as suggested by earlier investigators?

That methylation was not a detoxication step for arsenic was first suggested by studies showing that the marmoset, tamarin, and chimpanzee do not excrete MMA or DMA when given a dose of inorganic arsenic (Vahter et al., 1982, 1995); neither does the guinea pig (Healy et al., 1997). Our laboratory has demonstrated that these mammals and others do not have detectable arsenic methyltransferase activity in their livers (Table 2) (Zakharyan et al., 1996; Healy et al., 1997; Wildfang et al., 2001). It is obvious, therefore, that these animals, many of which are just as susceptible to inorganic arsenic toxicity, do not use methylation as a detoxication process.

5. MMA^V reductase

This enzyme (Fig. 1) appears to be the rate-limiting enzyme for the arsenic methylation pathway (Zakharyan and Aposhian, 1999). We have purified human liver MMA^V reductase (Zakharyan and Aposhian, 1999; Zakharyan et al., 2001) and sequenced it (Fig. 3). Sequences (92%) were analyzed and found to be identical to human glutathione-*S*-transferase-$\varnothing$ (GST-$\varnothing$) (Board et al., 2000; Zakharyan et al., 2001).

Table 1. Arsenite (As^{III}) and methylarsonous acid (MMA^{III}) methyltransferase purification summary.

Fraction	Volume (ml)	Protein (mg/ml)	Total protein (mg)	Activity (pmol/ml)		Specific activity (pmol/mg)		Recovery (%)		Purification (x-fold)	
				As^{III}	MMA^{III}	As^{III}	MMA^{III}	As^{III}	MMA^{III}	As^{III}	MMA^{III}
Cytosol	160	27.5	4400	1.35	2.90	0.049	0.105	n/a	n/a	n/a	n/a
DEAE Sepharose pool	14	15.4	216	31.87	5.26	2.07	0.341	206	15.9	42	3
Rotofor pool	4.8	1.23	5.9	25.92	2.48	21.02	2.01	57	2.6	429	19
Phenylsepharose pool	1.4	0.60	0.8	18.64	1.75	31.08	25.3^a	12	0.5	634	240
Red-120 dye agarose	2.5	0.003	0.008	2.798	n/a	932.6	n/a	3	n/a	19,033	n/a

[a] MMA^{III} methyltransferase specific activity value represents that of the peak fraction rather than the pool.

Table 2. Arsenite methyltransferase and arsenate reductase activities of primate livers.

Common name	Scientific name	n	Arsenate reductase activity (pmol As^{III}/mg protein/30 min)	Arsenite methyltransferase activity[a] (pmol/mg protein/h)
Great apes				
Gorilla	*Gorilla gorilla*	1	148	ND
Sumatran orangutan	*Pongo pygmaeus*	1	12	ND
Common chimpanzee	*Pan troglodytes*	4	190 ± 85	ND
Old world				
Rhesus macaque	*Macaca mulatta*	3	573 ± 127	0.224 ± 0.008
Pig-tailed macaque	*Macaca nemestrina*	2	943 ± 468	0.096 ± 0.049
Long-tailed macaque	*Macaca fascicularis*	1	502	0.071
Yellow baboon	*Papio cynocephalus*	3	191 ± 55	ND
New world				
Squirrel monkey	*Saimiri* sp.	3	439 ± 99	0.053 ($n = 1$); ND[b] ($n = 2$)
Owl monkey	*Aotus* sp.	4	527 ± 59	ND
Common marmoset	*Callithrix jacchus*	1	443	ND
Cotton-top tamarin	*Saguinus oedipus*	1	617	ND
Prosimians				
Aye-aye	*Daubentonia*	1	681	ND
Lesser bushbaby	*Galago senegalensis moholi*	1	211	ND
Slow loris	*Nycticebus coucang*	1	506	ND
Ring-tailed lemur	*Lemur catta*	1	281	ND
Verreaux's sifaka	*Propithecus verreauxi*	1	490	ND
Fat-tailed lemur	*Cheirogaleus medius*	1	368	ND

Note: Where $n \geq 2$, data are presented as the mean ± SEM.
[a] Golden Syrian hamster liver cytosol was assayed in parallel with all reactions as a positive control for assay and extraction procedures.
[b] ND, arsenite methyltransferase activity was not detected.

A Western blot of human liver MMA^V reductase gave one band with hGST-∅ antiserum. The evidence clearly demonstrates that human liver MMA^V reductase and human GST-∅ are identical proteins. In addition, they have the same competitive inhibitors (Zakharyan et al., 2001). MMA^V reductase has an absolute requirement for GSH. Other thiols are inactive.

Glutathione-*S*-transferases comprise a large family of enzymes in which the sulfur atom of GSH makes a nucleophilic attack on the electrophilic groups of endobiotics and xenobiotics. Members of this enzyme family are considered to be detoxifying enzymes and deficiencies of them can cause an increased risk of some diseases including cancer (Strange et al., 2000). A number of GST gene families such as alpha, mu, theta, pi, zeta, and omega have been identified (Hayes and Strange, 2000). The GST superfamily has an

Figure 3. Sequence identity between human MMA^V reductase and hGSTO 1-1. SDS-β-mercaptoethanol polyacrylamide gel band C was digested with trypsin and chymotrypsin separately and the respective peptide fragments analyzed by LC–MS/MS. The amino acid sequence of hGSTO 1-1(1) is aligned with MMA^V reductase peptides.

important function in protecting cells and there is a large interindividual variability in their expression (Clapper and Szarka, 1998). Oltipraz is an inducer of the alpha, mu, and pi members of the GST family (Salinas and Wong, 1999). Board et al. (2000) have pointed out that hGSTO 1-1 has properties unlike those of the other GSTs such as having (a) an active site cysteine that is able to form a disulfide bond with GSH. Reduction of MMA^V may be achieved *via* a one-electron donation by the thiolate of Cys-32 of MMA^V reductase/hGSTO 1-1 and the other electron from the thiolate of GSH; (b) a unique 19–20 residue N-terminal extension as compared to other GSTs; and (c) a novel structural unit formed by the N-terminal extension and the C-terminus (Board et al., 2000).

By structure, MMA^V reductase is a member of the GST superfamily, which includes alpha, mu, pi, sigma, theta, and zeta. By function, it is not. Alpha, mu, and pi do not reduce MMA^V. hGST-theta has 19% of GST-∅ activity and hGST-zeta has 2% of hGST-∅ activity when MMA^V is the substrate. If the 32 CySH of hGST-∅ is replaced by alanine, then MMA^V reductase activity is lost using MMA^V as the substrate; but 1-chloro-2,4-dinitrobenzene a GST substrate remains active (Table 3).

When one examines the K_m for human MMA^V reductase, a measure of the affinity of a substrate for its enzyme (the smaller the K_m, the tighter the binding), we find binding of DMA^V > arsenate > MMA^V. The K_m values are not that different, however.

Table 3. CY32 required for human MMA^V reductase activity.

	CDNB	MMA^V
	Product	
	μmol/min/mg	μmol/h/mg
GST01-1	0.40 ± 0.027	30.23 ± 1.06
GST01-1 Cy32/Ala	12.49 ± 0.350	ND

ND, not detected.

The velocity of the reaction is greater for MMA^V > arsenate = DMA^V. In the case of arsenate reductase, however, only arsenate, and not MMA^V or DMA^V, is a substrate (Table 4).

6. Toxicity of MMA^{III}

The product of the MMA^V reductase reaction when MMA^V is the substrate is MMA^{III}. MMA^{III} toxicity was first demonstrated in *Candida humicola* by Cullen et al. (1989) who found that MMA^{III} inhibited growth and decreased viability of this yeast. We (Petrick et al., 2000) and others (Styblo et al., 2002) have shown MMA^{III} to be more toxic than arsenite in cultured human cells. More importantly, our laboratory has shown (Fig. 4) that based on LD$_{50}$ determinations in hamsters, MMA^{III} given ip is about four times more toxic than arsenite (Petrick et al., 2001). The hamster is one of the few species that is believed to process inorganic arsenic in the same manner as humans. In addition, MMA^{III} is six times more inhibitory than sodium arsenite *in vitro* using partially purified porcine pyruvic acid dehydrogenase (Fig. 5), the classical biological target of arsenic (Petrick et al., 2001). Inhibition of other enzymes by MMA^{III} has been reported (Styblo et al., 1997). MMA^{III} has been found in human urine (Aposhian et al., 2000b), hamster tissues (Sampayo-Reyes et al., 2000), and rat bile (Gregus et al., 2000).

Table 4. Affinities and reaction rates.

Substrate	K_m	V_{max} μmol/mg/h
MMA^V reductase		
Arsenate	46.1×10^{-3} M	23.9
MMA^V	55.4×10^{-3} M	56.3
DMA^V	78.2×10^{-3} M	19.5
Arsenate reductase/PNP		
Arsenate	1.8×10^{-3} M	332.0

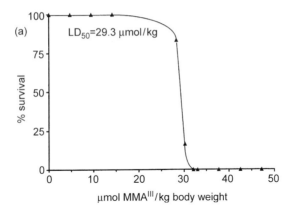

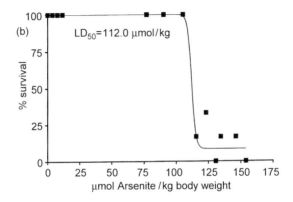

Figure 4. LD$_{50}$ of MMA^{III} or sodium arsenite in hamsters. Six animals per dose were injected ip with MMA^{III} or sodium arsenite. A total of 66 animals for MMA^{III} and 78 animals for sodium arsenite were used. (a) Percent survival for MMA^{III}; (b) percent survival for arsenite.

7. MMA^{III} methyltransferase

Using rabbit liver, the activities of arsenite methyltransferase and MMA^{III} methyltransferase copurified; and these two activities appear to be on the same protein (Zakharyan et al., 1995).

8. Inorganic arsenic toxicity

A great deal of research has been done recently and in the past on the mechanism of the inorganic arsenic toxicity (Aposhian, 1997; Aposhian et al., 1999). First of all, inorganic arsenic toxicity is not a simple problem. It involves, in most species, the accompanying and combined toxicities of arsenate, arsenite, MMA^{V}, MMA^{III}, DMA^{V}, and DMA^{III}, and in some other species, also TMAO and TMA. Excluding the latter two compounds, inorganic arsenic toxicity involves the interaction of a minimum of six recognized arsenic species. Very few, if any, other metals or metalloids have such a number of

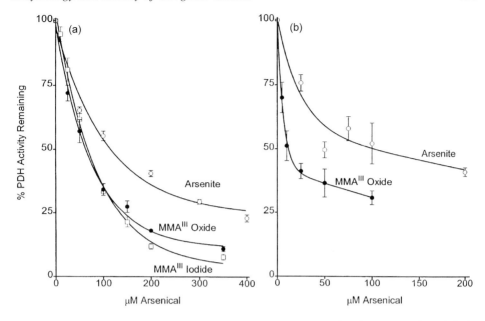

Figure 5. Inhibition of hamster kidney or purified porcine heart pyruvate dehydrogenase (PDH) activity by MMA^{III} or arsenite. (a) Inhibition of hamster kidney PDH activity. The concentrations needed for 50% inhibition of PDH were as follows: MMA^{III} oxide, 59.9 ± 6.5 μM; MMA^{III} iodide, 62.0 ± 1.8 μM; and arsenite, 115.7 ± 2.3 μM. Data are expressed as mean $\pm$ SE. (b) Inhibition of purified porcine heart PDH activity. The concentrations needed for 50% inhibition of PDH were 17.6 ± 4.1 μM MMA^{III} oxide or 106.1 ± 19.8 μM arsenite. Data are expressed as mean $\pm$ SE.

biotransformants involved in their metabolism and toxicity. After all, lead toxicity involves only Pb^{++} (Skerfving, 1986).

The toxicity of all these arsenic species must be considered and understood. It seems highly unlikely that the toxicity of arsenite does not involve the more reactive MMA^{III} and DMA^{III} species. It is also unlikely that the toxicity of arsenite, MMA^{III}, and DMA^{III} involves the inhibition of just one thiol or dithiol containing protein. These trivalent arsenic species are very reactive and have a high affinity for thiols of protein, especially proteins with vicinal thiols or two thiols very close in the three-dimensional structure of proteins. The reactions *in vivo* of the trivalent arsenic species with GSH are another and perhaps even more complex problem (Scott et al., 1993; Delnomdedieu et al., 1994). Some of these arsenic-GSH complexes are unstable and some may serve as intermediates in arsenic biotransformation. Others may be inhibitory as well.

Recent studies in our laboratory clearly show that there is a mechanism in cells that can oxidize arsenite, MMA^{III}, and DMA^{III} to arsenate, MMA^{V}, and DMA^{V}, respectively. It now appears that this mechanism is a means of decreasing the toxicity of trivalent arsenic species. The resulting pentavalent species are less toxic.

So, if we now step back and analyze the bigger picture of what happens to arsenate and arsenite in the body, we note two different fates (Figs. 6 and 7). Arsenate enters the cell *via* the phosphate carrier system. Once inside the cell, it can compete with phosphate. It can bind to polyphosphates, e.g. ADP, and is rapidly hydrolyzed. Some of it can be reduced

Metabolism of Arsenic (V) Species

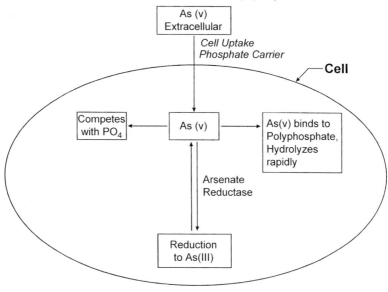

Figure 6.   Some interactions of arsenate.

Metabolism of Arsenic III Species

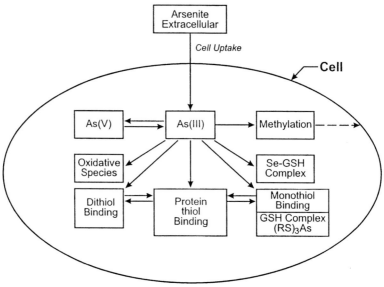

Figure 7.   Some interactions of arsenite.

enzymatically to arsenite by arsenate reductase (Radabaugh and Aposhian, 2000; Radabaugh et al., 2002) and some of it can be excreted in the urine.

The overall metabolism of the more reactive arsenite species is more complex (Fig. 7). At one time arsenite was believed to enter eukaryotic cells by diffusion because at the body's pH it is an uncharged molecule. Recent evidence suggests a transport system that brings it into the cell (Liu et al., 2002b). Once it is in the cell, there are many pathways it can follow. It can be oxidized to arsenate. A manuscript is being prepared on this mechanism in our lab. Because arsenite is so reactive, it can bind to thiols such as glutathione and proteins. It binds more strongly to dithiols such as lipoic acid and thiols that are in very close proximity in the three-dimensional structure of proteins.

We have shown that rabbits receiving arsenite and selenite excrete a GS–Se–As compound in the bile (Gailer et al., 2000) and have proposed that the formation of such a biliary complex may be responsible for the activity of selenium in reversing arsenic toxicity. Such a mechanism for reversing arsenic toxicity, however, is highly speculative and lacks the necessary *in vitro* and *in vivo* experimental evidence.

Much more needs to be done to understand arsenic metabolism and toxicity. In particular, the roles of glutathione adducts with arsenite, MMA[III], and DMA[III] need investigation in depth. Because of the species differences, human studies, both *in vitro* and *in vivo*, should be encouraged and emphasized. In this way perhaps, basic science approaches may help the 40 million people (Nordstrom, 2002) at risk of having arsenic-related cancers.

Acknowledgements

This work was supported in part by the Superfund Basic Research Program NIEHS Grant Number ES-04940 from the National Institute of Environmental Health Sciences and the Southwest Environmental Health Sciences Center P30-ES-06694.

References

Aposhian, H.V., 1997. Enzymatic methylation of arsenic species and other new approaches to arsenic toxicity. Annu. Rev. Pharmacol. Toxicol., 37, 397–419.

Aposhian, H.V., 2001. Environmental justice for children of the poor in developing countries: can we bring it about? Med. Americas, 2, 99–102.

Aposhian, H.V., Zakharyan, R.A., Wildfang, E.K., Healy, S.M., Gailer, J., Radabaugh, T.R., Bogdan, G.M., Powell, L.A., Aposhian, M.M., 1999. How is inorganic arsenic detoxified? In: Chappel, W.R., Abernathy, C.O., Calderon, R.L. (Eds), Arsenic Exposure and Health Effects: Proceedings of the Third International Conference on Arsenic Exposure and Health Effects, July 13–15, 1998, San Diego, California. Elsevier Science Ltd, Oxford, UK, pp. 289–297.

Aposhian, H.V., Zheng, B., Aposhian, M.M., Le, X.C., Cebrian, M.E., Cullen, W., Zakharyan, R.A., Ma, M., Dart, R.C., Cheng, Z., Andrews, P., Yip, L., O'Malley, G.F., Maiorino, R.M., Van Voorhies, W., Healy, S.M., Titcomb, A., 2000. DMPS-arsenic challenge test: II. Modulation of arsenic species, including monomethylarsonous acid (MMA[III]), excreted in human urine. Toxicol. Appl. Pharmacol., 165, 74–83.

Aposhian, H.V., Gurzau, E.S., Le, X.C., Gurzau, A., Healy, S.M., Lu, X., Ma, M., Yip, L., Zakharyan, R.A., Maiorino, R.M., Dart, R.C., Tircus, M.G., Gonzalez-Ramirez, D., Morgan, D.L., Avram, D., Aposhian, M.M., 2000. Occurrence of monomethylarsonous acid in urine of humans exposed to inorganic arsenic. Chem. Res. Toxicol., 13, 693–697.

Board, P.G., Coggan, M., Chelvanayagam, G., Easteal, S., Jermiin, L.S., Schulte, G.K., Danley, D.E., Hoth, L.R., Griffor, M.C., Kamath, A.V., Rosner, M.H., Chrunyk, B.A., Perregaux, D.E., Gabel, C.A., Geoghegan, K.F.,

Pandit, J., 2000. Identification, characterization, and crystal structure of the Omega class glutathione transferases. J. Biol. Chem., 275, 24798–24806.

Bzowska, A., Luic, M., Schroeder, W., Shugar, D., Saenger, W., Koellner, G., 1995. Calf spleen purine nucleoside phosphorylase: purification, sequence and crystal structure of its complex with an N(7)-acycloguanosine inhibitor. FEBS Lett., 367, 214–218.

Bzowska, A., Kulikowska, E., Shugar, D., 2000. Purine nucleoside phosphorylases: properties, functions, and clinical aspects. Pharmacol. Ther., 88, 349–425.

Chakraborti, D., Mohammed, M.R., Paul, K., Chowdhury, U.K., Sengupta, M.K., Lodh, D., Chanda, C.R., Saha, K.C., Mukherjee, S.C., 2002. Arsenic calamity in the Indian subcontinent. What lessons have been learned? Talanta, 58, 3–22.

Clapper, M.L., Szarka, C.E., 1998. Glutathione *S*-transferases: biomarkers of cancer risk and chemopreventive response. Chem.–Biol. Interact., 111–112, 377–388.

Cohen, S.M., Cano, M., St. John, M.K., Ryder, P.C., Uzvolgyi, E., Arnold, L.L., 2001. The carcinogenicity of dimethylarsinic acid (DMA) in rats. In: Chappel, W.R., Abernathy, C.O., Calderon, R.L. (Eds), Arsenic Exposure and Health Effects IV: Proceedings of the Fourth International Conference on Arsenic Exposure and Health Effects, June 18–22, 2000, San Diego, California. Elsevier Science Ltd, Oxford, UK, pp. 277–283.

Cullen, W.R., McBride, B.C., Manji, H., Pickett, A.W., Reglinski, J., 1989. The metabolism of methylarsine oxide and sulfide. Appl. Organomet. Chem., 3, 71–78.

Delnomdedieu, M., Basti, M.M., Otvos, J.D., Thomas, D.J., 1994. Reduction and binding of arsenate and dimethylarsinate by glutathione: a magnetic resonance study. Chem. Biol. Interact., 90, 139–155.

Del Razo, L.M., Styblo, M., Cullen, W.R., Thomas, D.J., 2001. Determination of trivalent methylated arsenicals in biological matrices. Toxicol. Appl. Pharmacol., 174, 282–293.

Erion, M.D., Takabayashi, K., Smith, H.B., Kessi, J., Wagner, S., Honger, S., Shames, S.L., Ealick, S.E., 1997. Purine nucleoside phosphorylase. 1. Structure–function studies. Biochemistry, 36, 11725–11734.

Gailer, J.G., George, G.N., Pickering, I.J., Prince, R.C., Ringwald, S.C., Pemberton, J.E., Glass, R.S., Younis, H.S., DeYoung, W., Aposhian, H.V., 2000. A metabolic link between arsenite and selenite: the seleno-bis (*S*-glutathionyl) arsinium ion. J. Am. Chem. Soc., 122, 4637–4639.

Gregus, Z., Németi, B., 2002. Purine nucleoside phosphorylase as a cytosolic arsenate reductase. Toxicol. Sci., 70, 13–19.

Gregus, Z., Gyurasics, A., Csanaky, I., 2000. Biliary and urinary excretion of inorganic arsenic: monomethylarsonous acid as a major biliary metabolite in rats. Toxicol. Sci., 56, 18–25.

Hayes, J.D., Strange, R.C., 2000. Glutathione *S*-transferase polymorphisms and their biological consequences. Pharmacology, 61, 154–166.

Healy, S.M., Zakharyan, R.A., Aposhian, H.V., 1997. Enzymatic methylation of arsenic compounds: IV. In vitro and in vivo deficiency of the methylation of arsenite and monomethylarsonic acid in the guinea pig. Mutat. Res., 386, 229–239.

Healy, S.M., Casarez, E.A., Ayala-Fierro, F., Aposhian, H.V., 1998. Enzymatic methylation of arsenic compounds. Toxicol. Appl. Pharmacol., 148, 65–70.

Kline, P.C., Schramm, V.L., 1993. Purine nucleoside phosphorylase. Catalytic mechanism and transition-state analysis of the arsenolysis reaction. Biochemistry, 32, 13212–13219.

Le, X.C., Lu, X., Ma, M., Cullen, W.R., Aposhian, H.V., Zheng, B., 2000. Speciation of key arsenic metabolic intermediates in human urine. Anal. Chem., 72, 5172–5177.

Le, X.C., Ma, M., Lu, X., Cullen, W.R., Aposhian, H.V., Zheng, B., 2000. Determination of monomethylarsonous acid, a key arsenic methylation intermediate, in human urine. Environ. Health Perspect., 108, 1015–1018.

Lin, S., Shi, Q., Nix, B., Styblo, M., Beck, M., Herbin-Davis, K.M., Hall, L.L., Simeonsson, J.B., Thomas, D.J., 2002. A novel *S*-adenosyl-ʟ-methionine:arsenic (III) methyltransferase from a rat liver cytosol. J. Biol. Chem., 277, 10795–10803.

Liu, J., Zheng, B., Aposhian, H.V., Zhou, Y., Chen, M.-L., Waalkes, M.P., 2002a. Chronic arsenic poisoning from burning high arsenic containing coal in Guiazhou, China. Environ. Health Perspect., 110, 119–122.

Liu, Z., Shen, J., Carbrey, J.M., Mukhopadhyay, R., Agre, P., Rosen, B.P., 2002b. Arsenite transport by mammalian aquaglyceroporins AQP7 and AQP9. Proc. Natl Acad. Sci. USA, 99, 6053–6058.

Markert, M.L., 1991. Purine nucleoside phosphorylase deficiency. Immunodefic. Rev., 3, 45–81.

Montgomery, J.A., Niwas, S., Rose, J.D., Secrist, J.A., III, Babu, Y.S., Bugg, C., Erion, M.D., Guida, W.C., Ealick, S.E., 1993. Structure-based design of inhibitors of purine nucleoside phosphorylase. 1. 9-(arylmethyl) derivatives of 9-deazaguanine. J. Med. Chem., 36, 55–69.

Németi, B., Gregus, Z., 2002. Reduction of arsenate to arsenite in hepatic cytosol. Toxicol. Sci., 70, 4–12.

Nordstrom, D.K., 2002. Worldwide occurrences of arsenic in ground water. Science, 296, 2143–2145.

Parks, R.E., Jr., Agarwall, R.P., 1972. Purine nucleoside phosphorylase. Methods Enzymol., 51, 483–514.

Parks, R.E., Stoeckler, J.D., Cambor, C., Savarese, T.M., Crabtree, G.W., Chu, S.H., 1981. In: Sartorelli, A., Lazo, J.S., Bertino, J.R. (Eds), Molecular Actions and Targets for Cancer Chemotherapeutic Agents. Academic Press, New York, pp. 229–252.

Petrick, J.S., Ayala-Fierro, F., Cullen, W.R., Carter, D.E., Aposhian, H.V., 2000. Monomethylarsonous acid (MMA^{III}) is more toxic than arsenite in Chang human hepatocytes. Toxicol. Appl. Pharmacol., 163, 203–207.

Petrick, J.S., Jagadish, B., Mash, E.A., Aposhian, H.V., 2001. Monomethylarsonous acid (MMA^{III}) and arsenite: LD$_{50}$ in hamsters and in vitro inhibition of pyruvate dehydrogenase. Chem. Res. Toxicol., 14, 651–656.

Radabaugh, T.R., Aposhian, H.V., 2000. Enzymatic reduction of arsenic compounds in mammalian systems: reduction of arsenate to arsenite by human liver arsenate reductase. Chem. Res. Toxicol., 13, 26–30.

Radabaugh, T.R., Sampayo-Reyes, A., Zakharyan, R.A., Aposhian, H.V., 2002. Arsenate reductase II. Purine nucleoside phosphorylase in the presence of dihydrolipoic acid is a route for reduction of arsenate to arsenite in mammalian systems. Chem. Res. Toxicol., 15, 692–698.

Rossman, T.G., Visalli, M.A., Uddin, A.N., Hu, Y., 2001. Human cell models for arsenic carcinogenicity and toxicity: transformation and genetic susceptibility. In: Chappel, W.R., Abernathy, C.O., Calderon, R.L. (Eds), Arsenic Exposure and Health Effects IV: Proceedings of the Fourth International Conference on Arsenic Exposure and Health Effects, June 18–22, 2000, San Diego, California. Elsevier Science Ltd, Oxford, UK, pp. 285–295.

Salinas, A.E., Wong, M.G., 1999. Glutathione S-transferase. Curr. Med. Chem., 6, 279–309.

Sampayo-Reyes, A., Zakharyan, R.A., Healy, S.M., Aposhian, H.V., 2000. Monomethylarsonic acid reductase and monomethylarsonous acid in hamster tissue. Chem. Res. Toxicol., 13, 1181–1186.

Scott, N., Hatlelid, K.M., MacKenzie, N.E., Carter, D.E., 1993. Reactions of arsenic (III) and arsenic (V) species with glutathione. Chem. Res. Toxicol., 6, 102–106.

Skerfving, S., 1986. Biological monitoring of exposure to inorganic lead. In: Clarkson, T.W., Friberg, L., Nordberg, G.F., Sager, P.R. (Eds), Biological Monitoring of Toxic Metals. Plenum Press, New York, pp. 169–197.

Stoeckler, J.D., 1984. In: Glazer, R.J. (Ed.), Developments in Cancer Chemotherapy. CRC Press, Boca Raton, FL, pp. 35–60.

Strange, R.C., Jones, P.W., Fryer, A.A., 2000. Glutathione S transferase: genetics and role in toxicology. Toxicol. Lett., 112–113, 357–363.

Styblo, M., Serves, S.V., Cullen, W.R., Thomas, D.J., 1997. Comparative inhibition of yeast glutathione reductase by arsenicals and arsenothiols. Chem. Res. Toxicol., 10, 27–33.

Styblo, M., Del Razo, L.M., Vega, L., Germolec, D.R., LeCluyse, E.L., Hamilton, G.A., Reed, W., Wang, C., Cullen, W.R., Thomas, D.J., 2000. Comparative toxicity of trivalent and pentavalent inorganic and methylated arsenicals in rat and human cells. Arch. Toxicol., 74, 289–299.

Styblo, M., Drobna, Z., Jaspers, I., Lin, S., Thomas, D.J., 2002. The role of biomethylation in toxicity and carcinogenicity of arsenic: a research update. Environ. Health Perspect., 110, 767–771.

Vahter, M., Marafante, E., Lindgren, A., Dencker, L., 1982. Tissue distribution and subcellular binding of arsenic in marmoset monkeys after injection of ^{74}As-arsenite. Arch. Toxicol., 51, 65–77.

Vahter, M., Couch, R., Nermell, B., Nilsson, R., 1995. Lack of methylation of inorganic arsenic in the chimpanzee. Toxicol. Appl. Pharmacol., 133, 262–268.

Waalkes, M.P., Keefer, L.K., Diwan, B.A., 2001. Induction of proliferative lesions of the uterus, testes and liver in Swiss mice given repeated injections of sodium arsenate. In: Chappel, W.R., Abernathy, C.O., Calderon, R.L. (Eds), Arsenic Exposure and Health Effects IV: Proceedings of the Fourth International Conference on Arsenic Exposure and Health Effects, June 18–22, 2000, San Diego, California. Elsevier Science Ltd, Oxford, UK, pp. 255–264.

Wildfang, E., Radabaugh, T.R., Aposhian, H.V., 2001. Enzymatic methylation of arsenic compounds. IX. Liver arsenite methyltransferase and arsenate reductase activities in primates. Toxicology, 168, 213–221.

Williams, S.R., Goddard, J.M., Martin, D.W., Jr., 1984. Human purine nucleoside phosphorylase cDNA sequence and genomic clone characterization. Nucleic Acids Res., 12, 5779–5787.

Zakharyan, R.A., Aposhian, H.V., 1999. Enzymatic reduction of arsenic compounds in mammalian systems: the rate-limiting enzyme of rabbit liver arsenic biotransformation is MMA^{V} reductase. Chem. Res. Toxicol., 12, 1278–1283.

Zakharyan, R.A., Wu, Y., Bogdan, G.M., Aposhian, H.V., 1995. Enzymatic methylation of arsenic compounds: assay, partial purification, and properties of arsenite methyltransferase and monomethylarsonic acid methyltransferase of rabbit liver. Chem. Res. Toxicol., 8, 1029–1038.

Zakharyan, R.A., Wildfang, E., Aposhian, H.V., 1996. Enzymatic methylation of arsenic compounds: III. The marmoset and tamarin, but not the rhesus, monkey are deficient in methyltransferase that methylate inorganic arsenic. Toxicol. Appl. Pharmacol., 140, 77–84.

Zakharyan, R.A., Ayala-Fierro, F., Cullen, W.R., Carter, D.E., Aposhian, H.V., 1999b. Enzymatic methylation of arsenic compounds. VII. Monomethylarsonous acid (MMA^{III}) is the substrate for MMA methyltransferase of rabbit liver and human hepatocytes. Toxicol. Appl. Pharamacol., 158, 9–15.

Zakharyan, R.A., Sampayo-Reyes, A., Healy, S.M., Tsaprailis, G., Board, P.G., Liebler, D.C., Aposhian, H.V., 2001. Human monomethylarsonic acid (MMA^{V}) reductase is a member of the glutathione-S-transferase superfamily. Chem. Res. Toxicol., 14, 1051–1057.

Arsenic Exposure and Health Effects V
W.R. Chappell, C.O. Abernathy, R.L. Calderon and D.J. Thomas, editors

241

Chapter 18

Structural proteomics of arsenic transport and detoxification

Zijuan Liu, Rita Mukhopadhyay, Jin Shi, Jun Ye and Barry P. Rosen

Abstract

All living organisms have systems for arsenic detoxification. Common features found in many organisms include: (a) arsenate uptake by phosphate transporters, (b) arsenite uptake by aquaglyceroporins, (c) enzymatic arsenate reduction to arsenite, the substrate of subsequent detoxification mechanisms and (d) arsenite detoxification by active extrusion out of cells or sequestration of As−thiol conjugates. In a number of organisms the genes and protein products for these transporters and enzymes have been identified, and the structures of some have been solved by X-ray crystallography.

Keywords: arsenic, aquaglyceroporin, transport ATPase, arsenate reductase

1. Introduction

Environmental ubiquity of arsenic has provided pressure for the evolution of detoxification mechanisms in organisms from *Escherichia coli* to man (Bhattacharjee et al., 1999) (Fig. 1). The metalloid arsenic has two oxidation states of biological relevance: As(V) and As(III). In solution these are found as oxyacids, arsenic acid (H_3AsO_4) or arsenous acid, also called arsenic trioxide (As_2O_3). At neutral pH, arsenic acid ionizes to the arsenate oxyanion, which is a substrate analogue of phosphate for many enzymes. However, the pK_a of arsenous acid is 9.2, which means that it is not ionized at neutral pH and would be a neutral species in solution, perhaps $As(OH)_3$. This means As(V) is transported into cells as the oxyanion arsenate, while As(III) is taken up as the neutral hydroxylated species.

2. Uptake systems for arsenate and arsenite

In *E. coli* there are two phosphate transporters, Pit and Pst (Rosenberg et al., 1977), both of which catalyze arsenate uptake (Willsky and Malamy, 1980). In the yeast *Saccharomyces cerevisiae* there are a number of phosphate transporters that also take up arsenate (Bun-ya et al., 1996; Yompakdee et al., 1996). It is likely that arsenate is taken up by phosphate transporters in most organisms, including humans.

Recently, the transporters for trivalent arsenite uptake into cells have been identified. We first identified a trivalent metalloid transporter as GlpF, the glycerol facilitator of *E. coli* (Sanders et al., 1997). GlpF is a member of the aquaporin superfamily (Fig. 1). While true aquaporins such as AQP1 are channels for only water, GlpF belongs to

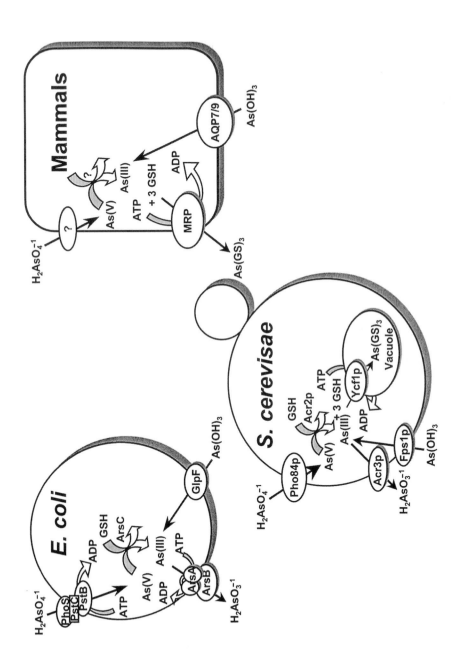

the group of aquaglyceroporins, multifunctional channels that transport neutral organic solutes such as glycerol and urea (Borgnia et al., 1999). What chemical species of As(III) do the aquaglyceroporins recognize? With a pK_a of 9.2, As(OH)$_3$ would predominate in solution at neutral pH. Since AQP7 and AQP9 conduct transmembrane movement of neutral species, As(OH)$_3$, which might be considered as an inorganic equivalent of glycerol, is the likely substrate to pass through the aquaglyceroporin channel.

In yeast, Fps1p, a homologue of GlpF, was recently shown to be the route of uptake of arsenite (Wysocki et al., 2001). An FPS1 deletion results in an increase in tolerance to arsenite, indicating that the toxic metal was not entering the cell. More recently we have shown that mammalian aquaglyceroporins catalyze uptake of the trivalent metalloids arsenite and antimonite (Liu et al., 2002). We had previously constructed a strain of *S. cerevisiae* with deletions in the genes for the two arsenite extrusion transporters, Acr3p and Ycf1p (Ghosh et al., 1999), which will be described in more detail below. Since a strain with disruptions of both ACR3 and YCF1 cannot get rid of arsenite, it becomes As(III) hypersensitive. Using that strain, we next constructed the triple deletion of ACR3, YCF1 and FPS1. This strain exhibited tolerance to arsenite compared to the parental double disrupted strain, consistent with a block in arsenite uptake. This strain was used to examine the arsenite transport properties of mammalian aquaglyceroporins. The genes for rodent AQP9 and AQP7 were cloned into a yeast expression vector, and their ability to complement the arsenite resistant phenotype of the triple yeast mutant studied. AQP9 was expressed well and restored arsenite sensitivity. Cells expressing AQP9 transported both [73]As(III) and [125]Sb(III), showing that AQP9 is a channel for both As(OH)$_3$ and Sb(OH)$_3$. AQP7 was not expressed in yeast. However, when either AQP7 or AQP9 cRNA was microinjected into *Xenopus laevis* oocytes, increased transport of [73]As(III) was observed (Liu et al., 2002). Thus, both AQP7 and AQP9 are arsenite and antimonite uptake channels.

Are all aquaglyceroporins able to facilitate As(III) transport? To date four mammalian aquaglyceroporins have been reported: AQP3 (Ishibashi et al., 1994), AQP7 (Ishibashi et al., 1997), AQP9 (Ishibashi et al., 1998) and AQP10 (Hatakeyama et al., 2001; Ishibashi et al., 2002). They share 36–45% identity with each other and are closely related to GlpF and Fps1p. Among them AQP9 has the broadest substrate range, able to transport water, glycerol and other uncharged solutes (Tsukaguchi et al., 1998). When the human forms of the aquaglyceroporins, AQP3, AQP9 and AQP10 were expressed in yeast, hAQP9 but not AQP3 or AQP10 complemented the arsenite tolerant phenotype (Z. Liu and B.P. Rosen, unpublished). In the aggregate, the results suggesting that, of the known aquaglyceroporins, only AQP7 and AQP9 aquaglyceroporins transport As(III).

Figure 1. Arsenical detoxification in prokaryotes and eukaryotes. Arsenate, As(V), is taken up by phosphate transporters (the Pst ABC transporter in *E. coli* and the Pho84p system in yeast), and As(III) is taken up by aquaglyceroporins (GlpF in *E. coli*, Fps1p in yeast and AQP7 and AQP9 in mammals). In both *E. coli* and *S. cerevisiae*, arsenate is reduced to arsenite by the bacterial ArsC or yeast Acr2p enzymes using GSH and glutaredoxin as reductants. The proteins responsible for arsenate uptake and reduction in mammals have not yet been identified. In *E. coli*, arsenite is extruded from the cells by the ArsAB ATPase. In yeast, Acr3p is a plasma membrane arsenite efflux protein, and Ycf1p, which is a member of the MRP family of the ABC superfamily of drug resistance pumps, transports As(GS)$_3$ into the vacuole. In mammals, MRP isoforms pump As(GS)$_3$ out of cells. For example, MRP2 extrudes As(III) into bile.

Recently the crystal structures of GlpF and AQP1 have been solved. GlpF was crystallized as a symmetric arrangement of four channels with three glycerol molecules in each (Fu et al., 2000). Each tetramer is circled by a hydrophobic surface matching the dimensions of the lipid bilayer. Like GlpF, AQP1 also exists as a tetramer with each subunit containing its own pore (Murata et al., 2000). Comparison of pore sizes of AQP1 and GlpF suggests that the pore of AQP1, with a diameter of 3 Å in the narrowest region, is too small for these molecules to pass through, while the 5 Å pore of GlpF is large enough for either to penetrate (Fig. 2). The As–O bond distance is 1.85 Å (Shi et al., 1996), and the O–O bond distance in unhydrated $As(OH)_3$ can be estimated at approximately 3 Å.

The finding that AQP7 and AQP9 are arsenite channels may have considerable relevance to human health and disease. Given the fact that AQP7 and AQP9 are expressed in adipocytes and hepatocytes, respectively, their glycerol permeation ability might be important during fasting or starvation states. On the one hand, differential expression of AQP7 or AQP9 might be responsible for the considerable individual variability in sensitivity to arsenic in drinking water, what has been observed in countries such as India and Bangladesh. Determination of the aquaglyceroporin status of exposed individuals may be useful for predicting their tolerance to arsenic in the water supplies.

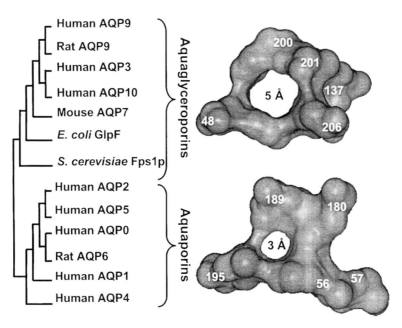

Figure 2. Aquaglyceroporins: $As(OH)_3$ channels. *Left*: A dendogram of representative members of the aquaporin family was made using the CLUSTAL4 algorithm (Higgins and Sharp, 1988). There are two subfamilies of aquaporins (Borgnia et al., 1999). The true aquaporins such as AQP1 are channels for water but not large molecules. The aquaglyceroporins have larger pores and are channels for uncharged solutes such as glycerol, urea and $As(OH)_3$. *Right*: The sizes of the AQP1 channel (bottom) and the GlpF (top) channel at their narrowest points can be visualized from the crystal structure of those two aquaglyceroporins. Residues in AQP1 surrounding the opening include Phe56, Gly57, His180, Cys189 and Arg195. Residues surrounding the channel of GlpF include Trp48, Thr137, Phe200, Ala201 and Arg206. AQP1 residue Arg195 and GlpF residue Arg206 are at the top of the channel, and the positive charge on the guanidinium side chain serves as a filter to prevent flow of protons or cations.

In clinical settings, arsenite is also an effective cancer chemotherapeutic agent. Its anhydrous form, arsenic trioxide (Trisenox) has been shown recently to be a high effective and relative safe drug in treating acute promyelocytic leukemia (APL) (Mushak and Crocetti, 1995; Chappell et al., 1997). Differences in expression of AQP7 or AQP9 could lead to variability in response to drug therapy in different patients. Rapid examination of expression of the aquaglyceroporin genes or the levels of their protein products in APL patients may be of value in tailoring treatment strategies to the individual.

3. Arsenate reduction

As described below, arsenic is removed from cells in the trivalent oxidation state. When arsenic is taken up as the pentavalent oxyanion arsenate, it must first be reduced to As(III) to become the substrate of the efflux or sequestration systems (Fig. 1). While arsenate can be reduced to arsenite nonenzymatically, this process is too slow to be physiologically significant, enzymatic reduction is required.

3.1. Three families of arsenate reductases arose through convergent evolution

Three families of arsenate reductase enzymes have been identified (Mukhopadhyay et al., 2002) (Fig. 3). Since these enzymes have little or no sequence similarity and quite different structures, it appears that these have each evolved separately. Why this should be the case has been the subject of speculation (Rosen, 1999; Mukhopadhyay et al., 2002). It is likely that arsenic resistance arose early following the origin of life. Since the original atmosphere lacked oxygen, As(III) would have been the major chemical species of arsenic in the primordial oceans. Resistance to As(III) arose, primarily through the evolution of membrane proteins that could remove arsenite from the cytosol (Rosen, 1999). Once photosynthesis arose, the atmosphere became oxidizing, and arsenite was oxidized to arsenate. This created a new evolutionary pressure to detoxify arsenate. Building on the existing platform of arsenite detoxification mechanisms, both prokaryotes and eukaryotes independently evolved thiol-linked enzymes to reduce arsenate to arsenite (Rosen, 1999; Mukhopadhyay et al., 2002) (Fig. 3). One family includes the *Staphylococcus aureus* plasmid pI258 encoded *arsC* gene product (Ji and Silver, 1992). This ArsC enzyme is homologous to low molecular weight protein phosphotyrosine phosphatases (lmw PTPases) (Bennett et al., 2001). A second family includes the *E. coli* plasmid R773 ArsC arsenate reductase (Gladysheva et al., 1994; Oden et al., 1994). The third family includes a eukaryotic arsenate reductase, Acr2p from *S. cerevisiae* (Bobrowicz et al., 1997; Mukhopadhyay and Rosen, 1998). Acr2p is homologous to the Cdc25a cell cycle protein tyrosine phosphatase (PTPase) (Fauman et al., 1998) and to rhodanase, a thiosulfate sulfurtransferase (Hofmann et al., 1998). It is interesting that the substrates of all three types of enzymes, arsenate reductases, phosphatases and sulfurtransferases, are anions. This evolutionary relatedness suggests that the common ancestor had an oxyanion-binding site. Furthermore, the fact that at least two of the arsenate reductases arose independently from ancestors of phosphoprotein phosphatases suggests that a large evolutionary leap is not required to produce an arsenate reductase. We suggest that all

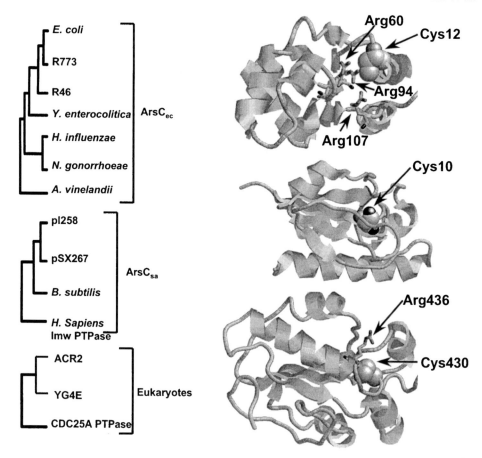

Figure 3. Three families of arsenate reductases. *Left*: A dendogram of representative members of the three families of arsenate reductases was made using the CLUSTAL4 algorithm. One family, found primarily in gram negative bacteria, includes the ArsC enzymes typified by the ArsC of *E. coli* plasmid R773 (ArsC$_{ec}$). A second bacterial family is typified by the ArsC of *S. aureus* plasmid pI258 (ArsC$_{sa}$). It is related to a family of lmw PTPases. The third family includes the ACR2 gene product of *S. cerevisiae*. Acr2p is related to the family of phosphoprotein phosphatases such as CDC25a. Opposite each dendogram is the structure of a member of the family. *Top*: R773 ArsC (Martin et al., 2001) (top), with the active site residues Cys12, Arg60, Arg94 and Arg107 identified. *Middle*: pI258 ArsC, with the active site Cys10 identified (Zegers et al., 2001). *Bottom*: A portion of Cdc25 is shown, including the Cys$_{430}$(X)$_5$Arg$_{436}$ active site (Fauman et al., 1998).

arsenate reductases, whether eukaryotic or prokaryotic, share a common evolutionary lineage with phosphatases.

3.2. *Prokaryotic arsenate reductases*

Unfortunately, the two families of bacterial arsenate reductases are both called ArsC, even though they have no sequence or structural homology. One ArsC is typified by the *arsC* gene product of *ars* operon of plasmid R773 from *E. coli* (Chen et al., 1986). The other ArsC is typified by the *arsC* gene product of plasmid pI258 of *S. aureus* (Ji and Silver, 1992).

Crystal structures of the 16 kDa R773 ArsC has been reported with either bound substrate (arsenate) or product (arsenite) in active site (Martin et al., 2001). In the reaction cycle, arsenate first binds to the anion site that consists of three basic residues, Arsg60, Arg94 and Arg107. Phosphate and sulfate are competitive for binding at this site. In the next step, arsenate (but not phosphate or sulfate) forms a covalent arsenate thioester intermediate with the active site Cys12. It is then reduced in two steps by glutaredoxin and glutathione, producing the Cys12-S-As(III) intermediate, which hydrolyzes to release arsenite. *E. coli* has three glutaredoxins, Grx1, Grx2 and Grx3, any one of which will serve as source of reducing potential for arsenate reduction, although Grx2 is preferred (Shi et al., 1999).

The pI258 enzyme does not use glutaredoxin but, instead, uses thioredoxin as the source of reducing potential (Ji et al., 1994). Thioredoxin will not substitute for glutaredoxin for the R773 ArsC, indicating catalytic differences between the two enzymes. The pI258 ArsC has three intramolecular cysteine residues that participate in the catalytic cycle (Messens et al., 1999). In the reaction pathway, reduction of arsenate to arsenite is through nucleophilic attack of Cys82 on Cys10 to form disulfide bond, which triggers a conformational change that allows Cys89 to move closely to Cys82 to form Cys82–Cys89 disulfide, and this oxidized pI258 is regenerated by thioredoxin and thioredoxin reductase (Messens et al., 2002). The intramolecular transport of electrons *via* thiol redox reaction is unique in pI258 enzyme, and distinguishes it from the reaction cycle of the R773 ArsC. The crystal structures of the pI258 ArsC and its homologue from *Bacillus subtilus* have recently been reported (Bennett et al., 2001; Zegers et al., 2001). Interestingly, the pI258 family of ArsC arsenate reductases is related to lmw PTPases and exhibits low-level phosphatase activity (Zegers et al., 2001).

3.3. Eukaryotic arsenate reductases

The first identified arsenate reductase enzyme in eukaryotes is Acr2p from *S. cerevisiae* (Bobrowicz et al., 1997; Mukhopadhyay and Rosen, 1998; Mukhopadhyay et al., 2002). Acr2p has a single active site cysteine residue and, like the R773 ArsC, uses glutaredoxin and glutathione as reductants (Mukhopadhyay and Rosen, 2001). Even though it is unrelated to either of the bacterial ArsC arsenate reductases, it can be heterologously expressed in *E. coli* and complements an *arsC* deletion (Mukhopadhyay and Rosen, 1998). All Grx proteins examined supported Acr2p activity, including a *S. cerevisiae* Grx and all three *E. coli* Grx's.

As mentioned, Acr2p is a member of the Cdc25A superfamily of PTPases such as human cell cycle dual specific phosphatase Cdc25a and is unrelated to the prokaryotic arsenate reductases. Members of this family have a consensus active site $HC(X)_5R$ motif (Fig. 3) (Denu and Dixon, 1995; Fauman et al., 1998). The consensus $C_{76}(X)_5R_{82}$ motif of Acr2p is likely to be part of the active site as well: either Cys76 or Arg82 mutations result in loss of arsenate resistance *in vivo* and arsenate reduction *in vitro* (Mukhopadhyay and Rosen, 2001). These results suggest that Cys76 is the equivalent of Cys12 in ArsC and may form As(V) and As(III) intermediates. Although Acr2p is a member of the PTPase family, it does not exhibit phosphatase activity. However, we have constructed an Acr2p mutant that has gained phosphatase activity and lost arsenate reductase activity (Mukhopadhyay *et al.*, 2003). The ease by which a reductase can be changed into a phosphatase has led us to propose that, under the selective pressure of ubiquitous arsenate in the environment, arsenate reductases may have evolved from phosphatases. Although no mammalian Acr2p homologues specific for arsenate have been identified, reductase activity has been reported in human liver (Radabaugh and Aposhian, 2000). Recently, Radabaugh and coworkers have reported that purified mammalian purine nucleotide phosphorylase reduces arsenate to arsenite in presence of dihydrolipoic acid (Radabaugh et al., 2002). Whether this *in vitro* activity reflects a physiological role for purine nucleotide phosphorylase in arsenic detoxification has not yet been examined.

4. Arsenite detoxification

Whether arsenite is taken up directly by aquaglyceroporins or is generated intracellularly by reduction of arsenate, subsequent detoxification occurs primarily by active extrusion from the cells or by sequestration into intracellular compartments. In either case, arsenite is removed from the cytosol, providing relief from As(III) toxicity.

4.1. Prokaryotic arsenic extrusion systems

In bacteria arsenite extrusion is catalyzed by arsenite carrier proteins that utilize the membrane potential for energy or by an arsenite-translocating ATPase, which uses ATP as the source of energy (Dey and Rosen, 1995). Two families of arsenite carriers have been identified. It is not clear whether the two groups share a common ancestor or arose through convergent evolution. Even though they show no sequence similarity, structure analysis might demonstrate relatedness. One group of arsenite carrier includes the arsenite resistance protein encoded by a gene in the SKIN element of *Bacillus subtilis* (Sato and Kobayashi, 1998). A larger group of bacterial arsenite carriers are the ArsB membrane proteins encoded by most *ars* operons. The ArsB encoded by the *E. coli* plasmid R773 is a 429-residue integral membrane protein with 12 membrane-spanning segments (Wu et al., 1992). From the energetics of efflux, ArsB is most likely a uniporter that extrudes arsenite using the membrane potential, positive exterior. As discussed above, in solution As(III) is predominately in the form of $As(OH)_3$, the substrate of the aquaglyceroporins. However, $As(OH)_3$ is in equilibrium with $As(OH)_2O^{-1}$. At neutral pH, a few percent of the total As(III) will always be in the form of the anion. If the affinity of ArsB for arsenite is high, this would be sufficient to reduce the intracellular arsenite concentration to subtoxic levels. In some organisms ArsB is made in conjunction with ArsA, and the two form an obligatorily ATP-coupled pump. Cells that make the ArsAB ATPase are more resistant to arsenite than those that have only ArsB because the ArsAB pump extrudes arsenite more efficiently than ArsB alone (Dey and Rosen, 1995).

ArsA is a member of a family of ATPases that has homologues in every organism, thus far sequenced, including eubacteria, archebacteria, fungi, plants and animals (Fig. 4) (Zhou and Rosen, 1997; Leipe et al., 2002). ArsA has two halves, A1 and A2, connected by a 25-residue linker (Li and Rosen, 2000). The two halves are homologous to each other, reflecting an ancestral gene duplication and fusion. The crystal structure of the enzyme has been determined (Fig. 4) (Zhou et al., 2000), and three types of domains can be resolved. In each half a nucleotide binding domain (NBD) can be observed. Second, there is

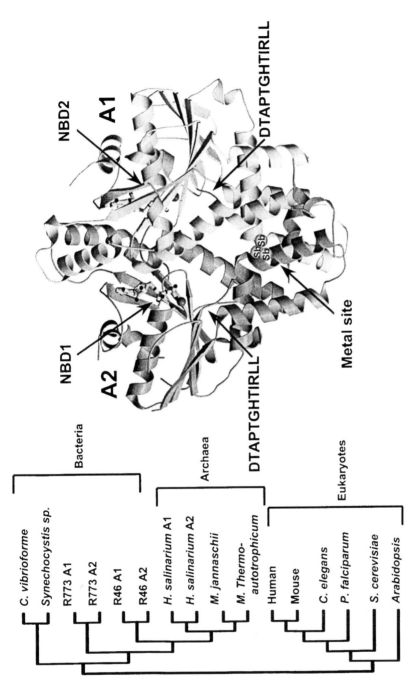

Figure 4. The ArsA family. Members of the ArsA ATPase family are identified by the presence of the 12-residue DTAPTGHTIRLL signature sequence. A dendogram of representative members is shown at left. The R773 ArsA is the catalytic subunit of the arsenite efflux pump of *E. coli*. At right is a ribbon structure of the R773 ArsA ATPase (Zhou et al., 2000). The two homologous halves are identified as A1 and A2. Also identified are the two NBDs filled with MgADP, the two signature sequences (DTAPTGHTIRLL) and the single As(III) binding site filled with three Sb(III).

a single As(III)/Sb(III) binding domain. The arsenic binding domain has three As(III) or Sb(III) coordinated to a number of residues including Cys113, Cys172 and Cys422 (Bhattacharjee et al., 1995; Bhattacharjee and Rosen, 1996) and His148 and His453 (Bhattacharjee and Rosen, 2000). We have proposed that A1 and A2 contact each other loosely in the absence of arsenic, but binding of As(III) acts as a "molecular glue" that brings the two halves of ArsA into tight contact, activating ATP hydrolysis at the two NBDs.

Connecting the single arsenic binding domain and the two NBDs are stretches of 12 residues in each half of ArsA (D_{142}TAPTGHTIRLL and D_{447}TAPTGHTIRLL) (Zhou and Rosen, 1997). This is the signature sequence that identifies the ArsA family. While the physiological roles of protein such as the human and mouse ArsA homologues are not known, they are probably unrelated to arsenic resistance (Bhattacharjee et al., 2001). In the R773 ArsA, the function of the two signature sequences is to physically link the catalytic NBDs through Asp142 and Asp447 with the allosteric arsenic binding domain at His148 and His453. This physical connection between the domains allows ATP hydrolysis at the NDBS to alter the affinity for As(III), and, reciprocally, allows As(III) binding to alter the affinity for ATP. It is hypothesized that, in the ArsAB complex, the metalloid binding site is at the interface with ArsB so that the bound As(III) is near the opening of the ArsB translocation pathway. ATP hydrolysis could then be coupled conformationally through the two signature sequences to a reduction in the affinity for As(III), which would be released into ArsB and through the membrane.

4.2. Eukaryotic arsenic extrusion systems

In humans and other mammals, arsenite is detoxified in the liver by conjugation with GSH followed by excretion into the bile (Fig. 1). Transport into the bile is catalyzed by MRP2, a member of the multidrug resistance-associated protein (MRP) group of the ABC superfamily of transport ATPases (Cole et al., 1994). The physiological roles of MRPs include export of GS-conjugates such as leukotriene C4 (LTC4) (Leier et al., 1994). In the liver, MRP2 pumps As(GS)$_3$ into bile (Kala et al., 2000).

In eukaryotic microorganisms such as *S. cerevisiae* and *Leishmania*, homologues of human MRP homologues confer arsenic resistance. Antimony-containing drugs are still treatment of choice for leishmaniasis, and about 25% of all cases show clinical resistance to these drugs. Arsenite- and antimonite-resistant *Leishmania* strains were selected *in vitro*, and some of them show increased expression of *pgpA*, which encodes an MRP homologue (Ouellette et al., 1990) that transports As(GS)$_3$ (Legare et al., 2001). The MRP homologue of *S. cerevisiae*, Ycf1p (Szczypka et al., 1994; Li et al., 1996), catalyzes transport of As(GS)$_3$ into the yeast vacuole (Fig. 1), conferring resistance by sequestration (Ghosh et al., 1999).

S. cerevisiae has a parallel pathway for arsenic detoxification that is encoded by a cluster of three genes, ACR1, ACR2 and ACR3 (Bobrowicz et al., 1997). Acr1p is a transcription factor. Acr2p, as described above, is an arsenate reductase. Acr3p is a plasma membrane arsenite transporter that catalyzes extrusion of arsenite from the cells (Fig. 1), conferring resistance (Wysocki et al., 1997; Ghosh et al., 1999). Acr3p is related to the bacterial homologues such as the *B. subtilis* SKIN element arsenite carrier protein (Sato and Kobayashi, 1998), but no Acr3p homologues have been found in mammals as yet.

Acknowledgements

This work was supported by grants GM55425, GM52216, AI45428 and ES10344 from the National Institutes of Health and a Wellcome Trust Biomedical Research Collaborative Grant.

References

Bennett, M.S., Guan, Z., Laurberg, M., Su, X.D., 2001. *Bacillus subtilis* arsenate reductase is structurally and functionally similar to low molecular weight protein tyrosine phosphatases. Proc. Natl Acad. Sci. USA, 98, 13577–13582.

Bhattacharjee, H., Rosen, B.P., 1996. Spatial proximity of Cys113, Cys172, and Cys422 in the metalloactivation domain of the ArsA ATPase. J. Biol. Chem., 271, 24465–24470.

Bhattacharjee, H., Rosen, B.P., 2000. Role of conserved histidine residues in metalloactivation of the ArsA ATPase. Biometals, 13, 281–288.

Bhattacharjee, H., Li, J., Ksenzenko, M.Y., Rosen, B.P., 1995. Role of cysteinyl residues in metalloactivation of the oxyanion-translocating ArsA ATPase. J. Biol. Chem., 270, 11245–11250.

Bhattacharjee, H., Ghosh, M., Mukhopadhyay, R., Rosen, B.P., 1999. Arsenic transporters from *E. coli* to humans. In: Broome-Smith, J.K., Baumberg, S., Sterling, C.J., Ward, F.B. (Eds), Transport of Molecules Across Microbial Membranes, Vol. 58. Society for General Micriobiology, Leeds, pp. 58–79.

Bhattacharjee, H., Ho, Y., Rosen, B.P., 2001. Genomic organization and chromosomal localization of the Asna1 gene, a mouse homologue of a bacterial arsenic-translocating ATPase gene. Gene, 272, 291–299.

Bobrowicz, P., Wysocki, R., Owsianik, G., Goffeau, A., Ulaszewski, S., 1997. Isolation of three contiguous genes, *ACR1*, *ACR2* and *ACR3*, involved in resistance to arsenic compounds in the yeast *Saccharomyces cerevisiae*. Yeast, 13, 819–828.

Borgnia, M., Nielsen, S., Engel, A., Agre, P., 1999. Cellular and molecular biology of the aquaporin water channels. Annu. Rev. Biochem., 68, 425–458.

Bun-ya, M., Shikata, K., Nakade, S., Yompakdee, C., Harashima, S., Oshima, Y., 1996. Two new genes, PHO86 and PHO87, involved in inorganic phosphate uptake in *Saccharomyces cerevisiae*. Curr. Genet., 29, 344–351.

Chappell, W.R., Beck, B.D., Brown, K.G., Chaney, R., Cothern, R., Cothern, C.R., Irgolic, K.J., North, D.W., Thornton, I., Tsongas, T.A., 1997. Inorganic arsenic: a need and an opportunity to improve risk assessment. Environ. Health Perspect., 105, 1060–1067.

Chen, C.M., Misra, T.K., Silver, S., Rosen, B.P., 1986. Nucleotide sequence of the structural genes for an anion pump. The plasmid-encoded arsenical resistance operon. J. Biol. Chem., 261, 15030–15038.

Cole, S.P., Sparks, K.E., Fraser, K., Loe, D.W., Grant, C.E., Wilson, G.M., Deeley, R.G., 1994. Pharmacological characterization of multidrug resistant MRP-transfected human tumor cells. Cancer Res., 54, 5902–5910.

Denu, J.M., Dixon, J.E., 1995. A catalytic mechanism for the dual-specific phosphatases. Proc. Natl Acad. Sci., 92, 5910–5914.

Dey, S., Rosen, B.P., 1995. Dual mode of energy coupling by the oxyanion-translocating ArsB protein. J. Bacteriol., 177, 385–389.

Fauman, E.B., Cogswell, J.P., Lovejoy, B., Rocque, W.J., Holmes, W., Montana, V.G., Piwnica-Worms, H., Rink, M.J., Saper, M.A., 1998. Crystal structure of the catalytic domain of the human cell cycle control phosphatase, Cdc25A. Cell, 93, 617–625.

Fu, D., Libson, A., Miercke, L.J., Weitzman, C., Nollert, P., Krucinski, J., Stroud, R.M., 2000. Structure of a glycerol-conducting channel and the basis for its selectivity. Science, 290, 481–486.

Ghosh, M., Shen, J., Rosen, B.P., 1999. Pathways of As(III) detoxification in *Saccharomyces cerevisiae*. Proc. Natl Acad. Sci. USA, 96, 5001–5006.

Gladysheva, T.B., Oden, K.L., Rosen, B.P., 1994. Properties of the arsenate reductase of plasmid R773. Biochemistry, 33, 7288–7293.

Hatakeyama, S., Yoshida, Y., Tani, T., Koyama, Y., Nihei, K., Ohshiro, K., Kamiie, J.I., Yaoita, E., Suda, T., Hatakeyama, K., Yamamoto, T., 2001. Cloning of a new aquaporin (aqp10) abundantly expressed in duodenum and jejunum. Biochem. Biophys. Res. Commun., 287, 814–819.

Higgins, D.G., Sharp, P.M., 1988. CLUSTAL: a package for performing multiple sequence alignment on a microcomputer. Gene, 73, 237–244.

Hofmann, K., Bucher, P., Kajava, A.V., 1998. A model of Cdc25 phosphatase catalytic domain and Cdk-interaction surface based on the presence of a rhodanese homology domain. J. Mol. Biol., 282, 195–208.

Ishibashi, K., Sasaki, S., Fushimi, K., Uchida, S., Kuwahara, M., Saito, H., Furukawa, T., Nakajima, K., Yamaguchi, Y., Gojobori, T., et al., 1994. Molecular cloning and expression of a member of the aquaporin family with permeability to glycerol and urea in addition to water expressed at the basolateral membrane of kidney collecting duct cells. Proc. Natl Acad. Sci. USA, 91, 6269–6273.

Ishibashi, K., Kuwahara, M., Gu, Y., Kageyama, Y., Tohsaka, A., Suzuki, F., Marumo, F., Sasaki, S., 1997. Cloning and functional expression of a new water channel abundantly expressed in the testis permeable to water, glycerol, and urea. J. Biol. Chem., 272, 20782–20786.

Ishibashi, K., Kuwahara, M., Gu, Y., Tanaka, Y., Marumo, F., Sasaki, S., 1998. Cloning and functional expression of a new aquaporin (AQP9) abundantly expressed in the peripheral leukocytes permeable to water and urea, but not to glycerol. Biochem. Biophys. Res. Commun., 244, 268–274.

Ishibashi, K., Morinaga, T., Kuwahara, M., Sasaki, S., Imai, M., 2002. Cloning and identification of a new member of water channel (AQP10) as an aquaglyceroporin. Biochim. Biophys. Acta, 1576, 335–340.

Ji, G., Silver, S., 1992. Reduction of arsenate to arsenite by the ArsC protein of the arsenic resistance operon of *Staphylococcus aureus* plasmid pI258. Proc. Natl Acad. Sci. USA, 89, 9474–9478.

Ji, G., Garber, E.A.E., Armes, L.G., Chen, C.M., Fuchs, J.A., Silver, S., 1994. Arsenate reductase of *Staphylococcus aureus* plasmid pI258. Biochemistry, 33, 7294–7299.

Kala, S.V., Neely, M.W., Kala, G., Prater, C.I., Atwood, D.W., Rice, J.S., Lieberman, M.W., 2000. The MRP2/cMOAT transporter and arsenic–glutathione complex formation are required for biliary excretion of arsenic. J. Biol. Chem., 275, 33404–33408.

Legare, D., Richard, D., Mukhopadhyay, R., Stierhof, Y.D., Rosen, B.P., Haimeur, A., Papadopoulou, B., Ouellette, M., 2001. The *Leishmania* ATP-binding cassette protein PGPA is an intracellular metal-thiol transporter ATPase. J. Biol. Chem., 276, 26301–26307.

Leier, I., Jedlitschky, G., Buchholz, U., Cole, S.P., Deeley, R.G., Keppler, D., 1994. The MRP gene encodes an ATP-dependent export pump for leukotriene C4 and structurally related conjugates. J. Biol. Chem., 269, 27807–27810.

Leipe, D.D., Wolf, Y.I., Koonin, E.V., Aravind, L., 2002. Classification and evolution of P-loop GTPases and related ATPases. J. Mol. Biol., 317, 41–72.

Li, J., Rosen, B.P., 2000. The linker peptide of the ArsA ATPase. Mol. Microbiol., 35, 361–367.

Li, Z.S., Szczypka, M., Lu, Y.P., Thiele, D.J., Rea, P.A., 1996. The yeast cadmium factor protein (YCF1) is a vacuolar glutathione S-conjugate pump. J. Biol. Chem., 271, 6509–6517.

Liu, Z., Shen, J., Carbrey, J.M., Mukhopadhyay, R., Agre, P., Rosen, B.P., 2002. Arsenite transport by mammalian aquaglyceroporins AQP7 and AQP9. Proc. Natl Acad. Sci. USA, 99, 6053–6058.

Martin, P., DeMel, S., Shi, J., Gladysheva, T., Gatti, D.L., Rosen, B.P., Edwards, B.F., 2001. Insights into the structure, solvation, and mechanism of ArsC arsenate reductase, a novel arsenic detoxification enzyme. Structure (Camb.), 9, 1071–1081.

Messens, J., Hayburn, G., Desmyter, A., Laus, G., Wyns, L., 1999. The essential catalytic redox couple in arsenate reductase from *Staphylococcus aureus*. Biochemistry, 38, 16857–16865.

Messens, J., Martins, J.C., Van Belle, K., Brosens, E., Desmyter, A., De Gieter, M., Wieruszeski, J.M., Willem, R., Wyns, L., Zegers, I., 2002. All intermediates of the arsenate reductase mechanism, including an intramolecular dynamic disulfide cascade. Proc. Natl Acad. Sci. USA, 99, 8506–8511.

Mukhopadhyay, R., Rosen, B.P., 1998. The *Saccharomyces cerevisiae* ACR2 gene encodes an arsenate reductase. FEMS Microbiol. Lett., 168, 127–136.

Mukhopadhyay, R., Rosen, B.P., 2001. The phosphatase C(X)5R motif is required for catalytic activity of the *Saccharomyces cerevisiae* Acr2p arsenate reductase. J. Biol. Chem., 18, 18.

Mukhopadhyay, R., Rosen, B., Phung, L., Silver, S., 2002. Microbial arsenic: from geocycles to genes and enzymes. FEMS Microbiol. Rev., 26, 311.

Mukhopadhyay, R., Zhou, Y., Rosen, B.P., 2003. Directed evolution of a yeast arsenate reductase into a protein tyrosine phosphatase. J. Biol. Chem., 278, 24476–24480.

Murata, K., Mitsuoka, K., Hirai, T., Walz, T., Agre, P., Heymann, J.B., Engel, A., Fujiyoshi, Y., 2000. Structural determinants of water permeation through aquaporin-1. Nature, 407, 599–605.

Mushak, P., Crocetti, A.F., 1995. Risk and revisionism in arsenic cancer risk assessment. Environ. Health Perspect., 103, 684–689.

Oden, K.L., Gladysheva, T.B., Rosen, B.P., 1994. Arsenate reduction mediated by the plasmid-encoded ArsC protein is coupled to glutathione. Mol. Microbiol., 12, 301–306.

Ouellette, M., Fase-Fowler, F., Borst, P., 1990. The amplified H circle of methotrexate-resistant *Leishmania tarentolae* contains a novel P-glycoprotein gene. EMBO J., 9, 1027–1033.

Radabaugh, T.R., Aposhian, H.V., 2000. Enzymatic reduction of arsenic compounds in mammalian systems: reduction of arsenate to arsenite by human liver arsenate reductase. Chem. Res. Toxicol., 13, 26–30.

Radabaugh, T.R., Sampayo-Reyes, A., Zakharyan, R.A., Aposhian, H.V., 2002. Arsenate reductase II. Purine nucleoside phosphorylase in the presence of dihydrolipoic acid is a route for reduction of arsenate to arsenite in mammalian systems. Chem. Res. Toxicol., 15, 692–698.

Rosen, B.P., 1999. Families of arsenic transporters. Trends Microbiol., 7, 207–212.

Rosenberg, H., Gerdes, R.G., Chegwidden, K., 1977. Two systems for the uptake of phosphate in *Escherichia coli.* J. Bacteriol., 131, 505–511.

Sanders, O.I., Rensing, C., Kuroda, M., Mitra, B., Rosen, B.P., 1997. Antimonite is accumulated by the glycerol facilitator GlpF in *Escherichia coli.* J. Bacteriol., 179, 3365–3367.

Sato, T., Kobayashi, Y., 1998. The *ars* operon in the skin element of *Bacillus subtilis* confers resistance to arsenate and arsenite. J. Bacteriol., 180, 1655–1661.

Shi, W., Dong, J., Scott, R.A., Ksenzenko, M.Y., Rosen, B.P., 1996. The role of arsenic–thiol interactions in metalloregulation of the *ars* operon. J. Biol. Chem., 271, 9291–9297.

Shi, J., Vlamis-Gardikas, A., Åslund, F., Holmgren, A., Rosen, B.P., 1999. Reactivity of glutaredoxins 1, 2, and 3 from *Escherichia coli* shows that glutaredoxin 2 is the primary hydrogen donor to ArsC-catalyzed arsenate reduction. J. Biol. Chem., 274, 36039–36042.

Szczypka, M.S., Wemmie, J.A., Moye-Rowley, W.S., Thiele, D.J., 1994. A yeast metal resistance protein similar to human cystic fibrosis transmembrane conductance regulator (CFTR) and multidrug resistance-associated protein. J. Biol. Chem., 269, 22853–22857.

Tsukaguchi, H., Shayakul, C., Berger, U.V., Mackenzie, B., Devidas, S., Guggino, W.B., van Hoek, A.N., Hediger, M.A., 1998. Molecular characterization of a broad selectivity neutral solute channel. J. Biol. Chem., 273, 24737–24743.

Willsky, G.R., Malamy, M.H., 1980. Characterization of two genetically separable inorganic phosphate transport systems in *Escherichia coli.* J. Bacteriol., 144, 356–365.

Wu, J., Tisa, L.S., Rosen, B.P., 1992. Membrane topology of the ArsB protein, the membrane subunit of an anion-translocating ATPase. J. Biol. Chem., 267, 12570–12576.

Wysocki, R., Bobrowicz, P., Ulaszewski, S., 1997. The *Saccharomyces cerevisiae ACR3* gene encodes a putative membrane protein involved in arsenite transport. J. Biol. Chem., 272, 30061–30066.

Wysocki, R., Chery, C.C., Wawrzycka, D., Van Hulle, M., Cornelis, R., Thevelein, J.M., Tamas, M.J., 2001. The glycerol channel Fps1p mediates the uptake of arsenite and antimonite in *Saccharomyces cerevisiae.* Mol. Microbiol., 40, 1391–1401.

Yompakdee, C., Bun-ya, M., Shikata, K., Ogawa, N., Harashima, S., Oshima, Y., 1996. A putative new membrane protein, Pho86p, in the inorganic phosphate uptake system of *Saccharomyces cerevisiae.* Gene, 171, 41–47.

Zegers, I., Martins, J.C., Willem, R., Wyns, L., Messens, J., 2001. Arsenate reductase from *S. aureus* plasmid pI258 is a phosphatase drafted for redox duty. Nat. Struct. Biol., 8, 843–847.

Zhou, T., Rosen, B.P., 1997. Tryptophan fluorescence reports nucleotide-induced conformational changes in a domain of the ArsA ATPase. J. Biol. Chem., 272, 19731–19737.

Zhou, T., Radaev, S., Rosen, B.P., Gatti, D.L., 2000. Structure of the ArsA ATPase: the catalytic subunit of a heavy metal resistance pump. EMBO J., 19, 1–8.

Arsenic Exposure and Health Effects V
W.R. Chappell, C.O. Abernathy, R.L. Calderon and D.J. Thomas, editors
Published by Elsevier B.V.

Chapter 19

A novel *S*-adenosylmethionine-dependent methyltransferase from rat liver cytosol catalyzes the formation of methylated arsenicals[*]

Stephen B. Waters, Shan Lin, Miroslav Styblo
and David J. Thomas

Abstract

Enzymatically catalyzed methylation of arsenic is part of a metabolic pathway that converts inorganic arsenic into methylated products. Hence, in humans chronically exposed to inorganic arsenic, methyl and dimethyl arsenic account for most of the arsenic that is excreted in the urine. Although methylation was originally thought to detoxify inorganic arsenic, methylated arsenicals containing trivalent arsenic formed as intermediates in the course of arsenic metabolism are more reactive than inorganic arsenic and may play important roles in toxicity and carcinogenicity associated with chronic arsenic exposure. In the work reported here, we isolated, purified, and characterized from rat liver cytosol a novel S-adenosylmethionine-dependent methyltransferase that converts arsenite to methyl and dimethyl arsenic. This 369-residue protein (molecular mass = 41,056) has features common to many non-nucleic acid methyltransferases. It is closely related to methyltransferases of unknown function that were identified by conceptual translations of the cyt19 genes in mouse and human genomes. Therefore, we designate rat liver arsenic methyltransferase as cyt19 and suggest that cyt19 is also an arsenic methyltransferase in mice and humans. Studies with either purified or recombinant rat cyt19 show that this protein efficiently catalyzes the conversion of arsenite to methyl and dimethyl arsenic. The activity of the protein is dependent on the availability of reducing equivalents. These may be supplied by dithiothreitol or tris-(2-carboxyethyl)phosphine. An enzymatic system consisting of thioredoxin, thioredoxin reductase, and NADPH can also provide reducing equivalents to support the catalytic function of cyt19.

Keywords: arsenic, methylated arsenicals, methylation, methyltransferase, rat

1. Introduction

A remarkable aspect of the metabolism of arsenic in humans and other species is transformation of inorganic arsenic into methylated forms. Braman and Foreback (1973) provided the first evidence that methyl arsenic and dimethylated arsenic could be detected in human urine. These investigators speculated that methylation of arsenic could be

* Disclaimer: This manuscript has been reviewed in accordance with the policy of the National Health and Environmental Effects Research Laboratory, US Environmental Protection Agency, and approved for publication. Approval does not signify that the contents necessarily reflect the views and policies of the Agency, nor does mention of trade names or commercial products constitute endorsement or recommendation for use.

attributed to methylcobalamin-methioneine (*sic*) reactions but admitted the site at which arsenic methylation occurs in humans "is not completely understood." The efforts of many investigators over the past three decades have been needed to better, if not completely, understand the processes involved in the conversion of inorganic arsenic into methylated arsenicals. Here, we report on recent identification of a novel enzyme involved in the production of methylated arsenicals. In the following paragraphs, we provide some historical perspective on the search for the source of methylated arsenicals and the rationale for our endeavor.

Peoples (1974) found that methyl arsenic in urine of cows treated with sodium arsenite or sodium arsenate and suggested that rumenal bacteria might be the source of this methylated product. However, the detection of methyl arsenic in urine of dogs treated with these inorganic arsenicals showed that non-rumenal species also methylated arsenic (Lakso and Peoples, 1975). Studies by Crecelius and associates in humans exposed to inorganic arsenic in occupational (Smith et al., 1977) or controlled experimental settings (Crecelius, 1977) confirmed the presence of methylated metabolites in urine and provided new impetus to elucidate the pathway for arsenic metabolism. Further studies in humans (Buchet et al., 1981) and other species (Tam et al., 1979a; Odanaka et al., 1980) indicated that methylation of arsenic was a common phenomenon. However, the site of or mechanism for arsenic methylation remained unknown. Tam et al. (1979b) reported a low rate of conversion of inorganic arsenic to dimethyl arsenic in human erythrocytes under *in vitro* assay conditions. Buchet and associates extended these studies using cytosol from rat liver as the source for the putative arsenic methyltransferase. In an *in vitro* assay system containing rat liver cytosol, they found that a methyl group donor, *S*-adenosylmethionine (AdoMet), and a thiol (e.g. glutathione (GSH)) were needed to support the conversion of inorganic arsenic to methyl and dimethyl arsenic (Buchet and Lauwerys, 1985, 1988). In an *in vitro* assay system, mouse liver was also shown to contain an arsenic methyltransferase activity (Hirata et al., 1989).

The next logical step in the study of the enzymology of arsenic methylation was purification of the protein that catalyzes the methylation reactions. Aposhian and his colleagues reported the purification of an enzyme from rabbit liver cytosol that methylates both arsenite and methylarsonous acid (Zakharyan et al., 1995, 1999). This ~60 kDa protein uses AdoMet as the methyl group donor. Its activity in methylation of arsenite is stimulated by a thiol (GSH) and its activity in methylation of methylarsonous acid is stimulated by a dithiol, dithiothreitol (DTT). These activities are designated arsenite methyltransferase (E.C. 2.1.1.137) and methylarsonite methyltransferase (E.C. 2.1.1.138), respectively. Although purification of these enzyme activities from rabbit liver provided much useful information on the characteristics of the methylation process, it did not provide protein sequence data that could be used to clone the gene encoding this protein or orthologous genes in other species, including humans. Therefore, we undertook the purification of an arsenic methyltransferase from rat liver cytosol with the goal of cloning its gene. This research has resulted in the identification of a novel arsenic methyltransferase that is conserved in rat, mouse, and human. Expression of the recombinant arsenic methyltransferase will facilitate future studies of the enzyme's properties and its role in the pathway leading from inorganic arsenic to methylated metabolites.

2. Results

2.1. Purification of arsenic methyltransferase from rat liver cytosol

The use of rat liver cytosol as the starting material for the purification of arsenic methyltransferase was based on the observation that in an *in vitro* assay system this tissue fraction rapidly converted inorganic arsenic into methylated metabolites (Styblo et al., 1996). Table 1a shows the scheme for the purification of arsenic methyltransferase activity from the cytosolic fraction prepared from livers of adult male Fischer 344 rats. Details of this purification scheme have been reported (Lin et al., 2001). A notable aspect of this purification scheme is the inclusion of both GSH (5 mM) and DTT (1 mM) in all buffer solutions used in the preparation of the cytosolic fraction and in all separatory steps. As noted above, both thiol reagents promote the methylation of arsenic. As summarized in Table 1b, this scheme yielded more than a 9000-fold increase in the specific activity for arsenic methyltransferase. SDS-polyacrylamide gel electrophoresis (SDS-PAGE) of the active fraction from *S*-adenosylhomocysteine affinity chromatography yielded a single Coomassie Blue-reactive band with an estimated molecular weight of 42 kDa.

Table 1a. Scheme for the purification of an arsenic methyltransferase activity from rat liver cytosol.

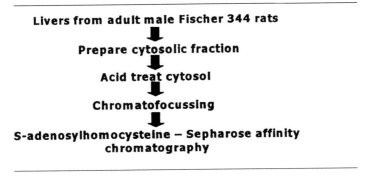

Livers from adult male Fischer 344 rats

⬇

Prepare cytosolic fraction

⬇

Acid treat cytosol

⬇

Chromatofocussing

⬇

S-adenosylhomocysteine – Sepharose affinity chromatography

Table 1b. Purification of an arsenic methyltransferase activity from rat liver cytosol.

Step	Total protein (mg)	Specific activity	Total activity	Purification
Cytosol	1200	0.16	192	1
Acidified cytosol	678	0.48	325	3
Chromatofocusing	22	216	4752	1347
S-AH affinity Chromatography	0.76	1490	1132	9312

Specific activity expressed as pmol of methylated and dimethylated arsenic formed per mg of protein in a reaction mixture containing 0.1 μM (^{73}As)-labeled sodium arsenite, 1 mM AdoMet, 5 mM GSH, and 1 mM DTT in 50 mM Na phosphate buffer that was incubated at 37 °C for 45 min. Total activity (pmol of methylated and dimethylated arsenic formed) is the product of the specific activity and the total protein content of that fraction.

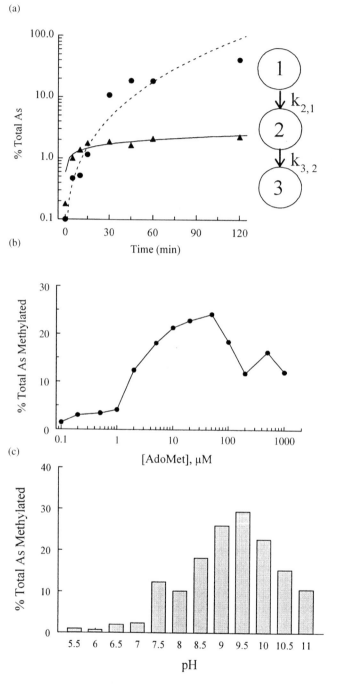

Figure 1. Characteristics of the methylation of arsenite by purified rat liver methyltransferase. (a) Time course of the production of methyl (▲) and dimethyl arsenic (•) in reaction mixtures containing 25 μg of purified enzyme, 0.1 μM [^{73}As]-labeled arsenite, 1 mM DTT, 5 mM GSH, 1 mM AdoMet in 50 mM phosphate buffer, pH 7.5, with 5% glycerol incubated at 37°C. Structure shown for the compartmental model for the sequential conversion of

2.2. Characterization of purified enzyme

The catalytic properties of the 42 kDa protein were examined using enzyme purified from rat liver cytosol. Arsenite was rapidly converted to methyl arsenic and dimethyl arsenic in assay systems incubated at 37°C for intervals up to 120 min (Fig. 1a). These data were fitted with a three-compartment kinetic model with the structure shown in Figure 1a using SAAMII (version 1.1.1, SAAM Institute, University of Washington) that provided estimates of the overall rates for the appearance of methylated and dimethylated arsenic in this assay system. Based on the reduction in the residual sum of squares, $k_{2,1}$ was estimated as 0.00443%/min and $k_{3,2}$ as 0.183%/min. Under similar conditions, neither arsenate (As^V) nor methylarsonic (As^V) acid was methylated (data not shown). By comparison, methylarsine (As^{III}) oxide was converted to dimethyl arsenic in this assay system. A K_m of 250 nM and a V_{max} of 68 pmol/mg of protein/min were estimated for the conversion of methylarsine oxide to dimethyl arsenic. Arsenic methyltransferase from rat liver cytosol was most active in the formation of methylated arsenicals from arsenite at concentrations of AdoMet below 100 μM (Fig. 1b). This enzyme had a broad pH range with maximal activity at or above pH 9 (Fig. 1c).

2.3. Sequencing of rat liver arsenic methyltransferase and cloning of its gene

The 42 kDa protein recovered from SDS-PAGE provided partial amino acid sequence data for rat liver arsenic methyltransferase. The scheme for the determination of the amino acid sequences of peptides prepared from the purified enzyme is shown in Table 2. To determine the mRNA sequence of the protein, a set of degenerate oligonucleotide primers were designed on the basis of the translation of peptide sequences obtained from rat liver arsenic methyltransferase. These primers were used in polymerase chain reactions (PCR) to amplify cDNA prepared from RNA isolated from the livers of adult male Fischer 344 rat. Several rounds of primer selection and sequencing of PCR amplified products permitted determination of the mRNA sequence encoding the complete protein. Table 3 summarizes the nucleotide sequence and its conceptual translation.

Arsenic methyltransferase was cloned by PCR amplification of rat liver cDNA using oligonucleotide primers encompassing the full-length nucleotide sequence (Waters, S.B., submitted). The amplified product was inserted into pRSETa vector and used to transform *E. coli* strain DH5α. Ampicillin-resistant colonies were PCR screened to identify positive (expressing) clones. A clone encoding the entire sequence of rat liver arsenic methyltransferase was selected and designated as pRSET-rcyt19. To produce recombinant protein, *E. coli* strain BL21(DE3)pLysS was transformed with pRSET-rcyt19. Expression of the His-tagged recombinant protein was driven by the addition of isopropyl thiogalactose to culture medium and recombinant protein was purified from cell lysate by chromatography on Ni-NTA resin.

---

arsenite (1) to methyl arsenic (2) to dimethyl arsenic (3). (b) Effect of AdoMet concentration on the production of methyl and dimethyl arsenic by purified rat liver methyltransferase under reaction conditions described above. Assays incubated for 12 min at 37°C. (c) Effect of pH on the production of methyl and dimethyl arsenic by purified rat liver methyltransferase in reaction mixtures containing 2.5 μg of purified enzyme, 0.1 μM [^{73}As]-labeled arsenite, 1 mM DTT, 5 mM GSH, 1 mM AdoMet that were incubated at 37°C incubated for 15 min.

Table 2.   Scheme for the determination and use of partial amino acid
scheme of As methyltransferase purified from rat liver cytosol.

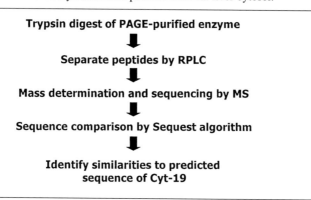

Trypsin digest of PAGE-purified enzyme
⬇
Separate peptides by RPLC
⬇
Mass determination and sequencing by MS
⬇
Sequence comparison by Sequest algorithm
⬇
**Identify similarities to predicted
sequence of Cyt-19**

2.4. Orthologous genes and proteins

Initial analysis of amino acid sequences obtained from a tryptic digest of purified rat liver
arsenic methyltransferase found similarities between this protein sequence and conceptual
translations of genes expressed in the mouse and human tissues that had been previously
designated as *cyt19*. Based on sequence similarity, we have designated rat liver arsenic
methyltransferase as cyt19. The alignments of the predicted sequence of rat liver arsenic
methyltransferase with the predicted sequences of the cyt19 proteins expressed in mouse
and human tissues identified common features shared by these proteins (Table 4). The
number of amino acid residues in the proteins were similar (rat, 369; mouse, 376; human,
375) and they shared a UbiE methylase-like domain containing sequence motifs (motif I,
post I, motif II, motif III) commonly found in non-nucleic acid methyltransferases. Twelve
cysteine residues were common to all proteins, including a carboxyl-terminal cysteine
residue. Although not unequivocally demonstrated, human or mouse cyt19 is likely to
function as arsenic methyltransferases. Expression of recombinant human cyt19 in *E. coli*
confers the capacity for the methylation of arsenite to cells that do not constitutively
methylate arsenite (Waters, S.B., unpublished results).

2.5. Expression of and mechanism for arsenic methyltransferase

The expression of cyt19 in the tissues of the rat was evaluated. As shown in Figure 2a,
reverse transcription (RT)-PCR analysis detected cyt19 mRNA in rat heart, adrenal,
urinary bladder, brain, kidney, lung, and liver. We also examined the expression of cyt19
in two human cell lines that differed in capacity to methylate arsenite (Fig. 2b). In HepG2
cells, a human hepatoma cell line, which efficiently methylates arsenite, cyt19 mRNA was
detected after 35 or 45 cycles of RT-PCR amplification. However, in UROtsa cells, an
immortalized cell line derived from normal human urothelium which does not methylate
arsenite, cyt19 mRNA was not detected after 45 cycles of RT-PCR amplification.

Table 3. Nucleotide sequence and conceptual translation of rat liver arsenic methyltransferase.

```
                                              Met Ala Ala Pro Arg Asp
   1 AGAGGGAATCCCTGGTTCTGGAAGTGGAGATCGTGAGTC ATG GCT GCT CCC CGA GAC

     Ala Glu Ile His Lys Asp Val Gln Asn Tyr Tyr Gly Asn Val Leu Lys
  58 GCA GAG ATC CAC AAG GAC GTT CAG AAC TAC TAT GGG AAT GTA CTG AAG

     Thr Ser Ala Asp Leu Gln Thr Asn Ala Cys Val Thr Pro Ala Lys Gly
 107 ACA TCT GCA GAC CTC CAG ACT AAT GCT TGT GTC ACC CCA GCC AAG GGG

     Val Pro Glu Tyr Ile Arg Lys Ser Leu Gln Asn Val His Glu Glu Val
 156 GTC CCT GAG TAC ATC CGG AAA AGT CTG CAG AAT GTA CAT GAA GAA GTT

     Ile Ser Arg Tyr Tyr Gly Cys Gly Leu Val Val Pro Glu His Leu Glu
 205 ATT TCC AGG TAT TAT GGC TGC GGT CTG GTG GTG CCT GAG CAT CTG GAA

     Asn Cys Arg Ile Leu Asp Leu Gly Ser Gly Ser Gly Arg Asp Cys Tyr
 253 AAC TGC CGG ATT TTG GAT CTG GGC AGT GGG AGT GGC AGA GAT TGC TAT

     Val Leu Ser Gln Leu Val Gly Gln Cys Gly His Ile Thr Gly Ile Asp
 302 GTG CTT AGC CAG CTG GTC GGC CAG AAG GGA CAC ATC ACC GGG ATA GAC

     Met Thr Lys Val Gln Val Glu Val Ala Lys Ala Tyr Leu Glu Tyr His
 351 ATG ACT AAG GTC CAG GTG GAA GTG GCT AAG GCC TAT CTT GAG TAC CAC

     Thr Glu Lys Phe Gly Phe Gln Thr Pro Asn Val Thr Phe Leu His Gly
 399 ACG GAA AAG TTC GGT TTC CAG ACA CCC AAT GTG ACT TTT CTT CAC GGC

     Gln Ile Glu Met Leu Ala Ala Ala Gly Ile Glr Lys Glu Ser Tyr Asp
 448 CAA ATT GAG ATG TTG GCA GAG GCC GGG ATC CAG AAG GAG AGC TAT GAT

     Ile Val Ile Ser Asn Cys Val Ile Asn Leu Val Pro Asp Lys Gln Lys
 497 ATC GTT ATA TCC AAT TGT GTG ATC AAC CTT GTT CCC GAC AAA CAA AAA

     Val Leu Arg Glu Val Tyr Gln Val Leu Lys Tyr Gly Gly Glu Leu Tyr
 546 GTC CTT CGG GAG GTC TAC CAA GTC CTG AAG TAC GGC GGG GAG CTC TAT

     Phe Ser Asp Val Tyr Ala Ser Leu Glu Val Ser Glu Asp Ile Lys Ser
 595 TTC AGT GAC GTC TAT GCT AGC CTT GAA GTG TCA GAA GAC ATC AAG TCA

     His Lys Val Leu Trp Gly Glu Cys Leu Gly Gly Ala Leu Tyr Trp Lys
 644 CAC AAG GTT TTA TGG GGG GAA TGC CTG GGT GGC GCT CTG TAC TGG AAG

     Asp Leu Ala Val Ile Ala Lys Lys Ile Gly Phe Cys Pro Pro Arg Leu
 693 GAC CTC GCC GTC ATT GCC AAA AAG ATT GGG TTC TGC CCT CCA CGT TTG

     Val Thr Ala Asn Ile Ile Thr Val Gly Asn Lys Glu Leu Glu Arg Val
 741 GTC ACT GCC AAT ATT ATT ACG GTT GGA AAC AAG GAA CTA GAA AGG GTT

     Leu Gly Asp Cys Arg Phe Val Ser Ala Thr Phe Arg Leu Phe Lys Leu
 790 CTT GGT GAC TGT CGC TTC GTG TCT GCC ACA TTT CGC CTC TTC AAA CTC

     Pro Lys Thr Glu Pro Ala Gly Arg Cys Gln Val Val Tyr Asn Gly Gly
 839 CCT AAG ACA GAG CCA GCC GGA AGA TGC CAA GTT GTT TAC AAT GGA GGA

     Ile Met Gly His Glu Lys Glu Leu Ile Phe Asp Ala Asn Phe Thr Phe
 888 ATC ATG GGG CAC GAA AAG GAA CTA ATT TTC GAT GCA AAT TTT ACA TTC

     Lys Glu Gly Glu Ala Val Glu Val Asp Glu Glu Thr Ala Ala Ile Leu
 937 AAG GAA GGT GAA GCT GTT GAA GTG GAT GAG GAG ACG GCA GCC ATC TTG

     Arg Asn Ser Arg Phe Ala his Asp Phe Leu Phe Thr Pro Val Glu Ala
 986 AGG AAC TCT CGG TTT GCT CAC GAT TTT CTC TTC ACA CCT GTT GAG GCC

     Ser Leu Leu Ala Pro Gln Thr Lys Val Ile Ile Arg Asp Pro Phe Lys
1035 TCC CTG TTG GCT CCC CAA ACA AAG GTT ATA ATC AGA GAT CCA TTC AAG

     Leu Ala Glu Glu Ser Asp Lys Met Lys Pro Arg Cys Ala Pro Glu Gly
1084 CTT GCA GAG GAA TCT GAC AAG ATG AAG CCG AGA TGT GCA CCA GAA GGC

     Thr Gly Gly Cys Cys Gly Gly Arg Lys Ser Cys Ter
1151 ACT GGA GGC TGC TGT GGC AAA AGG AAA AGC TGC TAG ACCTAGGGCCAGCGT

1203 AGAGCCCACC GAGCATGAGG GGGTGGCTAA AGGGCAGTCA CAAAGTCTTC TGAGCCTGCT
1264 CTTCACCAGA GCACAGACTA TGGGAAGATG GCAAAGCCAC TGCTAAGAAA AAGTGATTTT
1325 TAGAGGGTGT TAATTTAAGG TTCACAGCAA ACTTTTTCTA TATTTCAGAG TTCTGGCGCC
1386 ACCTAGTGGT CAGAAGTAGA ACTTGTACGT CTAAGGTTTA CAAATATCTT CCTTCTGGCC
1447 TACCACAGGA CACCTCTGGG TTTTTCTCTG TGGTTATTCA GGAAGCACAG TACTTACTGA
1508 ATTTATGCTG ACTATGCAAA AAGGTTGCCA ACTCAAATTT GGTAGGAGTA CTCTTTAGGT
1569 CGCTGTCTTC AAACTTTTTT CCCTGTAAAT GGAAATAAAT GAAAACAAAT GAAAAATAAA
1629 TTTGACTTCA TCCTGATAAA AAAAAAAAAA AAAAAAAAA A
```

Table 4. Multiple sequence alignment for rat, mouse, and human cytl9 with consensus sequence.

```
          1        10        20        30        40        50        60        70        80        90       100       110       120       130
rcyt.19   MAAPRDIA--EIHKDVANYYGNVLKTSADLQTNACYTPAKGYPEYTRKSLQNHHEEVISRYYGCGLVYPEHLENCRILDLGSGSGRDCYYLSQLVGQKGHITGIDMTKYQVEYAKGYLEYHTEKFGFQTPAV
mcyt.19   MAASRDADEIHKDVQNYYGNVLKTSADLQTHACYTRAKPYPSIYIRESLQNHHEDVSSRYYGCGLTVPEHLENCRILDLGSGSGRDCYYLSQLVGEKGHYTGIDMTKYQVEYAKGYLEHHNEKFGFQAPAV
hcyt.19   MAALRDA--EIQKDVQTYYGGVLKRSADLQTNGCYTRAPYYPKHIREALQNHHEVALRYYGCGLYTPEHLENCRILDLGSGSGRDCYYLSQLVGEKGHYTGIDMTKGQVEYAEKYLDYHHEKYGFQRSHV
Consensus MAA,RDA,EIHKDVQhYYG#VYLK,SADLQTHaCYT,Akp*P,yIResLQNYHE*Y,,sRYYGCGLv,rPEhLENC,ILDLGSGSGRDCYYLSQLVG#KGH!TGIDMTkyQVEYAk,,YL,#yhEK2GFQapHV
```

```
        131       140       150       160       170       180       190       200   Motif I  210       220       230   Post-I  240       250       260
rcyt.19   TFLHGQIEMLAERGIQKESYDIVISNCYINLVPDKQKYLREYYQVLKYGGELYFSDVYRSLEYSEDIKSHKYLAGECLGGALYIIKDLRYIDRKGIGFCPPRLVIANITIYGNKELERVLGBDCRFYSRITFRL
mcyt.19   TFLHGRIEKLAERGIQSESYDIVISNCYINLVPDKQQVLQEYYRYLKHGGELYFSDVYRSLEYPEDIKSHKYVLAGECLGGALYIAKDLRIIARGKCIGFCPPRLVIRDIITYENKELEGVLGBDCRFYSRITFRL
hcyt.19   TFLHGYIEKLGEAGKNESHDIYYSNCYINLVPDKQQVLQERYRYLKHGGELYFSDVYTSLELPEEIRTHKYLAGECLGGALYIAKELRYLAQKCIGFCPPRLVIANLITIQNKELERVIGBDCRFYSRITFRL
Consensus TFLHG,IEkLaERGIq,,ESyDIYY!SNCYINLVPDKQqvYLqEvYr*YLKhGGELYFSDVYaSLEvpE#IksHKYvLAGECLGGALYIaK#LR!iA,kCIGFCPPRLVIA#lIT!,NKELE*vLGBDCRFYSRITFRL
```

```
        261       270  Motif II  280       290       300  Motif III  310       320       330       340       350       360       370   376
rcyt.19   FKLPKTEPABRGQVYYNGGIDMGHEKELIFDANFTFKEGEAYEVDEETARILRNSRFAHDFLFTPVERSLLAPQ--------TKYIIRDPFKLAEESDKMKPCRPEGTGGCCGKRKSC
mcyt.19   FKLPKTEPAERCRVYYNGGIKGHEKELIFDANFTFKEGERVAYVDEETARYLKNSRFAPDFLFTPVDRSLPAPQGRSELETKYLTRIPFKLAEDSDKMKPAHPEGTGGCCGKRKNC
hcyt.19   FKHSKTGPTKRQQVYYNGGITGHEKELMFDANFTFKEGEYEVDEETARILKNSRFAQDFLIRPTGEKLPTSGGCSNLELKDIITDPFKLAEESDSHKSRCVPDAAGGCCGTKKSC
Consensus FKLpKTePa,RCqV!YHNGGI,GHEKEaYeVDEETAH!LkNSRFA,DFLftP!,asLpapvg,s,,letKv!I,rDPFKLAE#SDkMKpRcaP#gc6GCCGkrKsC
```

Multiple sequence alignment based on GenBank™ accession numbers for cyt19-rat, NP_543166; mouse, AAH13468; and human, XP_053690.

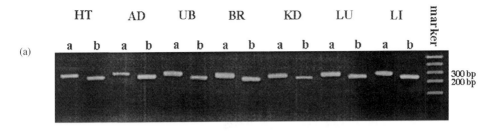

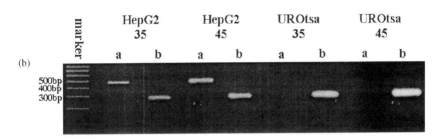

Figure 2. Detection of cyt19 mRNA in tissues and cells by RT-PCR. (a) Expression of cyt19 mRNA in tissues of rat. Predicted size of amplified cyt19 product was 250 bases; predicted size for β actin was 288 bases. For each tissue, lane "a" contains β actin product and lane "b" the cyt19 product. HT, heart; AD, adrenal; UB, urinary bladder; BR, brain; KD, kidney; LU, lung; LI, liver. φX174/*Hae* III DNA size markers shown in right lane. (b) Expression of cyt19 mRNA in human cell lines, HepG2 and UROtsa with 35 or 45 cycles of amplification. For cyt19, predicted size was 477 bases; predicted size for β actin was 288 bases. For each cell line, lane "a" contains the cyt19 product and lane "b" the β actin product. φX174/*Hae* III DNA size markers shown in left lane.

The role of reductants in the function of cyt19 was investigated. As noted above the inclusion of reductants (5 mM GSH and 1 mM DTT) preserved the arsenic methyltransferase activity of rat liver cytosol during the purification process, suggesting that the protein was susceptible to oxidation. Using recombinant rat cyt19 protein, we examined the function of two reductants in the catalytic function of this enzyme. Both DTT and *tris*-(2-carboxyethyl)phosphine (TCEP) support the formation by cyt19 of methylated arsenicals from arsenite (Fig. 3a and b). Without added reductant, arsenite was not methylated. A potential endogenous reductant for cyt19 was tested in assays in which a thioredoxin (Trx), thioredoxin reductase (TR), and NADPH system was added *in lieu* of a chemical reductant. Figure 3c compares the extent of arsenite methylation by cyt19 in the presence of 1 mM DTT or of various components of the TR/Trx/NADPH system. Only the reaction mixture containing TR, Trx, and NADPH produced methylated arsenicals from arsenite.

3. Discussion

A new enzyme has been purified from rat liver cytosol that converts inorganic arsenic into its methylated metabolites. This 42 kDa protein is not identical with an enzyme purified

S.B. Waters et al.

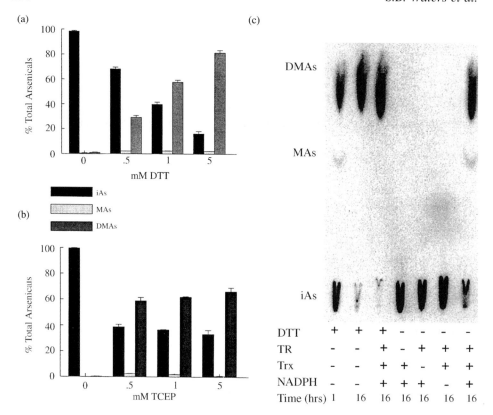

Figure 3. Reductants and the methylation of arsenite by recombinant rat cyt19 (a) Concentration dependence of DTT as a reductant in a reaction mixture containing 5 μg rCyt19, 1 mM AdoMet, and 1 μM [^{73}As]-labeled arsenite (iAsIII) in 100 mM Na phosphate, pH 7.4, that was incubated for 60 min at 37°C. (b) Concentration dependence of TCEP as a reductant in the above-described reaction mixture incubated for 60 min at 37°C. (c) TR/Trx/NADPH system supports methylation in a reaction mixture containing 5 μg rCyt19, 1 mM AdoMet, and 1 μM [^{73}As]-labeled with 10 μM *E. coli* Trx, 3 μM *E.coli* TR, 300 μM NADPH in 100 mM Na phosphate, pH 7.4, incubated up to 16 h at 37°C.

from rabbit liver cytosol (∼ 60 kDa) that catalyzes the methylation of arsenite and methylarsonous acid (Zakharyan et al., 1995, 1999). The low K_m value for the conversion of methylarsine oxide by the rat liver enzyme distinguishes it kinetically from arsenic methyltransferases purified from rabbit or hamster liver cytosol for which the estimated K_ms are orders of magnitude higher. Sequences for arsenic methyltransferases from rabbit or hamster liver and additional kinetic characterization of these enzymes are needed for further comparisons among these proteins. Methyltransferases from rat, rabbit, and hamster liver may be members of a group of related enzymes that are expressed in the tissues of different species. The sequence of rat liver arsenic methyltransferase contains motifs consistent with its function as a non-nucleic acid methyltransferase. This sequence data also permitted the identification of orthologous genes and proteins in the mouse and human. Based on the identity of these genes and gene products, the gene encoding arsenic methyltransferase is designated as *cyt19*. In human cell lines, levels of cyt19 mRNA are correlated with the capacity to convert arsenite into methylated metabolites.

The methyltransferase activity of cyt19 depended on the presence of a reductant in the reaction mixture. Both DTT and TCEP served as the chemical reductants; however, GSH, a monothiol reductant, did not support the methylation of arsenic by this enzyme. We found that reaction mixtures containing a Trx/TR/NADPH system were able to methylate arsenite. Hence, Trx may be an endogenous reductant for cyt19. Trx might reduce disulfide bonds formed between any two of the 13 cysteines in cyt19 that could alter the tertiary structure of the protein and affect its catalytic function. Alternatively, Trx could function as the reductant of pentavalent arsenicals formed as intermediates in the methylation process. Bacterial and yeast arsenate reductases that catalyze the reduction of pentavalent arsenic to trivalency use a Trx/TR/NADPH or a glutaredoxin/GSH reductase/GSH system to provide reducing equivalents to alter the oxidation state of arsenic (Messens et al., 2002; Mukhopadhyay and Rosen, 2002). Elucidation of the mechanistic basis of the reduction and the methylation of arsenicals by cyt19 is now underway.

4. Conclusions

Since the discovery that inorganic arsenic was converted into methylated metabolites in many species including humans, elucidating the pathway for arsenic metabolism has been a topic of considerable interest. Cyt19, a methyltransferase that catalyzes both monomethylation and dimethylation reactions, has been purified from rat liver cytosol and cloned to permit its expression as a recombinant protein. Cyt19 is also expressed in human tissues and is also an arsenic methyltransferases. Future research will probe the mechanism by which cyt19 catalyzes the oxidative methylation of arsenicals and the role of reductants in its function. Other studies will examine the relation between the cyt19 genotype and phenotypes including the profiles of arsenical metabolites in urine or susceptibility to adverse health effects of chronic arsenic exposure.

Acknowledgements

We thank our colleagues Dr Larry L. Hall, Dr Josef B. Simeonsson, and Ms Karen M. Herbin-Davis of the National Health Effects and Environmental Research Laboratory, and Dr Melinda A. Beck, Ms Qing Shi, and Mr F. Brent Nix of the Departments of Pediatrics and Nutrition, Schools of Medicine and of Public Health, University of North Carolina at Chapel Hill, for their many contributions to this research. We gratefully acknowledge the assistance of Dr Nicholas E. Sherman and his associates at the W.M. Keck Biomedical Mass Spectrometry Laboratory and the University of Virginia Biomedical Research Facility who obtained the peptide sequence data. The University of Virginia Pratt Committee funded this analytical facility. During the course of this research, SBW and SL were postdoctoral fellows supported by Training Grant T901915 of the US Environmental Protection Agency – University of North Carolina Toxicology Research Program with the Curriculum in Toxicology, University of North Carolina at Chapel Hill.

References

Braman, R.S., Foreback, C.C., 1973. Methylated forms of arsenic in the environment. Science, 182, 1247–1249.

Buchet, J.P., Lauwerys, R., 1985. Study of inorganic arsenic methylation by rat in vitro: relevance for the interpretation of observations in man. Arch. Toxicol., 57, 125–129.

Buchet, J.P., Lauwerys, R., 1988. Role of thiols in the in vitro methylation of inorganic arsenic by rat liver cytosol. Biochem. Pharmacol., 37, 3149–3153.

Buchet, J.P., Lauwerys, R., Roels, H., 1981. Urinary excretion of inorganic arsenic and its metabolites after repeated ingestion of sodium metaarsenite. Int. Arch. Occup. Environ., 48, 111–118.

Crecelius, E.A., 1977. Changes in the chemical speciation of arsenic following ingestion by man. Environ. Health Perspect., 19, 147–150.

Hirata, M., Mohri, T., Hisanaga, A., Ishinishi, N., 1989. Conversion of arsenite and arsenate to methylarsenic and dimethylarsenic compounds by homogenates prepared from livers and kidneys of rats and mice. Appl. Organomet. Chem., 3, 335–341.

Lakso, J.U., Peoples, S.A., 1975. Methylation of inorganic arsenic by mammals. J. Agric. Food Chem., 23, 674–676.

Lin, S., Shi, Q., Nix, F.B., Styblo, M., Beck, M.A., Herbin-Davis, K.M., Hall, L.L., Simeonsson, J.B., Thomas, D.J., 2001. A novel S-adenosyl-L-methionine: arsenic(III) methyltransferase from rat liver cytosol. J. Biol. Chem., 277, 10795–10803.

Messens, J., Martins, J.C., VanBelle, K., Brosens, E., Desmyter, A., DeGieter, M., Wieruszeski, J.-M., Willem, R., Wyns, L., Zegers, I., 2002. All intermediates of the arsenate reductase mechanism, including an intramolecular dynamic disulfide cascade. Proc. Natl. Acad. Sci. USA, 99, 8506–8511.

Mukhopadhyay, R., Rosen, B.P., 2002. Arsenate reductases in prokaryotes and eukaryotes. Environ. Health Perspect., 110 (Suppl. 5), 745–748.

Odanaka, Y., Matano, O., Goto, S., 1980. Biomethylation of inorganic arsenic by the rat and some laboratory animals. Bull. Environ. Contam. Toxicol., 24, 452–459.

Peoples, S.A., 1974. Review of arsenical pesticides. In: Woolson, E.A. (Ed.), Arsenical Pesticides. American Chemical Society, Washington, DC, pp. 1–12.

Smith, T.J., Crecelius, E.A., Reading, J.C., 1977. Airborne arsenic exposure and excretion of methylated arsenic compounds. Environ. Health Perspect., 19, 89–93.

Styblo, M., Delnomdedieu, M., Thomas, D.J., 1996. Mono- and dimethylation of arsenic in rat liver cytosol *in vitro*. Chem.–Biol. Interact., 99, 147–167.

Tam, G.K.H., Charbonneau, S.M., Lacroix, G., Bryce, F., 1979a. Confirmation of inorganic arsenic and dimethylarsenic in urine and plasma of dog by ion-exchange and TLC. Bull. Environ. Contam. Toxicol., 21, 371–374.

Tam, G.K.H., Charbonneau, S.M., Lacroix, G., Bryce, F., 1979b. *In vitro* methylation of [74]As in urine, plasma and red blood cells of human and dog. Bull. Environ. Contam. Toxicol., 22, 69–71.

Zakharyan, R., Wu, Y., Bogdan, G.M., Aposhian, H.V., 1995. Enzymatic methylation of arsenic compounds: assay, partial purification, and properties of arsenite methyltransferase and monomethylarsonic acid methyltransferase from rabbit liver. Chem. Res. Toxicol., 8, 1029–1038.

Zakharyan, R.A., Ayala-Fierro, F., Cullen, W.R., Carter, D.M., Aposhian, H.V., 1999. Enzymatic methylation of arsenic compounds. VII. Monomethylarsonous acid (MMAIII) is the substrate for MMA methyltransferase of rabbit liver and human hepatocytes. Toxicol. Appl. Pharmacol., 158, 9–15.

Arsenic Exposure and Health Effects V
W.R. Chappell, C.O. Abernathy, R.L. Calderon and D.J. Thomas, editors

Chapter 20

Metabolism of arsenic and gene transcription regulation: mechanism of AP-1 activation by methylated trivalent arsenicals

Zuzana Drobná, Ilona Jaspers and Miroslav Stýblo

Abstract

The enzymatic methylation of inorganic arsenic (iAs) in humans generates methylated arsenicals that contain either As^V or As^{III}. Unlike their pentavalent counterparts, methylated trivalent arsenicals, methylarsonous acid (MAs^{III}) and dimethylarsinous acid ($DMAs^{III}$), are more cytotoxic, genotoxic, and more potent enzyme inhibitors than iAs. Previous studies indicated that iAs does not act through classic genotoxic or mutagenic mechanisms, but rather is a co-carcinogen or a tumor promoter that interferes with signal transduction pathways. Trivalent iAs, arsenite, modifies expression and/or DNA binding activities of several transcription factors, thereby modulating cell growth and proliferation. However, effects of methylated arsenicals on gene transcription regulation have not been thoroughly characterized. We have examined the composition and DNA binding activity of one of the major transcription factors, activating protein-1 (AP-1), in human hepatic and urinary bladder cells exposed to arsenite, or to methylated trivalent arsenicals. Trivalent arsenicals modified phosphorylation of c-Jun oncoprotein and DNA binding activity of AP-1 in human primary hepatocytes, HepG2, UROtsa and T24 cells. Among cell types examined, UROtsa cells, a normal human urothelium cell line, were the most sensitive to trivalent arsenicals. In this cell line, methylarsine oxide ($MAs^{III}O$) and iododimethylarsine ($DMAs^{III}I$) were more potent inducers of c-Jun phosphorylation and AP-1 activation than was iAs^{III}. The examination of the upstream mechanisms suggest that activation of extracellular signal-regulated kinase (ERK), but not c-Jun N-terminal kinase (JNK) or p38 kinase, is responsible for the phosphorylation of c-Jun and induction of AP-1 DNA binding in UROtsa cells exposed to trivalent arsenicals. These results indicate that methylated trivalent arsenicals may be more potent than iAs^{III} in inducting AP-1-dependent gene transcription in human tissues, particularly in bladder urothelium.

Keywords: arsenic, methylation, AP-1, urinary bladder, cancer, human, cell

1. Introduction

Chronic exposures to iAs are associated with increased prevalence of skin cancer and cancers of internal organs, including lung, liver, kidney, and urinary bladder (US EPA, 1988; Bates et al., 1992). The mechanisms by which iAs induces cancers are not completely understood. In cultured cells and in hairless Skh1 mice exposed to ionizing radiation, arsenite (iAs^{III}) acts as a co-carcinogen (Rossman et al., 2001; Vogt and Rossman, 2001). In various cell types, iAs^{III} modifies signal transduction pathways that regulate cell growth and proliferation [reviewed by Simeonova and Luster (2000)]. iAs^{III} modulates the expression and/or DNA binding activities of several major transcription factors, including nuclear factor kappa B (NF-κB) (Barchowsky et al., 1996; Wiencke et al., 1997) and activating protein-1 (AP-1) (Burleson et al., 1996; Simeonova et al., 2000). The examination of mechanisms involved in the AP-1 activation indicate that iAs[III] stimulates the mitogen-activated protein kinase (MAPK) cascade, induces expression of *c-jun* and *c-fos* protooncogenes and enhances DNA binding activity of AP-1 (Burleson et al., 1996; Liu et al., 1996; Huang et al., 1999; Trouba et al., 2000; Simeonova et al., 2002). The modulation of AP-1-dependent gene transcription by iAs[III] may contribute to the induction of cell proliferation in tissues exposed to this arsenical.

Most studies of the molecular mechanisms involved in arsenic carcinogenesis have focused on iAs species, particularly iAs[III], and little attention has been given to methylated arsenicals which are metabolites of iAs. In humans, iAs is enzymatically methylated to methylarsenic species (MAs) and dimethylarsenic species (DMAs) [reviewed by Aposhian (1997) and Vahter (1998)]. Pentavalent and trivalent arsenicals that are intermediates or products in this pathway (Cullen et al., 1984) are found in urine of individuals chronically exposed to iAs (Aposhian et al., 2000; Le et al., 2000a,b; Del Razo et al., 2001; Mandal et al., 2001). Methylated trivalent arsenicals, methylarsonous acid (MAs[III]) and dimethylarsinous acid (DMAs[III]) are, unlike their pentavalent counterparts, methylarsonic acid (MAs[V]) and dimethylarsinic acid (DMAs[V]), potent cytotoxins, genotoxins, and enzyme inhibitors [reviewed by Thomas et al. (2001)]. In this study, we have examined effects of low micromolar concentrations of methylarsine oxide (MAs[III]O) and iododimethylarsine (DMAs[III]I), chemical derivatives of MAs[III] and DMAs[III], on the composition and DNA binding activity of AP-1 in cells derived from human liver, the major site of iAs methylation, and from urinary bladder, one of the targets for carcinogenic effects of iAs. We found that short-time exposures to micromolar concentrations of MAs[III]O or DMAs[III]I significantly increased the nuclear level of phospho-c-Jun (p-c-Jun) and the DNA binding activity of AP-1 in UROtsa, a cell line derived from normal human urothelium. Both methylated trivalent arsenicals were more potent inducers of c-Jun phosphorylation and activators of AP-1 DNA binding than iAs[III]. Analysis of MAPK activation patterns shows that extracellular signal-regulated kinase (ERK), but not c-Jun N-terminal kinase (JNK) or p38 kinase, is responsible for c-Jun phosphorylation and AP-1 activation in exposed cells. These results suggest that MAs[III] and DMAs[III], methylated trivalent metabolites produced in the course of iAs metabolism, are likely to contribute to the activation of AP-1 and to the induction of the AP-1 regulated gene transcription.

2. Methodology

2.1. Arsenic compounds

Arsenate (iAs[V]) and iAs[III] (sodium salts, at least 99% pure) were purchased from Sigma (St. Louis, MO). MAs[V] (sodium salt, 98% pure) was obtained from Chem Service (West Chester, PA) and DMAs[V] (98% pure) from Strem (Newburyport, MA). Methylated trivalent arsenicals, MAs[III]O and DMAs[III]I, were synthesized by Dr William R. Cullen (University of British Columbia, Vancouver, Canada) using previously described methods (Styblo et al., 1997). Radiolabeled [[73]As]iAs[V] was purchased from Los Alamos Meson Production Facility (Los Alamos, NM). [^{73}As]iAs^{III} was prepared from [^{73}As]iAs^{V} by reduction with metabisulfite/thiosulfate reagent (Reay and Asher, 1977).

2.2. Cell cultures and treatment

Primary human hepatocytes were isolated in collaboration with Dr Ed LeCluyse (School of Pharmacy, UNC, Chapel Hill) from a normal liver tissue sample obtained from a 65-year-old male donor. Hepatocyte monolayers were cultured in collagen coated 6-well culture plates (Falcon, Becton Dickinson Labware, Lincoln Park, NJ) in a modified William's medium E (Styblo et al., 1999). Cells derived from human bladder transient carcinoma (T24) and human hepatoma (HepG2) and SV40-immortalized normal human urothelial cells (UROtsa) were cultured in 60 or 100 mm diameter culture dishes (Falcon) as previously described (Del Razo et al., 2001; Styblo et al., 2002; Drobná et al., 2003). All cells were kept at 37°C in a humidified incubator in a 95% air and 5% CO_2 atmosphere. Cell cultures at 70–80% confluency (90% for hepatocytes) were used in experiments. Trivalent or pentavalent arsenicals were added into the cell cultures in sterile phosphate buffered saline (PBS). In some experiments, cells were pretreated with PD98059 (Promega Corporation, Madison, WI), a specific inhibitor of the ERK kinase (MEK), for 1 h before exposures to arsenicals.

2.3. Analysis of DNA binding activity and composition of AP-1

The methods used for analysis of DNA binding activity and composition of AP-1, as well as for immunoblot analyses of AP-1 constituents were described in detail by Drobná et al. (2003). Briefly, the DNA binding activity of AP-1 was analyzed in nuclear protein extracts from cells exposed to arsenicals and in control (untreated) cells by the electrophoretic mobility shift assay (EMSA) using a ^{32}P-labeled double-stranded oligonucleotide that contained AP-1 consensus sequence, 5'-TGAGTCAG-3' (Promega, Madison, WI). For supershift analyses, antibodies specific for the AP-1 constituents were added to the reaction mixture. Antibodies from two manufacturers were used, including rabbit polyclonal antibodies against c-Jun, c-Fos, Fra-1, ATF-2 and mouse monoclonal antibodies against p-c-Jun and phospho-ATF-2 (p-ATF-2) (all from Santa Cruz Biotechnology, Santa Cruz, CA), and rabbit polyclonal antibodies against human c-Jun, JunB, JunD, c-Fos, and FosB supplied in the Nushift Kit (Geneka Biotechnology, Inc., Montreal, Canada). A nuclear protein extract from 3T3 SR (serum responsive mouse embryo) cells (Nushift Kit, Geneka Biotechnology) served as a positive control for the supershift analysis. A neutralizing c-Fos peptide (Nushift Kit, Geneka Biotechnology) was used in the assay to confirm the specificity of antibodies against c-Fos. A 50-fold excess of unlabeled wild-type (wt) or mutant (mt) AP-1 probe was used to establish the binding specificity.

2.3.1. Immunoblot analysis

Immunoblot was used for analysis of AP-1 constituents and enzymes of the MAPK family in cells exposed to arsenicals and in control cells. Nuclear proteins were separated by SDS-PAGE and electroblotted on PVDF membranes. Blotted proteins were washed and blocked with 5% nonfat milk in PBS (Bio-Rad, Hercules, CA). Membranes were probed with rabbit polyclonal antibody specific to c-Jun p39 (*sc-44*, Santa Cruz Biotechnology), c-Fos p62 (*sc-52*; Santa Cruz), or Fra-1 (sc-605X; Santa Cruz); p-c-Jun was detected using mouse monoclonal antibodies *sc-822* (Santa Cruz). Enzymes of the MAPK family, JNK1 and 2, p38, and ERK1 and 2, were detected using rabbit polyclonal antibodies from Cell Signaling Technology (NEB, Beverly, MA). Mouse monoclonal antibodies from this supplier were used to detect phosphorylated forms of these kinases (p-JNK1,2, p-p38, and p-ERK1,2). Antigen–antibody complexes were stained with horseradish peroxidase-conjugated goat anti-rabbit or anti-mouse antibodies (Santa Cruz Biotechnology) and treated with an enhanced chemiluminescence reagent, Super Signal West Pico Chemiluminescent Substrate (Pierce, Rockford, IL). Immunoblot images were digitized and the optical densities of specific antigen–antibody complexes were quantified using the Gene Gnome imaging system supported with the Gene Tools software package (both from Syngene, Frederick, MD).

Note: All experiments were replicated to ensure the reproducibility of results. Representative findings are shown.

3. Results

3.1. DNA binding activity and composition of AP-1 in cells exposed to arsenicals

The DNA binding activity of AP-1 was examined by EMSA in cells treated with trivalent arsenicals (0.1–5 μM) for up to 2 h (Fig. 1). The most significant responses to these treatments were found in UROtsa cells. The AP-1 DNA binding activity was suppressed in UROtsa cells exposed for 1 h to 0.1 or 0.5 μM iAs^{III} or DMAs^{III}I but was strongly induced by 5 μM DMAs^{III}I. The AP-1 DNA binding activity remained unchanged in cells during 1 h exposures to 1 or 5 μM iAs^{III}, but increased slightly after 2 h exposures (data not shown). MAs^{III}O was a potent activator of AP-1 DNA binding regardless of the exposure time. One-hour exposure to MAs^{III}O activated AP-1 at concentrations as low as 0.1 μM; the most notable effect occurred at 1 μM. A lower DNA binding activity in UROtsa cells exposed to 5 μM MAs^{III}O was associated with decreased (by about 30%) cell viability. In T24 cells and in primary human hepatocytes, only exposures to 5 μM MAs^{III}O induced notable increases in the AP-1 DNA binding activity. In contrast, DNA binding activity of AP-1 in HepG2 cells was induced only by iAs^{III}, but not by MAs^{III}O or DMAs^{III}I. The components of AP-1–DNA binding complex were determined immunochemically by EMSA-supershift analysis in UROtsa cells exposed to iAs^{III}, MAs^{III}O, or DMAs^{III}I (Fig. 2). The concentrations of arsenicals that caused the most pronounced induction of AP-1 DNA binding activity (5 μM iAs^{III}, 1 μM MAs^{III}O, and 5 μM DMAs^{III}I) were used in these experiments. Incubation of the DNA–protein binding mixture in the presence of antibodies against p-c-Jun shifted almost entire AP-1–DNA complex, indicating that p-c-Jun is a major constituent of the activated AP-1 dimer (Fig. 2a). The p-c-Jun related supershifts were found in UROtsa cells treated with trivalent arsenicals and also in untreated (control) cells, indicating that

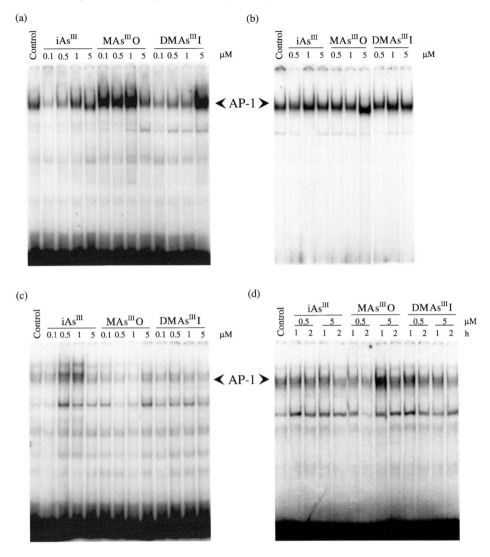

Figure 1. AP-1 DNA binding activity in cells treated with trivalent arsenicals and in control (untreated) cells: EMSA images for (a) UROtsa cells after 1 h treatment, (b) T24 and (c) HepG2 cells after 2 h treatments, and for (d) primary human hepatocytes after 1- and 2-h treatment. Six micrograms of nuclear proteins were analyzed from each treatment group.

p-c-Jun is present in AP-1 regardless of the treatment. Incubation with antibodies against c-Jun resulted in supershifts similar to those produced by antibodies against p-c-Jun. In contrast, antibodies against c-Fos had little or no effect on electrophoretic mobility of the AP-1 complex in all treatment groups, suggesting that under these conditions c-Fos is not a significant constituent. A nuclear protein extract from 3T3 SR cells was used as a positive control for c-Fos analysis. Additional EMSA-supershift analyses (Fig. 2b) showed presence of three other AP-1 constituents, Jun-B, JunD, and

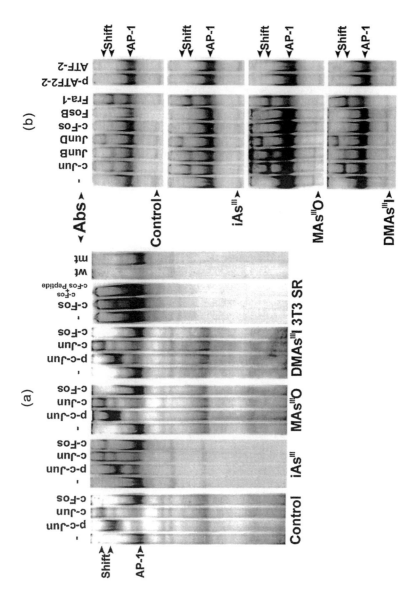

Figure 2. EMSA-supershift analysis of nuclear protein extracts from UROtsa cells treated for 1 h with 5 μM iAs^{III}, 1 μM MAs^{III}O, or 5 μM DMAs^{III}I and from control (untreated) cells. For immunochemical characterization of the AP-1 constituents, the DNA protein binding assay was carried out in the presence of antibodies specific for (a) p-c-Jun, c-Jun, or c-Fos (Santa Cruz Biotechnology) and for (b) c-Jun, JunB, JunD, c-Fos, or FosB (Geneka Biotechnology) and Fra-1, p-ATF-2, or ATF-2 (Santa Cruz Biotechnology). The specificity of the assay was established using a 50-fold excess of a wild-type (wt) or mutated (mt) AP-1 probe. A nuclear protein extract from 3T3 SR cells was used as a positive control for the analysis of c-Fos subunit. To confirm the specificity of the c-Fos antibodies, the 3T3 SR nuclear extract was incubated in the DNA binding assay mixture in the presence of an excess of c-Fos peptide.

especially Fra-1, in AP-1–DNA binding complex from cells exposed to trivalent arsenicals. On the other hand, FosB, ATF-2, and p-ATF-2 were not detected. Effects of pentavalent arsenicals on AP-1 DNA binding activity in UROtsa cells were also examined. No changes were detected in cells exposed to 1–100 μM iAs^V, MAs^V, or DMAs^V (data not shown).

3.2. Expression and activation of AP-1 constituents in cells exposed to arsenicals

The nuclear concentrations of p-c-Jun and Fra-1, proteins identified by the EMSA-supershift assay as the major components of the AP-1–DNA binding complex, were further analyzed by immunoblot. All trivalent arsenicals modified nuclear levels of p-c-Jun (Fig. 3). Treatments with low concentrations of iAs^{III} (0.1 and 0.5 μM) decreased nuclear p-c-Jun levels. However, exposure to 1 or 5 μM iAs^{III} increased p-c-Jun levels in a concentration-dependent manner. Both methylated trivalent arsenicals, MAs^{III}O and DMAs^{III}I, were more potent inducers of nuclear p-c-Jun than iAs^{III}. For all trivalent arsenicals, maximal effects occurred after a 1 h exposure. More than 20-fold increase of p-c-Jun was found in cells exposed to 5 μM MAs^{III}O as compared with 13- and 4-fold increases in cells treated with 5 μM DMAs^{III}I and 5 μM iAs^{III}, respectively. Notably, treatments with iAs^{III}, MAs^{III}O, or DMAs^{III}I (0.1–5 μM) had no effect on nuclear levels of c-Jun. In agreement with the results of EMSA-supershift analyses, no c-Fos was detected in nuclear protein extracts from UROtsa cells regardless of the treatment. However, cytoplasmic c-Fos was slightly increased in cells exposed to trivalent arsenicals, particularly in cells exposed to MAs^{III}O (data not shown). The immunoblot analysis of Fra-1 in nuclear protein extracts from untreated cells showed a triplet of immunoreactive protein bands in a 35–43 kDa molecular mass range (Fig. 4). Exposures to MAs^{III}O increased the intensity of all three bands and shifted the rapidly migrating immunoreactive proteins to the slowly migrating 43 kDa band of the triplet. DMAs^{III}I was a less potent inducer of Fra-1. No significant changes in the Fra-1 immunoblot profiles were detected in cells exposed to iAs^{III}.

3.3. Effects of trivalent arsenicals on enzymes of the MAPK family

To identify upstream mechanisms responsible for c-Jun phosphorylation and AP-1 activation by trivalent arsenicals, protein extracts from control and exposed cells were analyzed for the presence of the parent and activated (phosphorylated) forms of enzymes of the MAPK family, JNK, p38, and ERK. No changes in the concentrations of JNK1, JNK2, and p38, nor induction of the corresponding phosphorylated forms (p-JNK1, p-JNK2, and p-p38) were detected in cells treated with iAs^{III} or DMAs^{III}I (0.1–5 μM) for up to 2 h (data not shown). Similarly, exposures to MAs^{III}O, the most potent inducer of c-Jun phosphorylation, did not alter the immunoblot patterns for either JNK or p38 kinases (Fig. 5). The protein extract from cells exposed to a cytotoxic concentration of iAs^{III} (400 μM) was used in these experiments as a positive control. Treatments with iAs^{III}, MAs^{III}O, or DMAs^{III}I did not increase basal levels of ERK1 or ERK2 (Fig. 6). However, all three arsenicals induced phosphorylation of ERK2. Increased levels of p-ERK1 were found in cells

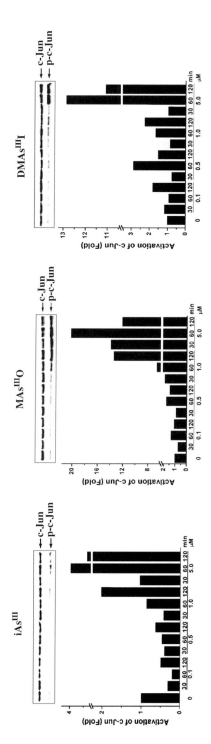

Figure 3.   Immunoblot analysis of c-Jun and p-c-Jun in nuclear protein extracts from UROtsa cells exposed to iAs[III], MAs[III]O, or DMAs[III]I for 1/2, 1, or 2 h and from control (untreated) cells. Immunoblot images for c-Jun and p-c-Jun (upper panels) and results of the quantitative analysis for p-c-Jun (lower panels) are shown.

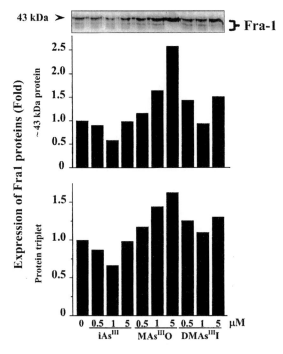

Figure 4. Immunoblot analysis of Fra-1 in nuclear protein extracts from UROtsa cells treated for 1 h with trivalent arsenicals and from untreated cells (upper panel). Results of the quantitative analyses of ~43 kDa Fra-1 fraction (middle panel) and of the entire Fra-1 triplet (lower panel).

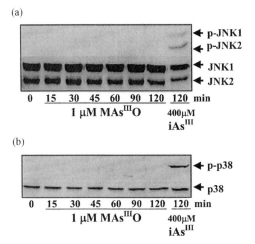

Figure 5. Immunoblot analyses of (a) JNK1,2 and p-JNK1,2 and (b) p38 and p-p38 in whole cell lysates from UROtsa cells exposed to 1 μM MAs^{III}O for up to 2 h. UROtsa cells exposed to 400 μM iAs^{III} were used as positive controls for p-JNK1,2 and p-p38 analyses.

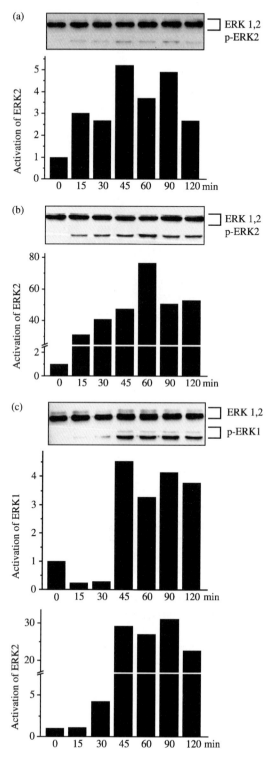

exposed to DMAs^{III}I, but not to iAs^{III} or MAs^{III}O (Fig. 6c). A specific inhibitor of MEK, PD98059, was used in further experiments to evaluate the relationship between ERK activation and c-Jun phosphorylation in UROtsa cells exposed to MAs^{III}O or DMAs^{III}I. Treatment with PD98059 (5–100 μM) suppressed entirely ERK phosphorylation in exposed cells (Fig. 7) and caused a concentration-dependent decrease in the production of p-c-Jun. No cytotoxicity was observed in cells co-treated with PD98059 and arsenicals.

4. Discussion

Recent reports on adverse biological effects of methylated trivalent arsenicals as potent enzyme inhibitors, cytotoxins and genotoxins (Thomas et al., 2001) indicate that these metabolites play a significant role in the overall toxicity and carcinogenicity of iAs. Modulation of MAPK activities that results in changes in the DNA binding patterns of major transcription factors, including NF-κB and AP-1, are believed to underlie some of the toxic and cancer promoting effects associated with exposures to iAs. However, the role of methylated metabolites in this process has not been characterized. In this study, we compare the effects of inorganic and methylated arsenicals on the composition and DNA binding activity of AP-1 in human hepatic and urinary bladder cells. Based on the activation pattern of AP-1, UROtsa cells that are derived from normal bladder epithelium were most sensitive to exposures to trivalent arsenicals. Notably, urinary bladder is one of the major targets for carcinogenic effects of iAs in humans. Among arsenicals examined, MAs^{III}O was the most potent inducer of AP-1 DNA binding activity in UROtsa cells. The immunochemical analysis showed that p-c-Jun is the major component of the AP-1–DNA binding complex in UROtsa cells before and after the exposure to arsenicals. However, treatment with trivalent arsenicals, particularly MAs^{III}O, dramatically increased the presence of p-c-Jun in the nuclear protein extracts, indicating that c-Jun is actively phosphorylated in response to the treatment. Unlike in previous studies that were carried out in cultured cells exposed to relatively high concentrations of iAs^{III}, no significant changes in c-Jun or c-Fos expression were found in UROtsa cells treated with 0.1–5 μM iAs^{III}, MAs^{III}O, or DMAs^{III}I. Thus, the phosphorylation of c-Jun, rather than induction of c-Jun or c-Fos expression, may be a sensitive marker of the exposure to low concentrations of trivalent arsenicals in human urinary bladder cells. In addition to p-c-Jun, other constituents, including Fra-1 and to a lesser extent JunB and JunD, were detected in the AP-1–DNA binding complex. A distinct electrophoretic shift from low- to high-molecular weight protein bands was found during the immunoblot analysis of Fra-1. A similar shift has been described previously by Gruda et al. (1994) in fibroblasts exposed to O-tetradecanoylphorbol 13-acetate (TPA). This shift was associated with an increased presence of the phosphorylated form of Fra-1 (p-Fra-1) in exposed cells. Thus, exposures

---

Figure 6. Immunoblot analysis of ERK1,2 and p-ERK1,2 in UROtsa cells exposed to (a) 5 μM iAs^{III}, (b) 1 μM MAs^{III}O, or (c) 5 μM DMAs^{III}I for up to 2 h. Each panel shows immunoblot images and results of the quantitative analysis of these images (bar charts). The data in the bar charts are shown as folds of ERK activation.

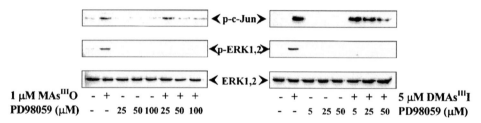

$1\ \mu M\ MAs^{III}O$ - + - - - + + +   - + - - - + + + $5\ \mu M\ DMAs^{III}I$
PD98059 (μM) - - 25 50 100 25 50 100   - - 5 25 50 5 25 50   PD98059 (μM)

Figure 7. Immunoblot analysis of p-c-Jun, p-ERK1,2, and ERK1,2 in UROtsa cells pretreated with PD98059 for 1 h before exposure to 1 μM MAs^{III}O (left panels) or 5 μM DMAs^{III}I (right panels).

to trivalent arsenicals induced Fra-1 phosphorylation in UROtsa cells. Taken together, these data suggest that in UROtsa cells the AP-1 dimers are composed mostly of the p-c-Jun and Fra-1 proteins, yielding p-c-Jun:p-c-Jun homodimers or p-c-Jun:Fra-1 heterodimers. In general, c-Jun-containing AP-1 dimers have been associated with the induction of cell proliferation (Karin et al., 1997; Mechta-Grigoriou et al., 2001; Shaulian and Karin, 2001; Vogt, 2001).

Several studies have shown that phosphorylation of JNK and/or p38 is required for AP-1 activation in cultured cells exposed to 40–300 μM iAs^{III} (Cavigelli et al., 1996; Huang et al., 1999; Simeonova et al., 2001). In this study, increased levels of p-ERK1 and/ or 2, but not p-JNK1 and 2 or p-p38 were detected in UROtsa cells exposed to much lower concentrations of trivalent arsenicals. The kinetic pattern of ERK1,2 phosphorylation in cells exposed to arsenicals resembles that of p-c-Jun formation, suggesting ERK1 and/or 2 to be responsible for c-Jun phosphorylation. Indeed, inhibition of ERK1,2 phosphorylation by PD98059 resulted in a significant decrease in the production of p-c-Jun in cells treated with MAs^{III}O or DMAs^{III}I. Thus, phosphorylation of c-Jun in response to exposures to trivalent methylated arsenicals is directly mediated by active, phosphorylated, forms of ERK1 and/or 2. Although JNKs are regarded as the most specific and efficient enzymes involved in phosphorylation of c-Jun (Mechta-Grigoriou et al., 2001), other protein kinases, including Abl tyrosine protein kinase (Barila et al., 2000) and ERK (Alvarez et al., 1991; Frost et al., 1994), have been shown to phosphorylate this transcription factor. For example, ERK was involved in c-Jun phosphorylation and AP-1 activation in p53-deficient human lung cancer cells exposed to hydrogen peroxide (Chung et al., 2002). Notably, exposures to inorganic and methylated arsenicals are known to induce oxidative stress in cultured cells (Kitchin, 2001; Thomas et al., 2001). It is possible, that the activation of ERK/c-Jun pathway in UROtsa cells exposed to iAs^{III}, MAs^{III}O, or DMAs^{III}I is mediated by an increased production of reactive oxygen species. Activation of ERK1,2 by trivalent arsenicals may also be responsible for increased phosphorylation of Fra-1 that was detected in exposed cells as a shift in the electrophoretic mobility of the Fra-1 protein fractions during the immunoblot analysis. The transactivation of Fra-1 by ERK has been reported by Young et al. (2002) in JB6 cells. In these cells, the ERK-dependent transactivation of Fra-1 was a prerequisite for a mitogen-induced activation of AP-1. Activation of ERK1 and/or 2 by iAs^{III} has previously been described in several types of animal and human cells (Ludwig et al., 1998; Huang et al., 1999). Notably, activation of ERK1,2 but not JNK, was required for iAs^{III}-induced cell transformation (Huang et al., 1999). In this study, we show that methylated trivalent arsenicals are at least equipotent with iAs[III] as inducers of ERK phosphorylation in human urothelial cells. Simeonova et al. (2002) have reported that activation of the epidermal growth factor receptor provides the initial signal for ERK phosphorylation in UROtsa cells exposed to iAs[III]. The upstream mechanisms as well as quantitative and qualitative aspects of the ERK activation by methylated trivalent arsenicals remain to be examined.

In conclusion, exposures to trivalent arsenicals, iAs[III], MAs[III]O, or DMAs[III]I, increase AP-1 DNA binding activity in UROtsa cells through the ERK-dependent induction of c-Jun and Fra-1 phosphorylation. Among these arsenicals, MAs[III]O is the most potent inducer of c-Jun/Fra-1 phosphorylation and AP-1 activation, followed by DMAs[III]I and iAs[III]. It is possible that methyl groups in MAs[III]O and DMAs[III]I facilitate interactions of these arsenicals with critical lipophilic moieties in cellular receptors or in components of the signaling cascade that is responsible for c-Jun/AP-1 activation in UROtsa cells. Thus, methylated trivalent arsenicals that are chemically consistent with methylated metabolites of iAs, MAs[III], and DMAs[III], are considerably more potent AP-1 activators than is the parent compound in cells derived from urinary bladder, one of the target tissues for carcinogenesis in humans chronically exposed to iAs. These findings support the hypothesis that the biomethylation is a process that activates iAs as a toxin and human carcinogen. Further studies of the downstream events associated with the AP-1 activation induced by arsenic metabolites in human urinary bladder will elucidate the mechanisms by which arsenic produce their toxic and carcinogenic effects.

Acknowledgements

The work presented here was supported by NIH grant ES09941 to M.S. and NIH Clinical Nutrition Research Center Grant DK 56350. The authors would like to thank Dr Nyseo Unimye (West Virginia University, Morgantown, WV) for providing UROtsa cell line and Dr William R. Cullen (University of British Columbia, Vancouver, BC, Canada) for supplying custom synthesized methylated trivalent arsenicals for this study.

References

Alvarez, E., Northwood, I.C., Gonzales, F.A., Latour, D.A., Seth, A., Abate, C., Curran, T., Davis, R.J., 1991. Pro–Leu–Ser/Thr–Pro is a consensus primary sequence for substrate protein phosphorylation. Characterization of the phosphorylation of c-myc and c-jun proteins by an epidermal growth factor receptor threonin 669 protein kinase. J. Biol. Chem., 266, 15277–15285.

Aposhian, H.V., 1997. Enzymatic methylation of arsenic species and other new approaches to arsenic toxicity. Annu. Rev. Pharmacol. Toxicol., 37, 397–419.

Aposhian, H.V., Gurzau, E.S., Le, X.C., Gurzau, A., Healy, S.M., Lu, X., Ma, M., Yip, L., Zakharyan, R.A., Maiorino, R.M., Dart, R.C., Tircus, M.G., Gonzales-Ramirez, D., Morgan, D.L., Avram, D., Aposhian, M.M., 2000. Occurrence of monomethylarsonous acid in urine of humans exposed to inorganic arsenic. Chem. Res. Toxicol., 13, 693–697.

Barchowsky, A., Dudek, E.J., Treadwell, M.D., Wetterhahn, K.E., 1996. Arsenic induces oxidant stress and NF-κB activation in cultured aortic endothelial cells. Free Radic. Biol. Med., 21, 783–790.

Barila, D., Mangano, R., Gomfloni, S., Kretzschmar, J., Moro, M., Bohmann, D., Superti-Furga, G., 2000. A nuclear tyrosine phosphorylation circuit: c-Jun as an activator and substrate of c-Abl and JNK. EMBO J., 19, 273–281.

Bates, M.N., Smith, A.H., Hopenhayn-Rich, C., 1992. Arsenic ingestion and internal cancers: a review. Am. J. Epidemiol., 135, 462–476.

Burleson, F.G., Simeonova, P.P., Germolec, D.R., Luster, M.I., 1996. Dermatotoxic chemical stimulate of *c-jun* and *c-fos* transcription and AP-1 DNA binding in human keratinocytes. Res. Commun. Mol. Pathol. Pharmacol., 93, 131–148.

Cavigelli, M., Li, W.W., Lin, A., Su, B., Yoshioka, K., Karin, M., 1996. The tumor promoter arsenite stimulates AP-1 activity by inhibiting a JNK phosphatase. EMBO J., 15, 6269–6279.

Chung, Y.W., Jeong, D.-W., Won, J.Y., Choi, E.-J., Choi, Y.H., Kim, I.Y., 2002. H_2O_2-induced AP-1 activation and effect on p21$^{WAF1/CIP1}$-mediated G2/M arrest in a p53-deficient human lung cancer cell. Biochem. Biophys. Res. Commun., 293, 1248–1253.

Cullen, W.R., McBride, B.C., Reglinski, J., 1984. The reaction of methylarsenicals with thiols: some biological implications. J. Inorg. Biochem., 21, 179–194.

Del Razo, L.M., Styblo, M., Cullen, W.R., Thomas, D.J., 2001. Determination of trivalent methylated arsenicals in biological matrices. Toxicol. Appl. Pharmacol., 174, 282–293.

Drobná, Z., Jaspers, I., Thomas, D.J., Styblo, M., 2003. Differential activation of AP-1 in human bladder epithelial cells by inorganic and methylated arsenicals. FASEB J., 17, 67–69.

Frost, J.A., Geppert, T.D., Cobb, M.H., Feramisco, J.R., 1994. A requirement for extracellular signal-regulated kinase (ERK) function in the activation of AP-1 by Ha-Ras, phorbol 12-myristate 14-acetate, and serum. Proc. Natl Acad. Sci. USA, 91, 3844–3848.

Gruda, M.C., Kovary, K., Metz, R., Bravo, R., 1994. Regulation of Fra-1 and Fra-2 phosphorylation differs during the cell cycle of fibroblasts and phosphorylation in vitro by MAP kinase affects DNA binding activity. Oncogene, 9, 2537–2547.

Huang, C., Ma, W.Y., Li, J., Goranson, A., Dong, Z., 1999. Requirement of ERK, but not JNK, for arsenite-induced cell transformation. J. Biol. Chem., 274, 14595–14601.

Karin, M., Liu, Z.-G., Zandi, E., 1997. AP-1 function and regulation. Curr. Opin. Cell Biol., 9, 240–246.

Kitchin, K.T., 2001. Recent advances in arsenic carcinogenesis: modes of action, animal model systems, and methylated arsenic metabolites. Toxicol. Appl. Pharmacol., 172, 249–261.

Le, X.C., Le, X.C., Lu, X., Ma, M., Cullen, W.R., Aposhian, H.V., Zheng, B., 2000a. Speciation of key arsenic metabolic intermediates in human urine. Anal. Chem., 72, 5172–5177.

Le, X.C., Ma, M., Cullen, W.R., Aposhian, H.V., Lu, X., Zheng, B., 2000b. Determination of monomethylarsonous acid, a key arsenic methylation intermediate, in human urine. Environ. Health Perspect., 108, 1015–1018.

Liu, Y., Guyton, K.Z., Gorospe, M., Xu, Q., Lee, J.C., Holbrook, N.J., 1996. Differential activation of ERK, JNK/ SAPK, and P38/CSBP/RK map kinase family members during the cellular response to arsenite. Free Radic. Biol. Med., 21, 771–781.

Ludwig, S., Hoffmeyer, A., Goebeler, M., Kilian, K., Hafner, H., Neufeld, B., Han, J., Rapp, U.R., 1998. The stress inducer arsenite activates mitogen-activated protein kinases extracellular signal-regulated kinases 1 and 2 via a MAPK kinase 6/p38-dependent pathway. J. Biol. Chem., 273, 1917–1922.

Mandal, B.K., Ogra, Y., Suzuki, K.T., 2001. Identification of dimethylarsinous and monomethylarsonous acids in human urine of the arsenic-affected areas in West Bengal, India. Chem. Res. Toxicol., 14, 371–378.

Mechta-Grigoriou, F., Gerald, D., Yaniv, M., 2001. The mammalian Jun proteins: redundancy and specificity. Oncogene, 20, 2378–2389.

Reay, P.F., Asher, C.J., 1977. Preparation and purification of ^{74}As-labeled arsenate and arsenite for use in biological experiments. Anal. Biochem., 78, 557–560.

Rossman, T.G., Uddin, A.N., Burns, F.J., Bosland, M.C., 2001. Arsenite is a cocarcinogen with solar ultraviolet radiation for mouse skin: an animal model for arsenic carcinogenesis. Toxicol. Appl. Pharmacol., 176, 64–71.

Shaulian, E., Karin, M., 2001. AP-1 in cell proliferation and survival. Oncogene, 20, 2390–2400.

Simeonova, P.P., Luster, M.I., 2000. Mechanisms of arsenic carcinogenicity: genetic or epigenetic mechanisms? J. Environ. Pathol. Toxicol. Oncol., 19, 281–286.

Simeonova, P.P., Wang, S., Toriumi, W., Kommineni, C., Matheson, J., Unimye, N., Kayama, F., Harki, D., Ding, M., Vallyathan, V., Luster, M.I., 2000. Arsenic mediates cell proliferation and gene expression in the bladder epithelium: association with AP-1 transactivation. Cancer Res., 60, 3445–3453.

Simeonova, P.P., Wang, S., Kashon, M.L., Kommineni, C., Creselius, E., Luster, M.I., 2001. Quantitative relationship between arsenic exposure and AP-1 activity in mouse urinary bladder epithelium. Toxicol. Sci., 60, 279–284.

Simeonova, P.P., Wang, S., Hulderman, T., Luster, M.I., 2002. c-Src-dependent activation of the epidermal growth factor receptor and mitogen-activated protein kinase pathway by arsenic. Role in carcinogenesis. J. Biol. Chem., 277, 2945–2950.

Styblo, M., Serves, S.V., Cullen, W.R., Thomas, D.J., 1997. Comparative inhibition of yeast glutathione reductase by arsenicals and arsenothiols. Chem. Res. Toxicol., 10, 27–33.

Styblo, M., Del Razo, L.M., LeCluyse, E.L., Hamilton, G.A., Wang, C., Cullen, W.R., Thomas, D.J., 1999. Metabolism of arsenic in primary cultures of human and rat hepatocytes. Chem. Res. Toxicol., 12, 560–565.

Styblo, M., Drobná, Z., Jaspers, I., Lin, S., Thomas, D.J., 2002. The role of biomethylation in toxicity and carcinogenicity of arsenic. A research update. Environ. Health Perspect., 110 (Suppl. 5), 767–771.

Thomas, D.J., Styblo, M., Lin, S., 2001. The cellular metabolism and systemic toxicity of arsenic. Toxicol. Appl. Pharmacol., 176, 127–144.

Trouba, K.J., Wauson, E.M., Vorce, R.L., 2000. Sodium arsenite-induced dysregulation of proteins involved in proliferation signaling. Toxicol. Appl. Pharmacol., 164, 161–170.

US EPA (United States Environmental Protection Agency), 1988. Risk Assessment Forum. Special Report on Ingested Inorganic Arsenic: Skin Cancer, Nutrition Essentiality. Washington, DC.

Vahter, M., 1998. Methylation of inorganic arsenic in different mammalian species and population groups. Sci. Prog., 82, 69–88.

Vogt, P.K., 2001. Jun, the oncoprotein. Oncogene, 20, 2365–2377.

Vogt, B.L., Rossman, T.G., 2001. Effects of arsenite on p53, p21 and cyclin D expression in normal human fibroblasts – a possible mechanism for arsenite's comutagenicity. Mutat. Res., 478, 159–168.

Wiencke, J.K., Yager, J.W., Varkonyi, A., Hultner, M., Lutze, L.H., 1997. Study of arsenic mutagenesis using the plasmid shuttle vector pZ189 propagated in DNA repair proficient human cells. Mutat. Res., 386, 335–344.

Young, M.R., Nair, R., Bucheimer, N., Tulsian, P., Brown, N., Chapp, C., Hsu, T.Ch., Colburn, N.H., 2002. Transactivation of Fra-1 and consequent activation of AP-1 occur extracellular signal-regulated kinase dependently. Mol. Cell. Biol., 22, 587–598.

Arsenic Exposure and Health Effects V
W.R. Chappell, C.O. Abernathy, R.L. Calderon and D.J. Thomas, editors
© 2003 Published by Elsevier B.V.

Chapter 21

Effect of antioxidants on the papilloma response and liver glutathione modulation mediated by arsenic in Tg.AC transgenic mice

K. Trouba, A. Nyska, M. Styblo, D. Dunson, L. Lomnitski,
S. Grossman, G. Moser, A. Suttie, R. Patterson, F. Walton
and D. Germolec

Abstract

Epidemiological studies indicate that inorganic arsenicals produce various skin lesions as well as skin, lung, bladder, liver, prostate, and renal cancer. Our laboratory previously demonstrated that low-dose 12-O-tetradecanoylphorbol-13-acetate (TPA) increased the number of skin papillomas in Tg.AC transgenic mice that received sodium arsenite in drinking water, an effect dependent on proinflammatory cytokines. Because proinflammatory cytokine expression can be modulated by free radicals and oxidative stress, we hypothesized that oxidative stress contributes to TPA-promoted papilloma development in Tg.AC mice exposed to sodium arsenite. To evaluate the contribution of oxidative stress to arsenic skin carcinogenesis, two free-radical scavengers were tested for their ability to suppress papilloma responses (e.g. induction, latency, and multiplicity) modulated by arsenite in Tg.AC mice. Data indicate that arsenite increased papilloma responses in TPA-promoted Tg.AC mice as compared to control animals (no arsenite). The antioxidant vitamin E or a water-soluble natural antioxidant fraction from spinach had no inhibitory effect on TPA-promoted papilloma responses following arsenite exposure. Although not conclusively defined by our studies, oxidative stress generated by arsenic may contribute to skin carcinogenesis; however, it is not likely to be the sole or primary mechanism that enhances papilloma responses following arsenite exposure and TPA promotion.

Keywords: antioxidants, papilloma, Tg.AC, glutathione, NAO, vitamin E

1. Introduction

Arsenic, a ubiquitous metalloid, is a human carcinogen and exposure *via* drinking water is linked to skin cancer, hyperkeratoses, and pigmentation abnormalities (Goering et al., 1999). Several theories have been proposed to explain the mechanism(s) of arsenic carcinogenesis including alterations in DNA methylation, DNA repair, chromosome structure/number, cell proliferation, and free-radical/oxidative stress generation (Kitchin, 2001). Free-radical generation induced by arsenic is associated with cellular stress, DNA damage, chromosomal aberrations, protein expression alterations, and it influences inflammatory events involved in carcinogenesis (Matsui et al., 1999; Bernstam and Nriagu, 2000; Liu et al., 2001a,b).

Tumor initiation, promotion, and progression are causally linked to free-radical action as demonstrated by the capacity of 12-*O*-tetradecanoylphorbol-13-acetate (TPA) to stimulate free-radical release from inflammatory cells and generate oxidative species within epidermal cells (Marks and Furstenberger, 1983; Slaga and Butler, 1983; Schwarz et al., 1984). Compounds that induce oxidative stress *via* free-radical action enhance the transition of papillomas, a population of precancerous lesions, to carcinomas during the progression stage of carcinogenesis (Athar et al., 1989). Papilloma outgrowth also is sensitive to free radicals generated by the numerous inflammatory cells and proliferating keratinocytes that reside in and around these lesions (Rotstein and Slaga, 1988).

The transgenic Tg.AC mouse, carrying the v-Ha-ras structural gene linked to a ζ-globin promoter (Leder et al., 1990), has been used for the study of papilloma responses (e.g. induction, latency, and multiplicity) induced by chemical carcinogens that generate free radicals and oxidative stress (Spalding et al., 1993; Trempus et al., 1998). Tg.AC mice exhibit keratinocyte proliferation when challenged with TPA (Owens et al., 1995), and display an elevated papilloma response following arsenic exposure and TPA promotion (Germolec et al., 1998). Keratinocytes are the primary target of arsenic in skin (Germolec et al., 1996; Schwartz, 1996; Germolec et al., 1997; Bernstam et al., 2002), and are central to papilloma development (Krieg et al., 1993; Saez et al., 1995; Yuspa, 1998). Keratinocytes also produce free radicals in response to many environmental stimuli (Timmins and Davies, 1993; Shvedova et al., 2000). The generation of free radicals following arsenic exposure is due in part to the transfer of electrons by this metalloid during oxidative metabolism, leading to the production of unstable radical species such as the superoxide radical (Ercal et al., 2001; Liu et al., 2001a,b). The superoxide radical is converted, enzymatically, into hydrogen peroxide and subsequently into highly reactive hydroxyl radicals, all of which contribute to increased oxidative stress (Trouba et al., 2002).

Antioxidants have the capacity to suppress the effects of keratinocyte-derived free radicals (Peus et al., 2001), and they modulate skin tumor initiation, promotion, and progression *in vivo* (Slaga, 1995). Antioxidants like vitamin E, a naturally occurring lipid-soluble antioxidant, can protect skin from the damaging effects of radiation-induced oxidative stress (Nachbar and Korting, 1995), and non-enzymatic antioxidants, including ascorbic acid and selenium, ameliorate the toxicity of arsenic *in vivo* and *in vitro* (Schrauzer, 1992; Chen and Whanger, 1994; Lee and Ho, 1994; Davis et al., 2000; Chattopadhyay et al., 2001). The latter suggests that antioxidants are useful in mitigating the toxic effects of arsenic. However, research addressing antioxidant effects on arsenic skin carcinogenesis *in vivo* has not been examined in detail. Recently, it was demonstrated that natural antioxidant (NAO) fractions reduced papilloma responses in Tg.AC mice (Nyska et al., 2001). NAO, extracted from spinach leaves, contains an effective lipooxygenase inhibitor and free-radical scavenger composed primarily of aromatic polyphenols, glucurinated flavonoids, *cis-* and *trans*-coumaric acid derivatives, and uridine (Grossman et al., 1994; Bergman et al., 2001, 2003).

The purpose of this study was to determine the effect of NAO or vitamin E on arsenite-induced, TPA-promoted papilloma responses and the modulation of liver glutathione levels in Tg.AC mice. It was anticipated that these studies would help define the epidermal response (e.g. papilloma induction) to antioxidant supplementation under conditions permissive for cancer development (i.e. following arsenic exposure).

2. Methodology

2.1. Reagents

NAO, a water-soluble antioxidant composed of a mixture of natural molecules extracted and purified from spinach leaves, was prepared and characterized as described previously (Bergman et al., 2001). TPA and sodium arsenite were obtained from Sigma Chemical Co. (St. Louis, MO).

2.2. Animal treatment

Female, homozygous and hemizygous Tg.AC mice containing the fetal ζ-globin promoter fused to the v-Ha-ras structural gene (with mutations at codons 12 and 59) and linked to a simian virus 40 polyadenylylation/splice sequence were obtained from Taconic Farms (Germantown, NY). Homozygous mice, compared to hemizygotes, are generally more susceptible to arsenic-induced carcinogenecity and toxicity. Mice were maintained in the Integrated Laboratory Systems (ILS) animal facility in compliance with approved guidelines for the humane treatment of laboratory animals and were fed Purina Pico Chow 5058 and water *ad libitum*. Groups of 8-week-old, age-matched Tg.AC transgenic mice were provided arsenic (150 mg/l) as sodium arsenite in their drinking water for a period of 20 weeks. Seven weeks after initiation of arsenic treatment, the dorsal skin was shaved with electric clippers, and the first of four TPA doses (1.25 μg in 200 μl of acetone) was applied topically twice a week for 2 weeks. Animals were shaved thereafter during TPA dosing as needed. Papilloma occurrence and regression were recorded three times per week during the 25-week study. Vitamin E acetate in corn oil vehicle was used at a dose of 200 U/kg BW in a volume of 10 ml/kg, and NAO was dissolved in phosphate buffered saline to a dose level of 80 mg/kg BW in a volume of 10 ml/kg. Antioxidant administration was performed by gavage and was initiated 2 weeks prior to sodium arsenite treatment, and maintained for 2 weeks after initiation of TPA promotion (2 weeks prior to terminal sacrifice). The NAO dose and route of administration were based on previous studies with Tg.AC mice where 100 mg/kg by gavage was found to be safe and effective (i.e. administration of this NAO dose for 13 weeks to Tg.AC mice promoted with TPA caused a significant delay in skin hyperplasia progression and reduced tumor multiplicity) (Nyska et al., 2001).

2.3. Glutathione measurements

Glutathione (GSH) and glutathione disulfide (GSSG) concentrations in the livers of treated mice were determined using the previously described glutathione reductase (GR) recycling assay (Anderson, 1985). Fresh tissue was homogenized in ice-cold 5% 5-sulfosalicylic acid (SSA) on ice. Twenty percent (w/v) homogenates were centrifuged at $4°C$, $10,000g$, for 10 min and protein-free extracts frozen at $-80°C$ until analysis. For the assay, an aliquot of the SSA-extract was treated with 5,5′-dithiobis(2-nitrobenzoic acid) (DTNB) to oxidize GSH to GSSG. Total GSSG in the extract was then converted to GSH in a reaction catalyzed by yeast GR in the presence of NADPH. DTNB re-oxidizes GSH formed in the GR reactions, closing the cycle. In this assay, the sulfhydryl group of

GSH reacts with DTNB and produces a yellow-colored 5-thio-2-nitrobenzoic acid (TNB). The mixed disulfide, GSTNB (between GSH and TNB) that is concomitantly produced, is reduced by GR to recycle the GSH and produce more TNB. The rate of TNB production is directly proportional to this recycling reaction that in turn is directly proportional to the concentration of GSH in the sample. Measurement of TNB absorbance at 412 nM provides an accurate estimation of GSH in the sample. GSH is easily oxidized to the disulfide dimer GSSG and to limit this reaction, an aliquot of SSA-extract was treated with 2-vinylpyridine to block GSH thiols, preventing their oxidation to GSSG. Endogenous GSSG was then analyzed as described above, and the calibration curve for GSSG was used to quantify results of both assays. GSH concentration was calculated as the difference between GSH plus GSSG and GSSG as determined in 2-vinylpyridine-treated samples.

2.4. Statistical analysis

Individual animal papilloma data consisted of weekly observations of the number of skin papillomas in the test site over 25 weeks. To assess differences between groups in papilloma latency and multiplicity, accounting for within-animal and serial dependency in the papilloma counts, the Bayesian approach proposed by Dunson was employed (Dunson et al., 2000). Results are based on a model that accounts for heterogeneity among mice in (1) the rate of developing the first papilloma and (2) the rate at which new papillomas occur after the initial latency period. The model also allows for differential expression of the v-Ha-ras transgene, which can result in differences in sensitivity to a test agent. Groups differences are assessed based on posterior probabilities (PPs), with a PP >0.95 considered unlikely to be due to chance. In testing hypotheses, 1-PP is often used as a Bayesian version of the p-value. Analysis of variance (ANOVA) was performed on data obtained in liver GSH experiments. Following ANOVA, significant differences were determined by application of Dunnett's or the Tukey–Kramer multiple comparisons tests.

3. Results

To address the effects of antioxidants on arsenic skin toxicity *in vivo*, papilloma incidence, latency, and multiplicity were quantified in liver from Tg.AC mice treated with sodium arsenite/TPA alone or in combination with vitamin E or NAO. The level of reduced (GSH) and oxidized (GSSG) glutathione in liver from Tg.AC mice treated with sodium arsenite alone or in combination with vitamin E or NAO also was quantified.

Histology revealed no difference in tumor appearance among treatment groups, indicating that all tumors were papillomas (data not shown). Papilloma incidence (i.e. % animals with tumors), latency (i.e. time to development), and multiplicity (i.e. peak number of tumors/animal) are presented in Table 1. As indicated, papilloma incidence in Tg.AC mice following arsenite exposure and TPA promotion was 68.8% for homozygous mice and 57.1% for hemizygous mice. Exposure to arsenite significantly increased the incidence of papillomas compared to the control group (TPA only). Mean latency to first papilloma was reduced as compared to control (18 weeks), with 16.4 weeks for the homozygous and 16.0 weeks for the hemizygous mice. Data in Table 1 show that mice receiving NAO alone displayed a papilloma incidence that increased slightly compared to

Table 1. Papilloma incidence, latency, and peak number in homozygous Tg.AC mice treated with sodium arsenite ± antioxidants.

Group no.	Treatment[a]	Incidence (%)[b]		Mean latency (weeks)		Peak no. of Papillomas[c]		
		%	N	To first papilloma	To max. papillomas[d]	Mean	S.E.	N
1	Control	16.0	25	18.0	20.0	0.2	0.14	25
2	As (150 mg/l)	68.8	16	16.4	16.9	1.2	0.42	16
3[e]	As (150 mg/l)	57.1	14	16.0	17.4	1.3	0.73	14
4	As + NAO (80 mg/kg)	64.0	25	15.7	17.6	1.6	0.59	25
5	As + vit. E (200 U/kg)	68.2	22	15.1	17.3	2.0	0.69	22
6	NAO (80 mg/kg)	20.0	15	16.0	17.7	0.3	0.21	15
7	Vit. E (200 U/kg)	30.8	13	16.3	17.0	0.2	0.21	13

[a] All animals were TPA promoted.
[b] Incidence = (number of animals with at least one papilloma/number of animals in group) × 100; animals not surviving past week 13 excluded.
[c] Peak no. of papillomas = maximum number of papillomas sustained for three consecutive observations; mean is for all animals surviving past week 13.
[d] Max. papillomas = maximum number of papillomas observed.
[e] Hemizygotes.

the control group (20.0 compared to 16.0%). The incidence of papillomas in the group treated with arsenite and NAO was 64.0%, and was similar to animals receiving arsenite alone. The mean latency to papilloma was similar to the arsenite-treated group and was modestly decreased compared to the control group in both cases (15.7 weeks for the arsenite/NAO group and 16.0 weeks for the NAO group). While mean latency to maximum papillomas for both groups (17.6 weeks for the arsenite/NAO group and 17.7 weeks for the NAO group) increased slightly compared to the group receiving arsenite (16.9 weeks), it was reduced for both groups when compared to control. Combined, these data indicate that treatment with NAO did not significantly affect TPA-promoted papilloma responses in Tg.AC mice exposed to sodium arsenite.

Vitamin E effects on papilloma responses are presented in Table 1. Papilloma incidence in mice receiving arsenite and vitamin E was 68.2%. This value was 4.3-fold higher than that of control mice (16.0%) and similar to the group receiving arsenite alone. The group receiving only vitamin E displayed a papilloma incidence of 30.8%, which is 1.9-fold greater than that of control. The vitamin E group displayed a mean latency to first papilloma of 16.3 weeks, which was similar to the arsenite-treated animals and decreased as compared to control. In mice exposed to arsenite and vitamin E, the mean latency to first papilloma was 15.1 weeks, slightly lower than that of mice exposed to arsenite alone. Mean latency to maximum papillomas (17.3 weeks for the arsenite/vitamin E group and 17.0 weeks for the vitamin E group) was decreased compared to the control, but similar to the arsenite-treated group. These results demonstrate that vitamin E has little or no inhibitory effect on the papilloma response in Tg.AC mice treated concurrently with sodium arsenite.

Table 2. Results from Bayesian analysis assessing differences between groups in papilloma incidence and multiplicity. Posterior probabilities >0.95 are considered significant.

Contrast	Posterior probabilities (PPs)	
	Latency	Multiplicity
Group 1 vs. Group 2[a]	>0.999	>0.999
Group 1 vs. Group 4[a]	>0.999	0.998
Group 1 vs. Group 5[a]	>0.999	0.999
Group 1 vs. Group 6[a]	>0.999	0.998
Group 1 vs. Group 7[a]	0.836	0.190
Group 2 vs. Group 4[b]	0.152	0.995
Group 2 vs. Group 5[b]	0.166	0.886

[a] PPs for decrease in latency and increase in multiplicity.
[b] PPs for increase in latency and decrease in multiplicity.

Additional analyses were performed to determine if group comparisons (as presented in Table 2), along with PPs were present. Statistical analysis indicates that there was a significant decrease in the time to first papilloma (i.e. latency) and a significant increase in the number of new papillomas for mice treated with arsenite (group 2), arsenite and NAO (group 4), or arsenite and vitamin E (group 5) relative to control (group 1). These data confirm that arsenite induces papillomas in Tg.AC mice. There also were significant decreases in latency and increases in multiplicity among mice treated with NAO alone (group 6) relative to control (group 1). Evidence of a decrease in multiplicity among mice treated with arsenite and NAO (group 4) compared with mice treated with arsenite alone also was observed (group 2).

Glutathione (GSH) is important in the arsenic detoxification scheme *via* its involvement in the reduction of arsenic (V) to arsenic (III) in preparation for enzyme-catalyzed oxidative methylation, free-radical scavenging, and direct binding of this metalloid (Thompson, 1993; Wildfang et al., 1998). To further define the effect of NAO or vitamin E on biological changes induced by arsenite in Tg.AC mice, we quantified liver GSH and GSSG following arsenite exposure. Table 3 shows that there is a dose-dependent decrease in liver GSH in mice treated with arsenite; subsequent experiments examining the effect of antioxidants on GSH used an arsenite concentration of 150 mg/l. Table 4

Table 3. Dose-dependent changes in glutathione (mean ± S.D.) in livers of homozygous Tg.AC mice treated with sodium arsenite for 20 weeks.

Treatment	GSH[a] (µmol/g)	GSSG (µmol/g)	GSH/GSSG ratio	N
Control	6.26 ± 0.8	0.305 ± 0.06	21.56 ± 6.4	8
20 mg/l As	5.56 ± 1.1	0.284 ± 0.09	21.31 ± 7.6	9
50 mg/l As	5.13 ± 0.7	0.285 ± 0.11	19.90 ± 6.3	8
150 mg/l As	5.06 ± 0.6	0.261 ± 0.06	20.36 ± 5.0	8

[a] Statistical significance based on linear regression (trend analysis) ($p < 0.02$).

Table 4. Glutathione (mean ± S.D.) in livers of homozygous Tg.AC mice treated with sodium arsenite (150 mg/l) for 20 weeks ± antioxidants.

Group no.	Treatment	GSH (μmol/g)	GSSG (μmol/g)	GSH/GSSG ratio	N
1	Control	8.21 ± 1.72	0.186 ± 0.045	45.5 ± 10.33	16
2	As	5.91 ± 1.54[a]	0.184 ± 0.045	33.3 ± 9.67[a]	16
3[b]	As	7.26 ± 1.12	0.180 ± 0.027	41.0 ± 7.05	6
4	As + NAO	6.49 ± 1.49[a]	0.183 ± 0.051	37.5 ± 11.1	17
5	As + vit. E	6.57 ± 1.55[a]	0.186 ± 0.040	36.4 ± 11.5	13
6	NAO	7.82 ± 1.80	0.175 ± 0.041	46.2 ± 11.29	9
7	Vit. E	8.45 ± 2.07	0.174 ± 0.049	49.4 ± 10.12	7

[a] Significantly ($p < 0.05$) different from Control (Group no. 1) by Dunnett's multiple comparison test.
[b] Hemizygotes.

shows that control mice displayed a GSH level of 8.21 ± 1.7 μmol/g and a ratio of GSG/GSSG that was 45.5 ± 10.3 μmol/g. As determined using Dunnett's test, arsenite treatment resulted in a statistically significant decrease in GSH to 5.91 ± 1.5 μmol/g and lowered the GSH/GSSG ratio to 33.3 ± 9.67 μmol/g, respectively. Significant differences among all groups in Table 4 were determined using the Tukey–Kramer multiple comparisons test. Analysis indicated that in animals exposed to NAO plus arsenite (6.49 ± 1.49) or vitamin E plus arsenite (6.57 ± 1.55), there was no significant increase in GSH concentration compared to animals treated with arsenite alone (5.91 ± 1.54). In addition, no significant change occurred in GSH/GSSG ratio in animals treated with arsenite versus arsenite in combination with either antioxidant. It should be noted that the range of variation in GSSG concentration (compare 150 mg/l arsenite in Tables 3 and 4) is common in biological samples, and reflects a short-term change in redox metabolism of GSH dependent on immediate redox status of tissues, timing of dissections, sample handling, etc. GSSG concentrations also are almost 100 times lower than GSH concentrations in most cell types (tissues) making it difficult to measure GSSG concentrations, leading to greater inter-assay variation.

4. Discussion

Our previous studies demonstrating that sodium arsenite enhances the papilloma response *in vivo* were confirmed by the current studies using Tg.AC transgenic mice (Germolec et al., 1998). Vitamin E and NAO, however, had little effect on papilloma response induced following arsenite treatment/TPA promotion. Evidence in the literature indicating that vitamin E (Rotstein and Slaga, 1988) and NAO (Nyska et al., 2001) are capable of attenuating papilloma development is not supported by data in our study.

In general, vitamin E is an effective scavenger of singlet oxygen and peroxyl radicals (Chaudiere and Ferrari-Iliou, 1999; Wang and Quinn, 1999), whereas NAO is a scavenger of several reactive oxygen species such as the superoxide and hydroxyl anion radical, and it also is capable of preventing lipid peroxidation in mouse skin (Grossman et al., 1994; Bergman et al., 2001, 2003). The inability of NAO or vitamin E to inhibit the papilloma response in our study suggests that arsenite toxicity in skin is not dependent exclusively on the generation/action of free radicals. Alternatively, it is possible that the dose or route of administration of vitamin E or NAO was not optimal under our experimental conditions, or that these antioxidants do not target the primary free-radical species produced by arsenite that might enhance the papilloma response in Tg.AC mice. We feel the latter may not be a significant factor in our study because vitamin E has a strong propensity for reacting with a wide range of free radicals and can reduce tumor development in skin carcinogenesis models (Slaga, 1995). Interestingly, mice receiving vitamin E alone displayed a papilloma incidence 1.9-fold greater than that of control, an effect that was unanticipated but not without precedence as vitamin E has been shown to have tumor promoting properties (Perchellet et al., 1987; Mitchel and McCann, 1993). The pro-oxidant characteristic of vitamin E and/or the disruption of negative regulatory mechanisms that control tumor development also are possible explanations for this result (Yamashita et al., 1998). NAO, similar to vitamin E, has anti-carcinogenic effects in several experimental models including skin (Nyska et al., 2001) and prostate (Nyska et al., 2003). NAO prevents hyperoxia induced by reperfusion in the rat (Zurovsky et al., 1995), hepatic injury induced by LPS (Ben-Shaul et al., 2000; Lomnitski et al., 2000), and cardiotoxicity-induced by several anti-tumor agents (Breitbart et al., 2001). In our hands, NAO had little or no ability to counter the papilloma responses induced by arsenite.

Antioxidants mitigate arsenic toxicity associated with free-radical generation/action both *in vivo* (Ramanathan et al., 2002) and *in vitro*. For example, elevated superoxide dismutase activity reduces the frequency of arsenic-induced sister chromatid exchanges in human lymphocytes *in vitro* (Nordenson and Beckman, 1991), and vitamin E protects human fibroblasts from arsenic-induced cytotoxicity (Lee and Ho, 1994). Arsenic exposure also results in elevated levels of metallothionein and heme oxygenase (Keyse and Tyrrell, 1989; Keyse et al., 1990; Albores et al., 1992) in fibroblasts *in vitro*, GSH-related gene and enzyme expression in human cells *in vitro* (Schuliga et al., 2002), and GSH synthesis in mice *in vivo* (Santra et al., 2000); all of which may contribute to the attenuation of free-radical action. Paradoxically, arsenic reduces GSH level under certain conditions both *in vitro* and *in vivo* (Lee and Ho, 1995; Flora, 1999; Liu et al., 2000), demonstrating the complex action of arsenic on GSH and GSH-related enzyme systems. In our study, arsenite significantly reduced liver GSH; however, the reduction in liver GSH level was not rescued/inhibited by either vitamin E or NAO. Nonetheless, long-term liver GSH depletion may have important implications in chronic arsenic toxicity and cancer, and deserves further investigation.

5. Conclusions

In summary, the results of this study indicate the antioxidants vitamin E and a water-soluble antioxidant-containing fraction-NAO had no significant inhibitory effect on the papilloma response enhanced by sodium arsenite in Tg.AC mice. Both antioxidants also were incapable of attenuating the reduction in liver GSH level that occurred following arsenite exposure. These results support the concept that arsenic modulates multiple biological events involved in the carcinogenic process, and that free-radical action may not be the sole or primary mechanism by which arsenic stimulates TPA-promoted papillomagenesis in Tg.AC mice.

Acknowledgements

The authors would like to thank the many individuals at ILS who provided support for the study presented herein. We also are grateful to Yvette Rebolloso, Drs Joseph Wachsman, Maria Kadiiska, and Jie Liu for their insightful comments concerning the manuscript.

References

Albores, A., Koropatnick, J., Cherian, M.G., Zelazowski, A.J., 1992. Arsenic induces and enhances rat hepatic metallothionein production in vivo. Chem. Biol. Interact., 85 (2–3), 127–140.

Anderson, M., 1985. Determination of glutathione and glutathione disulfide in biological samples. Methods in Enzymology, Vol. 113. Academic Press, pp. 548–555.

Athar, M., Lloyd, J.R., Bickers, D.R., Mukhtar, H., 1989. Malignant conversion of UV radiation and chemically induced mouse skin benign tumors by free-radical-generating compounds. Carcinogenesis, 10 (10), 1841–1845.

Ben-Shaul, V., Lomnitski, L., Nyska, A., Carbonatto, M., Peano, S., Zurovsky, Y., Bergman, M., Eldridge, S.R., Grossman, S., 2000. Effect of natural antioxidants and apocynin on LPS-induced endotoxemia in rabbit. Hum. Exp. Toxicol., 19 (11), 604–614.

Bergman, M., Varshavsky, L., Gottlieb, H., Grossman, S., 2001. The antioxidant activity of aqueous spinach extract: chemical identification of active fractions. Phytochemistry, 58, 143–152.

Bergman, M., Perelman, A., Dubinsky, Z., Grossman, S., 2003. Scavenging of reactive oxygen species by a novel glucurinated flavonoid antioxidant isolated and purified from spinach. Phytochemistry, 62 (5), 753–762.

Bernstam, L., Nriagu, J., 2000. Molecular aspects of arsenic stress. J. Toxicol. Environ. Health B Crit. Rev., 3 (4), 293–322.

Bernstam, L., Lan, C.H., Lee, J., Nriagu, J.O., 2002. Effects of arsenic on human keratinocytes: morphological, physiological, and precursor incorporation studies. Environ. Res., 89 (3), 220–235.

Breitbart, E., Lomnitski, L., Nyska, A., Malik, Z., Bergman, M., Sofer, Y., Haseman, J.K., Grossman, S., 2001. Effects of water-soluble antioxidant from spinach, NAO, on doxorubicin-induced heart injury. Hum. Exp. Toxicol., 20 (7), 337–345.

Chattopadhyay, S., Ghosh, S., Debnath, J., Ghosh, D., 2001. Protection of sodium arsenite-induced ovarian toxicity by coadministration of L-ascorbate (vitamin C) in mature Wistar strain rat. Arch. Environ. Contam. Toxicol., 41 (1), 83–89.

Chaudiere, J., Ferrari-Iliou, R., 1999. Intracellular antioxidants: from chemical to biochemical mechanisms. Food Chem. Toxicol., 37 (9–10), 949–962.

Chen, C.L., Whanger, P.D., 1994. Interaction of selenium and arsenic with metallothionein: effect of vitamin B12. J. Inorg. Biochem., 54 (4), 267–276.

Davis, C.D., Uthus, E.O., Finley, J.W., 2000. Dietary selenium and arsenic affect DNA methylation in vitro in Caco-2 cells and in vivo in rat liver and colon. J. Nutr., 130 (12), 2903–2909.

Dunson, D.B., Haseman, J.K., van Birgelen, A.P., Stasiewicz, S., Tennant, R.W., 2000. Statistical analysis of skin tumor data from Tg.AC mouse bioassays. Toxicol. Sci., 55 (2), 293–302.

Ercal, N., Gurer-Orhan, H., Aykin-Burns, N., 2001. Toxic metals and oxidative stress part I: mechanisms involved in metal-induced oxidative damage. Curr. Top. Med. Chem., 1 (6), 529–539.

Flora, S.J., 1999. Arsenic-induced oxidative stress and its reversibility following combined administration of *N*-acetylcysteine and meso 2,3-dimercaptosuccinic acid in rats. Clin. Exp. Pharmacol. Physiol., 26 (11), 865–869.

Germolec, D.R., Yoshida, T., Gaido, K., Wilmer, J.L., Simeonova, P.P., Kayama, F., Burleson, F., Dong, W., Lange, R.W., Luster, M.I., 1996. Arsenic induces overexpression of growth factors in human keratinocytes. Toxicol. Appl. Pharmacol., 141 (1), 308–318.

Germolec, D.R., Spalding, J., Boorman, G.A., Wilmer, J.L., Yoshida, T., Simeonova, P.P., Bruccoleri, A., Kayama, F., Gaido, K., Tennant, R., Burleson, F., Dong, W., Lang, R.W., Luster, M.I., 1997. Arsenic can mediate skin neoplasia by chronic stimulation of keratinocyte-derived growth factors. Mutat. Res., 386 (3), 209–218.

Germolec, D.R., Spalding, J., Yu, H.S., Chen, G.S., Simeonova, P.P., Humble, M.C., Bruccoleri, A., Boorman, G.A., Foley, J.F., Yoshida, T., Luster, M.I., 1998. Arsenic enhancement of skin neoplasia by chronic stimulation of growth factors. Am. J. Pathol., 153 (6), 1775–1785.

Goering, P.L., Aposhian, H.V., Mass, M.J., Cebrian, M., Beck, B.D., Waalkes, M.P., 1999. The enigma of arsenic carcinogenesis: role of metabolism. Toxicol. Sci., 49 (1), 5–14.

Grossman, S., Reznik, R., Tamari, T., Albeck, M., 1994. New Plant Water Soluble Antioxidant (NAO) from Spinach. Elsevier Science, Amsterdam.

Keyse, S.M., Tyrrell, R.M., 1989. Heme oxygenase is the major 32-kDa stress protein induced in human skin fibroblasts by UVA radiation, hydrogen peroxide, and sodium arsenite. Proc. Natl Acad. Sci. USA, 86 (1), 99–103.

Keyse, S.M., Applegate, L.A., Tromvoukis, Y., Tyrrell, R.M., 1990. Oxidant stress leads to transcriptional activation of the human heme oxygenase gene in cultured skin fibroblasts. Mol. Cell. Biol., 10 (9), 4967–4969.

Kitchin, K.T., 2001. Recent advances in arsenic carcinogenesis: modes of action, animal model systems, and methylated arsenic metabolites. Toxicol. Appl. Pharmacol., 172 (3), 249–261.

Krieg, P., Feil, S., Furstenberger, G., Bowden, G.T., 1993. Tumor-specific overexpression of a novel keratinocyte lipid-binding protein. Identification and characterization of a cloned sequence activated during multistage carcinogenesis in mouse skin. J. Biol. Chem., 268 (23), 17362–17369.

Leder, A., Kuo, A., Cardiff, R.D., Sinn, E., Leder, P., 1990. v-Ha-ras transgene abrogates the initiation step in mouse skin tumorigenesis: effects of phorbol esters and retinoic acid. Proc. Natl Acad. Sci. USA, 87 (23), 9178–9182.

Lee, T.C., Ho, I.C., 1994. Differential cytotoxic effects of arsenic on human and animal cells. Environ. Health Perspect., 102 (Suppl. 3), 101–105.

Lee, T.C., Ho, I.C., 1995. Modulation of cellular antioxidant defense activities by sodium arsenite in human fibroblasts. Arch. Toxicol., 69 (7), 498–504.

Liu, J., Liu, Y., Goyer, R.A., Achanzar, W., Waalkes, M.P., 2000. Metallothionein-I/II null mice are more sensitive than wild-type mice to the hepatotoxic and nephrotoxic effects of chronic oral or injected inorganic arsenicals. Toxicol. Sci., 55 (2), 460–467.

Liu, J., Kadiiska, M.B., Liu, Y., Lu, T., Qu, W., Waalkes, M.P., 2001. Stress-related gene expression in mice treated with inorganic arsenicals. Toxicol. Sci., 61 (2), 314–320.

Liu, S.X., Athar, M., Lippai, I., Waldren, C., Hei, T.K., 2001. Induction of oxyradicals by arsenic: implication for mechanism of genotoxicity. Proc. Natl Acad. Sci. USA, 98 (4), 1643–1648.

Lomnitski, L., Nyska, A., Ben-Shaul, V., Maronpot, R.R., Haseman, J.K., Harrus, T.L., Bergman, M., Grossman, S., 2000. Effects of antioxidants apocynin and the natural water-soluble antioxidant from spinach on cellular damage induced by lipopolysaccharide in the rat. Toxicol. Pathol., 28 (4), 580–587.

Marks, F., Furstenberger, G., 1983. Multistage tumor promotion in skin. Princess Takamatsu Symp., 14, 273–287.

Matsui, M., Nishigori, C., Toyokuni, S., Takada, J., Akaboshi, M., Ishikawa, M., Imamura, S., Miyachi, Y., 1999. The role of oxidative DNA damage in human arsenic carcinogenesis: detection of 8-hydroxy-2'-deoxyguanosine in arsenic-related Bowen's disease. J. Invest. Dermatol., 113 (1), 26–31.

Mitchel, R.E., McCann, R., 1993. Vitamin E is a complete tumor promoter in mouse skin. Carcinogenesis, 14 (4), 659–662.

Nachbar, F., Korting, H.C., 1995. The role of vitamin E in normal and damaged skin. J. Mol. Med., 73 (1), 7–17.

Nordenson, I., Beckman, L., 1991. Is the genotoxic effect of arsenic mediated by oxygen free radicals? Hum. Hered., 41 (1), 71–73.

Nyska, A., Lomnitski, L., Spalding, J., Dunson, D.B., Goldsworthy, T.L., Ben-Shaul, V., Grossman, S., Bergman, M., Boorman, G., 2001. Topical and oral administration of the natural water-soluble antioxidant from spinach reduces the multiplicity of papillomas in the Tg.AC mouse model. Toxicol. Lett., 122 (1), 33–44.

Nyska, A., Suttie, A., Bakshi, S., Lomnitski, L., Grossman, S., 2003. Slowing tumorigenic progression in TRAMP mice and prostatic carcinoma cell lines using natural antioxidant from spinach, NAO – a comparative study of three antioxidants. Toxicol. Pathol., 31 (1), 1–13.

Owens, D.M., Spalding, J.W., Tennant, R.W., Smart, R.C., 1995. Genetic alterations cooperate with v-Ha-ras to accelerate multistage carcinogenesis in Tg.AC transgenic mouse skin. Cancer Res., 55 (14), 3171–3178.

Perchellet, J.P., Abney, N.L., Thomas, R.M., Guislain, Y.L., Perchellet, E.M., 1987. Effects of combined treatments with selenium, glutathione, and vitamin E on glutathione peroxidase activity, ornithine decarboxylase induction, and complete and multistage carcinogenesis in mouse skin. Cancer Res., 47 (2), 477–485.

Peus, D., Meves, A., Pott, M., Beyerle, A., Pittelkow, M.R., 2001. Vitamin E analog modulates UVB-induced signaling pathway activation and enhances cell survival. Free Radic. Biol. Med., 30 (4), 425–432.

Ramanathan, K., Balakumar, B.S., Panneerselvam, C., 2002. Effects of ascorbic acid and alpha-tocopherol on arsenic-induced oxidative stress. Hum. Exp. Toxicol., 21 (12), 675–680.

Rotstein, J.B., Slaga, T.J., 1988. Effect of exogenous glutathione on tumor progression in the murine skin multistage carcinogenesis model. Carcinogenesis, 9 (9), 1547–1551.

Saez, E., Rutberg, S.E., Mueller, E., Oppenheim, H., Smoluk, J., Yuspa, S.H., Spiegelman, B.M., 1995. c-fos is required for malignant progression of skin tumors. Cell, 82 (5), 721–732.

Santra, A., Maiti, A., Chowdhury, A., Mazumder, D.N., 2000. Oxidative stress in liver of mice exposed to arsenic-contaminated water. Indian J. Gastroenterol., 19 (3), 112–115.

Schrauzer, G.N., 1992. Selenium. Mechanistic aspects of anticarcinogenic action. Biol. Trace Elem. Res., 33, 51–62.

Schuliga, M., Chouchane, S., Snow, E.T., 2002. Upregulation of glutathione-related genes and enzyme activities in cultured human cells by sublethal concentrations of inorganic arsenic. Toxicol. Sci., 70 (2), 183–192.

Schwartz, R.A., 1996. Premalignant keratinocytic neoplasms. J. Am. Acad. Dermatol., 35 (2 Pt 1), 223–242.

Schwarz, M., Peres, G., Kunz, W., Furstenberger, G., Kittstein, W., Marks, F., 1984. On the role of superoxide anion radicals in skin tumour promotion. Carcinogenesis, 5 (12), 1663–1670.

Shvedova, A.A., Kommineni, C., Jeffries, B.A., Castranova, V., Tyurina, Y.Y., Tyurin, V.A., Serbinova, E.A., Fabisiak, J.P., Kagan, V.E., 2000. Redox cycling of phenol induces oxidative stress in human epidermal keratinocytes. J. Invest. Dermatol., 114 (2), 354–364.

Slaga, T.J., 1995. Inhibition of the induction of cancer by antioxidants. Adv. Exp. Med. Biol., 369, 167–174.

Slaga, T.J., Butler, A.P., 1983. Cellular and biochemical changes during multistage skin tumor promotion. Princess Takamatsu Symp., 14, 291–301.

Spalding, J.W., Momma, J., Elwell, M.R., Tennant, R.W., 1993. Chemically induced skin carcinogenesis in a transgenic mouse line (Tg.AC) carrying a v-Ha-ras gene. Carcinogenesis, 14 (7), 1335–1341.

Thompson, D.J., 1993. A chemical hypothesis for arsenic methylation in mammals. Chem. Biol. Interact., 88 (2–3), 89–114.

Timmins, G.S., Davies, M.J., 1993. Free radical formation in isolated murine keratinocytes treated with organic peroxides and its modulation by antioxidants. Carcinogenesis, 14 (8), 1615–1620.

Trempus, C.S., Mahler, J.F., Ananthaswamy, H.N., Loughlin, S.M., French, J.E., Tennant, R.W., 1998. Photocarcinogenesis and susceptibility to UV radiation in the v-Ha-ras transgenic Tg.AC mouse. J. Invest. Dermatol., 111 (3), 445–451.

Trouba, K.J., Hamadeh, H.K., Amin, R.P., Germolec, D.R., 2002. Oxidative stress and its role in skin disease. Antioxid. Redox Signal., 4 (4), 665–673.

Wang, X., Quinn, P.J., 1999. Vitamin E and its function in membranes. Prog. Lipid Res., 38 (4), 309–336.

Wildfang, E., Zakharyan, R.A., Aposhian, H.V., 1998. Enzymatic methylation of arsenic compounds. VI. Characterization of hamster liver arsenite and methylarsonic acid methyltransferase activities in vitro. Toxicol. Appl. Pharmacol., 152 (2), 366–375.

Yamashita, N., Murata, M., Inoue, S., Burkitt, M.J., Milne, L., Kawanishi, S., 1998. Alpha-tocopherol induces oxidative damage to DNA in the presence of copper(II) ions. Chem. Res. Toxicol., 11 (8), 855–862.

Yuspa, S.H., 1998. The pathogenesis of squamous cell cancer: lessons learned from studies of skin carcinogenesis. J. Dermatol. Sci., 17 (1), 1–7.

Zurovsky, Y., Eligal, Z., Grossman, S., 1995. Unilateral renal ischemia reperfusion in the rat: effect of blood volume trapped in the kidney, sucrose infusion, and antioxidant treatments. Exp. Toxicol. Pathol., 47 (6), 471–478.

Arsenic Exposure and Health Effects V
W.R. Chappell, C.O. Abernathy, R.L. Calderon and D.J. Thomas, editors

Chapter 22

Application of filter arrays to the study of arsenic toxicity and carcinogenesis

Jie Liu, Hua Chen, Maria Kadiiska, Yaxiong Xie
and Michael P. Waalkes

Abstract

Arsenic is a known human carcinogen. Acute arsenic exposure produces toxicity to multiple organs, while chronic exposure to arsenic causes tumors of the skin, lung, bladder, liver, and kidney. Since alterations in gene expression following arsenic exposure that are linked to toxic effects are incompletely understood, microarray technology was utilized to analyze gene expression changes associated with arsenic toxicity and carcinogenesis. Total RNA was extracted from cells transformed by chronic arsenic exposure, from mouse livers of acute and chronic arsenic exposure, and from liver tumors induced by transplacental arsenic exposure. RNA was then converted to cDNA probes using reverse transcriptase and cDNA synthesis primers. The ^{32}P-labeled cDNA probes were hybridized to array membranes on which hundreds of known cDNA fragments are fixed. The hybridization intensity was quantified using AtlasImage software, and the selected gene expression was further confirmed by RT-PCR or Western blot analysis. Induction of oxidative stress-related genes, such as heme oxygenase-1 and heat-shock proteins, is associated with acute arsenic exposure, while the activation of oncogenes, such as c-myc, H-ras, c-met, and Wilm's tumor protein, is associated with chronic arsenic exposure and/or arsenic transformation. Dysregulation of cell cycles, such as overexpression of estrogen receptor-α, cyclin D1, proliferative cell nuclear antigen (PCNA), and downregulation of p27 is observed following chronic exposure to arsenic. Overexpression of α-fetoprotein and downregulation of syndecan-1 were mainly seen in arsenic-induced tumors. Thus, microarray is a useful tool to profile genetic events associated with arsenic toxicity and carcinogenesis.

Keywords: cDNA microarray, arsenic, acute and chronic toxicity, carcinogenesis

1. Introduction

Inorganic arsenic is a known human carcinogen and exerts many other acute and chronic adverse effects. Epidemiology studies show that chronic arsenic exposure produces tumors of the skin, bladder, lung, liver, prostate, and kidney (Abernathy et al., 1999; NRC, 1999; ATSDR, 2000). Environmental arsenic exposure is mainly from consumption of drinking water contaminated with inorganic arsenic (NRC, 1999), but can also result from burning of coal containing high levels of inorganic arsenic (Liu et al., 2002). The alterations in gene expression that are linked to arsenic toxicity and carcinogenicity are poorly defined (Abernathy et al., 1999; Goering et al., 1999; Kitchin, 2001).

There are a variety of ways in which arsenic can alter gene expression (Kitchin, 2001; Liu et al., 2001a), which can be associated with aberrant cellular phenotypes. We have utilized cDNA microarray technology to profile inorganic arsenic-induced malignant transformation in rat liver epithelial cells (Chen et al., 2001a), inorganic arsenic-induced acute toxicity in the mouse liver (Liu et al., 2001a), chronic arsenic-induced liver disorders in human population of Guizhou, China, using liver biopsy samples (Lu et al., 2001), and arsenic-induced hepatocellular carcinomas following a brief transplacental exposure in mice (Waalkes et al., 2003; Liu et al., 2003). The results clearly showed that arsenic toxicity and carcinogenicity are associated with a variety of aberrant gene expressions, including oxidative stress-related genes, oncogens, cell-cycle regulators, DNA damage/repair responsive genes, and genes encoding growth factors, hormone receptors, and metabolic enzymes. This initial gene profiling could advance our understanding of the mechanism of arsenic toxicity and carcinogenesis.

2. Materials and methods

2.1. Microarray analysis

The Clontech filter arrays (140–1200 genes, Clontech, Palo Alto, CA) were used for cDNA microarray analysis. Total RNA was isolated from tissues or cell cultures with RNeasy columns (Qiagen, Valencia, CA). 2–5 μg of total RNA was converted to [α-^{32}P]-dATP-labeled cDNA probe using MMLV reverse transcriptase and the Atlas array specific cDNA synthesis primer mix, and then purified with a NucleoSpin column. The membranes were prehybridized with Expresshyb from Clontech for 2 h at 68°C, followed by hybridization with the cDNA probe overnight at 68°C. The membranes were then washed four times in 2 × SSC/1% SDS, 30 min each, and two times in 0.1 × SSC/0.5% SDS for 30 min. The membranes were then sealed with plastic wrap and exposed to a Molecular Dynamics Phosphoimage Screen. The images were analyzed densitometrically using AtlasImage software (Clontech, version 2.01). The gene expression intensities were first corrected with the background and then normalized with the sum of all nine housekeeping genes on the array (40S ribosomal protein S29, 45 kDa calcium-binding protein, β-actin, ornithine decarboxylase, myosin, G3PDH, hypoxanthine–guanine phosphoribosyltransferase, ubiquitin, and phospholipase A2).

2.2. RT-PCR analysis and statistics

The levels of expression of the selected genes were quantified by real-time RT-PCR analysis with the SYBR green DNA PCR kit (Applied Biosystems, Foster City, CA) or by conventional RT-PCR analysis with the One-step RT-PCR kit (Clontech). Pooled samples ($n = 4$–6) were used. To ensure the reproducibility, mean and SEM of 3–6 replicates were calculated. For the comparisons of gene expression between two groups, Student's *t*-test was performed. The level of significance was set at $p < 0.05$.

3. Results and discussion

3.1. Microarray analysis of chronic As-transformed cells

An *in vitro* model of arsenic-induced malignant transformation was developed by chronic (18 weeks or more) exposure of the rat liver epithelial cell line (TRL1215) to low concentrations of sodium arsenite (0–500 nM, equivalent to 0–37.5 ppb that is relevant to environmental exposure). After 18 weeks of exposure, the diploid, non-tumorigenic TRL1215 cells became morphologically transformed, and produced malignant tumors capable of metastasis upon inoculation into Nude mice (Zhao et al., 1997), and were termed as CAsE cells. Genomic DNA hypomethylation (Zhao et al., 1997) and acquired tolerance to arsenic (Romach et al., 2000) occurred concomitantly in CAsE cells. Altered DNA methylation status could affect the expression of a variety of genes, and is a potential event in chemical carcinogenesis (Wachsman, 1997). Microarrays were therefore used to profile the aberrant gene expression patterns associated with arsenic-induced malignant transformation using CAsE cell model.

Reproducibility is a major concern in microarray analysis. Thus, hybridizations were repeated four times using separate cell preparations. After background subtraction and normalization, we obtained hybrid intensity signals which allowed the calculation of mean and SEM of each gene. Using this approach, we reported significant alterations in the expression of approximately 80 genes associated with arsenic-induced malignant transformation (Chen et al., 2001a). These included oncogenes, cell-cycle regulator genes, signaling pathway genes, genes in response to oxidative stress and DNA damage, and genes encoding hormone receptors and metabolic enzymes (Chen et al., 2001a). From this initial gene expression analysis, follow-up confirmatory studies were conducted to verify the microarray results.

3.1.1. Follow-up study #1: cell-cycle dysregulation

Among the altered gene expressions seen in the array studies with the arsenic trans-formants were the overexpression of cell-cycle regulators such as c-*myc*, cyclin D1, and PCNA. To confirm these results, RT-PCR analysis, which measures gene expression at the transcription level, and Western blot analysis, which measures levels of the translation products were performed (Fig. 1). In all cases, the RT-PCR and Western blot analyses confirmed the microarray results.

The prominent overexpression of c-*myc*, a gene frequently activated during hepatocarcinogenesis, as well as the overexpression of cyclin D1 and PCNA, implies

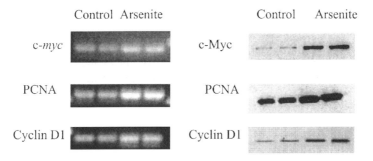

Figure 1. Representative RT-PCR (left panel) and Western blot (right panel) analysis of c-*myc*, PCNA, and cyclin D1, respectively, in control and As-transformed cells exposed to 500 nM arsenite for 24 weeks. Each left two lanes are from controls, and each right two lanes are from As-transformed cells. All of them were significantly increased in As-transformed cells (Adapted from Chen et al. 2001b).

that cell-cycle dysregulation is an important event in arsenite-induced malignant transformation. Indeed, these arsenic transformed cells are characterized by a much higher proliferation rate in a fashion related to the level of arsenic exposure. The expression of c-*myc* was highly correlated with cellular hyperproliferation ($r = 0.961$) (Chen et al., 2001b). Thus, the follow-up analysis provides convincing evidence that overexpression of c-*myc*, cyclin D1, and PCNA are mechanistically linked to arsenite-induced malignant transformation in this model system.

3.1.2. Follow-up study #2: GST-π, MRP and MDR overexpression

An important feature of these arsenic-transformed cells is the development of self-tolerance (Romach et al., 2000). There are several possible mechanisms for arsenic-induced self-tolerance in these transformed cells. One possible mechanism is that arsenite exposure leads to the increased expression of arsenic efflux transporters. Indeed, one of the most dramatic changes in these arsenic-transformed cells is the marked reduction in cellular arsenic accumulation, apparently due to an increase in arsenic efflux (Romach et al., 2000). The initial gene array screening revealed overexpression of glutathione *S*-transferase pi (GST-π), multidrug resistance transporters and P-glycoproteins in these arsenic-transformed cells. Based on these results, we hypothesized that multidrug-resistance transporters could play an important role in acquired arsenic tolerance. In support of this hypothesis, RT-PCR and Western blot analysis confirmed increases in GST-π, MRP, and P-glycoproteins in arsenite-transformed cells (Liu et al., 2001b). We next tested the biological function of these proteins in cellular arsenic accumulation by utilizing the specific transporter inhibitors, MK571 for MRP and PSC833 for P-glycoproteins (Fig. 2).

The transporter inhibitors greatly increased cellular arsenic accumulation, and reversed arsenic tolerance. Thus, this follow-up study revealed that enhanced arsenic transport is an important mechanism for acquired tolerance to arsenic. In CAsE cells, arsenic self-tolerance is associated with increased expression of GST-π, multidrug resistance transporters and P-glycoproteins), which together contribute to arsenic tolerance by reducing cellular arsenic content (Liu et al., 2001b). These data have important implications in the toxicology and pharmacology of arsenic, as acquired arsenic tolerance through altered toxicokinetics is important during the long-term arsenic chemotherapy or chronic arsenic exposure.

3.2. Microarray analysis of acute arsenic-exposed animals

High local concentrations of arsenic could lead to cell death and carcinogenesis. It was thus appropriate to profile acute arsenic-induced oxidative stress-related gene expression *in vivo*. Mice were injected sc with inorganic arsenite (As(III), 100 μmol/kg) or arsenate (As(V), 300 μmol/kg), and livers were removed 30 min later to detect radical production *via* ESR (Fig. 3). As(III) produced higher ESR signals as compared to As(V) and controls.

To profile the gene expression associated with arsenic-induced oxidative stress, livers were removed 3 h after arsenic injection for RNA isolation, and subjected to microarray analysis. A number of functional categories of genes were altered following acute arsenic

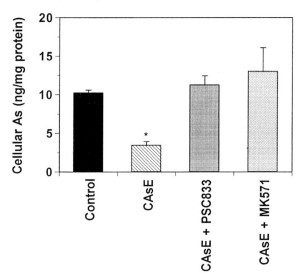

Figure 2. Effect of specific transporter inhibitors on cellular arsenic accumulation after 24 h treatment with 10 μM of arsenite to control and As-transformed CAsE cells. The P-glycoprotein inhibitor PSC833 (3 μM) and multidrug resistance protein inhibitor MK571 (15 μM) were added simultaneously with arsenic. *Significantly different from control cells $p < 0.05$.

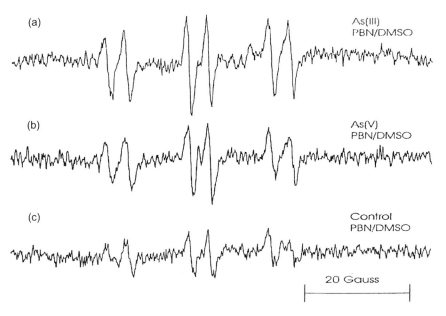

Figure 3. Electron spin resonance (ESR) spectrum for PBN-radical adducts in mouse liver lipid extracts 30 min after As(III) (100 μmol/kg, sc), As(V) (300 μmol/kg, sc), or saline administration.

exposure in mice, including stress proteins, such as heme oxygenase-1 and DNA damage/repair genes. To confirm this array profile, multiprobe RNase protection assays were used in follow-up studies. The activation of the c-Jun/AP-1 transcription complex coincides with the induction of stress-related genes. Western blot analysis further confirmed the enhanced production of arsenic-induced stress proteins such as heme oxygenase-1, heat shock protein-70, heat shock protein-90, metallothionein, the metal-responsive transcription factor metal transcription factor-1 (MTF-1), nuclear factor kappa B, and c-Jun/AP-1 following acute arsenic exposure. Increases in caspase-1 and cytokines such as tumor necrosis factor-α and macrophage inflammatory protein-2 were also evident (Liu et al., 2001a). In general, arsenite is more effective than arsenate in inducing these gene alterations following acute exposure.

3.3. Microarray analysis of As-induced tumors following transplacental exposure

The development of rodent models of inorganic arsenic carcinogenesis is critical to elucidate the carcinogenic potential of inorganic arsenic, and to understand its molecular mechanisms. We have recently shown that short-term transplacental arsenic exposure during gestation, a period of high sensitivity to chemical carcinogenesis in rodents, produced a variety of tumors in the offspring (Waalkes et al., 2003). This includes aggressive epithelial malignancies, such as hepatocellular carcinoma in males and pulmonary adenocarcinoma in females. These tumors occurred in the absence of any other treatment. Transplacental arsenic exposure also induced neoplasms of the adrenal gland, ovary, and proliferative lesions in uterus and oviduct (Waalkes et al., 2003). Thus, inorganic arsenic can act as a "complete carcinogen" during brief exposure in the gestational period.

Microarray analysis was conducted to profile gene expression associated with transplacental arsenic carcinogenesis. Total RNA was extracted from liver tumors and surrounding tissues of mice transplacentally exposed to sodium arsenite (As-85 ppm) or As-0 (control) in the drinking water, and subjected to microarray and quantitative RT-PCR analysis. Among 600 customer-designed genes, arsenic-induced tumors had higher rates of aberrant gene expression (>2-fold, 31%) than spontaneous tumors (20%). Selected gene expression changes are shown in Figure 4.

Overexpression of estrogen receptor-α, cyclin D1, c-*myc* and suppression of p27 were associated with arsenic exposure. The overexpression of α-fetoprotein and decreased expression of syndecan-1, markers for malignancy of hepatocellular carcinoma, were seen in arsenic-induced tumors only. The non-tumor tissues of arsenic-exposed offspring also showed alterations in the expression of genes associated with cell proliferation, oxidative stress, signal transduction, DNA damage/repair, and metabolism enzymes. Real time RT-PCR largely confirmed these findings. These aberrantly expressed genes could play integrated roles in arsenic carcinogenesis.

3.4. Microarray analysis of chronic arsenic-exposed human liver biopsy samples

To extrapolate the findings in cultured cells and rodents to human situations, analysis of available human samples is necessary. There are areas in Guizhou, China where

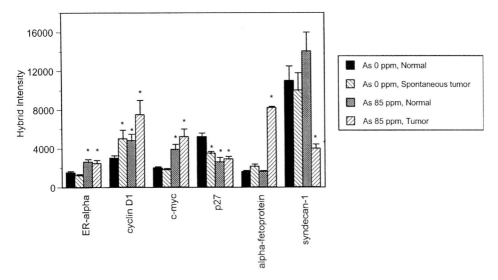

Figure 4. Expression of genes in liver tumors and surrounding tissues of mice transplacentally exposed to As-0 (control) and As-85 ppm. Total RNA from pooled samples ($n = 4$) was subjected to microarray analysis. Data are mean ± SEM of 3–6 replicates. *Significantly different from controls, $p < 0.05$.

arsenic exposure occurs through the burning of high arsenic-containing coal. The coal is burned inside the home in open pits for daily cooking and crop drying, resulting in high concentration of arsenic in indoor air. Arsenic in the air coats and permeates the food being dried, resulting in high concentrations on foodstuffs (Liu et al., 2002). At least 3000 arsenic exposed patients were found in the Southwest Prefecture of Guizhou, and approximately 200,000 people are at risk for such exposures. Skin lesions are common, including keratosis of the hand and feet, pigmentation on the truck, ulceration, and skin cancers. Toxicities to internal organs, including lung dysfunction, neuropathy, and nephrotoxicity, are clinically evident. The prevalence of hepatomegaly was 20%, and cirrhosis, ascites, and liver cancer are the most serious outcomes of arsenic poisoning. To evaluate the therapeutic effects of treating the patients with Chinese medicinal preparations, needle liver biopsies were performed in selected patients and very small amounts (~5 mg) of liver biopsy samples were used for microarray analysis.

Approximately 60 genes (10%) were differentially expressed in arsenic-exposed human livers as compared to controls. Similar to what was seen in rodents, the differentially expressed genes included those involved in cell-cycle regulation, apoptosis, DNA damage response, intermediate filaments, and hormone receptors (Lu et al., 2001). The observed gene alterations appear to be reflective of hepatic degenerative lesions seen in these arsenic-exposed patients. Clearly, a variety of gene expression changes may play an integral role in arsenic hepatotoxicity and possibly carcinogenesis. Table 1 shows a comparison of selected gene alterations following chronic arsenic exposure in rat liver cells, in mouse liver tumors, and in chronic arsenic-exposed human liver biopsy samples.

Table 1.   A comparison of selected gene alterations following chronic arsenic treatments.

	As-transformed rat liver CAsE cells	As-induced mouse liver tumors	As-exposed human liver biopsy samples
Cyclin D1	Up	Up	Up
Estrogen receptor-α	N/A	Up	N/A
p27	Down	Down	N/A
Cu–Zn SOD	Up	Up	Up
c-*myc*	Up	Up	Up
α-fetoprotein	N/A	Up	N/A
Insulin-like growth factor	Down	Down	No change
IGF-1 binding protein	Up	Up	Up
Fibrosis growth factor	Up	Up	Up

N/A: Not available on the array membranes.

4. Challenges and prospectives

Based on the expression profile analysis and other classic approaches, multiple mechanisms that lead to arsenic toxicity and carcinogenicity can be invoked. The major challenge for investigators using gene arrays is to formulate valid hypotheses based on the alteration of potentially large numbers of genes. It should be kept in mind that many of the gene alterations are unexpected because the function and the regulation of most genes are not entirely understood. It is anticipated that as more investigators apply gene expression profiling to either toxicogenomic or disease analysis, the signaling pathways important in perturbations of cellular homeostasis will be better characterized.

One of the major challenges for the effective use of expression profile data is the inherent platform and biological variability. Platform variability may come from differences in hybridization and washing conditions from experiment to experiment as well as the amount or quality of the DNA spotted onto the array membranes. Biological variability may come from differences in the way in which individual animals and/or separate cell preparations respond to treatment. Additional biological studies to verify array findings with known functions are essential for hypothesis confirmation, and are an important aspect of toxicogenomics investigations.

Both dose–response and time-course studies coupled with appropriate toxicogenomic analysis are crucial for making linkage between the gene expression changes and toxicity. However, because of the significant costs, it is not realistic to conduct comprehensive array studies. Until the costs for performing array experiments decreases substantially, it should be kept in mind that arrays may best be used in focused initial screening followed by more complete studies on a few genes in detail by standard techniques. This strategy will be important to create a comprehensive and reliable understanding of the mechanisms of chemical toxicity and allow an extrapolation of the data to human risk assessment.

References

Abernathy, C.O., Liu, Y.-P., Longfellow, D., Aposhian, H.V., Beck, B., Fowler, B., Goyer, R., Menzer, R., Rossman, T., Thompson, C., Waalkes, M.P., 1999. Arsenic: health effects, mechanisms of actions, research issues. Environ. Health Perspect., 107, 593–597.

Agency for Toxic Substances and Disease Registry (ATSDR), 2000. Toxicological Profiles for Arsenic.

Chen, H., Liu, J., Merrick, A., Waalkes, M.P., 2001a. Genetic events associated with arsenite-induced malignant transformation: application of cDNA microarray technology. Mol. Carcinog., 30, 79–87.

Chen, H., Liu, J., Zhao, C., Merrick, A.B., Waalkes, M.P., 2001b. Association of oncogene activation and cell-cycle dysregulation with arsenite-induced malignant transformation. Toxicol. Appl. Pharmacol., 175, 260–268.

Goering, P.L., Aposhian, H.V., Mass, M.J., Cebrian, M., Beck, B.D., Waalkes, M.P., 1999. The enigma of arsenic carcinogenesis: role of metabolism. Toxicol. Sci., 49, 5–14.

Kitchin, K.T., 2001. Recent advances in arsenic carcinogenesis: modes of action, animal model systems, and methylated arsenic metabolites. Toxicol. Appl. Pharmacol., 172, 249–261.

Liu, J., Kadiiska, M., Liu, Y., Lu, T., Qu, W., Waalkes, M.P., 2001a. Stress-related gene expression in mice treated with inorganic arsenicals. Toxicol. Sci., 61, 314–320.

Liu, J., Chen, H., Miller, D.S., Saavedra, J.E., Keefer, L.K., Johnson, D.R., Klaassen, C.D., Waalkes, M.P., 2001b. Overexpression of glutathione *S*-transferase Pi and multidrug resistance transport proteins is associated with acquired tolerance to inorganic arsenic. Mol. Pharmacol., 60, 302–309.

Liu, J., Zheng, B., Aposhian, H.V., Zhou, Y., Chen, M.-L., Waalkes, M.P., 2002. Chronic arsenic poisoning from burning arsenic-containing coal in Guizhou, China. Environ. Health Perspect., 110, 119–122.

Liu, J., Xie, Y., Ward, J.M., Diwan, B.A., Waalkes, M.P., 2003. Toxicogenomic analysis of aberrant gene expression in liver and liver tumors induced by transplacental exposure to inorganic arsenic in mice. Toxicologist, 68, 1038.

Lu, T., Liu, J., LeCluyse, E.L., Zhou, Y.-S., Cheng, M.-L., Waalkes, M.P., 2001. Application of cDNA microarray to the study of arsenic-induced liver diseases in the population of Guizhou, China. Toxicol. Sci., 59, 185–192.

National Research Council (NRC), 1999. Arsenic in Drinking Water. National Academy Press, Washington, DC.

Romach, E.H., Zhao, C.Q., Del Razo, L.M., Cebrian, M.E., Waalkes, M.P., 2000. Studies on the mechanism of arsenic-induced self tolerance in liver epithelial cells through continuous low-level arsenite exposure. Toxicol. Sci., 54, 500–508.

Waalkes, M.P., Ward, J.M., Liu, J., Diwan, B.A., 2003. Transplacental carcinogenicity of inorganic arsenic in mice: induction of hepatic, ovarian, pulmonary and adrenal tumors and associated overexpression of cyclin D1 and estrogen receptor-α. Toxicol. Appl. Pharmacol, 186, 7–17.

Wachsman, J.T., 1997. DNA methylation and the association between genetic and epigenetic changes: relation to carcinogenesis. Mutat. Res., 375, 1–8.

Zhao, C.Q., Young, M.R., Diwan, B.A., Coogan, T.P., Waalkes, M.P., 1997. Association of arsenic-induced malignant transformation with DNA hypomethylation and aberrant gene expression. Proc. Natl Acad. Sci. USA, 94, 10907–10912.

Arsenic Exposure and Health Effects V
W.R. Chappell, C.O. Abernathy, R.L. Calderon and D.J. Thomas, editors

Chapter 23

Regulation of redox and DNA repair genes by arsenic: low dose protection against oxidative stress?

Elizabeth T. Snow, Yu Hu, Catherine B. Klein, Kate L. McCluskey, Michael Schuliga and Peter Sykora

Abstract

We have evaluated the molecular responses of human epithelial cells to low dose arsenic to ascertain how target cells may respond to physiologically relevant concentrations of arsenic. Data gathered in numerous experiments in different cell types all point to the same conclusion: low dose arsenic induces what appears to be a protective response against subsequent exposure to oxidative stress or DNA damage, whereas higher doses often provoke synergistic toxicity. In particular, exposure to low, sub-toxic doses of arsenite, As(III), causes coordinate up-regulation of multiple redox and redox-related genes including thioredoxin (Trx) and glutathione reductase (GR). Glutathione peroxidase (GPx) is down-regulated in fibroblasts, but up-regulated in keratinocytes, as is glutathione S-transferase (GST). The maximum effect on these redox genes occurs after 24 h exposure to $5-10$ μM As(III). This is 10-fold higher than the maximum As(III) concentrations required for induction of DNA repair genes, but within the dose region where DNA repair genes are coordinately down-regulated. These changes in gene regulation are brought about in part by changes in DNA binding activity of the transcription factors activating protein-1 (AP-1), nuclear factor kappa-B, and cAMP response element binding protein (CREB). Although sub-acute exposure to micromolar As(III) up-regulates transcription factor binding, chronic exposure to sub-micromolar As(III) causes persistent down-regulation of this response. Similar long-term exposure to micromolar concentrations of arsenate in drinking water results in a decrease in skin tumour formation in dimethylbenzanthracene (DMBA)/phorbol 12-tetradecanoate 13-acetate (TPA) treated mice. Altered response patterns after long exposure to As(III) may play a significant role in As(III) toxicology in ways that may not be predicted by experimental protocols using short-term exposures.

Keywords: arsenic, gene expression, redox, DNA repair, human fibroblasts

1. Introduction

Arsenic is a toxic metalloid often found in association with heavy metals such as gold and copper. It has been used as a pesticide, as an herbicide, and as a poison, and continues to be used as a pharmaceutical. Recently, environmental exposure to arsenic in drinking water has become recognised as a serious world health problem. Chronic exposure to environmental arsenic is known to cause skin lesions, lung disease (including cancer), other internal cancers (particularly of the bladder), vascular disease, peripheral neuropathies, and increased risk of diabetes [reviewed in Kitchin (2001)]. The risk of increased tumour formation in humans due to environmental arsenic exposure appears quite high. It has been estimated that arsenic toxicity is the ultimate cause of up to 7% of all deaths in Region II of Chile (Smith et al., 1998). Nevertheless, questions remain regarding the shape of the dose response curve, particularly at low doses. In some animal models, relatively high doses of arsenic (> 100 μM As(III) in drinking water) appear to be carcinogenic or co-carcinogenic (Rossman et al., 2001; Wang et al., 2002). There is only one, unverified, report of lower doses causing an increase in cancer in a rodent model (Ng et al., 1999). However, as reported below, we have found that low doses of arsenate in drinking water, similar to levels to which human populations can be exposed, can also decrease papilloma formation in mice exposed to a classic initiation/promotion protocol using dimethylbenzanthracene (DMBA) as an initiator and phorbol acetate (TPA) as a promoter.

The molecular mechanism(s) of arsenic carcinogenesis in humans are not well understood. There are numerous hypotheses related to different possible mechanisms: from induction of hyperproliferation by activation of keratinocyte growth factors (Germolec et al., 1998), to oxidative damage caused by increased production of reactive oxygen species (ROS) (Kessel et al., 2002), to alteration of transcription factor activation (Hu et al., 2002), and inhibition of DNA repair (Li and Rossman, 1989; Hartmann and Speit, 1996; Yager and Wiencke, 1997; Snow et al., 1999). Whatever the mechanism, or suite of mechanisms, it is clear that chronic health effects are the result of "low dose" arsenic exposure. Doses sufficient to kill large numbers of cells provoke acute toxicity and are not generally relevant to tumour promotion or progression. Doses of arsenite, As(III), that produce molecular responses relevant to carcinogenesis seem to be in the range of $0.01-10$ μM, i.e. doses that can cause genetic or proliferative responses in human cells, without being overtly toxic (Germolec et al., 1998; Simeonova et al., 2000; Kitchin, 2001; Hu et al., 2002). This is below the range found in some environmental sources of As, but within the range of total arsenic species found in the blood of people exposed to high levels of As over long periods of time (Concha et al., 1998; Wu et al., 2001). It is also in the sub-toxic range in which arsenic causes measurable and relevant changes in cellular response such as hyperproliferation and altered gene expression (Germolec et al., 1996; Hu et al., 2002; Schuliga et al., 2002). It is also well below the range in which most enzymes are directly inhibited by arsenic species (Hu et al., 1998; Chouchane and Snow, 2001). Thus, chronic disease that develops with time occurs in living tissues and is the result of cellular exposure to doses that are sub- or minimally toxic. It is also clear that epithelial cells predominate in target tissues for arsenic-induced chronic disease end-points, such as the skin, lung, bladder, and vascular tissues. It follows that the study of these cell types is critical to our understanding of both the mechanism and dose dependence of cellular responses to arsenic species that may ultimately result in cancer or other arsenic-related diseases.

Another parameter critical for our understanding of molecular responses to arsenic in humans is chemical speciation. Environmental arsenic exposure is generally in the form of inorganic arsenic species, either as arsenous acid [arsenite, As(III)] or arsenic acid [arsenate, As(V)]. Arsenite, which reacts preferentially with sulfur (thiols), is significantly more toxic than arsenate. The binding of As(III) to critical or high affinity cellular thiols, particularly dithiols, can cause cross-linkages. This mimics normal oxidative processes and may serve to trigger responses as if to oxidative stress. Thus, cellular processes related to redox control may be shifted in unusual or long-lasting ways. Higher concentrations of As(III) or As(III)-species may also inhibit critical thiol-containing enzymes and thereby cause cell death. It is generally believed that most of the toxicity of arsenate is due to its reduction *in vivo* to arsenite. Organic forms of arsenic may also be found in the environment, particularly in food products. In general, however, these metabolites are less toxic than inorganic arsenic and are of less concern for human health (Kojima et al., 2002). Arsenic is metabolised *in vivo* to methylated derivatives and the trivalent forms such as monomethylarsenic(III) and dimethylarsenic(III) may also be important for carcinogenesis in some tissues (Styblo et al., 2000). However, it remains to be established whether these metabolites are present in reasonable quantities in target tissues such as the skin, lung, or even bladder.

Over the past several years we have examined the molecular response of human epidermal keratinocytes and fibroblasts to low, micromolar doses of arsenite. In particular, we have examined the expression of genes and proteins involved in cellular redox control and genes involved in base excision DNA repair.

Low doses of arsenic are known to affect several cellular systems that might be involved in the carcinogenic process. Foremost among these is the oxidative stress response. Arsenic can induce several stress response genes, including heme oxygenase and heat shock proteins (Del Razo et al., 2001). Low doses of arsenic also cause a rapid and transient burst of ROS (Wang and Huang, 1994; Liu et al., 2001; Kessel et al., 2002). In addition to confirming some of these responses using human epithelial fibroblasts, we have found that micromolar concentrations of arsenic alter the expression of a battery of genes involved in cellular redox control, including a series of thioredoxin (Trx) and glutathione-related enzymes. Thioredoxin is a small cysteine-rich protein involved in the redox regulation of numerous critical cellular proteins, including ribonucleotide reductase and several transcription factors. Thioredoxin reductase (TR) is the enzyme that maintains the redox status of Trx. Glutathione (GSH), a cysteine-containing tripeptide, is the most abundant single redox-active thiol within the cell. Glutathione reductase (GR) reduces oxidised GSH (GSSG) maintaining the GSH/GSSG balance within the cell. Glutathione peroxidase (GPx) uses GSH to reduce and thereby detoxify reactive peroxides. Glutathione *S*-transferase (GST) consists of a series of closely related enzymes that conjugate endogenous and exogenous metabolites to GSH facilitating their transport out of the cell.

In addition to the effects on cellular redox control processes, we have found that the inhibition of DNA repair by arsenic is the consequence of down-regulation of multiple DNA repair genes. However, sub-micromolar concentrations of arsenic actually enhance levels of DNA base excision repair (BER). These unexpected findings are consistent with an adaptive response that would protect against subsequent exposure to higher doses of arsenic or other agents that produce oxidative DNA damage and are seen at concentrations of arsenic that are below the levels necessary to inhibit enzyme activity of all but the most sensitive proteins. These results are significant both with respect to possible mechanisms of As-induced carcinogenesis, but also for determining a mechanistic dose response for arsenic toxicity.

2. Materials and methods

Caution: Inorganic arsenic is toxic and classified as a human carcinogen. It must be handled with appropriate care and caution.

2.1. Materials and chemicals

Sodium arsenite (NaAsO$_2$), reduced GSH, oxidised GSH (GSSG), baker's yeast GR, β-nicotinamide adenine dinucleotide phosphate (NADPH) tetrasodium salt, sulfosalicylic acid, 5,5′-dithio-bis(2-nitrobenzoic acid) (DTNB), 1-chloro-2,4-dinitrobenzene (CDNB), neutral red (NR), monoclonal anti-β-actin IgG and oligonucleotide primers for polymerase chain reaction (PCR) were obtained from Sigma (Australia). Anti-GR polyclonal antibodies were kindly provided by R.H. Schirmer (University of Heidelberg, Germany), whereas anti-rabbit and mouse IgG horseradish peroxidase (HRP) conjugated secondary antibodies (raised in sheep) were obtained from Silenus (Australia). Media and sera for cell culture were obtained from Trace Elements (Australia). Hybond-N nylon membranes and Hybond-P PVDF membranes were obtained from Amersham (USA and Australia). α[^{32}P]-dCTP was supplied by Amersham.

2.2. Cell culture

Normal diploid WI-38 lung and immortalised GM847 human skin fibroblasts were obtained from CSL (Geelong, Australia) and the Murdoch Institute (Melbourne, Australia), respectively. Cells were maintained in either basal medium Eagle's (WI-38) or Dulbecco's modified Eagle's medium, 2 mM glutamate, 100 Units penicillin, and 100 μg/ml streptomycin at 37°C. All cell lines were maintained in a humidified atmosphere of 5% CO$_2$ and were routinely split at a 1:4 ratio. Sodium arsenite in PBS was added to complete medium of pre-confluent cell cultures in the logarithmic phase of growth 24 h after seeding.

2.3. Neutral red assay

The viability of cells grown in 96-well microtitre plates was assessed by the uptake of NR dye according to the protocol of Babich and Borenfreund (1987).

2.4. Dichlorofluorescein (DCF) assay

The production of ROS (peroxides) was measured using dichlorofluorescein diacetate (DCF-dAc) as described by Huang et al. (1994). Cultured cells, maintained under normal conditions, were loaded with 20 μM DCF-dAc for 10 min prior to the addition of As(III). After a further 30 min, cells were harvested in PBS using a rubber policeman, pelleted by centrifugation, and then re-suspended in PBS at 2×10^5 cells/ml. DCF fluorescence was determined using a Versa Fluor fluorometer (Bio-Rad) with excitation and emission filters of 480 and 520 nm, respectively.

2.5. Western analysis

Protein extracts for western analysis were obtained according to the protocol described by Tan et al. (1999). Equal amounts of denatured supernatant protein, along with protein standard markers (Bio-Rad), were resolved on 12% SDS−polyacrylamide gels and then electroblotted onto PVDF membranes. Membranes were blocked in 5% w/v non-fat dry milk powder in 10 mM Tris–HCl (pH 7.5) with 0.15 M NaCl and 0.1% v/v Tween 20 and then incubated with a mouse primary antibody. Bound antibody was detected with HRP-conjugated secondary antibodies (1:1000 dilution) and chemiluminescence using the ECL kit (Amersham). Western blots were re-probed with a mouse antibody raised against β-actin (1:5000 dilution) to verify equal protein loading.

2.6. DNA binding

Transcription factor DNA binding was carried out using band shift assays with Promega reagents as described in Hu et al. (2002).

2.7. Enzyme assays

Enzymatic assays of GR and GST activity were performed as described in Schuliga et al. (2002). The GST enzyme assay was based on that described by Lee et al. (1989). Reactions, conducted at room temperature, were initiated by adding 50 μl of cell lysate to 950 μl of 0.1 M sodium phosphate buffer (pH 6.8) with 1 mM EDTA, 1 mM reduced GSH, and 1 mM CDNB. The formation of conjugates of GSH and CDNB were monitored spectrophotometrically at 340 nm.

The GR enzyme assay was based on that described previously by Styblo et al. (1997), and involved monitoring the reduction in NADPH absorbance at 340 nm. Cell extracts were pre-incubated in 0.15 M sodium phosphate buffer (pH 7) containing 6 mM EDTA and 0.1 mM oxidised GSH at 37°C. Reactions were initiated by the addition of 0.23 mM NADPH.

The DNA ligase assay was performed using a poly dA·oligo dT substrate as described by Arrand and Willis (1986).

2.8. Northern blot probes

Northern analysis of mRNA was performed as described previously (Snow et al., 2001; Schuliga et al., 2002) using PCR amplified regions of human cDNA as probes. All values for mRNA concentrations were calculated relative to the amount of GAPDH in the same lane. Developed X-ray films were scanned using a Bio-Rad G710 densitometer, and analysed using Bio-Rad Quantity One software.

2.9. Animal experiments

A mouse skin papilloma assay was used to investigate the effect of chronic arsenic exposure on polyaromatic hydrocarbon-induced skin cancer. Papilloma formation was initiated in C57Bl/6 × CBA F1 mice by a single topical treatment with DMBA, applied in 100 μl acetone at a concentration of 200 nmol per mouse, and was promoted by topical application of phorbol 12-tetradecanoate 13-acetate (TPA) (Frenkel et al., 1993; Lu et al., 1997). The effect of arsenate on papilloma formation was tested by continued exposure to low (0.2 mg/l) and high (2 mg/l) dose As(V) (as Na_2HAsO_4) in the drinking water starting 2 days before topical DMBA treatment. TPA (5 μg/ml in acetone) was administered topically twice weekly starting 1 week after DMBA treatment. Papillomas were counted as a measure of formation of pre-neoplastic lesions.

2.10. Statistical analysis

Where appropriate, results are expressed as the mean and standard error of the mean (SEM) of at least three separate experiments. Statistical analyses were performed using the Student's *t*-test. Values of $p < 0.05$ were considered to represent statistically significant differences.

3. Results

3.1. Reactive oxygen species

As(III) causes a rapid but transient burst in ROS in WI-38 cells, resolving to baseline within an hour when assayed by DCF fluorescence. Thirty-minute exposure to doses of As(III) up to 10 µM resulted in a steady increase in ROS, as detected *via* the fluorescence of DCF (Fig. 1). This rapid burst of ROS is resolved promptly but, especially at higher doses, may contribute to As-induced cytotoxicity. The change in the redox environment of the cell may also initiate cell signalling that mediates a longer-term response.

3.2. Cell viability

As(III) toxicity in WI-38 fibroblasts was determined *via* NR uptake of viable cells. Short-term (24 h) exposure to As(III) showed a 50% decrease in dye uptake (EC$_{50}$) in the range of 15–20 µM (Fig. 1). All subsequent experiments were performed within this dose range to ensure cell viability. Long-term exposure to sub-toxic concentrations of As(III)

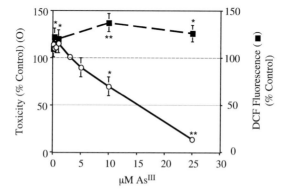

Figure 1. Toxicity and ROS production in WI-38 cells after exposure to As(III). Toxicity was measured using NR dye uptake after 24 h exposure, whereas ROS was measured using the DCF fluorescence assay after 30 min exposure, as described in Section 2. *Indicates significantly different from the control, $p < 0.05$. **Indicates $p < 0.01$.

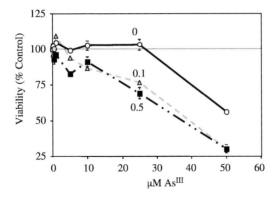

Figure 2. Long-term exposure (7 weeks) to low dose (0.1 or 0.5 μM) As increases the sensitivity of human fibroblast cells to subsequent acute (24 h) exposure to As(III). GM847 fibroblasts were grown for 50 days in control media (○) or in media containing 0.1 μM (△) or 0.5 μM (■) As(III), as described previously (Hu et al., 2002). The cultured cells were then tested for their response to an acute dose of As(III). To do this, cells from each of the long-term cultures were plated in 96-well plates, allowed to attach for 24 h, then exposed for 24 h to 0–50 μM As(III) and viability assessed by NR dye uptake, as described in Section 2.

over periods of 50 days increased the sensitivity of the GM847 fibroblast cell line to subsequent short-term (24 h) incubation with As(III) (Fig. 2) Short-term incubation led to an EC_{50} of approximately 50 μM in the control group that had been cultured for 50 days in the absence of exogenous As(III). Significant cell death did not occur at doses lower than 30 μM. Chronic exposure to 0.1 or 0.5 μM As(III) resulted in an increased sensitivity to a challenge dose of As(III), with 50% cytotoxicity occurring after 24 h exposure to < 40 μM As(III).

3.3. Cellular redox systems

The two main redox systems studied were the glutathione system and the Trx system. Both of these systems involve numerous proteins and enzymes that maintain the cell in its reduced state. Twenty-four hour exposure to As(III) at concentrations of up to 10 μM resulted in coordinated up-regulation of the steady-state mRNA levels of several key proteins. There was a threefold increase in GR mRNA and a twofold increase in Trx mRNA in exposed WI-38 fibroblasts (Fig. 3). GPx, however, was down-regulated by almost 50% in fibroblasts exposed to the same conditions (Fig. 3). The PMC42 breast tumour cell line responded similarly with regards to GPx (Fig. 4A). In contrast, GPx mRNA was up-regulated to almost twice its baseline level in the HaCaT keratinocyte cell line after 24 h exposure to As (Fig. 4A). GST activity was also up-regulated in keratinocytes, while being down-regulated in the breast tumour and fibroblast cell lines (Fig. 4B).

As we have seen previously in GM847 cells (Schuliga et al., 2002), GR protein levels and enzyme activity are also up-regulated in parallel with mRNA levels in the WI-38 cells (Fig. 5). The relative increase in enzyme activity is not as great as that seen for the mRNA levels, but still shows a significant increase ($p < 0.01$) after exposure of the cells to greater than 0.25 μM As(III).

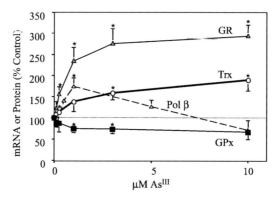

Figure 3. Relative gene expression measured by mRNA levels or protein levels (of DNA polβ) in WI-38 cells after 24 h exposure to As(III). GR, Trx, and GPx mRNA levels were assayed by northern blot analysis after 24 h As(III) exposure as described previously (Schuliga et al., 2002). DNA polβ protein levels were measured by western blot analysis of nuclear protein extracts using affinity purified polβ antibody obtained from Dr S.H. Wilson (NIEHS). All assays were done in quadruplicate. *Indicates values significantly different from the unexposed controls with $p < 0.05$.

3.4. DNA repair

The BER pathway is responsible for the repair of most DNA damage resulting from cellular metabolism, including damage caused by ROS. In retrospect, it is not therefore surprising that in addition to causing up-regulation of redox enzymes to deal with the increased production of ROS, low doses of As(III) also cause up-regulation of the BER pathway. At least three enzymes in this pathway have been found to be up-regulated by sub-micromolar As(III): apurinic/apyrimidinic (AP) endonuclease, DNA polymerase β (polβ), and DNA ligase I. After either spontaneous or enzymatic removal of a damaged

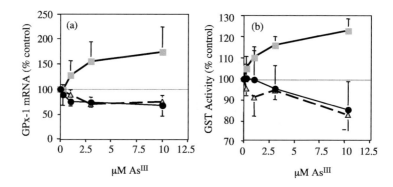

Figure 4. GPx expression (a) and GST activity (b) co-regulate in different cell types after exposure to As(III). Levels of GPx mRNA were assessed by northern blots and GST activity was measured by enzyme assay using total cell extracts, as described in Section 2. Each data point represents an average of four measurements ± standard errors. HaCaT keratinocytes are indicated with a square, GM847 fibroblasts with a filled circle, and PMC42 breast tumour cells with a triangle.

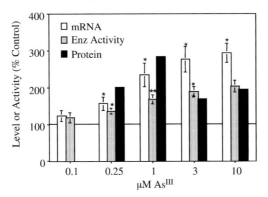

Figure 5. GR varies in a similar dose-dependent manner in WI-38 cells after 24 h exposure to As(III) whether measured by mRNA levels (using northern blots), by enzyme activity, or by protein levels (using western blots). All assays were done in quadruplicate and *indicates values significantly different from the control with $p < 0.05$.

base, AP endonuclease cleaves the backbone of the DNA. Polβ is then responsible for removing the 5′-deoxyribose and incorporating the appropriate base in the resulting single nucleotide gap using the undamaged complementary strand as a template. DNA ligase then joins the newly incorporated nucleotide to the adjacent 5′-end of the existing DNA. We have shown previously that AP endonuclease mRNA is up-regulated by treatment of human keratinocytes with 0.2–1 μM As(III) (Snow et al., 2001), Figure 6 shows that both DNA polβ and DNA ligase I are also up-regulated by similar concentrations of As(III) in human WI-38 and GM847 fibroblasts. Polβ is maximally up-regulated by low micromolar concentrations of As(III), reaching a peak of more than 150% of control at 1 μM As(III). The protein concentration of ligase I also increases to approximately 150% of control at low micromolar concentrations of As(III) (Fig. 6). The DNA ligase enzyme is twice

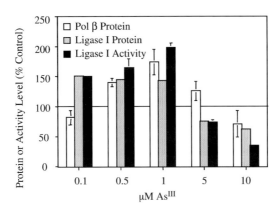

Figure 6. Parallel up- and down-regulation of DNA repair proteins and DNA ligase activity after low dose exposure to As(III) in fibroblasts (GM847 and WI-38 cells). Polβ protein levels were measured by western blot in WI-38 cells ($n = 4$). DNA ligase I protein ($n = 1$) and enzyme activity levels ($n = 3$) were assayed in GM847 fibroblasts.

as active in cells treated with 1 μM As(III) than in control cells. It should be noted that the maximal up-regulation of the BER proteins is seen at concentrations 10-fold lower than those resulting in maximal up-regulation of the redox proteins (compare Figure 6 with Figures 3 and 5). Doses beyond this point, however, result in significant down-regulation of BER protein concentrations and enzyme activity below that of the control. At 10 μM As(III), the concentration at which many of the redox proteins peak, the relative concentrations of the BER proteins are reduced to substantially below baseline.

3.5. *Transcription factors*

Transcription factors alter the activation of early response genes. Such genes are often of a protective or reparative nature. Stress (including oxidative stress) induces the activation of transcription factors that trigger the expression of these genes. Two such transcription factors are activating protein-1 (AP-1) and cAMP response element binding protein (CREB). Both AP-1 and CREB are up-regulated by low (0.1–5 μM) concentrations of arsenic in GM847 fibroblasts (data not shown). To evaluate cellular responses after long-term, chronic exposure to As, GM847 fibroblasts were treated with 0.1 or 0.5 μM As(III) semi-continuously for up to 10 weeks. As shown in Figure 7, this long-term exposure to low concentrations of As caused a significant dose-dependent decrease in the basal levels of AP-1 and CREB binding in the As treated cells compared to cells grown for the same amount of time in the absence of added As(III).

The short-term response to oxidative stress was determined by exposing the arsenic-exposed cells to the tumour promoter TPA and the oxidant H_2O_2. Although TPA treatment

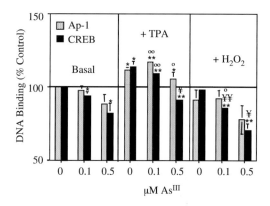

Figure 7.   Long-term (10 week) exposure of GM847 fibroblasts to low dose As down-regulates AP-1 and CREB transcription factor binding activity and ameliorates cellular responses to subsequent short exposures to oxidative stress induced by TPA (phorbol ester) or H_2O_2. GM847 cells were exposed to low dose As for 10 weeks as described in Hu et al. (2002), then challenged with 100 ng/ml TPA for 3 h or 200 μM H_2O_2 for 30 min. After an additional 60 min of incubation the cells were harvested and tested for AP-1 or CREB binding activity by an electromobility band shift assay. An asterisk indicates results that are significantly different from the untreated (0 As) controls ($^* = p < 0.5$, $^{**} = p < 0.01$). Circles indicate significantly different from control cells maintained in the same level of As prior to treatment ($^\circ = p < 0.5$, $^{\circ\circ} = p < 0.01$). ¥ symbols indicate that the response of the 0.1 or 0.5 μM As-treated cells was significantly different than the other As-treated cells to the same TPA or H_2O_2 treatment ($¥ = p < 0.05$, $¥¥ = p < 0.01$).

induced both AP-1 and CREB binding in control and 0.1 μM As exposed cells, the response was significantly decreased in cells that had been chronically exposed to 0.5 μM As(III). H_2O_2 also caused a decrease in transcription factor binding that was much more pronounced in the cells that had been chronically exposed to As. A similar decrease in the DNA binding capacity of the transcription factor NF-κB after long-term exposure to 0.1 and 0.5 μM As(III) has previously been reported (Hu et al., 2002).

Together, these results indicate that there is a lowered response to oxidative stress after long-term exposure to As(III). This reduced activation of early response genes and resultant lessening of the cellular response to oxidative stress induced by TPA and H_2O_2 would be expected to decrease the effect of these agents as tumour promoters and therefore possibly protect against some forms of cancer.

3.6. Animal experiments

In order to evaluate the effect of long-term exposure to arsenic as a tumour promoter, or anti-promoter, we performed a two-stage mouse skin carcinogenesis experiment using DMBA as the initiator and TPA as the promoter while simultaneously exposing the mice to environmentally relevant concentrations of arsenate in their drinking water. The doses of As(V) used in these experiments were non-toxic to the mice as evidenced by the fact that there were no differences in weight gain among the groups and no differences in water intake between groups exposed to different concentrations of As (data not shown). The results of these experiments were rather surprising: the arsenate significantly decreased the formation of papillomas on the skin of the mice. The high dose of arsenic (2 mg/l arsenate in the drinking water for the duration of the experiment) was twice as effective as the low dose (0.2 mg/l arsenate) (Fig. 8). Papilloma formation was not induced in the presence of

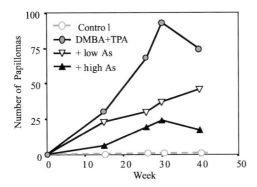

Figure 8. Exposure to relatively low concentrations of As(V) in drinking water reduces papilloma formation induced by a DMBA/TPA tumour initiation protocol. Mice were exposed *ad libitum* to As(V) at concentrations of 0, 0.2 (low As), or 2 mg/l (high As) in their drinking water starting 2 days prior to DMBA treatment and continuing for the duration of the experiment. Starting 1 week after DMBA the mice were also treated with twice weekly applications of TPA for 30 weeks, as described in Section 2. Papilloma formation on the dorsal skin of the mice was assessed visually at various times after initiation with DMBA and the total number of papillomas in each group of 18 mice per treatment is presented. Similar results were obtained when the average number of papillomas per mouse, or the number of mice with papillomas was plotted.

arsenic alone, if DMBA was omitted, or if the TPA promoter was eliminated or reduced to 3 μg/ml (not shown). These results differ from those reported by others (Rossman et al., 2001; Wang et al., 2002), however, it should be noted that we used As(V) rather than As(III) in the drinking water and that the doses of As(V) used in our experiments were significantly lower than that for As(III) used in most other experiments (Wang et al., 2002). There is only one report of inorganic arsenic in any form at concentrations of less than 10 mg/l increasing cancer risk (Ng et al., 1999).

4. Discussion

As mentioned previously, exposure to As(III) causes a rapid and transient burst of ROS. This burst of ROS is fully dose dependent, observable at doses as low as 0.1 μM As(III) and increasing up to doses of 20 μM or more, and can be prevented or decreased by adding antioxidants such as β-hydroxytoluene or *N*-acetylcysteine to the cell culture medium before treating the cells with As(III) (data not shown).

Our results show that micromolar doses of As(III) cause a coordinated two- to threefold up-regulation of Trx and GR. GPx is also up-regulated in keratinocyte cells, but is down-regulated in human fibroblasts. The altered expression of GR is seen not only with respect to changes in levels of mRNA, but also in increased (or decreased) protein levels and enzyme activity. Interestingly, GST activity also increases in keratinocytes and decreases in fibroblasts, in parallel with GPx, but not the other GSH-related activities. These findings are consistent with an adaptive response to increased oxidative stress caused by the observed burst of ROS and suggests that the redox system response may confer greater protection against oxidative damage in skin keratinocyte cells compared with fibroblast cells. It should also be noted that while the WI-38 fibroblast cells used here were derived from human lung cells, we obtained similar results with GM847 skin fibroblasts (Snow et al., 2001; Schuliga et al., 2002). Our results clearly show that human cells derived from different organ systems exhibit different molecular responses after exposure to low doses of arsenic. While this may be related to the tissue specificity of arsenic-induced carcinogenesis, we do not yet know how changes in gene expression after short (24 h) exposures to As compare with the cellular response in keratinocytes after long-term exposure to sub-micromolar concentrations of arsenite. Changes in molecular response should also be evaluated in the context of intact organ systems, rather than in isolated cell types.

Redox genes are not the only category of genes affected by arsenic. DNA repair genes are also strongly modulated by low concentrations of As(III). The adverse affects of arsenic on DNA repair have been known for over 10 years. In particular, moderately toxic concentrations of As(III) have been shown to inhibit DNA ligase activity (Li and Rossman, 1989), decrease the repair of UV-induced DNA damage (Hartwig et al., 1997), and decrease the activity of poly(ADP-ribose)polymerase (PARP) (Yager and Wiencke, 1997). It was previously thought that arsenic inhibited these DNA repair enzymes directly, such that DNA ligase activity in particular was inhibited by arsenic binding to critical protein SH groups. In 1998, we showed that direct inhibition of enzyme activity did not occur (Hu et al., 1998). We have now found that although micromolar concentrations of As(III) can cause significant down-regulation of DNA repair activity, lower sub-micromolar concentrations can actually cause a significant increase in DNA polymerase and DNA ligase activity. AP endonuclease mRNA (APE/Ref-1) is also up-regulated in keratinocytes by low sub-micromolar As(III) (Snow et al., 2001). The repair enzymes we have studied are part of the BER pathway, which acts on small, often spontaneous, lesions such as oxidative damage to DNA bases and alkylation damage. Because As(III) can promote oxidative damage by the production of ROS, activation of this repair pathway by low doses of As(III) also constitutes an adaptive response. This adaptive up-regulation of BER activity is maximal after 24 h exposure to approximately 1 μM As(III). Higher doses of As(III) cause a loss of BER activity and after exposure to 10 μM As(III) and above, mRNA levels and enzyme activity are decreased to 50–70% of the untreated control values. Thus, the response of DNA repair genes to As(III) peaks at a 10-fold lower As(III) concentration than the response of the redox genes.

Our findings with respect to arsenic modulation of cellular redox control and DNA repair are important because the biphasic dose response (the enzyme activities are up-regulated by arsenic at doses 10-fold below those required to down-regulate the same processes) implies that any adverse response to arsenic must exhibit a distinct threshold. They show that arsenic, like many other trace elements, has distinctly beneficial effects at very low doses.

Regulation of gene expression at the level of mRNA transcription is controlled in part by transcription factors that bind to the $5'$ promoter region of the gene. The DNA binding activities of at least three common transcription factors that are known to be redox regulated, AP-1, CREB, and NF-κB (Hu et al., 2002) are modulated by low dose As(III). Short-term (approximately 24 h) exposure to As(III) generally causes a dose-dependent up-regulation of these transcription factors, with AP-1 responding slightly more strongly than CREB (data not shown). However, after long-term (up to 10 weeks) exposure to low dose As(III), the binding activities of both AP-1 and CREB are decreased in a dose-dependent manner, with CREB being more strongly down-regulated than AP-1. Cellular response to other forms of exogenous oxidative stress, such as TPA or H_2O_2 is mediated in part by binding of either or both AP-1 and CREB to appropriate response elements within the promoters of stress response genes and is also strongly down-regulated by long-term exposure to As(III). The decreased response of AP-1 after long-term As(III) exposure is due primarily to down-regulation of the component proteins, c-Fos and c-Jun (Hu et al., 2002). We conclude from these experiments that chronic exposure to As(III) causes a decreased responsiveness of epithelial cells to oxidative stress. Because many tumour promoters such as TPA act by causing oxidative stress to the cells, a decreased responsiveness to TPA can lead to a decrease in tumour production. In support of this, when we exposed mice to As(V) in drinking water at the same time as we exposed them to a DMBA/TPA-dependent tumour regimen, the mice exposed to the arsenic exhibited significantly fewer papillomas than the mice that received As-free drinking water (as shown in Figure 8).

In summary, we have found that sub-micromolar doses of As(III) cause an adaptive response in epithelial fibroblasts that is generally cytoprotective in nature rather than cytotoxic. Even low doses of As(III), especially after long-term exposure, can cause significant changes in gene expression at the level of transcription factor binding and steady-state mRNA levels. These changes can lead to a decreased cellular responsiveness to various forms of oxidative stress and may be initially anti-carcinogenic in nature.

We would suggest that for risk assessment purposes, it is important to undertake critical evaluation of dose response curves, especially at low doses. Human tissue dose and cellular responses must also be carefully evaluated after chronic exposure to low doses of As(III) or As(V). Because of their different dose response patterns, arsenic-related gene induction and toxicity are likely to represent different intracellular or molecular targets, especially in different target cell types such as fibroblasts versus keratinocytes.

Acknowledgements

This work was supported in part by the US Environmental Protection Agency's Science Achieve Results (STAR) program (EPA R 826135); the Electric Power Research Institute, contract no. EP-P4898/C2396; the New York University School of Medicine's NIEHS Center (ES00260); and the Centre for Cellular and Molecular Biology, School of Biological and Chemical Sciences, Deakin University, Australia.

References

Arrand, J., Willis, A., 1986. Different substrate specificities of the two DNA ligases of mammalian cells. J. Biol. Chem., 261, 9079–9082.

Babich, H., Borenfreund, E.E., 1987. Structure activity relationship (SAR) model established *in vitro* with the neutral red cytotoxicity assay. Toxicol. In Vitro, 1, 3–9.

Chouchane, S., Snow, E.T., 2001. *In vitro* effect of arsenical compounds on glutathione related enzymes. Chem. Res. Toxicol., 14, 517–522.

Concha, G., Vogler, G., Nermell, B., Vahter, M., 1998. Low-level arsenic excretion in breast milk of native Andean women exposed to high levels of arsenic in the drinking water. Int. Arch. Occup. Environ. Health, 71, 42–46.

Del Razo, L.M., Quintanilla Vega, B., Brambila Colombres, E., Calderon Aranda, E.S., Manno, M., Albores, A., 2001. Stress proteins induced by arsenic. Toxicol. Appl. Pharmacol., 177, 132–148.

Frenkel, K., Wei, H., Bhimani, R., Ye, J., Zadunaisky, J., Huang, M., Ferraro, T., Conney, A., Grunberger, D., 1993. Inhibition of tumor promoter-mediated processes in mouse skin and bovine lens by caffeic acid phenethyl ester. Cancer Res., 53, 1255–1261.

Germolec, D.R., Yoshida, T., Gaido, K., Wilmer, J.L., Simeonova, P.P., Kayama, F., Burleson, F., Dong, W., Lange, R.W., Luster, M.I., 1996. Arsenic induces overexpression of growth factors in human keratinocytes. Toxicol. Appl. Pharmacol., 141, 308–318.

Germolec, D.R., Spalding, J., Yu, H.S., Chen, G.S., Simeonova, P.P., Humble, M.C., Bruccoleri, A., Boorman, G.A., Foley, J.F., Yoshida, T., Luster, M.I., 1998. Arsenic enhancement of skin neoplasia by chronic stimulation of growth factors. Am. J. Pathol., 153, 1775–1785.

Hartmann, A., Speit, G., 1996. Effect of arsenic and cadmium on the persistence of mutagen-induced DNA lesions in human cells. Environ. Mol. Mutagen., 27, 98–104.

Hartwig, A., Groblinghoff, U.D., Beyersmann, D., Natarajan, A.T., Filon, R., Mullenders, L.H.F., 1997. Interaction of arsenic(III) with nucleotide excision repair in UV-irradiated human fibroblasts. Carcinogenesis, 18, 399–405.

Hu, Y., Su, L., Snow, E.T., 1998. Arsenic toxicity is enzyme specific and its affects on ligation are not caused by the direct inhibition of DNA repair enzymes. Mutat. Res., 408, 203–218.

Hu, Y., Jin, X., Snow, E., 2002. Effect of arsenic on transcription factor AP-1 and NF-kappa B DNA binding activity and related gene expression. Toxicol. Lett., 133, 33–45.

Huang, X., Zhuang, Z., Frenkel, K., Klein, C., Costa, M., 1994. The role of nickel and nickel-mediated reactive oxygen species in the mechanism of nickel carcinogenesis. Environ. Health Perspect., 102, 281–284.

Kessel, M., Liu, S., Xu, A., Santella, R., Hei, T., 2002. Arsenic induces oxidative DNA damage in mammalian cells. Mol. Cell. Biochem., 234, 301–308.

Kitchin, K.T., 2001. Recent advances in arsenic carcinogenesis: modes of action, animal model systems, and methylated arsenic metabolites. Toxicol. Appl. Pharmacol., 172, 249–261.

Kojima, C., Sakurai, T., Ochiai, M., Kumata, H., Qu, W., Waalkes, M.P., Fujiwara, K., 2002. Cytotoxicological aspects of the organic arsenic compound arsenobetaine in marine animals. Appl. Organomet. Chem., 16, 421–426.

Lee, T.C., Wei, M.L., Chang, W.J., Ho, I.C., Lo, J.F., Jan, K.Y., Huang, H., 1989. Elevation of glutathione levels and glutathione S-transferase activity in arsenic-resistant Chinese hamster ovary cells. In Vitro Cell. Dev. Biol., 25, 442–448.

Li, J.-H., Rossman, T.G., 1989. Inhibition of DNA ligase activity by arsenite: a possible mechanism of its comutagenesis. Mol. Toxicol., 2, 1–9.

Liu, S.X., Athar, M., Lippai, I., Waldren, C., Hei, T.K., 2001. Induction of oxyradicals by arsenic: implication for mechanism of genotoxicity. Proc. Natl Acad. Sci. USA, 98, 1643–1648.

Lu, Y.-P., Lou, Y.-R., Yen, P., Newmark, H., Mirochnitchenko, O.I., Inouye, M., 1997. Enhanced skin carcinogenesis in transgenic mice with high expression of glutathione peroxidase or both glutathione peroxidase and superoxide dismutase. Cancer Res., 57, 1468–1474.

Ng, J., Seawright, A., Qi, L.X., Garnett, C., Chiswell, B., Moore, M., 1999. Tumours in mice exposed to sodium arsenate in drinking water. In: Chappell, W., Abernathy, C. and Calderon, R. (Eds), Arsenic Exposure and Health Effects: Proceedings of the Third International Conference, July 12–15, 1998. Elsevier, New York, pp. 217–223.

Rossman, T.G., Uddin, A.N., Burns, F.J., Bosland, M.C., 2001. Arsenite is a cocarcinogen with solar ultraviolet radiation for mouse skin: an animal model for arsenic carcinogenesis. Toxicol. Appl. Pharmacol., 176, 64–71.

Schuliga, M., Chouchane, S., Snow, E.T., 2002. Upregulation of glutathione-related genes and enzyme activities in cultured human cells by sublethal concentrations of inorganic arsenic. Toxicol. Sci., 70, 183–192.

Simeonava, P.P., Wang, S., Toriuma, W., Kommineni, V., Matheson, J., Unimye, N., Kayama, F., Harki, D., Ding, M., Vallyathan, V., Luster, M.I., 2000. Arsenic mediates cell proliferation and gene expression in the bladder epithelium: association with activating protein-1 transactivation. Cancer Res., 60, 3445–3453.

Smith, A.H., Goycolea, M., Hague, R., Biggs, M.L., 1998. Marked increase in bladder and lung cancer mortality in a region of Northern Chile due to arsenic in drinking water. Am. J. Epidemiol., 147, 660–669.

Snow, E., Hu, Y., Yan, C., Chouchane, S., 1999. Modulation of DNA repair and glutathione levels in human keratinocytes by micromolar arsenite. In: Chappell, W.R., Abernathy, C.O., Calderon, R.L. (Eds), Arsenic Exposure and Health Effects. Elsevier, Oxford, pp. 243–251.

Snow, E.T., Schuliga, M., Chouchane, S., Hu, Y., 2001. Sub-toxic arsenite induces a multi-component protective response against oxidative stress in human cells. In: Chappell, W.R., Abernathy, C.O., Calderon, R.L. (Eds), Arsenic Exposure and Health Effects IV. Elsevier, Oxford, pp. 265–276.

Styblo, M., Serves, S.V., Cullen, W.R., Thomas, D.J., 1997. Comparative inhibition of yeast glutathione reductase by arsenicals and arsenothiols. Chem. Res. Toxicol., 10, 27–33.

Styblo, M., Del Razo, L.M., Vega, L., Germolec, D.R., LeCluyse, E.L., Hamilton, G.A., Reed, W., Wang, C., Cullen, W.R., Thomas, D.J., 2000. Comparative toxicity of trivalent and pentavalent inorganic and methylated arsenicals in rat and human cells. Arch. Toxicol., 74, 289–299.

Tan, M., Li, S., Swaroop, M., Guan, K., Oberley, L., Sun, Y., 1999. Transcriptional activation of the human glutathione peroxidase promoter by p53. J. Biol. Chem., 274, 12061–12066.

Wang, T.S., Huang, H., 1994. Active oxygen species are involved in the induction of micronuclei by arsenite in XRS-5 cells. Mutagenesis, 9, 253–257.

Wang, J.P., Qi, L.X., Moore, M.R., Ng, J.C., 2002. A review of animal models for the study of arsenic carcinogenesis. Toxicol. Lett., 133, 17–31.

Wu, M.M., Chiou, H.Y., Wang, T.W., Hsueh, Y.M., Wang, I.H., Chen, C.J., Lee, T.C., 2001. Association of blood arsenic levels with increased reactive oxidants and decreased antioxidant capacity in a human population of northeastern Taiwan. Environ. Health Perspect., 109, 1011–1017.

Yager, J.W., Wiencke, J.K., 1997. Inhibition of poly(ADP-ribose) polymerase by arsenite. Mutat. Res., 386, 345–351.

Arsenic Exposure and Health Effects V
W.R. Chappell, C.O. Abernathy, R.L. Calderon and D.J. Thomas, editors
© 2003 Elsevier B.V. All rights reserved.

Chapter 24

Carcinogenicity of dimethylarsinic acid (DMA^V)

Samuel M. Cohen, Chris Le, Xiufen Lu, Marty Cano and Lora L. Arnold

Abstract

Dimethylarsinic acid (DMA^V) administered at high doses in the diet or drinking water produces bladder tumors in rats, with females being more susceptible than males. In contrast, tumors are not produced when even higher doses are administered to mice, and in short-term studies (10 weeks) there are no toxic or preneoplastic changes in the urinary tract of hamsters. Cytotoxicity of the urothelium is induced within 6 h of administration, with consequent regenerative hyperplasia beginning after 3 days of administration. The toxicity and regeneration follows a dose response similar to the carcinogenic effects seen in a 2-year bioassay, and the hyperplasia is reversible after 10 weeks of administration. The two major urinary products of rats treated with DMA^V are DMA^V itself and trimethylarsine oxide (TMAO). However, micromolar concentrations of dimethylarsinous acid (DMA^{III}) are also produced. DMA^{III} is a highly cytotoxic organic trivalent metabolite of DMA^V and other arsenicals, and is a possible proximate form of DMA^V that produces the cytotoxicity of the urothelium. In vitro studies show comparable cytotoxicity of rat and human urothelial cell lines at micromolar concentrations for the trivalent arsenicals, but require millimolar concentrations of organic pentavalent arsenicals, such as DMA^V and TMAO. Extrapolation of these findings in rodents to human risk must be made cautiously, because the doses used in these studies are several orders of magnitude greater than potential human exposures, and the metabolism of DMA^V appears to differ significantly among species, especially in the rat.

Keywords: dimethylarsinic acid, dimethylarsinous acid, urinary bladder, urinary metabolites, cytotoxicity, cell proliferation

1. DMA^V carcinogenesis in rats

Dimethylarsinic acid (DMA^V) was evaluated in a standard 2-year dietary carcinogenicity study in rats and mice and was found to produce bladder tumors in rats, with a greater response in female rats than in male rats (van Gemert and Eldan, 1998). Hyperplasia was identified at doses of 40 ppm and tumors at 100 ppm. Wei et al. (1999) subsequently reported induction of bladder tumors in male rats administered DMA^V in the drinking water at similar doses. In addition, Fukushima and his colleagues demonstrated an enhancing effect on rat bladder carcinogenesis when DMA^V was administered in the drinking water following brief exposure to *N*-butyl-*N*-(4-hydroxybutyl)nitrosamine (BBN) (Wanibuchi et al., 1996). In contrast, there was no carcinogenic effect of dietary treatment with DMA^V on the mouse urinary tract, including the bladder (van Gemert and Eldan, 1998). An increased incidence of tumors was also not observed at any other site in mice.

In the mouse study of DMA^V, an unusual intracytoplasmic inclusion was observed in the superficial cells of the urinary bladder urothelium, but this was not accompanied by cytotoxicity, necrosis or evidence of regeneration (unpublished observation). These inclusions have been identified in studies with a variety of chemicals and in untreated control mice (Cohen et al., 2002b). It is not yet clear what they represent, but there has been some evidence that they may represent lipid inclusions. They do not pose a deleterious effect on the functioning of the cells or the bladder as a whole, as there is no evidence of cytotoxicity or regeneration and the mice appeared to have normal bladder function. They are not precancerous lesions, either with DMA^V or following administration of other chemicals leading to this effect.

Based on these observations in rats and mice, DMA^V represents a relatively weak bladder carcinogen in rats and is not carcinogenic in mice. Inorganic arsenicals are known to be carcinogenic to humans. In humans, high levels of inorganic arsenic in the drinking water are associated with cancers of the skin, lung and urinary bladder and possibly with kidney and liver (Bates et al., 1992). However, the relationship of organic arsenicals, such as DMA^V, to human cancer remains unknown.

An important determinant of mode of action of DMA^V or any arsenical is its potential for genotoxicity. More specifically, ascertainment of the possibility that any of these arsenicals chemically react with DNA to form adducts which potentially could be mutagenic is essential in evaluating potential risk to the human. The majority of evidence to date, based on chemistry and results of numerous *in vitro* and *in vivo* studies, suggests that arsenicals do not covalently bind to DNA (Rossman, 1998; Basu et al., 2001; Environmental Protection Agency, 2001; Kenyon and Hughes, 2001; Kitchin, 2001). This conclusion was based primarily on studies with inorganic arsenic, especially arsenite, but there have been some studies with organic arsenicals such as DMA^V.

Arsenic is metabolized primarily by a series of redox reactions and methylations progressing from arsenate to arsenite to methylarsonic acid (MMA^V) to methylarsonous acid (MMA^{III}) to DMA^V to dimethylarsinous acid (DMA^{III}) to trimethylarsine oxide (TMAO). This sequence of oxidative methylation was postulated several years ago, but evidence in support of it has been provided only recently, both in humans and in animal models (Aposhian et al., 2000; Le et al., 2000). As expected, the trivalent forms of the organic arsenicals, MMA^{III} and DMA^{III}, are less stable chemicals, which are highly cytotoxic *in vitro* and *in vivo*. However, their role in the carcinogenicity of arsenite itself remains unclear.

The potential genotoxicity of these trivalent methylated arsenicals has only recently been evaluated. Similar to the results with arsenite, they tend to produce abnormalities in a variety of *in vitro* and *in vivo* assays, but they do not appear to directly bind or interact with DNA. In a recent study published by Mass et al. (2001), there was some evidence of direct damage to DNA in an *in vitro* assay involving the interaction of MMA^{III} and DMA^{III} with φX174 DNA. However, the effects were only seen at extremely high concentrations (MMA^{III}, 30 mM; DMA^{III}, 150 μM), whereas these chemicals are cytotoxic at micromolar concentrations and lower (Styblo et al., 2000; Vega et al., 2001). Thus, it is untenable for these chemicals to attain such high concentrations in an *in vivo* setting. There is no evidence from *in vivo* studies that any of these arsenicals directly interact with DNA.

The other tests for genotoxicity all involve a variety of indirect mechanisms. These potential mechanisms involve interactions of the arsenicals with non-nucleotide cellular components, most likely various proteins such as tubulin, the major component of the mitotic spindle, DNA repair enzymes or structural components of the cell membrane or cytoskeleton. The implications of these indirect genotoxic mechanisms are quite different than the implications of direct DNA reactivity, particularly with respect to low dose extrapolation.

Another possibility that has been suggested as an indirect mechanism for arsenicals to damage DNA is oxidative damage, most likely occurring through formation of free radical arsenicals (Kitchin, 2001). Evidence for such a mechanism has been identified in *in vitro* studies, but evidence has yet to be found supporting such a mechanism in *in vivo* studies. However, this represents a plausible mechanism of action that needs to be more fully evaluated and is quite compatible with a cytotoxicity-regeneration mode of action.

One of the major difficulties in evaluating arsenicals in animal models, particularly rats, has been the significant differences between the metabolism of arsenicals in rats compared to humans and even to other rodents (Aposhian, 1997). These differences are generally more quantitative than qualitative. Significant differences exist with respect to the toxicokinetics of arsenicals because of retention of dimethyl arsenicals in rat red blood cells. This does not occur in other rodents and apparently does not occur in humans.

More recently, another potential confounding factor in interpreting and extrapolating results from animals to humans has been identified by Oremland and his colleagues with whom we are collaborating (Herbel et al., 2002). This is based on the anaerobic dissimilatory reduction of arsenicals by various gastrointestinal bacteria. This could contribute significantly to variations among species as well as within species in the amount of arsenate that is reduced to arsenite, and therefore, the amount that is more readily bioavailable for absorption across the gastrointestinal membrane and into the systemic circulation. We have preliminary evidence that administration of arsenicals, such as arsenate, can actually modify the extent of this gastrointestinal metabolism.

2. Cytotoxicity and regeneration

Because it appeared most unlikely that DMA^V was producing bladder cancer in rats by a mechanism involving direct DNA reactivity, we hypothesized that a major contributing factor could be cytotoxicity of the urothelium produced by DMA^V administration, leading to significant regenerative hyperplasia and ultimately to low incidences of bladder tumors. Such a mode of action involving cytotoxicity and regeneration has been identified in numerous tissues, including the urinary bladder (Cohen, 1998), but it has been most extensively evaluated with respect to regulatory implications for the chemical chloroform, which produces kidney and liver tumors (Andersen et al., 2000). Based on this hypothesis, we have developed and evaluated a short-term model of bladder toxicity and carcinogenesis in rats that we have utilized for evaluation of a variety of chemicals (Cohen et al., 1990; Cohen, 1998). We have found that most chemicals that produce bladder cytotoxicity and regeneration do so within the 10-week time frame of these short-term studies. We used this 10-week protocol in our initial study of DMA^V. We also tried to evaluate DMA^V utilizing a protocol as close as possible to that with which the 2-year bioassay was conducted. Thus, we used the same strain of animals as well as the same doses used in the 2-year bioassay.

The experiment (Arnold et al., 1999) was performed administering DMA^V in Purina 5002 diet at doses of 0, 2, 10, 40 or 100 ppm to female rats, because they were the most susceptible of the sexes in the 2-year bioassay. Groups of male rats were also fed either 0 or 100 ppm DMA^V for comparison with the females, and groups of female rats were fed 0 or 100 ppm DMA^V in Altromin 1321 diet to evaluate the possible effects of elevated urinary pH on DMA^V-induced urothelial changes and because this was the diet used in the 2-year bioassay. Lastly, the reversibility of the changes produced in the female rats at the highest dose, 100 ppm, was evaluated in a group fed DMA^V in Purina 5002 for 10 weeks, then fed Purina diet without added DMA^V for an additional 10 weeks. A group of rats were fed DMA^V continuously for 20 weeks for comparison. Ten rats were evaluated in each group at each time point. The urine was analyzed for chemical changes and for the possible production of urinary solids. The bladders were evaluated by light and scanning electron microscopy (SEM) and by bromodeoxyuridine (BrdU) labeling index, utilizing procedures established in our laboratory. In addition, the kidneys were evaluated because of an indication of calcification in the kidneys in the 2-year bioassay. To evaluate calcification in the kidneys and in the bladder, sections were evaluated by hematoxylin and eosin stain and by von Kossa stain for calcium. The various treatments are summarized in Table 1.

DMA^V administration did not have an appreciable effect on body weight gain, but there were variable, slight effects on food consumption (data not shown). Water consumption was increased with the higher doses of DMA^V (data not shown).

The significant findings in the urine were primarily the absence of the formation of urinary solids as detected by SEM. There was no evidence of precipitate, abnormal microcrystalluria or calculus formation in any of the treatment groups. Urinalysis showed a slight increase in urinary volume with increasing dose associated with the increases in

Table 1.   Study design for Experiment 1.

Group	Sex	Treatment	Basal diet
1a	F	0 ppm DMA^V – 9 weeks	Purina 5002
1b	F	0 ppm DMA^V – 10 weeks	Purina 5002
1c	F	0 ppm DMA^V – 20 weeks	Purina 5002
2	F	2 ppm DMA^V – 10 weeks	Purina 5002
3	F	10 ppm DMA^V – 10 weeks	Purina 5002
4	F	40 ppm DMA^V – 10 weeks	Purina 5002
5a	F	100 ppm DMA^V – 9 weeks	Purina 5002
5b	F	100 ppm DMA^V – 10 weeks	Purina 5002
5c	F	100 ppm DMA^V – 10 weeks 0 ppm DMA^V – 10 weeks	Purina 5002
5d	F	100 ppm DMA^V – 20 weeks	Purina 5002
6	F	0 ppm DMA^V – 10 weeks	Altromin 1321
7	F	100 ppm DMA^V – 10 weeks	Altromin 1321
8	M	0 ppm DMA^V – 10 weeks	Purina 5002
9	M	100 ppm DMA^V – 10 weeks	Purina 5002

Table 2. Effect of dietary treatment with DMA^V on volume, creatinine and calcium of 24-h urines collected during week 5 of the study.

Group	Treatment	Sex	Volume (ml)	Creatinine (mg/dl)	Calcium (mg/dl)	Calcium/creatinine (mg/mg)
1	0 ppm DMA^V – Purina	F	6 ± 0	68 ± 3	15.6 ± 1.3	0.25 ± 0.03
2	2 ppm DMA^V – Purina	F	7 ± 1	66 ± 6	15.3 ± 0.9	0.26 ± 0.02
3	10 ppm DMA^V – Purina	F	7 ± 0	65 ± 4	15.5 ± 0.9	0.25 ± 0.02
4	40 ppm DMA^V – Purina	F	9 ± 1[a]	52 ± 2[a]	17.7 ± 1.4	0.35 ± 0.03[a]
5	100 ppm DMA^V – Purina	F	10 ± 1[a]	48 ± 2[a]	21.5 ± 2.0[a]	0.47 ± 0.05[a]
6	0 ppm DMA^V – Altromin	F	7 ± 1	49 ± 2[a]	26.3 ± 1.3[a]	0.55 ± 0.03[a]
7	100 ppm DMA^V – Altromin	F	8 ± 0	44 ± 2	24.9 ± 1.4	0.58 ± 0.04
8	0 ppm DMA^V – Purina	M	8 ± 1	89 ± 4[a]	15.7 ± 0.5	0.18 ± 0.01
9	100 ppm DMA^V – Purina	M	12 ± 1[b]	59 ± 4[b]	14.7 ± 2.0	0.29 ± 0.05[b]

Values expressed as the mean ±SE.
[a] Significantly different from female Purina 0 ppm group, $p < 0.05$.
[b] Significantly different from male Purina 0 ppm group, $p < 0.05$.

water consumption (Table 2). Correspondingly, the urinary creatinine was reduced in the groups where the urinary volume was increased, indicating a dilutional effect from the increased volume. Interestingly, urinary calcium was elevated at the higher doses in females administered DMA^V in the Purina diet, but not in the Altromin diet nor in the males. However, when expressed as unit of calcium per unit of creatinine, the urinary calcium was elevated at 40 and 100 ppm of the diet in female rats fed Purina diet and in the male rats. Occasional other changes were sporadically present, but there were no significant effects that could be clearly related to the treatment.

The effects on the urinary bladder after 10 weeks of administration are listed in Table 3. As can be seen from this table, there is a significant proliferative effect at 40 and 100 ppm in the female rats, as evidenced by light microscopy, labeling index and SEM. Evidence of cytotoxicity and cell necrosis were present at 40 and 100 ppm of the diet with a suggestion of an effect at 10 ppm. There were similar changes in the male rats, although the extent of the changes appeared to be somewhat less than in the females, similar to the results in the 2-year bioassay. Also, there did not appear to be a significant difference in the effect between DMA^V administered in Altromin diet compared to Purina diet, although the elevated urinary pH seen with the Altromin control diet complicated the interpretation to some extent.

In the reversibility study, the effects on the urothelium appeared to be nearly completely reversible within the 10 weeks on the control diet. The light microscopic histology and labeling index were the same in the control group as in the group fed DMA^V for 10 weeks and then control diet for 10 weeks. Minor changes were present in the reversibility group, as seen by SEM, although the effect was considerably less than that seen in the rats administered DMA^V for either 10 weeks or 20 weeks by itself. Interestingly, there was less evidence of cytotoxicity and hyperplasia in the rats administered DMA^V for 20 weeks compared to those sacrificed after 10 weeks of DMA^V administration. We have recently extended these studies in female rats to 26 weeks of

Table 3. Effects of dietary administration of DMA^V for 10 weeks on the urinary bladder.

Group	Treatment	Sex	Bladder weight (g) Mean ± SE	Bladder histology Normal	Simple hyperplasia	Labeling index (%) Mean ± SE (n)	SEM classification 1	2	3	4	5
1	0 ppm DMA^V – Purina	F	0.071 ± 0.005	9	1	0.22 ± 0.05 (8)	5	5	–	–	–
2	2 ppm DMA^V – Purina	F	0.070 ± 0.005	10	0	0.20 ± 0.03 (9)	–	4	5	1	–
3	10 ppm DMA^V – Purina	F	0.075 ± 0.005	10	0	0.33 ± 0.08 (10)	–	2	5	3	–
4	40 ppm DMA^V – Purina	F	0.097 ± 0.009[a]	6	4	0.95 ± 0.15 (8)[a]	–	5	3	2	–
5	100 ppm DMA^V – Purina	F	0.086 ± 0.005	1	9[a]	0.93 ± 0.11 (7)[a]	–	–	–	4	6[b]
6	0 ppm DMA^V – Altromin	F	0.085 ± 0.009	6	4	0.19 ± 0.03 (10)	–	4	4	2	–
7	100 ppm DMA^V – Altromin	F	0.084 ± 0.007	4	6	0.87 ± 0.09 (10)[c]	–	–	1	3	6[d]
8	0 ppm DMA^V – Purina	M	0.099 ± 0.007[a]	10	0	0.23 ± 0.03 (10)	–	3	7	–	–
9	100 ppm DMA^V – Purina	M	0.122 ± 0.008[e]	8	2	0.95 ± 0.05 (10)[e]	–	1	4	5	–

[a] Significantly different from female Purina 0 ppm group, $p < 0.05$.
[b] Number of class 5 bladders significantly different from female Purina 0 ppm group, $p < 0.05$.
[c] Significantly different from female Altromin 0 ppm group, $p < 0.05$.
[d] Number of class 5 bladders significantly different from female Altromin 0 ppm group, $p < 0.05$.
[e] Significantly different from male Purina 0 ppm group, $p < 0.05$.

treatment with 100 ppm DMA^V. We observed a continued decrease in the proliferative effect that we had seen previously at 20 weeks compared to 10 weeks. Thus, the proliferative effect, although it continues over time, appears to be attenuated to some extent. Some confusion in the interpretation of the SEM classification has arisen. In the classification system that we have used, classes 1, 2 and 3 are considered within normal limits, whereas classes 4 and 5 are abnormal. In the group administered DMA^V for 20 weeks, seven of the 10 rats had abnormal bladders by SEM, whereas the reversibility group showed abnormalities in only four out of 10. This is in contrast to the rats administered DMA^V for 10 weeks where all 10 animals showed abnormal bladders by SEM.

Examination of the kidneys showed an increase in the deposition of calcification at the corticomedullary junction, most readily observable using von Kossa stain (Table 4). These changes were present at 40 and 100 ppm of the diet in females, but it was not seen in the female rats administered DMA^V in Altromin 1321 diet or in the male rats. The urinary bladder showed no evidence of calcification with any of the treatments. Based on these observations, we suggested that urinary calcium and tissue calcification might in some way be contributing to the mode of action of DMA^V-induced urothelial changes. However, subsequent studies have indicated that such a hypothesis is unlikely.

This first study clearly demonstrated that DMA^V administered in the diet showed a dose-related effect in producing cytotoxicity and regeneration, and that the effect was greater in the female rat than in the male rat, corresponding to the results of the 2-year bioassay. In addition, it appeared that urinary pH, although variably altered following DMA^V treatment, did not significantly affect the changes in the bladder, as demonstrated in the portion of the study involving Altromin 1321 diet. It was also apparent that the changes produced by 10 weeks of administration of DMA^V to female rats at a dose of 100 ppm were nearly completely reversible within 10 weeks of change of administration to control diet without DMA^V. Slight changes were still detectable by SEM, but the histology and labeling index values had returned to the same as controls.

Table 4. Effects of dietary administration of DMA^V for 10 weeks on the kidneys.

Group	Treatment	Sex	Corticomedullary junction calcification				
			Negative	+1	+2	+3	+4
1	0 ppm DMA^V – Purina	F	4	4	2		
2	2 ppm DMA^V – Purina	F	7	3			
3	10 ppm DMA^V – Purina	F	5	4	1		
4	40 ppm DMA^V – Purina	F			9	1	
5	100 ppm DMA^V – Purina	F		1	4	4	1
6	0 ppm DMA^V – Altromin	F	10				
7	100 ppm DMA^V – Altromin	F	9				
8	0 ppm DMA^V – Purina	M	8	2			
9	100 ppm DMA^V – Purina	M	8	1			

No statistical analysis of results.

The major observation from this study, however, was that the toxicity and regeneration were not related to the formation of any urinary solids, whether precipitate, microcrystalluria or calculus formation. Thus, we were left with the possibility that the cytotoxicity and regeneration were due either to an effect from DMA^V and/or metabolite directly on the bladder epithelium or an alteration in the urine itself, such as urinary calcium, increased volume and decreased osmolality that might produce the changes. As indicated below, the evidence now strongly supports a mode of action involving cytotoxicity secondary to a metabolite of DMA^V in the urine. It should be noted that other studies have demonstrated that orally administered DMA^V is excreted predominately in the urine rather than in the feces (Aposhian, 1997).

Based on the observations in this first experiment, we conducted two additional experiments evaluating the time course of the urinary and urothelial changes following administration of DMA^V to female F344 rats at a dose of 100 ppm of the diet for up to 10 weeks (Cohen et al., 2001). Bladders were examined after 6 and 24 h, 3, 7 and 14 days and 10 weeks of treatment. Again, urinary examinations were performed in addition to evaluation of the urothelium by light microscopy, SEM and BrdU labeling index.

In these experiments evaluating the time course, we were able to show that the urinary volume again was somewhat increased but the urinary pH was not significantly altered, although there was a suggestion of a slight decrease in urinary pH. Importantly, we were able to show that the urinary calcium was comparable to control levels after 7 days of treatment with DMA^V, was elevated at four weeks, but had actually decreased compared to control levels by 8 weeks.

By SEM, we were able to detect evidence of cytotoxicity as early as 6 h after administration of DMA^V began. This increased in severity by 24 h, and even further through 3 days and 7 days of administration. It was also significantly present at 2 weeks, and by 10 weeks of administration all 10 bladders were classified as class 4 and 5. In contrast, an increase in labeling index did not become evident until 7 days of administration, whereas at 3 days, it was still at control levels. This indicates that at some time between 3 and 7 days of administration of DMA^V, there was an increased proliferative response. Thus, this experiment demonstrates that cytotoxicity and necrosis appear first, and are followed by a regenerative hyperplastic response. It does not appear that the necrosis is secondary to the increased proliferation. Importantly, the changes by labeling index were more sensitive in detecting the proliferative response than light microscopy. Similarly, SEM is by far more sensitive for detecting cellular necrosis, because it is predominantly present only in superficial cells. We did not see extensive necrosis involving the full thickness of the bladder urothelium, nor was there evidence of full thickness ulceration with an inflammatory response.

Based on this second set of experiments, we can conclude that the sequence of events in the bladder is urothelial cytotoxicity that is present rapidly (by 6 h of administration), and this is followed some days later by a regenerative hyperplastic response. It is also clear that this cytotoxicity is neither due to the formation of abnormal urothelial solids, nor it is due to abnormalities in urinary calcium or calcification of urinary tract tissues. Thus, the most likely mode of action is cytotoxicity, secondary to DMA^V and/or a metabolite present in the urine.

Based on these studies, we were also able to demonstrate that the urothelial proliferation appeared to be maximum at approximately 2 weeks of administration, and that the amount of proliferation had actually declined by 10 weeks. This gave us an opportunity to evaluate the other aspects of DMA^V in a shorter time course, rather than having to rely on experiments conducted for 10 weeks because changes were present nearly immediately following administration of DMA^V in the diet.

3. Inhibition by DMPS and possible role of DMA^{III}

Because these studies strongly suggested that DMA^V produced urothelial cytotoxicity by a direct chemical interaction of either DMA^V or, more likely, one or more of its metabolites, we began measuring urinary metabolites of DMA^V. At this point, we hypothesized that it was most likely that trivalent arsenicals were forming and concentrating in the urine and that these were probably responsible for the urothelial cytotoxicity produced by DMA^V administration. 2,3-Dimercaptopropane-1-sulfonic acid (DMPS) is known to be an effective treatment for arsenic poisoning predominantly through chelation of trivalent arsenicals. We postulated that if trivalent arsenicals were involved with the cytotoxicity following DMA^V administration to rats, DMPS should inhibit this cytotoxicity and thus the regeneration.

To evaluate this hypothesis, we administered DMA^V in the diet to female F344 rats at a dose of 100 ppm and DMPS at a dose of 5600 ppm (dose established in a separate experiment) (Cohen et al., 2002a). These were administered in the diet for up to 2 weeks and the bladders evaluated as above. Twenty-four hour urines were analyzed for arsenical species and various endogenous components.

We again were able to demonstrate that 2 weeks of DMA^V administration at 100 ppm to female rats produced significant cytotoxicity and regenerative hyperplasia (Table 5). In this experiment, DMPS almost completely inhibited both the cytotoxicity and the increased proliferation.

Initial examination of the urinary specimens for different arsenical species suggested that there was an increase in arsenite in the DMA^V-treated rats as well as high levels of DMA^V. However, further development of the HPLC procedure demonstrated that arsenite was not increased and that the major arsenicals found in the urine were the trimethyl arsenical, TMAO, and unchanged DMA^V (Table 6). MMA^{III} was also detected in the 24-h urine of rats co-administered DMA^V and DMPS, but at extremely low concentrations (0.21 ± 0.03 μM). Although metabolism of DMA^V by demethylation is theoretically possible, it is unlikely due to the strong bond between carbon and arsenic. Thus, it is much more likely that it either remains as the dimethyl parent compound (DMA^V) or it is further methylated, which is what we observed. Interestingly, co-administration of DMA^V with DMPS led to a marked increase in the urinary concentration of DMA^V and a marked decrease in TMAO urinary concentration.

The presence of TMAO in high concentrations in the urine of rats fed high levels of DMA^V strongly suggested that DMA^{III} would be present as a less stable intermediate, because it is required in the metabolism of DMA^V to TMAO. However, collection of urines over a 24-h period was inadequate for preserving DMA^{III} and we were unable to detect any DMA^{III}; therefore, we attempted to identify the presence of DMA^{III} utilizing fresh voided urines, immediately frozen and analyzed. Following such a procedure, we were able to demonstrate that DMA^{III} was present in the urine of rats

Table 5. Effects of treatment for 2 weeks with DMA^V, DMPS or DMA^V + DMPS on the female rat bladder.

Group	Treatment	Bladder weight (g) Mean ± SE	Bladder histology Normal	Simple hyperplasia	Labeling index (%) Mean ± SE (n)	SEM classification 1	2	3	4	5
1	Control	0.078 ± 0.003	10	0	0.16 ± 0.02 (9)	7	3	0	0	0
2	100 ppm DMA^{Va}	0.090 ± 0.005	9	1	0.63 ± 0.10 (10)[b]	0	2	2	0	6
3	5600 ppm DMPS^a	0.066 ± 0.003	8	2	0.12 ± 0.02 (9)	0	2	2	3	3
4	100 ppm DMA^V + 5600 ppm DMPS^{a,c,d}	0.073 ± 0.005[e]	10	0	0.09 ± 0.01 (8)[e]	0	8	0	1	1

[a] SEM classification significantly different from control, $p < 0.05$.
[b] Significantly different from control, $p < 0.05$.
[c] SEM classification significantly different from 100 ppm DMA^V group, $p < 0.05$.
[d] SEM classification significantly different from 5600 ppm DMPS group, $p < 0.05$.
[e] Significantly different from 100 ppm DMA^V group, $p < 0.05$.

Table 6.   Quantitation of urinary species present in 24-h urines collected during week 2 from female rats treated with DMA^V, DMPS or DMA^V + DMPS.

Treatment	Arsenite (μM)	Arsenate (μM)	MMA^V (μM)	DMA^V (μM)	TMAO (μM)
Control	0.04 ± 0.00	0.02 ± 0.00	0.01 ± 0.00	0.2 ± 0.0	0.3 ± 0.1
100 ppm DMA^V	ND	ND	ND	66.4 ± 2.7[a]	73.2 ± 9.5[a]
5600 ppm DMPS	0.03 ± 0.00	0.02 ± 0.00	0.04 ± 0.00	0.2 ± 0.0	0.1 ± 0.0
100 ppm DMA^V + 5600 ppm DMPS	0.11 ± 0.01[a,b]	DL	5.35 ± 0.53[a,b]	506.8 ± 31.1[a,b,c]	2.8 ± 0.4[c]

Values expressed as the mean ±SD. DL: below detection limit. Detection limits: As^{III}, 0.007 μM; As^V, 0.014 μM; MMA^{III}, 0.026 μM; MMA^V, 0.007 μM; DMA^V, 0.014 μM; TMAO, 0.1 μM. ND: not determined due to large dilution of sample required to quantitate DMA^V.
[a] Significantly different from control group, $p < 0.05$.
[b] Significantly different from 5600 ppm DMPS group, $p < 0.05$.
[c] Significantly different from 100 ppm DMA^V group, $p < 0.05$.

administered 100 ppm of DMA^V in the diet, with or without co-administration of DMPS, at approximately micromolar concentrations (Table 7). Although the levels of DMA^{III} in the rats administered DMA^V plus DMPS were somewhat lower than in the rats administered DMA^V, they were at comparable levels.

In addition to the above arsenicals, there were two additional arsenic-containing peaks identified in the 24-h urines during the HPLC separation, but we have not yet been able to identify them. Two unknown metabolites were reported previously in the urine of rats administered DMA^V in the drinking water from the laboratory of Fukushima (Li et al., 1998). They also have not yet been able to identify what these two compounds are.

These analyses suggest that DMA^V is extensively metabolized, predominantly to TMAO through a DMA^{III} intermediate. Although TMAO itself might be producing cytotoxicity, it is more likely that DMA^{III}, especially at these concentrations, is acting as a cytotoxic metabolite. However, such a conclusion is not clear from these data, particularly because the urinary levels of DMA^{III} were similar in the DMA^V group and in the DMA^V plus DMPS group, despite significant differences in biological response. Also, the potential role that the two unidentified metabolites play in the cytotoxic response to

Table 7.   Concentration of DMA^{III} (μM) in fresh void urines collected from female rats.

Treatment	Day 1	Day 71	Day 175
Control	DL	DL	DL
100 ppm DMA^V	1.38 ± 0.44[a]	5.05 ± 1.19[a]	0.80 ± 0.31[a]
100 ppm DMA^V + 5600 ppm DMPS	0.92 ± 0.25[a]	3.05 ± 0.21[a,b]	0.67 ± 0.07[a]

Values expressed as the mean ±SD. DL: below detection limit of 0.026 μM.
[a] Significantly different from control group, $p < 0.05$.
[b] Significantly different from 100 ppm DMA^V group, $p < 0.05$.

DMA^V administration remains unknown and cannot be evaluated until these metabolites have been identified.

4. *In vitro* cytotoxicity

To gain some indication of the cytotoxic potency of various forms of arsenic, we established an *in vitro* cytotoxicity assay utilizing rat (MYP3) and human (1T1) urothelial cells (Cohen et al., 2002a). With both cell lines, trivalent arsenicals were cytotoxic at concentrations of approximately 0.4–4.8 μM. Sodium arsenate was cytotoxic at approximately 5 μM (rat) and 30 μM (human), but the pentavalent methylated arsenicals, MMA^V, DMA^V and TMAO, were cytotoxic only at millimolar concentrations.

Although the results from *in vitro* studies cannot be directly extrapolated to the *in vivo* setting, these *in vitro* studies give an indication that the trivalent methylated arsenicals are cytotoxic at essentially micromolar concentrations, whereas the pentavalent methylated arsenicals are only cytotoxic at millimolar concentrations. More importantly, the concentration of DMA^{III} in the urine of rats fed 100 ppm in the diet attains a level of approximately 1–5 μM, which potentially could be cytotoxic in the *in vivo* setting. In contrast, the urinary concentrations of MMA^V, DMA^V and TMAO are not likely to be sufficient to produce cytotoxicity in the *in vivo* setting, although there might be some interaction between the different methylated derivatives. This is yet to be evaluated. Again, we have not been able to evaluate the cytotoxicity of two unknown metabolites, but we will do so as soon as their identity is known.

Based on these experiments, our working hypothesis is that DMA^V administered at relatively high doses in the diet produces a sufficient concentration of a cytotoxic metabolite or metabolites, which rapidly produces cytotoxicity and consequent regenerative hyperplasia, which, if present for a prolonged period of time, is sufficient to lead to the development of tumors. A cytotoxicity-regenerative hyperplasia mode of action is well established in chemical carcinogenesis, and is a plausible mode of action for DMA^V in the rat. This would suggest the potential for a non-linear dose response, possibly even a threshold effect.

5. DMA^V in the hamster

As indicated before, the mouse has proven to be unresponsive to the urothelial effects of DMA^V administration, whereas the rat shows cytotoxicity with regeneration and tumor formation in the bladder urothelium. However, the hamster is considered a more relevant species to human metabolism of arsenicals than either the mouse or the rat (Aposhian, 1997). We thus wanted to evaluate the potential of DMA^V to produce an effect on the urothelium of the hamster. This was evaluated in Syrian Golden hamsters in an experiment involving administration of DMA^V at a dose of 100 ppm for 10 weeks (Cano et al., 2001). An initial experiment showed that diet can greatly influence the type of response that occurs in the hamster urothelium, even without DMA^V added. We were able to identify a diet, NIH-07, that by itself did not produce an effect on the hamster urothelium. Using that diet, 10 weeks of DMA^V administration showed no evidence of a cytotoxic or proliferative effect on the hamster bladder. This was not a definitive 2-year bioassay, but the lack of response in the hamster to treatment with DMA^V for 10 weeks is in contrast to the very rapid changes that occurred in the DMA^V-treated rat. This strongly suggests that the hamster urothelium does not respond to treatment with DMA^V.

6. Summary and conclusions

DMA^V has been shown to be carcinogenic to the rat urinary bladder when administered at relatively high doses in either the diet or drinking water, with the female somewhat more responsive than the male (van Gemert and Eldan, 1998; Wei et al., 1999). We demonstrated that this occurs through a process of urothelial cytotoxicity followed by regenerative hyperplasia (Cohen et al., 2001). The effect appears to be specific for the rat, because there are no tumorigenic effects in the mouse (van Gemert and Eldan, 1998), and there is no effect on the hamster urinary tract when administered for up to 10 weeks (Cano et al., 2001).

The cytotoxic effect appears rapidly, by no later than 6 h after administration of DMA^V begins (Cohen et al., 2001). There is also a clear dose response effect with changes present at 40 and 100 ppm of the diet and marginal changes at 10 ppm of the diet (Arnold et al., 1999). The cytotoxic and regenerative changes are greater in the female than in the male rat, similar to the tumorigenic effects in the 2-year bioassay. Also, the urothelial changes produced by DMA^V appear to be reversible, at least to a significant extent within a 10-week period of control diet administered after 10 weeks of DMA^V.

Based on examination of the urine of animals fed DMA^V, it is clear that there are variable effects. There is an increase in urinary volume with a corresponding decrease in osmolality and creatinine concentration in rats administered DMA^V. There are variable effects on urinary pH and calcium, with increases and decreases present, but these do not appear to be related to the effects of DMA^V administration on the urothelium. Importantly, we could find no evidence of the formation of abnormal urinary solids in the rats administered DMA^V for periods of up to 20 weeks.

Based on *in vivo* analytical analyses of urinary arsenicals and *in vitro* cytotoxicity assays, we have found evidence that there is formation of DMA^{III} in the urines of rats administered high doses of DMA^V. The major arsenicals present in the urine of DMA^V-treated rats appear to be unchanged DMA^V and TMAO. Although DMA^{III} is a minor metabolite, it is present at micromolar concentrations, which, at least in *in vitro* studies, are adequate for production of cytotoxicity. Similarly, co-administration of DMPS with DMA^V inhibits the cytotoxicity and regenerative hyperplasia, supporting a hypothesis of involvement of trivalent arsenicals in the mechanism of action of DMA^V on the rat urothelium. Considerably more research will be necessary to further identify the responsible agent or agents that produce the cytotoxicity following administration of high doses of DMA^V, and to also evaluate the dose response.

Extrapolating the results in the rat to humans is difficult. The doses used in these animal experiments are several orders of magnitude greater than the exposure levels of humans under natural conditions or following utilization of DMA^V as a commercial product. Similarly, the concentrations of DMA^V in humans have not been anywhere near the levels attained in these studies. In addition, the effect appears to be species specific, at least with respect to mice and hamsters. Where the human fits into this is unknown. Complicating this extrapolation further is the fact that metabolism of arsenicals is significantly different in rodents than in humans, particularly in the rat. In rats, the major metabolite of DMA^V is TMAO. TMAO is a minor metabolite in mice and hamsters and it is unclear whether it forms at all in humans, although it has been reported in one experiment in which high levels of arsenicals were administered to humans.

It is also most likely, based on the weight of the evidence, that arsenicals are not DNA-reactive substances, although genotoxicity might be produced by indirect mechanisms. Nevertheless, whether indirect genotoxicity and/or involvement of cytotoxicity and regeneration, the dose response is expected to be non-linear. Based on these considerations, a reasonable risk assessment for DMA^V at this time would be to apply a margin of exposure, non-linear approach, based on the most recent version of the revised cancer risk assessment guidelines of the United States Environmental Protection Agency.

Acknowledgements

We appreciate the efforts of Earline Wayne-Titsworth in the preparation of this manuscript. We gratefully acknowledge the assistance of Dr William Cullen, University of British Columbia, for preparation of specific arsenicals, and for his advice concerning our research.

References

Andersen, M.E., Meed, E., Boorman, G.A., Brusick, D.J., Cohen, S.M., Dragan, Y.P., Frederick, C.B., Goodman, J.I., Hard, G.C., O'Flaherty, E.J., Robinson, D.E., 2000. Lessons learned in applying the U.S. EPA's proposed cancer guidelines to specific compounds. Toxicol. Sci., 53, 159–172.

Aposhian, H.V., 1997. Enzymatic methylation of arsenic species and other new approaches to arsenic toxicity. Annu. Rev. Pharmacol. Toxicol., 37, 397–419.

Aposhian, H.V., Zheng, B., Aposhian, M.M., Le, X.C., Cebrian, M.E., Cullen, W., Zakharyan, R.A., Dart, R.C., Cheng, Z., Andres, P., Yip, L., O'Malley, G.F., Maiorino, R.M., Van Voorhies, W., Healy, S.M., Titcomb, A., 2000. DMPS–arsenic challenge test. Toxicol. Appl. Pharmacol., 165, 74–83.

Arnold, L.L., Cano, M., St. John, M., Eldan, M., van Gemert, M., Cohen, S.M., 1999. Effects of dietary dimethylarsinic acid on the urine and urothelium of rats. Carcinogenesis, 20, 2171–2179.

Basu, A., Mahata, J., Gupta, S., Giri, A.K., 2001. Genetic toxicology of a paradoxical human carcinogen, arsenic: a review. Mutat. Res., 488, 171–194.

Bates, M.N., Smith, A.H., Hopenhayn-Rich, C., 1992. Arsenic ingestion and internal cancers: a review. Am. J. Epidemiol., 135, 462–476.

Cano, M., Arnold, L.L., Cohen, S.M., 2001. Evaluation of diet and dimethylarsinic acid on the urothelium of Syrian golden hamsters. Toxicol. Pathol., 29, 600–606.

Cohen, S.M., 1998. Urinary bladder carcinogenesis. Toxicol. Pathol., 26, 121–127.

Cohen, S.M., Fisher, M.J., Sakata, T., Cano, M., Schoenig, G.P., Chappel, C.I., Garland, E.M., 1990. Comparative analysis of the proliferative response of the rat urinary bladder to sodium saccharin by light and scanning electron microscopy and autoradiography. Scanning Microsc., 4, 135–142.

Cohen, S.M., Yamamoto, S., Cano, M., Arnold, L.L., 2001. Urothelial cytotoxicity and regeneration induced by dimethylarsinic acid in rats. Toxicol. Sci., 59, 68–74.

Cohen, S.M., Arnold, L.L., Uzvolgyi, E., Cano, M., St. John, M., Yamamoto, S., Lu, X., Le, X.C., 2002a. Possible role of dimethylarsinous acid in dimethylarsinic acid-induced urothelial toxicity and regeneration in the rat. Chem. Res. Toxicol., 15, 1150–1157.

Cohen, S.M., Wanibuchi, H., Fukushima, S., 2002b. Lower urinary tract. In: Haschek, W.M., Rousseaux, C.G., Wallig, M.A. (Eds), Handbook of Toxicologic Pathology, Vol. II. Academic Press, San Diego, CA, pp. 337–362.

Environmental Protection Agency, 2001. National primary drinking water regulations: arsenic and clarifications to compliance and new source contaminants monitoring; Final Rule. Fed. Reg., 66, 6976–7066a.

Herbel, M.J., Switzer Blum, J., Hoeft, S.E., Cohen, S.M., Arnold, L.L., Lisak, J., Stolz, J.F., Oremland, R.S., 2002. Dissimilatory arsenate reductase activity and arsenate-respiring bacteria in bovine rumen fluid, hamster feces, and the termite hindgut. FEMS Microbiol. Ecol., 41, 59–67.

Kenyon, E.M., Hughes, M.F., 2001. A concise review of the toxicity and carcinogenicity of dimethylarsinic acid. Toxicology, 160, 227–236.

Kitchin, K.T., 2001. Recent advances in arsenic carcinogenesis: modes of action, animal model systems, and methylated arsenic metabolites. Toxicol. Appl. Pharmacol., 172, 249–261.

Le, X.C., Lu, X., Ma, M., Cullen, W.R., Aposhian, H.V., Zheng, B., 2000. Speciation of key arsenic metabolic intermediates in human urine. Anal. Chem., 72, 5172–5177.

Li, W., Wanibuchi, H., Salim, E.I., Yamamoto, S., Yoshida, K., Endo, G., Fukushima, S., 1998. Promotion of NCI-Black-Reiter male rat bladder carcinogenesis by dimethylarsinic acid an organic arsenic compound. Cancer Lett., 134, 29–36.

Mass, M.J., Tennant, A., Roop, R.C., Cullen, W.R., Styblo, M., Thomas, D.J., Kligerman, A.D., 2001. Methylated trivalent arsenic species are genotoxic. Chem. Res. Toxicol., 14, 355–361.

Rossman, T.G., 1998. Molecular and genetic toxicology of arsenic. In: Rose, J. (Ed.), Environmental Toxicology: Current Developments. Gordon and Breach, Amsterdam, pp. 171–187.

Styblo, M., Del Razo, L.M., Vega, L., Germolec, D.R., LeCluyse, E.L., Hamilton, G.A., Reed, W., Wang, C., Cullen, W.R., Thomas, D.J., 2000. Comparative toxicity of trivalent and pentavalent inorganic and methylated arsenicals in rat and human cells. Arch. Toxicol., 74, 289–299.

van Gemert, M., Eldan, M., 1998. Chronic carcinogenicity assessment of cacodylic acid. 3rd Abstract, International Conference on Arsenic Exposure and Health Effects, San Diego, CA.

Vega, L., Styblo, M., Patterson, R., Cullen, W., Wang, C., Germolec, D., 2001. Differential effects of trivalent and pentavalent arsenicals on cell proliferation and cytokine secretion in normal human epidermal keratinocytes. Toxicol. Appl. Pharmacol., 172, 225–232.

Wanibuchi, H., Yamamoto, S., Chen, H., Yoshida, K., Endo, G., Hori, T., Fukushima, S., 1996. Promoting effects of dimethylarsinic acid on N-butyl-N-(4-hydroxybutyl)nitrosamine-induced urinary bladder carcinogenesis in rats. Carcinogenesis, 17, 2435–2439.

Wei, M., Wanibuchi, H., Yamamoto, S., Li, W., Fukushima, S., 1999. Urinary bladder carcinogenicity of dimethylarsinic acid in male F344 rats. Carcinogenesis, 20, 1873–1876.

Arsenic Exposure and Health Effects V
W.R. Chappell, C.O. Abernathy, R.L. Calderon and D.J. Thomas, editors
© 2003 Elsevier B.V. All rights reserved.

Chapter 25

Urinary speciation of sodium arsenate in folate receptor knockout mice

Ofer Spiegelstein, Xiufen Lu, X. Chris Le and Richard H. Finnell

Abstract

Previously, we have shown that folate binding protein-2 nullizygous (Folbp2$^{-/-}$) mice were more sensitive to in utero sodium arsenate exposure resulting in higher rates of neural tube defects compared to wildtype control mice. These differences in response frequencies were further exacerbated when the mice were maintained on a reduced folate diet. We have also determined that arsenic biotransformation was not the cause of the increased susceptibility in Folbp2$^{-/-}$ embryos; however, a definite conclusion could not be made due to the high dose of arsenic used in that study (30 mg/kg). In the present study, we performed urinary arsenic speciation in Folbp2$^{-/-}$ and folate binding protein-1 nullizygous (Folbp1$^{-/-}$) adult male mice using a single 1 mg/kg intraperitoneal injection of sodium arsenate. Results of the present study confirmed our previous findings in that Folbp2$^{-/-}$ mice have similar arsenic biotransformation, as do wildtype control mice. In addition, Folbp1$^{-/-}$ mice were also found to have similar arsenic speciation characteristics compared to the wildtype controls. The conclusions from this study are that the absence of a functional Folbp1 or Folbp2 protein does not seem to have a significant effect on arsenic biotransformation, and that the increased susceptibility of Folbp2$^{-/-}$ mice to in utero arsenic exposure is not likely due to differences in biotransformation.

Keywords: arsenic, Folbp1, Folbp2, urine

1. Introduction

Inorganic arsenic has long been suspected to be a human teratogen; however, the existing published literature is limited to a few case reports and questionably designed epidemiological studies. As a result, the association between human fetal exposure and adverse pregnancy outcome remains controversial (Shalat et al., 1996; DeSesso et al., 1998; Holson et al., 2000; DeSesso, 2001). In stark contrast to human studies, arsenic has consistently been shown to be teratogenic in a number of laboratory animal species, primarily inducing exencephaly, the major form of neural tube defect (NTD) seen in laboratory animals (Shalat et al., 1996; Golub et al., 1998; Holson et al., 2000). However, arsenic-induced teratogenicity in animals has been achieved almost exclusively by high-dose injections, which is a much different route of exposure compared to the human situation. Thus, extrapolating from animal studies to humans is not straightforward, and warrants additional high-quality human epidemiological and animal studies.

A considerable number of published studies have shown the protective role of periconceptional folic acid supplementation in preventing congenital malformations in humans. Daily supplementation with folic acid in the range of 0.4–5.0 mg has been shown to reduce the incidence of NTDs by up to 70% (Mulinare et al., 1988; Czeizel and Dudas, 1992; Werler et al., 1993; Shaw et al., 1994, 1995). However, despite the accumulating evidence regarding the embryo-protective effect of folic acid supplementation, the mechanism by which this is achieved remains unknown.

Folic acid enters cells by way of the folate receptors (FRs), also known as folate binding proteins (Folbps), and the reduced folate carrier. While the reduced folate carrier is a membrane transporter ubiquitously located in all tissues and cells, FRs are membrane bound and have both tissue- and cell-specific expression patterns (Ross et al., 1994). FR-α is primarily localized in the placenta, kidneys, choroids plexus and epithelial cells (Elwood, 1989), whereas FR-β is variably expressed in most tissues, and its binding affinity to folates are lower than FR-α (Ratnam et al., 1989; Ross et al., 1994). FR-γ/γ' are specific for hematopoietic cells (Shen et al., 1994) and the expression of FR-δ is yet to be determined (Spiegelstein et al., 2000). In the mouse, Folbp1 and Folbp2 are considered to be the homologs of FR-α and FR-β, respectively (Barber et al., 1999). To assess the role of murine FRs during embryogenesis, mice lacking a functional Folbp1 or Folbp2 alleles were generated by homologous recombination in embryonic stem cells (Piedrahita et al., 1999). Folbp1 nullizygous (Folbp1$^{-/-}$) embryos were found to die *in utero* by gestational day (GD) $10\frac{1}{2}$ due to multiple malformations, but could be rescued by high-dose folate supplementation throughout gestation (Piedrahita et al., 1999; Finnell et al., 2002). Thus far, we have been able to raise several Folbp1$^{-/-}$ mice, which do not require constant folate supplementation for survival beyond weaning and have even been able to reproduce. In contrast to Folbp1$^{-/-}$ mice, Folbp2$^{-/-}$ embryos seemed to develop normally, are free of any congenital malformations and are able to reproduce (Piedrahita et al., 1999).

Based on the hypothesis that Folbp2$^{-/-}$ mice may be more susceptible to teratogenic insults, especially those that are involved with the folate biosynthetic pathways, pregnant Folbp2$^{-/-}$ and Folbp2$^{+/+}$ dams were mated with sires of the same genotype. Sodium arsenate (40 mg/kg) was injected intraperitoneally (ip) on GDs 7½ and 8½, the critical period of neural tube closure (Wlodarczyk et al., 2001). As hypothesized, on a regular rodent chow (~ 2.7 mg folate/kg diet), Folbp2$^{-/-}$ mice had higher rates of NTDs compared to control Folbp2$^{+/+}$ mice (40.6 and 24.0%, respectively). Under conditions of reduced dietary folate intake (0.3 mg folate/kg diet), the rate of NTDs remained unchanged in control animals (25.7%) but significantly increased to 64.0% in Folbp2$^{-/-}$ mice. In order to test whether the differences in susceptibility are due to differences in arsenate biotransformation and exposure, we analyzed the 24-h urinary excretion of inorganic and methylated arsenicals following a single 30 mg/kg ip injection of sodium arsenate in adult Folbp2$^{-/-}$ and Folbp2$^{+/+}$ mice (Wlodarczyk et al., 2001). The results from this study failed to reveal any differences in the excretion of arsenicals between the two genotypes. However, the data that were obtained suggested that the dose that was used might have been too high to allow subtle distinctions between the genotypes to be detected.

The purpose of this study was to test whether there are differences in arsenate biotransformation between the two FR knockout mouse models. In this study, we performed a 24-h urinary speciation following a single 1 mg/kg ip injection of sodium arsenate in adult male Folbp1$^{-/-}$, Folbp2$^{-/-}$ and control wildtype mice.

2. Methodology

2.1. Animal study

Five wildtype, five Folbp2$^{-/-}$ and six Folbp1$^{-/-}$ adult male mice (3–5 months old) were placed on an amino-acid defined diet containing 2.7 mg folate/kg diet (Dyets #517839) for 2 weeks. After this 2-week conditioning period, the mice received a single 1 mg/kg ip injection of sodium arsenate (Sigma, Saint Louis, MO) dissolved in sterile water for injection (Abbott Laboratories, Chicago, IL), at a volume of 10 μl/g body weight. Each mouse was individually placed in a metabolic cage (Nalge Nunc International, Rochester, NY) with the powdered diet and a 10% sucrose solution provided *ad libitum*. Into each urinary collection tube, 100 mg of sodium ascorbate (Sigma, Saint Louis, MO) and 7.5 mg of sodium azide (Acros, NJ) were added. Twenty-four hours post-injection, the animal was removed and the cage interior thoroughly rinsed with approximately 10 ml of water to ensure collection of all the dried-up urine. The total volume of diluted urine was recorded and an aliquot was sent frozen on dry ice on the same day to the lab of XCL for arsenic speciation analysis.

2.2. Urinary speciation

Once samples were received in the analysis lab (less than 48-h post urine collection), they were immediately processed and prepared for analysis, in order to increase the chances of detecting unstable trivalent arsenicals, such as MMA[III]. Analysis was preformed by using ion pair chromatographic separation and anion exchange chromatographic separation with hydride generation atomic fluorescence detection (Le et al., 2000a). A high performance liquid chromatography (HPLC) system consisted of a Gilson (Middletone, WI) HPLC pump (Model 307), a Rheodyne (Cotati, CA) 6-port sample injector (Model 7725i) with a 20 μL sample loop and an HPLC column. The ion pair chromatographic separation was performed on a reversed-phase C18 column (ODS-3, 150 mm × 4.6 mm, 3 μm particle size (Phenomenex, Torrance, CA). A mobile phase solution (pH 5.9) containing 5 mM tetrabutylammonium hydroxide, 3 mM malonic acid and 5% methanol (flow rate was 1.2 ml/min) was used for the separation. The column temperature was maintained at 50°C. For the determination of trimethylarsine oxide (TMAO), an anion exchange column (PRP-X100, 150 mm × 4.1 mm, 3 μm particle size, Hamilton, Reno, NV) was used, and separation was performed at room temperature. The mobile phase solution (pH 8.2) contained 5 mM Na_2HPO_4 and 5% methanol, and its flow rate was 0.8 ml/min. A hydride generation atomic fluorescence detector (Model Excalibur 10.003, P.S. Analytical, Kent, UK) was used for the detection of arsenic.

2.3. Statistical analysis

Analysis for significance (*P* value set at 0.05, two sided) in excretion of each individual arsenic species and the cumulative amount (total) of arsenic excreted in the urine between the two knockout mice strains and controls was performed using the non-parametric Mann–Whitney test.

3. Results

The 24-h urinary arsenic speciation data of individual mice and groups' averages are presented both in Table 1 and Figure 1. On average, dimethylarsinic acid (DMA[V]) was the primary arsenic species excreted among all genotypes, and accounted for 53–61% of the administered dose. Arsenate (As[V]) accounted for 19–26%, arsenite (As[III]) 8–12%, whereas approximately only 1% of monomethylarsonic acid (MMA[V]) was detected in the urine of all groups. The total amount of arsenic excreted in 24 h was 87–93% of the administered dose. No significant differences in

Table 1.   Twenty-four hour urinary excretion of arsenicals in Folbp1[−/−] and Folbp2[−/−] mice, following a 1 mg/kg ip injection of sodium arsenate.

Animal no.	As[III] (%)	As[V] (%)	MMA[V] (%)	DMA[V] (%)	Total[a] (%)
Controls					
1	8.2	20.3	0.62	41.0	70.1
2	10.2	26.5	1.11	68.3	106.1
3	17.0	16.8	1.06	67.8	102.7
4	10.3	24.5	0.50	49.5	84.7
5	8.2	30.1	1.03	38.6	77.9
Mean	10.8	23.6	0.87	53.0	88.3
SD[b]	3.6	5.2	0.28	14.3	15.6
CV (%)[c]	33.8	22.1	32.6	27.0	17.7
Folbp1[−/−]					
1	7.2	9.9	0.68	44.8	62.6
2	16.7	22.9	1.30	67.1	108.0
3	11.6	13.2	1.29	75.8	101.9
4	19.6	7.8	1.80	88.4	117.6
5	8.2	34.7	0.90	50.0	93.9
6	10.9	22.9	0.80	42.1	76.6
Mean	12.4	18.6	1.13	61.4	93.4
SD[b]	4.9	10.2	0.42	18.7	20.5
CV (%)[c]	39.2	54.8	36.9	30.5	21.9
Folbp2[−/−]					
1	10.9	24.6	0.00	66.2	101.6
2	11.3	18.1	1.20	67.9	98.5
3	4.0	36.6	0.30	40.1	81.0
4	10.1	27.0	0.69	43.6	81.3
5	4.1	23.7	0.70	45.4	73.9
Mean	8.1	26.0	0.58	52.6	87.3
SD[b]	3.7	6.8	0.45	13.3	12.1
CV (%)[c]	45.5	26.0	78.5	25.3	13.9

All excretion values represent the percentage from the administered dose.
[a] Cumulative amount of arsenicals excreted in the urine.
[b] Standard deviation.
[c] Percent coefficient of variation.

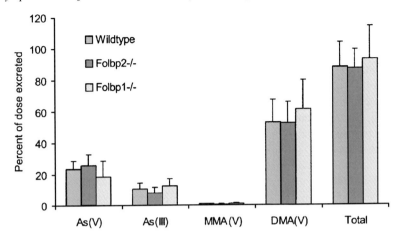

Figure 1. Twenty-four hour urinary excretion of arsenicals in wildtype controls, Folbp1$^{-/-}$ and Folbp2$^{-/-}$ knockout mice. Data represents the average ± standard deviation of the percentage from the administered dose excreted in the urine. As[V], arsenate; As[III], arsenite; MMA[V], monomethylarsonic acid; DMA[V], dimethylarsinic acid.

excretion of individual or total arsenicals were detected between Folbp2$^{-/-}$ or Folbp1$^{-/-}$ mice and the control group ($P > 0.05$).

Using the system described, the following limits of detection were obtained: 0.5 μg/l for As[III] and MMA[V]; 1 μg/l for DMA[V] and As[V]; 2 μg/l for MMA[III] and DMA[III]; 7 μg/l for TMAO. The presence of MMA[III], DMA[III] and TMAO was not detected in all urine samples.

4. Discussion

Metabolism of arsenate in most mammalian species proceeds *via* reduction to As[III] and biomethylation to MMA[V]. MMA[V] is subsequently reduced to monomethylarsonous acid (MMA[III]) and then biomethylated to DMA[V], which is the major arsenic metabolite excreted in the urine of most mammalian species (Vahter and Marafante, 1988). Both methylation reactions are catalyzed by methyltransferases that require *S*-adenosylmethionine (SAM) as the methyl donor (Vahter and Enval, 1983; Aposhian, 1997; Vahter, 2000). SAM is eventually regenerated by remethylation through the homocysteine remethylation cycle, a process that requires (6*S*)-5-methyl tetrahydrofolate as a co-factor, and thus it seems as if arsenic biomethylation would be dependent on an adequate folate supply.

In this study, we hypothesized that mice deficient of the Folbp1 or Folbp2 gene products would have perturbed folate homeostasis, and as a result would demonstrate altered arsenic metabolism and altered urinary excretion of arsenicals. Specifically, we hypothesized that the key effects would be reduced excretion of DMA[V], as seen with methyl-deficient rabbits (Vahter and Marafante, 1987). However, after a single ip injection of 1 mg/kg sodium arsenate, no significant differences in excretion of neither organic nor inorganic arsenicals were observed between the two knockout mice strains and controls.

The data obtained in the present study support our previous findings regarding the insignificant role of Folbp2 in arsenic biotransformation (Wlodarczyk et al., 2001). Moreover, it supports our previous suggestion that the increased *in utero* susceptibility of Folbp2$^{-/-}$ mice to arsenate is not due to differences in metabolism and exposure. An alternative explanation would be that other, yet unknown genetic determinants play an important role in conferring increased susceptibility to the Folbp2$^{-/-}$ mice. Currently, we are exploring several such options using cDNA microarrays in gene expression studies, and other experiments aimed at examining the potential role of DNA methylation.

On the other hand, it is possible that the ultimate teratogenic arsenical was not detected in this study, and thus, no differences in arsenic speciation between Folbp2$^{-/-}$ and control mice were feasible. For example, trivalent arsenicals may have been excreted into the urine; however, being unstable and the lack of an effective preservative at the time this study was performed, it was impossible to detect them. For example, MMA[III] as well as DMA[III] have been previously detected in human urine (Aposhian et al., 2000; Le et al., 2000a,b; Mandal et al., 2001; Styblo et al., 2001), and are known to be equal to or more toxic than As[V] or As[III] (Styblo et al., 1997; Lin et al., 1999, 2001; Petrick et al., 2000, 2001; Mass et al., 2001). Although it has been shown that detection of trivalent arsenicals requires the analysis of fresh urine samples that readily oxidize to their respective pentavalent forms (Le et al., 2000a,b; Gong et al., 2001), we were unable to detect these arsenic species despite the relatively prompt shipping process at very low temperatures. Thus, it is possible that trivalent arsenical(s) are the ultimate teratogen(s), and that only our technical inability to detect them prevented us from deciphering a likely explanation for the increased susceptibility of Folbp2$^{-/-}$ mice. Only recently, Le and colleagues have reported on the effectiveness of diethylammonium diethyldithiocarbamate in preventing oxidation of MMA[III] and DMA[III] to their respective pentavalent species (Le et al., 2002). Thus, in the future we will be able to repeat this study in order to re-examine the potential role of trivalent arsenicals in arsenic-induced teratogenicity.

On the other hand, it may also be that examining the urine of adult male mice was not the most appropriate method to detect differences in arsenic exposure of mid-gestation embryos. Perhaps, analysis of maternal blood, or the embryonic tissue itself, may prove to be much more instructive. However, despite our inability to establish a causative effect of altered arsenic metabolism in Folbp2$^{-/-}$ mice, we feel confident that the experimental paradigm used in this study was appropriate and sensitive enough to detect such differences had they been present.

The important role of FR-1 has been demonstrated in our previous studies, where we found that Folbp1$^{-/-}$ mice die *in utero* by GD 10½, but can be rescued and brought to adulthood by high-dose folate supplementation of dams prior to and throughout pregnancy (Piedrahita et al., 1999; Finnell et al., 2002). To date, Folbp1 mice have not been utilized in teratological studies with arsenic as the teratogen. However, because we had several living Folbp1$^{-/-}$ mice in our laboratory, we were able to utilize them in this study in order to assess the role of Folbp1 in arsenic biotransformation. As presented in Table 1 and Figure 1, it seems that the Folbp1 gene product lacks an important role in the metabolism of arsenic, because no significant differences in arsenicals excretion were detected.

In conclusion, the data we present herein demonstrate that no significant differences in arsenic biotransformation and excretion occur between Folbp2$^{-/-}$ and wildtype control mice. Thus, we suggest that the increased sensitivity of Folbp2$^{-/-}$ mice to arsenic exposure during gestation may not be due to increased exposure to arsenicals. In addition, while Folbp1 has an important role in maintaining folate homeostasis especially during embryonic development, it does not seem to have any significant effect on the biotransformation of arsenic.

Acknowledgements

This project was supported in part by grants ES 04917, ES 09106 and ES 11775 from the National Institute of Environmental Health to RHF, and grants from the Natural Sciences and Engineering Research Council of Canada and Canadian Water Networks NCE to XCL. Its contents are solely the responsibility of the authors and do not necessarily represent the official views of the NIEHS, NIH. The authors would like to thank Ms Michelle Merriweather and Mr Joe Wicker from the Institute of Biosciences and Technology at Texas A&M University Health Science Center for the care, well-being and for genotyping of the mice used in this study.

References

Aposhian, H.V., 1997. Enzymatic methylation of arsenic species and other new approaches to arsenic toxicity. Annu. Rev. Pharmacol. Toxicol., 37, 397–419.

Aposhian, H.V., Gurzau, E.S., Le, X.C., Gurzau, A., Healy, S.M., Lu, X., Ma, M., Yip, L., Zakharyan, R.A., Maiorino, R.M., Dart, R.C., Tirus, M.G., Gonzalez-Ramirez, D., Morgan, D.L., Avram, D., Aposhian, M.M., 2000. Occurrence of monomethylarsonous acid (MMAIII) in urine of humans exposed to inorganic arsenic. Chem. Res. Toxicol., 13, 693–697.

Barber, R.C., Bennett, G.D., Greer, K.A., Finnell, R.H., 1999. Expression patterns of folate binding proteins one and two in the developing mouse embryo. Mol. Genet. Metab., 66, 31–39.

Czeizel, A.E., Dudas, I., 1992. Prevention of the first occurrence of neural-tube defects by periconceptional vitamin supplementation. N. Engl. J. Med., 327, 1832–1835.

DeSesso, J.M., 2001. Teratogen update: inorganic arsenic. Teratology, 63, 170–173.

DeSesso, J.M., Jacobson, C.F., Scialli, A.R., Farr, C.H., Holson, J.F., 1998. An assessment of the developmental toxicity of inorganic arsenic. Reprod. Toxicol., 12, 385–433.

Elwood, P.C., 1989. Molecular cloning and characterization of the human folate binding protein cDNA from placenta and malignant tissue culture. J. Biol. Chem., 264, 14893–14901.

Finnell, R.H., Wlodarczyk, B., Spiegelstein, O., Triplett, A., Gelineau-vanWaes, J., 2002. Folate transport abnormalities and congenital defects. In: Milstein, S., Kapatos, G., Levine, R.A., Shane, B. (Eds), Chemistry and Biology of Pteridines and Folates. Kluwer Academic Publishers, Norwich, MA, pp. 637–642.

Golub, M.S., Macintosh, M.S., Baumrind, N., 1998. Developmental and reproductive toxicity of inorganic arsenic: animal studies and human concerns. J. Toxicol. Environ. Health B Crit. Rev., 1, 199–241.

Gong, Z., Lu, X., Cullen, W.R., Le, X.C., 2001. Stability of monomethylarsonous acid and dimethylarsinous acid in human urine. J. Anal. At. Spectrom., 16, 1409–1413.

Holson, F.H., Desesso, J.M., Jacobson, C.F., Farr, C.F., 2000. Appropriate use of animal models in the assessment of risk during prenatal development: an illustration using inorganic arsenic. Teratology, 62, 51–71.

Le, X.C., Lu, X., Ma, M., Cullen, W.R., Aposhian, V., Zheng, B., 2000a. Speciation of key arsenic metabolic intermediates in human urine. Anal. Chem., 72, 5172–5177.

Le, X.C., Ma, M., Lu, X., Cullen, W.R., Aposhian, V., Zheng, B., 2000b. Determination of monomethylarsonous acid, a key arsenic methylation intermediate, in human urine. Environ. Health Perspect., 108, 1015–1018.

Le, X.C., Gong, Z., Jiang, G., Lu, X., Cullen, W.R., 2002. Trivalent arsenic compounds: speciation, preservation, and interaction with proteins. 5th International Conference on Arsenic Exposure and Health Effects, San Diego, CA, July 14–18, 2002, p. 24.

Lin, S., Cullen, W.R., Thomas, D.J., 1999. Methylarsenicals and arsinothiols are potent inhibitors of mouse liver thioredoxin reductase. Chem. Res. Toxicol., 12, 924–930.

Lin, S., Del Razo, L.M., Styblo, M., Wang, C., Cullen, W.R., Thomas, D.J., 2001. Arsenicals inhibit thioredoxin reductase in cultured rat hepatocytes. Chem. Res. Toxicol., 14, 305–311.

Mandal, B.K., Ogara, Y., Suzuki, K.T., 2001. Identification of dimethylarsinous and monomethylarsonous acids in human urine of the arsenic-affected areas in West Bengal, India. Chem. Res. Toxicol., 14, 371–378.

Mass, M.J., Tennant, A., Roop, R.C., Cullen, W.R., Styblo, M., Thomas, D.J., Kligerman, A.D., 2001. Methylated trivalent arsenic species are genotoxic. Chem. Res. Toxicol., 14, 355–361.

Mulinare, J., Cordero, J.F., Erickson, J.D., Berry, R.J., 1988. Periconceptional use of multivitamins and the occurrence of neural tube defects. J. Am. Med. Assoc., 260, 3141–3145.

Petrick, J.S., Ayala-Fierro, F., Cullen, W.R., Carter, D.E., Aposhian, H.V., 2000. Monomethylarsonous acid (MMAIII) is more toxic than arsenite in Chang human hepatocytes. Toxicol. Appl. Pharmacol., 163, 203–207.

Petrick, J.S., Jagadish, B., Mash, E.A., Aposhian, H.V., 2001. Methylarsonous acid (MMAIII) and arsenite: LD50 in hamsters and in vitro inhibition of pyruvate dehydrogenase. Chem. Res. Toxicol., 14, 651–656.

Piedrahita, J.A., Oetama, B., Bennett, G.D., van Waes, J., Kamen, B.A., Richardson, J., Lacey, S.W., Anderson, R.G., Finnell, R.H., 1999. Mice lacking the folic acid-binding protein Folbp1 are defective in early embryonic development. Nat. Genet., 23, 228–232.

Ratnam, M., Marquardt, H., Duhring, J.L., Freisheim, J.H., 1989. Homologous membrane folate binding proteins in human placenta: cloning and sequence of a cDNA. Biochemistry, 28, 8249–8254.

Ross, J.F., Chaudhuri, P.K., Ratnam, M., 1994. Differential regulation of folate receptor isoforms in normal and malignant tissues in vivo and in established cell lines. Physiologic and clinical implications. Cancer, 73, 2432–2443.

Shalat, S.L., Walker, D.B., Finnell, R.H., 1996. Role of arsenic as a reproductive toxin with particular attention to neural tube defects. J. Toxicol. Environ. Health, 48, 253–272.

Shaw, G.M., Jensvold, N.J., Wasserman, C.R., Lammer, E.J., 1994. Epidemiologic characteristics of phenotypically distinct neural tube defects among 0.7 million California births, 1983–1987. Teratology, 49, 143–149.

Shaw, G.M., Lammer, E.J., Wasserman, C.R., O'Malley, C.D., Tolarova, M.M., 1995. Risks of orofacial clefts in children born to women using multivitamins containing folic acid periconceptionally. Lancet, 346, 393–396.

Shen, F., Ross, J.F., Wang, X., Ratnam, M., 1994. Identification of a novel folate receptor, a truncated receptor, and receptor type beta in hematopoietic cells: cDNA cloning, expression, immunoreactivity, and tissue specificity. Biochemistry, 8, 1209–1215.

Spiegelstein, O., Eudy, J.D., Finnell, R.H., 2000. Identification of two putative novel folate receptor genes in humans and mouse. Gene, 27, 117–125.

Styblo, M., Serves, S.V., Cullen, W.R., Thomas, D.J., 1997. Comparative inhibition of yeast glutathione reductase by arsenicals and arsenothiols. Chem. Res. Toxicol., 10, 27–33.

Styblo, M., Lin, S., Del Razo, L.M., Thomas, D.J., 2001. Trivalent methylated arsenicals: toxic products of the metabolism of inorganic arsenic. In: Chappell, W.R., Abernathy, C.C., Calderon, R.L. (Eds), Arsenic Exposure and Health Effects. Elsevier, New York, pp. 325–338.

Vahter, M., 2000. Genetic polymorphisms in the biotransformation of inorganic arsenic and its role in toxicity. Toxicol. Lett., 112–113, 209–217.

Vahter, M., Enval, J., 1983. In vivo reduction of arsenate in mice and rabbits. Environ. Res., 32, 14–24.

Vahter, M., Marafante, E., 1987. Effects of low dietary intake of methionine, choline or proteins on the biotransformation of arsenite in the rabbit. Toxicol. Lett., 37, 41–46.

Vahter, M., Marafante, E., 1988. In vivo methylation and detoxification of arsenic. In: Craig, P.J., Glockling, F. (Eds), The Biological Alkylation of Heavy Elements. Royal Society of Chemistry, London, UK, 105 pp.

Werler, M.M., Shapiro, S., Mitchell, A.A., 1993. Periconceptional folic acid exposure and risk of occurrent neural tube defects. J. Am. Med. Assoc., 269, 1257–1261.

Wlodarczyk, B., Spiegelstein, O., Gelineau van-Waes, J., Vorce, R.L., Lu, X., Le, X.C., Finnell, R.H., 2001. Arsenic-induced congenital malformations in genetically susceptible folate binding protein-2 knockout mice. Toxicol. Appl. Pharmacol., 177, 238–246.

Arsenic Exposure and Health Effects V
W.R. Chappell, C.O. Abernathy, R.L. Calderon and D.J. Thomas, editors
© 2003 Published by Elsevier B.V.

Chapter 26

Some chemical properties underlying arsenic's biological activity[*]

Kirk T. Kitchin, Kathleen Wallace and Paul Andrewes

Abstract

In this paper, some of the chemical properties of arsenicals (atomic and molecular orbitals, electronegativity, valence state, changes between valence state, nucleophilicity, the hard/soft acid/base principle) that may account for some of the biological activity of arsenicals are presented. Trivalent arsenicals can act as soft acids while pentavalent arsenicals tend to act as borderline soft to hard acids. As soft acids, trivalent arsenicals have a high propensity to bond with soft bases, which in biological systems are usually sulfur (e.g. cysteine moieties of proteins) and selenium containing compounds. This soft acid property of trivalent arsenicals can lead to many of the proposed modes of carcinogenesis (e.g. altered DNA repair, altered growth factors, cell proliferation, altered DNA methylation patterns and promotion of carcinogenesis). Trivalent arsenicals can act as nucleophiles, whereas pentavalent arsenicals usually do not. Because of less electrostatic repulsion, nonionized trivalent arsenicals are more capable of interacting with negatively charged DNA than negatively charged pentavalent arsenicals. Experimental binding studies with [73]As arsenite to a model peptide of 25 amino acids and 4 free sulfhydryl groups (based on a zinc finger region of the human estrogen receptor alpha) gave a Kd (dissociation equilibrium constant) of 2.2 µM and a Bmax (concentration of binding sites for the ligand) of 89 nmol/mg protein. This arsenite binding was dependent on (a) the presence of free sulfhydryls in the peptide and (b) trivalency of the arsenical (i.e. arsenite binds well, arsenate does not).

Keywords: arsenic, arsenite, arsenate, binding, sulfhydryl, carcinogenesis, monomethyarsonic acid, dimethylarsinic acid

Abbreviations:

Bmax, concentration of binding sites for the ligand; DMA(V), dimethylarsinic acid; DMA(III), dimethylarsinous acid; Kd, dissociation equilibrium constant; MMA(V), monomethylarsonic acid; MMA(III), monomethylarsonous acid.

1. Introduction

Three of the chemical properties of arsenic most likely to account for its biological activity are the soft acid/soft base principle (trivalent arsenicals and sulfhydryls), nucleophilicity of trivalent arsenicals and the formation of free radicals and/or

---

[*] Disclaimer: This manuscript has been reviewed in accordance with the policy of the National Health and Environmental Effects Research Laboratory, U.S. Environmental Protection Agency, and approved for publication. Approval does not signify that the contents necessarily reflect the views and policies of the Agency, nor does mention of trade names of commercial products constitute endorsement or recommendation for use.

reactive oxygen species (ROS) by arsenicals. Other chemical properties of arsenic (the bonding electrons of the individual arsenic atom, the electronegativities of arsenic, the ionization energy of arsenic, the molecular orbitals of an arsenic compound and the pKa values for arsenic acid and arsenious acid) will be presented later in this article.

1.1. Soft acid/soft base (trivalent arsenicals and sulfhydryls)

The hard/soft acid/base principle is that soft acids tend to form complexes with soft bases and hard acids tend to form complexes to hard bases (Pearson, 1963). Four properties useful in placing a chemical species into the right hard/soft acid/base category are electronegativity, molecular size, charge and polarizability. Examples of soft acids include Cu^+, Hg^{2+} and Pb^{2+}. The anions I^-, Br^- and S^{2-} are soft bases. Examples of hard acids include H^+, Ca^{2+}, Fe^{3+} and Al^{3+}. The chemical species F^-, O^{2-}, PO_4^{3-} and H_2O are hard bases. The hard/soft acid/base principle is useful to distinguish the properties of arsenite (soft acid) and arsenate (a borderline soft to hard Lewis acid). Thus, As(V) is often found, in nature, bound to oxygen in arsenate compounds. In contrast, arsenic(III) is a soft acid due to its lower oxidation state and the ease with which its electron cloud can be polarized or distorted. Thus, due to its soft acid nature arsenic(III) has a high propensity to bond with soft bases, which in biological systems are usually sulfur (e.g. cysteine) and selenium containing compounds. Arsenate can be reduced to arsenite by reduced glutathione (Winski and Carter, 1995). Arsenite can complex with three additional glutathione moieties to form As(SG)$_3$ (Scott et al., 1993; Winski and Carter, 1995). In protein binding experiments using rabbit erythrocyte lysates, rapid protein binding was demonstrated by arsenite, slower binding by arsenate and far slower binding by either MMA(V) or DMA(V) (Delnomdedieu et al., 1995). As a soft acid, arsenic(III) is comparable to mercury(II), another highly toxic sulfhydryl seeking species.

Arsenic has long been thought of as attacking protein (via sulfhydryls) and not DNA. Several of the possible protein targets of trivalent arsenicals include DNA repair enzymes, transcription factors, steroid hormone receptors, zinc finger proteins and tubulin (National Research Council, 1999). When methylarsenic species are considered, the same nucleophilicity and hard/soft acid/base principle rules hold true. The number of sites available for binding of sulfur is reduced by each methyl group that is attached to the arsenic. Thus, monomethylarsenic(III) can react with two sulfur containing ligands whereas dimethylarsenic(III) can react with just one such ligand. In biological systems, the sulfhydryl groups of cysteine in peptides and proteins are important in the binding of trivalent arsenicals (Aposhian, 1989). If trivalent arsenicals acting as soft acids are causally important, then the likely modes of action of arsenic carcinogenesis may include altered DNA repair, altered growth factors, cell proliferation, altered DNA methylation patterns and promotion of carcinogenesis. It is no surprise that so many possible modes of carcinogenesis for arsenic classify reasonably well under the soft acid and soft base, and the trivalent arsenical and sulfhydryl poison banners.

1.2. Nucleophilicity of trivalent arsenicals

By virtue of their lone electron pair, arsenic(III) species are nucleophiles. As a nucleophile, arsenic(III) behaves as a base and demonstrates its amphoteric metalloid nature.

Depending on the biological molecule with which it interacts, the biological activity of arsenic(III) may be due to either its soft acid nature [like Hg(II)] and/or its nucleophilic behavior. The nucleophilicity of As(III) is illustrated by the Meyer reaction where arsenic(III) species will react with methyl iodide in an oxidative addition reaction. In biological systems, this is seen in the Challenger mechanism for arsenic biomethylation (Challenger, 1945) where the source of methyl carbocations is *S*-adenosyl methionine and methyl transferase enzymes catalyze the nucleophilic addition of arsenic(III) species to a methyl carbocation. However, based on theoretical grounds DNA cannot be realistically considered as a good substrate for nucleophilic attack because it contains no electrophilic functional groups. In contrast, for an electrophile there are many possible sites for interaction with the bases of DNA. For example, on deoxyguanosine, possible electrophile attack sites would include the 06, N7, N3, N1, N2 and C8 atoms.

1.3. Free radicals/ROS

As arsenicals exist in both the trivalent and pentavalent oxidation states *in vivo*, there might be a redox cycle with substantial rates of simultaneous reduction and oxidation of arsenic compounds [e.g. arsenate ↔ arsenite, MMA(V) ↔ MMA(III) and DMA(V) ↔ DMA(III)]. If redox cycling of arsenicals occurs *in vivo* to a substantial degree, and if this redox cycling generates other ROS or free radical species in some manner, then redox cycling between the two valence states of arsenic could be one of the chemical properties underlying arsenic's biological actions (Kitchin and Ahmad, 2003).

The mono- and dimethylation of inorganic arsenic requires an oxidative methylation step from the trivalent to pentavalent oxidation state [e.g. MMA(III) (monomethylarsonous acid) → DMA(V) (dimethylarsinic acid)]. This oxidative methylation step might be a source of reactive species or free radicals. However, this possible source of reactive species is probably less likely than redox cycling of arsenicals *per se*.

1.4. Binding studies

In this paper, the soft acid activity of a trivalent arsenical is demonstrated by the binding of arsenite (as [73]As) to a peptide with four free sulfhydryl groups. In this experimental system, arsenate binds poorly. This type of interaction requires both an available sulfhydryl on the peptide and a trivalent arsenic species.

2. Methodology

2.1. Binding studies

A model peptide (named peptide 15) containing 25 amino acids (with 4 free sulfhydryls) with an isothiocyanate linked fluorescent tag (fluorescein-RYCAVCNDYA SGYHYGVWSCEGCKA, molecular weight 3164) was studied in binding experiments with [73]As arsenite and [73]As arsenate. Peptides were synthesized and mass spectroscopy and HPLC purity determinations were run by a commercial laboratory (Alpha Diagnostics). This peptide is part of the Zn finger region of the human estrogen receptor alpha (Green et al., 1986). This protein was selected because of results indicating that arsenite is a "potential nonsteroidal environmental estrogen" (Stoica et al., 2000). In MCF-7 cells (an estrogen receptor positive breast cancer cell), arsenite displayed substantial endocrine-like activity by: (1) decreasing the estrogen receptor alpha concentration by 60%, (2) increasing the progesterone receptor concentration 22-fold, (3) stimulating cell growth in estrogen-depleted media and (4) blocking the binding of estrogen to its receptor [the K_i value (50% inhibitory concentration) for arsenite was 5 nM] (Stoica et al., 2000). Two other interesting connections exist between arsenicals and estrogenicity. Swiss mice given weekly intravenous treatment with 0.5 mg/kg of arsenate for 20 weeks exhibited proliferative preneoplastic lesions of the uterus, testes and liver, uterine cystic hyperplasia and changes in the uterine concentration of estrogen receptor protein (Waalkes et al., 2000). In female C3H mice drinking water containing either 42.5 or 85 ppm arsenite during days 8–18 of gestation, treatment effects included hepatocellular carcinoma and adrenal tumors in male offspring and ovarian tumors, lung carcinoma and proliferative lesions in the uterus and oviduct in female offspring (Waalkes et al., 2003).

[73]As (arsenate) was obtained from Los Alamos National Laboratories and reduced to arsenite by SO_2. Binding experiments used the test peptide or protein diluted in cold water, [73]As arsenite diluted in cold 150 mM NaCl, pH 7.5 buffer containing 100 mM Tris–HCl, 0.45 μM nitrocellulose filters soaked in a pH 7.5 solution containing 150 mM NaCl, 0.3% polyethylenimide and 100 mM Tris–HCl and a Brandel Model M-24C Membrane Harvestor. Solutions were deoxygenated by bubbling nitrogen gas through them. Peptide and arsenite were incubated for 60 min at 2–8°C prior to filtration. To reduce nonspecific binding, filtered peptides were washed three times with 2 ml of cold 150 mM NaCl, 100 mM Tris–HCl, pH 7.5 solution. The wash solution was kept on ice. Nonspecific binding was determined by the addition of 100-fold excess of unlabeled arsenite to [73]As arsenite. Gamma counting was done in a Packard Minaxi Auto gamma 5000. Protein was stripped from the filters in a pH 10 buffer of 62.5 mM $NaCO_3$ containing 5% sodium lauryl sulfate. The concentration of protein was determined fluorometrically via the fluorescein tag on the peptide with a Wallac Victor 1420 multilabel counter. Kd and Bmax values were estimated from saturation binding experiments by use of the program Graph Pad Prism 3.

3. Results

3.1. Binding studies

A model peptide based on the human estrogen receptor alpha, which contained 25 amino acids (4 of them are free sulfhydryls), was studied in binding experiments with [73]As arsenite and [73]As arsenate. Overall, the saturation binding curves with arsenite gave an estimate for the Kd of 2.2 μM and a Bmax of 89 nmol/mg protein (Fig. 1). The 95% confidence intervals for the Kd were 1.64–2.70 μM and for the Bmax were 83–95 nmol/mg. The goodness of fit r^2 value was 0.98. The binding curve shows the expected higher

Arsenite Binding to Peptide 15

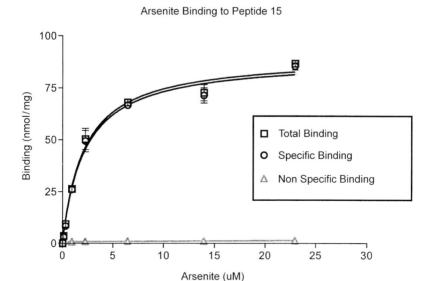

Figure 1. Results from a saturation binding experiment with [73]As arsenite and peptide 15 (fluorescein-RYCAVCNDYASGYHYGVWSCEGCKA). There are 4 free sulfhydryls among the 25 amino acids of this synthetic peptide which is based on a Zn finger region of the human estrogen receptor alpha protein.

slope at low arsenite concentrations in the range of 0.4–3.0 μM and then a much lower slope closer to the upper asymptote in the concentration range of about 6–25 μM.

When [73]As arsenate was substituted for arsenite in the binding studies, far less binding to peptide 15 occurred [only 2.5% of the binding observed with equimolar

(40 µM) arsenite]. If treatment with 400 µM N-ethylmaleimide (a reagent which inactivates free sulfhydryl groups) for 10 min precedes a binding study of arsenite and peptide 15, then the binding is reduced by 99.6%. Proteins and peptides lacking free sulfhydryl groups bind [73]As arsenite poorly (only 5.0% or less of the binding observed with peptide 15) (data not shown).

4. Discussion

4.1. Binding experiments with arsenicals

As soft acids, trivalent arsenicals have a well known propensity to interact with elements of group 16 of the periodic chart – oxygen, selenium and sulfur [e.g. the sulfhydryl groups of proteins (Aposhian, 1989; Stoica et al., 2000)]. Proteins and peptides contain cysteine both in reduced and oxidized (and hence unavailable) forms. Trivalent arsenicals are not known for strongly binding to methionine, serine, threonine or tyrosine moieties in peptides or proteins. Methionine contains a CH_3-S-CH_2-moiety whereas serine, threonine and tyrosine all contain hydroxy groups.

In our binding studies with arsenite and a model peptide containing 4 sulfhydryl groups (Fig. 1), a Kd of 2.2 µM for specific binding and a Bmax of 89 nmol/mg protein were observed. This specific binding demonstrated the expected properties of an interaction between a trivalent arsenical and a sulfhydryl containing peptide: (a) the majority of the specific binding of arsenite could be eliminated by prior treatment of the peptide with the sulfhydryl agent N-ethylmaleimide, (b) arsenate did not bind well to peptide 15, (c) arsenite did not bind well to other proteins that did not contain free sulfhydryls and (d) arsenite did not bind well to a peptide containing several positively charged amino acids (His, Arg and Lys). When proteins and DNA are compared on a nmol arsenical/mg of target macromolecule basis, arsenite binds much better to a sulfhydryl containing peptide (89 nmol/mg protein, data of Fig. 1) than any arsenical is known to bind to DNA.

4.2. Chemical properties of arsenicals

Some additional chemical properties that could underlay arsenic's biological actions will now be presented. These chemical properties include the bonding electrons of the individual arsenic atom, the electronegativities of arsenic and some other elements, the ionization energy of arsenic, the molecular orbitals of an arsenic compound and finally the pKa values for arsenic acid and arsenious acid. As a group 15 member of the periodic chart, the important bonding electron orbitals of arsenic are $4s^2$ and $4p^3$ (Hill and Matrone, 1970). When arsenic is bonded to the atoms H, C, O and S, the common valences for arsenic are either three or five (Cotton and Wilkinson, 1972).

On the Pauling electronegativity scale, arsenic has a value of 2.18. The electronegativity of the elements arsenic often combines with are 2.20 for H, 2.55 for C, 2.58 for S and 3.44 for O. The electronegativity values for group 15 elements are 3.04 for N, 2.19 for P, 2.18 for As, 2.05 for Sb and 2.02 for Bi. Electronegativity values alone cannot totally explain arsenic's bonding to other atoms and the strength of these bonds. For the first six ionizable electrons of arsenic, the ionization energies are 947, 1798, 2735, 4837, 6043 and

12310 kJ per mole. The bond enthalpies for several diatomic molecules containing arsenic are 481 for AsO, 449 for AsN, 448 for AsCl, 433 for AsP, 410 for AsF, 382 for AsAs, 379 for AsS, 330 for AsSb, 297 for AsI, 274 for AsH and 96 kJ/mole for AsSe. The relatively large bond enthalpies for the As–O bond (481 kJ/mole) and the As–S bond (379 kJ/mole) are part of the reason arsenic is so often found bonded to these two elements.

A more sophisticated approach to arsenic bonding to other atoms must use molecular orbitals and not atomic orbitals. Arsenic can promote one of its 4s electrons into a 4d orbital to form sp^3 hybrid orbitals and thus form a tetrahedral molecular structure (Hill and Matrone, 1970). The molecular orbitals of As(V)O$_4$ and P(V)O$_4$ are identical except that arsenic uses 4s, 4p and 4d molecular orbitals whereas phosphorus uses 3s, 3p and 3d molecular orbitals. For both arsenate and phosphate, the double bond to oxygen involves one sp^3 orbital and one d orbital. Pentavalent arsenicals often act similarly to phosphate analogs whereas trivalent arsenicals can better form additional bonds with other molecules.

4.3. How the chemical properties of arsenic can underlay some possible modes/mechanisms of arsenic's biological action

From chemical properties such as atomic and molecular orbitals, electronegativity, enthalpies and ionization energies, we can get a picture of some of the properties of arsenic's bonding electrons, the generally tetrahedral structure of arsenicals and the chemical bonds arsenic makes with other atoms. From the hard/soft acid/base principle, nucleophilicity of trivalent arsenicals and the possibility that arsenic containing free radicals and/or ROS can be formed, we get a more useful set of tools with which to approach arsenic's mode of biological action. However, it is difficult, at present, to first go from these chemical parameters to any of the nine proposed modes of arsenic carcinogenesis and then secondarily to arsenic exposed human populations with interindividual variation of responses and disease outcomes. Future progress can come from (i) biological and experimental animal studies which include chronic exposures, later developing stages of arsenic-induced diseases and study parameters that allow one to connect to earlier causal events and chemical properties of arsenicals and (ii) epidemiological studies which include biomarkers of effect which can be related to possible modes of actions and/or underlying chemical properties of arsenic. Three of the proposed nine different possible modes of action of arsenic carcinogenesis (progression of carcinogenesis, gene amplification and p53 suppression) are somewhat difficult to classify under one of the predominant chemical properties (soft acid, nucleophilicity and free radical/ROS) of arsenic.

4.4. Free radicals and ROS

There is one pathway of arsenic metabolism already known to produce free radicals (Yamanaka et al., 1990). Dimethylarsine, a reduced metabolite of DMA(V), can react with molecular oxygen forming a $(CH_3)_2As^\bullet$ radical and superoxide. This $(CH_3)_2As^\bullet$ radical can add a molecule of oxygen and form the $(CH_3)_2AsOO^\bullet$ radical. A second pathway that can produce superoxide is in the oxidation of ferrous to ferric iron. Arsenic exposure can lead to ferrous iron release by (i) the release of iron from ferritin (Ahmad et al., 2000) and (ii) the induction of the enzyme heme oxygenase by arsenite (Sunderman, 1987; Kitchin et al., 1999). Superoxide can lead to hydrogen peroxide, which can generate hydroxyl radical via iron or copper-dependent processes. From exposure to these arsenic-free radicals and the ROS they can generate, DNA damage such as DNA single strand breaks can occur. Among the nine proposed modes of action of arsenic carcinogenesis (Kitchin, 2001), oxidative stress (e.g. base oxidation), chromosomal abnormalities and promotion of carcinogenesis are three of the modes of action which may be partly mediated by ROS.

4.5. Arsenical interactions with DNA

To electrically interact with DNA, it is advantageous for a molecule to have minimal negative charge, be uncharged or even have a positive charge. At physiological pH any of the nonionized trivalent arsenicals, arsenite, MMA(III) (monomethylarsonous acid) and DMA(III) (dimethylarsinous acid) are excellent candidates for interacting with DNA without electrostatic repulsion preventing close proximity of the two molecules (the pKa for H_3AsO_3 is 9.23). Arsenate would have one or two hydroxy groups ionized at physiological pH (the three pKa values for H_3AsO_4 are 2.25, 6.77 and 11.6). For the methylated pentavalent arsenicals the pKa values are 4.1 and 8.7 for MMA(V) and 6.9 for DMA(V) (Aposhian, 1989). As they do not have an ionizable hydroxy group, the arsenicals dimethylarsine, trimethylarsine oxide and trimethylarsine do not have any negative charges in a cell and thus are also candidates for interacting with DNA. Dimethylarsine has been shown to generate free radicals in vivo (Yamanaka et al., 1990). Trimethylarsine oxide has been shown to exert some degree of promotion of carcinogenesis in rats (Wanibuchi et al., 2000; Nishikawa et al., 2002). To date, trimethylarsine has not been well tested for possible promotion or carcinogenicity.

DNA damage caused by oral administration of DMA(V) in both mice and rats suggested that DMA(V) was unusually genotoxic (Yamanaka et al., 1989; Brown et al., 1997; Kitchin and Ahmad, 2003). The first observations that a methylated trivalent arsenical (i) damages DNA directly (Ahmad et al., 2000, 2002; Mass et al., 2001), (ii) damages DNA indirectly (via iron released from ferritin) (Ahmad et al., 2000, 2002) and (iii) that this DNA damage was partly mediated by ROS species (Ahmad et al., 2002; Nesnow et al., 2002; Kitchin and Ahmad, 2003) were made by our research group at the US EPA.

4.6. Methylated trivalent arsenicals

Recent experimental evidence has suggested that methylated trivalent metabolites of inorganic arsenic not only occur in biological systems (Le et al., 2000; Sampayo-Reyes et al., 2000), but also have strong biological activity (Styblo et al., 1997; Lin et al., 1999; Petrick et al., 2001; Kitchin, 2001). Other studies showed an unusually high degree of biological activity of dimethylarsinic acid [DMA(V)] (Yamanaka et al., 1989; Yamamoto et al., 1995; Brown et al., 1997). In vivo, DMA(V) can be reduced to DMA(III) by enzymatic (glutathionine S-transferase omega form) and probably by nonenzymatic means. Methylated trivalent arsenicals are expected to act as soft acids, nucleophiles and they may also generate free radical/ROS species although it is not clear exactly how.

5. Conclusions

At present, chemical properties (soft acid and sulfhydryls, nucleophilicity, free radicals and ROS, atomic and molecular orbitals, electronegativity, valence state, enthalpies and ionization energies) can explain only part of the biological activity of arsenicals. What we really need to know are the important, causal targets (e.g. DNA, proteins, peptides, transcription factors, etc.) in arsenic's modes/mechanisms of action. Interaction studies between various arsenicals and the appropriate target biological molecules can then be expected to be much more fruitful. It may eventually be found that one or more of the chemical properties of arsenicals may enable several arsenic species to act through several modes of carcinogenic action at many stages of multistage carcinogenesis. Important future research questions include (1) which or how many of the known arsenic chemical species are causally active in toxicity/carcinogenesis and (2) is there a single predominant mode or mechanism of arsenic toxicity/carcinogenesis or do many modes/mechanisms participate. Identifying biological targets that respond at low tissue concentrations of arsenicals is a matter of great importance. A scientific and risk assessment issue of major importance is how should one mathematically extrapolate from the higher arsenic exposure concentrations associated with known toxicity and carcinogenicity to the lower arsenic concentration region below that in which human epidemiology and experimental biological experimentation can give useful data.

Acknowledgements

This work was partially supported by a grant from the National Research Council, Washington, DC, to a postdoctoral fellow (Dr Paul Andrewes) at U.S. Environmental Protection Agency, North Carolina. We thank Drs Mike Hughes and Russell D. Owen for reviewing this manuscript for in-house EPA clearance.

References

Ahmad, S., Kitchin, K.T., Cullen, W.R., 2000. Arsenic species that cause release of iron from ferritin and generation of activated oxygen. Arch. Biochem. Biophys., 382, 195–202.

Ahmad, S., Kitchin, K.T., Cullen, W.R., 2002. Plasmid DNA damage caused by methylated arsenicals, ascorbic acid and human liver ferritin. Toxicol. Lett., 133, 47–57.

Aposhian, H.V., 1989. Biochemical toxicology of arsenic. In: Hodgson, E., Bend, J., Philpot, R. (Eds), Reviews in Biochemical Toxicology, Vol. 10. Elsevier Science, New York, pp. 265–300.

Brown, J.L., Kitchin, K.T., George, M., 1997. Dimethylarsinic acid treatment alters six different rat biochemical parameters: relevance to arsenic carcinogenesis. Teratog. Carcinog. Mutagen., 17, 71–84.

Challenger, F., 1945. Biological methylation. Chem. Rev., 36, 315–361.

Cotton, F.A., Wilkinson, G., 1972. Advanced Inorganic Chemistry. Interscience Publishers, New York, pp. 394–402.

Delnomdedieu, M., Styblo, M., Thomas, D.J., 1995. Time dependence of accumulation and binding of inorganic and organic arsenic species in rabbit erythrocytes. Chem.-Biol. Interact., 98, 69–83.

Green, S., Walter, P., Kumar, V., Krust, A., Bornet, J., Argos, P., Chambon, P., 1986. Human oestrogen receptor cDNA: sequence, expression and homology to v-erb-A. Nature, 320, 134–139.

Hill, C.H., Matrone, G., 1970. Chemical parameters in the study of *in vivo* and *in vitro* interactions of transition elements. Fed. Proc., 29, 1474–1481.

Kitchin, K.T., 2001. Recent advances in arsenic carcinogenesis: modes of action, animal model system and methylated arsenic metabolites. Toxicol. Appl. Pharmacol., 172, 249–261.

Kitchin, K.T., Ahmad, S., 2003. Oxidative stress as a possible mode of action for arsenic. Toxicol. Lett., 137, 3–13.

Kitchin, K.T., Del Razo, L.M., Brown, J.L., Anderson, W.L., Kenyon, E.M., 1999. An integrated pharmacokinetic and pharmacodynamic study of arsenite action. 1. Heme oxygenase induction in rats. Teratog. Carcinog. Mutagen., 19, 385–402.

Le, X.C., Lu, X., Ma, M., Cullen, W.R., Aposhian, H.V., Zheng, B., 2000. Speciation of key arsenic metabolic intermediates in human urine. Anal. Chem., 72, 5172–5177.

Lin, S., Cullen, W.R., Thomas, D.J., 1999. Methylarsenicals and arsinothiols are potent inhibitors of mouse liver thioredoxin reductase. Chem. Res. Toxicol., 12, 924–930.

Mass, M.J., Tennant, A., Roop, B., Cullen, W., Styblo, M., Thomas, D., Kligerman, A., 2001. Methylated trivalent arsenic species are genotoxic. Chem. Res. Toxicol., 14, 355–361.

National Research Council, 1999. Mechanism of toxicity. Arsenic in Drinking Water. National Academy Press, Washington, DC, pp. 194–196.

Nesnow, S.N., Roop, B.C., Lambert, G., Kadiiska, M., Mason, R.P., Mass, M.J., 2002. DNA damage induced by trivalent arsenicals is mediated by reactive oxygen species. Chem. Res. Toxicol., 15, 1627–1634.

Nishikawa, T., Wanibuchi, H., Oawa, M., Kinoshita, A., Morimura, K., Hiroi, T., Funae, Y., Kishida, H., Nakae, D., Fukushima, S., 2002. Promoting effects on rat hepatocarcinogenesis of monomethylarsinic acid, dimethylarsinic acid and trimethylarsine oxide: evidence for a possible reactive oxygen species mechanism. Int. J. Cancer, 100, 136–139.

Pearson, R.G., 1963. Hard and soft acids and bases. J. Am. Chem. Soc., 85, 3533–3536.

Petrick, J.S., Jagadish, B., Mash, E.A., Aposhian, H.V., 2001. Monomethylarsonous acid (MMA(III)) and arsenite: LD(50) in hamsters and *in vitro* inhibition of pyruvate dehydrogenase. Chem. Res. Toxicol., 14, 651–656.

Sampayo-Reyes, A., Zakharyan, R.A., Healy, S.H., Aposhian, H.V., 2000. Monomethylarsonic acid reductase and monomethylarsonous acid in hamster tissue. Chem. Res. Toxicol., 13, 1181–1186.

Scott, N., Hatlelid, K.M., MacKenzie, N.E., Carter, D.E., 1993. Reactions of arsenic(III) and arsenic(V) species with glutathione. Chem. Res. Toxicol., 6, 102–106.

Stoica, A., Pentecost, E., Martin, M.B., 2000. Effects of arsenite on estrogen receptor-alpha expression and activity in MCF-7 breast cancer cells. Endocrinology, 141, 3595–3602.

Styblo, M., Serves, S.V., Cullen, W.R., Thomas, D.J., 1997. Comparative inhibition of yeast glutathione reductase by arsenicals and arsenothiols. Chem. Res. Toxicol., 10, 27–33.

Sunderman, W., 1987. Metal induction of heme oxygenase. Ann. N. Y. Acad. Sci., 514, 65–80.

Waalkes, M., Keefer, L., Diwan, B., 2000. Induction of proliferative lesions of the uterus, testes and liver in Swiss mice given repeated injections of sodium arsenate: possible estrogenic mode of action. Toxicol. Appl. Pharmacol., 166, 24–35.

Waalkes, M., Ward, J., Liu, J., Diwan, B., 2003. Transplacental carcinogenicity of inorganic arsenic in the drinking water: induction of hepatic, ovarian, pulmonary and adrenal tumors in mice. Toxicol. Appl. Pharmacol., 186, 7–17.

Wanibuchi, H., Wei, M., Yamamoto, S., Li, W., Fukushima, S., 2000. Carcinogenicity of an organic arsenical, dimethylarsenic acid and related arsenicals in rat urinary bladder. Proc. Am. Assoc. Cancer Res., 40 (Abstract No. 2309).

Winski, S.L., Carter, D., 1995. Interactions of rat blood cell sulfhydryls with arsenate and arsenite. J.Toxicol. Environ. Health, 46, 379–397.

Yamamoto, S., Konishi, Y., Matsuda, T., Murai, T., Shibata, M., Matsui-Yuasa, I., Otani, S., Kuroda, K., Endo, G., Fukushima, S., 1995. Cancer induction by an organic arsenic compound, dimethylarsinic acid (cacodylic acid), in F344/DuCrj rats after pretreatment with five carcinogens. Cancer Res., 55, 1271–1275.

Yamanaka, K., Hasegawa, A., Sawamura, R., Okada, S., 1989. Dimethylated arsenics induce DNA strand breaks in lung via the production of active oxygen in mice. Biochem. Biophys. Res. Commun., 165, 43–50.

Yamanaka, K., Hoshino, M., Okanoto, M., Sawamura, R., Hasegawa, A., Okada, S., 1990. Induction of DNA damage by dimethylarsine, a metabolite of inorganic arsenics, is for the major part likely due to its peroxyl radical. Biochem. Biophys. Res. Commun., 168, 58–64.

Chapter 27

Arsenic metabolism in hyperbilirubinemic rats: distribution and excretion in relation to transformation

Kazuo T. Suzuki, Takayuki Tomita, Yasumitsu Ogra
and Masayoshi Ohmichi

Abstract

Metabolic pathway for arsenic was studied in Eisai hyperbilirubinemic (EHB) rats to estimate the participation of hepato-enteric circulation during the transformation and distribution. Arsenite was injected intravenously into male EHB rats of 8 weeks of age at a single dose of 0.5 mg As/kg body weight, and time-dependent changes in the concentration of arsenic in organs and body fluids were determined. Arsenic disappeared from the bloodstream within 10 min after the injection, and appeared in the liver (40% of the dose), kidneys, lungs, spleen, testes, skin (in total 15%) and urine (10%). The rest (35%) was not detected in any of these organs/tissues/body fluids. However, arsenic accumulating in the erythrocytes amounted to 90% of the dose after 6 h, indicating that the hepato-enteric circulation can explain only a half of the missing 70% of the dose in normal rats in our previous study. Although the arsenite once distributed in the liver was redistributed in the erythrocytes by 1 h, it took 6 h for the missing 35% to reappear in the erythrocytes, suggesting that the missing arsenic be transformed more slowly in the hidden organ. Although the concentration was low, the whole arsenic in the muscle was calculated to amount approximately to the missing one (31%). Arsenite injected intravenously into normal rats was estimated to distribute at first in the liver (40%), muscle (35%), other organs (15%) and urine (10%) in its original form in an organ-specific manner, and then redistributed in the erythrocytes after being transformed to dimethylated arsenic with different reduction/methylation efficiency, the liver being most active in the transformation and in the excretion into the hepato-enteric circulation.

Keywords: arsenic, arsenite, dimethylated arsenic, hyperbilirubinemic rat, rat, metabolism of arsenic

Abbreviations:

iAs^V, arsenate; iAs^{III}, arsenite; AsB, arsenobetaine; DMA, dimethylated arsenical; DMA^{III}, dimethylarsinous acid; DMA^V, dimethylarsinic acid; MMA, monomethylated arsenical; MMA^{III}, monomethylarsonous acid; MMA^V, monomethylarsonic acid; TMA, trimethylated arsenical; HPLC, high performance liquid chromatography; ICP, MS, inductively coupled argon plasma mass spectrometry; GSH, glutathione; EHB rats, Eisai hyperbilirubinemic rats; MRP2/cMOAT, multidrug resistance protein 2/canalicular multi-specific organic anion transporter.

1. Introduction

Diverse chemical species of arsenic are taken up by humans through dietary foods and water. However, the most important chemical species from the viewpoint of the toxicity and exposure to humans are arsenite (iAs^{III}) and arsenate (iAs^V) (National Research Council Report, 1999). Because the inorganic arsenic absorbed in the body is excreted in the urine mostly in the form of dimethylated arsenic (DMA), the toxicity of arsenic must be caused by one of the metabolites during its transformation. It is of our interest to reveal the toxic form of arsenic (proximate toxic form), and to explain the metabolic pathway including the way in which the arsenic is delivered to the target organs.

Arsenite absorbed by the body is believed to be transferred to the liver, and then transformed by consecutive methylation and reduction reactions to DMA (Vahter and Marafante, 1983; Vahter et al., 1984a,b; Fischer et al., 1985; Goyer, 1996; Vahter, 1999). Namely, iAs^{III} is oxidatively methylated to monomethylarsonic acid (MMA^V), and then reduced to monomethylarsonous acid (MMA^{III}) for further oxidative methylation to dimethylarsinic acid (DMA^V) (Cullen et al., 1984; Aposhian et al., 2000; Le et al., 2000a, b; Mandal et al., 2001). In rats, DMA^V is further reduced efficiently to dimethylarsinous acid (DMA^{III}), and then excreted into the bloodstream, where DMA^{III} is sequestered selectively by erythrocytes, resulting in the selective accumulation of arsenic in erythrocytes in rats (Shiobara et al., 2001).

During the metabolic transformation of arsenic of iAs^{III} origin injected intravenously (iv) into rats, substantial amount of the arsenic was shown to disappear from both the bloodstream and major organs, and then reappear with time in the erythrocytes (Suzuki et al., 2001). The arsenic disappeared from the bloodstream and major organs was assumed to be excreted into the bile, i.e. get into the hepato-enteric circulation in the form conjugated with glutathione (GSH) (iAs^{III}(GS)$_3$) during the metabolic transformation. However, further examinations in the balance study such as the absorption rate of free and GSH-conjugated forms of iAs^{III} from the intestine raised a question to explain the hidden arsenic only by the hepato-enteric circulation.

The balance study of arsenic was carried out using Eisai hyperbilirubinemic (EHB) rats, in which the excretion of anionic substances through the multidrug resistance protein 2/canalicular multi-specific organic anion transporter (MRP2/cMOAT) is not active owing to genetic defection of the expression of the transporter (Gregus et al., 2000; Kala et al., 2000; Leslie et al., 2001; Cherrington et al., 2002). Although more arsenic was distributed in the liver of EHB rats than of normal SD rats, a substantial amount of arsenic was not detected again in the major organs even in EHB rats where arsenic is not excreted into bile. In response to the inhibition of biliary excretion, the arsenic accumulating in the erythrocytes was much higher in EHB rats than in SD rats after 6 h post-injection.

The arsenic disappeared from the bloodstream and hidden in organs was traced by determining arsenic concentrations in most organs/tissues and body fluids. The recovery of arsenic present in the body immediately after the injection was calculated to be 90% on the assumption of some data such as percentages of blood volume (7.0%), skin (7.0%) and muscle (50%) relative to body weight. The missing arsenic was explained mostly by the distribution in the muscle though the concentration (μg/g organ) was 100 times lower than that in the liver. Based on these results, the distribution of arsenic among organs/tissues/ body fluids was estimated, and then the chemical forms during the transfer to each target sites were discussed.

2. Experimental procedures

2.1. Reagents

All reagents were of analytical grade. Milli-Q SP water (Millipore) was used throughout. Sodium arsenite (NaAsO$_2$) (iAs^{III}), dimethylarsinic acid [(CH$_3$)$_2$AsO(OH)] (DMA^V) and sulfuric and nitric acids were purchased from Wako Pure Chemical Industries, Ltd. (Osaka, Japan), while Trizma® HCl and Base from Sigma (St. Louis, MO, USA). The arsenic standard solution (1000 μg/ml) for ICP MS was purchased from SPEX CentiPrep (Metuchen, NJ, USA). Arsenobetaine (AsB) was a gift from Professor T. Kaise (Tokyo University of Pharmacy and Life Science, Tokyo, Japan). Stock solutions of all arsenic compounds (10 mmol/l) were prepared from the respective standard compounds, and stored in the dark at 0° C. Diluted standard solutions for analysis were prepared daily prior to use.

2.2. Animal experiments

Male EHB and SD rats were purchased at 7 weeks of age from a breeder (Japan SLC Co., Hamamatsu, Japan). They were housed in a humidity-controlled room, maintained at 22–25° C with a 12 h light–dark cycle. The animals were fed a commercial diet (CE-2; Clea Japan Co.) and tap water *ad libitum*. Following a 1-week acclimation period, the animals, at 8 weeks old and weighing 220–270 g, were used for experiments.

iAs^{III} was dissolved in saline (Otsuka Pharmaceutical Co., Ltd., Naruto, Japan) immediately before injection. Arsenic solutions were adjusted to 1.0 ml/kg body weight. Control rats were injected with the same volume of saline as that in the experimental group. Rats were sacrificed at 10 and 30 min, 1, 6 and 12 h after the injection.

Bile collection was performed according to Gregus et al. (2000). The rats were anesthetized with sodium pentobarbital (Dainippon Pharmaceutical Co., Ltd., Osaka), and then the bile duct was cannulated with a polyethylene tube (o.d. 0.61 mm × i.d. 0.28 mm., Natume, Tokyo).

2.3. Whole body perfusion

As arsenic accumulates preferentially in erythrocytes in rats (Charbonneau et al., 1980; Lerman and Clarkson, 1983; Vahter et al., 1984a,b; Marafante and Vahter, 1987), the concentrations of arsenic in organs and tissues have to be determined after completely removing blood (erythrocytes) by whole body perfusion. In the present study, perfusion was performed according to the method developed by Professor H. Yamauchi, St. Marianna University School of Medicine as described in our previous communication (Shiobara et al., 2001).

2.4. Instruments

Speciation and quantification experiments for arsenic were carried out using the HPLC–ICP MS and ICP MS systems as reported elsewhere (Mandal et al., 2001).

2.5. Analytical procedures

The concentrations of arsenic in whole blood, liver, kidneys, spleen, testes, lungs, muscle, urine and bile were determined by ICP MS after wet-ashing with nitric acid and hydrogen peroxide. A 20 µl aliquot of 6 h urine was also applied to an anion exchange column (Shodex Asahipak ES-502N 7C, Showa Denko, Tokyo), and the column was eluted with 15 mM citric acid (pH 2.0 adjusted with nitric acid at 25° C) at the flow rate of 1.0 ml/min (Suzuki et al., 2001).

Reduction of standard DMA^V compound was carried out to study trimethylated arsenical (DMA^{III}) using metabisulfite and thiosulfate reagents as described by Reay and Asher (1977).

3. Results

The concentration of total bilirubin in the plasma of EHB rats was much higher than that of normal SD rats before and after an injection of arsenite, as shown in Figure 1, suggesting that bilirubin was not secreted into the bile, and that the animal is hyperbilirubinemic.

In response to inhibition of the secretion of bilirubin into the bile, arsenic was not excreted into the bile in EHB rats, as shown in Figure 2A. On the other hand, the concentration of arsenic in the bile of SD rats was increased immediately after the injection of arsenic, and the excretion amounted to 20% of the dose within 2 h, as shown in Figure 2B. At the same time, the present results indicated that arsenic is excreted into the bile only through the MRP2/cMOAT transporter at least under the present dose condition.

As arsenic is not excreted into the bile in EHB rats, it was supposed that arsenic should be retained more in the liver of EHB rats, which might cause a higher toxicity in EHB rats than in normal rats. Therefore, the dose was lowered in the present experiment, i.e. 0.5 mg As/kg body weight in the present study compared with 2.0 mg As/kg body weight in normal Wistar rats in the previous study (Suzuki et al., 2001).

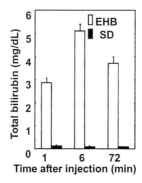

Figure 1. Changes in the concentration of total bililubin in the plasma of EHB and SD rats. Male EHB and SD rats (3 rats/group, 8 weeks of age) were injected arsenite intravenously at a single dose of 0.5 mg As/kg body weight, and were blooded under sodium pentobarbital anesthesia 10, 60 and 720 min later. The concentrations of total bilirubin are expressed as means ± SE.

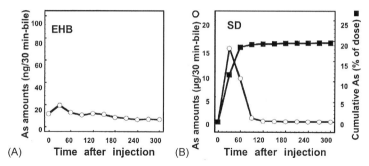

Figure 2. Changes in the amount of arsenic in the bile with time after a single intravenous injection of arsenite into EHB and SD rats. Male EHB and SD rats (3 rats/group, 8 weeks of age) were cannulated for the collection of normal bile for 30 min without injecting arsenite under sodium pentobarbital anesthesia, and then, injected arsenite intravenously at a single dose of 0.5 mg As/kg body weight. Thirty-minute bile was collected for up to 300 min after the injection. The amount of arsenic in 30 min bile and its recovery relative to the dose were plotted against time after the injection.

The arsenic injected in the form of arsenite was expected to distribute more in the organs/tissues and to be excreted more into the urine in EHB rats than in normal rats as a result of inhibition of the excretion into the bile. The distribution of the arsenic in the organs/tissues and bloodstream was determined to reveal the material balance of arsenic in the body, as shown in Figure 3. The concentration of arsenic in the whole blood (Fig. 3A) indicates that arsenite injected into the plasma was cleared rapidly from the bloodstream within 10 min after the injection, as also observed in normal rats with the higher dose (Suzuki et al., 2001). Then, the arsenic concentration in the bloodstream started to increase with time after 30 min, amounting up to approximately 90% of the dose after 6 h post-injection, as shown in Figure 3A. The arsenic accumulating in the bloodstream with time was primarily present in the erythrocytes but not in the plasma, as also observed previously in normal rats (Aposhian, 1997; Shiobara et al., 2001).

The concentration of arsenic in the liver (Fig. 3B) was highest at the earliest time point of 10 min, and then, decreased with time after the injection. The maximum concentration at 10 min amounted only to approximately 40% of the dose despite of the inhibition of excretion into the bile in EHB rats.

The concentration of arsenic in the kidneys (Fig. 3C) was also highest at 10 min after the injection, and decreased thereafter similarly to that in the liver. The maximum concentration in the kidneys amounted to approximately 7% of the dose at 10 min.

The distributions of arsenic in the lungs (Fig. 3D), spleen (Fig. 3E) and testes (Fig. 3F) were lower than those in the liver and kidneys. The concentration of arsenic in the lungs and testes was highest at 10 min, and then decreased with time after the injection as in the case in the liver and kidneys. However, the concentration of arsenic in the spleen was at the same level as those in the liver and kidneys, and it was maintained at a constant level throughout, as shown in Figure 3E, suggesting a different role for spleen from other organs.

The concentration of arsenic in the skin (Fig. 3G) was low and at the same level as those in the lungs and testes, though it decreased with time more slowly than in the liver and

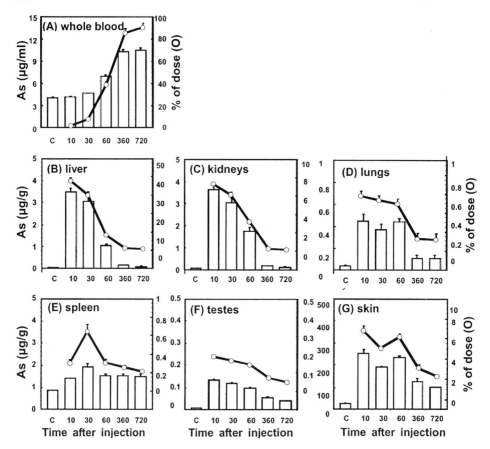

Figure 3. Changes in the concentration and recovery of arsenic in the blood and organs/tissues of EHB rats after a single intravenous injection of arsenite. Male EHB rats (3 rats/group, 8 weeks of age) were injected arsenite intravenously at a single dose of 0.5 mg As/kg body weight, and were sacrificed 10, 30, 60, 360 and 720 min later. The concentrations of arsenic in the whole blood (A), liver (B), kidneys (C), lungs (D), spleen (E), testes (F) and skin (G) were determined after wet-ashing. The concentration (open column) and recovery as to the dose (open circle) were plotted against time after the injection. Whole blood was assumed to be 7% of the body weight. "C" in each panel denotes the corresponding control.

kidneys. The amount of arsenic retained in the skin was greater than 1% of the dose at 6 h after the injection, and it was higher than those in the other five organs examined in the present experiment, suggesting that arsenic is metabolized more slowly or retained more efficiently in the skin than in other organs.

The arsenic recovered in the whole blood and the six organs/tissues amounted only to 50–57% of the dose during 10–30 min after the injection, as shown in Figure 5. However, the distribution of arsenic in these organs changed dramatically with time, the distribution in the liver being decreased and that in the whole blood increased with time. This tendency was continued during the observation period. The arsenic in the bloodstream was present mostly in the erythrocytes except for a short period after the injection of arsenite.

Arsenic was excreted more efficiently into the urine before 6 h after the injection, and then, more slowly thereafter, as shown in Table 1. The amount of urinary arsenic was significantly higher in EHB rats than in SD rats, reflecting the higher distribution to the liver and other organs/tissues owing to the inhibition of the biliary excretion in EHB rats and to the higher excretion into urine.

The chemical species of arsenic in the urine was determined on an anion exchange column by the HPLC–ICP MS method, as shown in Figure 4. Arsenic was excreted into the urine mostly in the two forms during the first 6 h after the injection, i.e. arsenite and DMA^V accompanied by a small amount of MMA^V. The higher urinary excretion of arsenite at an early time suggests a direct excretion of arsenite immediately after the injection. However, in 6–12 h urine, arsenic was detected mostly in the form of DMA^V accompanied by two minor peaks of arsenite and MMA^V. Arsenobetaine (AsB) of diet origin was also detected. The distribution profile of arsenic in the urine was not different between EHB and SD rats, as shown in Figure 4(ii)–(v).

The amounts of arsenic recovered in the whole blood and six organs/tissues were summed up and plotted against time after the injection, as shown in Figure 5. The summed amount was 50–57% of the dose before 1 h after the injection, and then, was around 90% after 6 h. As approximately 10% of the dose was recovered in the urine within 12 h, mostly within the first 6 h, as shown in Table 1, the material balance of the injected arsenic in the form of arsenite can be estimated. Although the arsenic appeared in the kidneys at 1 h after the injection may be excreted into the urine and it was counted doubly, the material balance in Figure 5 indicates that approximately 35% of the dose distributed in organs other than the six organs/tissues was examined in the present study.

Then, the missing 35% of the dose was traced with several prerequisites: (1) the arsenic disappears from the plasma within several minutes in the form of arsenite; (2) the amount is 35% of the dose, an equivalent amount to that distributed in the liver; (3) the missing arsenic accumulates finally in the erythrocytes after being transformed to DMA^{III};

Table 1. Changes in the amount of arsenic in the urine with time after a single intravenous injection of arsenite into EHB and SD rats.

	As (μg/6 h)	As (% of dose)
EHB rat		
0–6 h	14.80 ± 0.33	9.26 ± 0.23
6–12 h	1.31 ± 0.37	–
0–12 h	16.11 ± 0.55	10.06 ± 0.17
SD rat		
0–6 h	11.59 ± 0.31	7.45 ± 0.21
6–12 h	1.39 ± 0.16	–
0–12 h	12.99 ± 0.40	8.07 ± 0.19

Male EHB and SD rats (3 rats/group, 8 weeks of age) were housed individually in a metabolic cage for the collection of urine after the intravenous injection of arsenite at a single dose of 0.5 mg As/kg body weight. Six-hour urine was collected for up 12 h after the injection. The amount of arsenic in 6 h urine and its recovery relative to the dose were expressed as means ± SE.

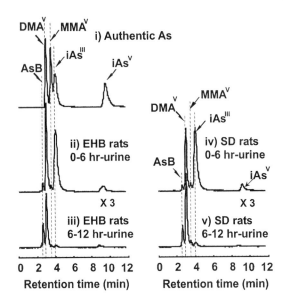

Figure 4. Changes in the distribution of arsenic in the urine of EHB and SD rats on an anion exchange column by the HPLC–ICP MS method after a single intravenous injection of arsenite. The 6 h urine collected from the EHB (ii and iii) and SD rats (iv and v) in Figure 3 was subjected to the HPLC–ICP MS analysis on an anion exchange column (Shodex Asahipak ES-502N 7C). The intensity of arsenic for the profile of the first 6 h urine is reduced threefold less than that of the second 6 h urine. The standard arsenicals for arsenobetaine (AsB), DMA^V, MMA^V, arsenite (iAs^{III}) and arsenate (iAs^V) are drawn in the panel (i).

(4) the transformation takes place in the organ/tissue more slowly than in the liver. As a candidate organ/tissue, muscle was nominated and the concentration of arsenic was determined in a separate experiment with normal SD rats. Although the concentration was low (200 ± 15 ng/g, mean ± SD), the amount of arsenic in the whole muscle was estimated to be 31.2 ± 2.3% of the dose on the assumption that muscle weight is 50% of the body weight.

4. Discussion

Inorganic forms of arsenic of foods and drinks are of more concern from the toxicological viewpoint than organic forms of arsenic, as it is well documented for the on-going worldwide health problems caused by drinking of arsenic-contaminated groundwater (National Research Council Report, 1999; WHO, 2001). Arsenite and arsenate are transformed to methylated metabolites by consecutive reduction and oxidative methylation reactions, and the metabolic pathway leading to the methylated metabolites has been considered to be the detoxification process for arsenic (Tam et al., 1979; Fischer et al., 1985; Yamauchi and Fowler, 1994; Goyer, 1996; Vahter, 1999, 2000). However, methylated metabolites, especially the trivalent forms, MMA^{III} and DMA^{III}, have been proposed to be the toxic forms in the metabolic pathway for arsenic administered in the inorganic form (Styblo et al., 1997, 2000; Goering et al., 1999; Lin et al., 1999; Petrick

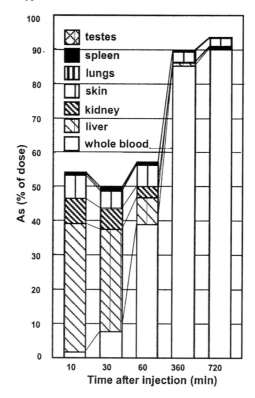

Figure 5. Changes in the recovery of arsenic summed up in the six organs/tissues and whole blood of EHB rats after a single intravenous injection of arsenite. The mean amounts of arsenic in the whole blood (which is assumed to be 7% of the body weight), liver, kidneys, skin, lungs, spleen and testes in Figure 3 were summed up, and the recovery relative to the dose was plotted against time after the injection.

et al., 2000; Del Razo et al., 2001). Nevertheless, it is not known where the toxic forms of arsenic are produced and/or how those are transferred to the target sites.

Our interest is to correlate the toxicity of arsenic with its chemical forms or to find the proximate toxic arsenicals, as several groups have also tried with similar intention (Vahter, 1988, 1994; Mitchell et al., 2000; Shibata and Morita, 2000; Kitchin, 2001). As the first approach, we tried to reveal how arsenic is distributed to the target sites during the transformation leading to the excretion form, DMA in rats. We observed that a substantial amount of arsenic gets into the hepato-enteric circulation when arsenite was administered intravenously (Suzuki et al., 2001). However, further investigation on the metabolic balance of arsenic raised a doubt to explain the hidden arsenic immediately after an intravenous injection only by the hepato-enteric circulation. Namely, the role of arsenic in the hepato-enteric circulation may be overestimated and the hidden arsenic immediately after the injection from the bloodstream may not be explained simply by the arsenic circulating in the intestine.

The present results with EHB rats indicate that arsenite distributes 40% in the liver, 15% in total in the kidneys, lungs, spleen, testes and skin, and 10% in the urine. However,

35% of the dose was not detected in any of these compartments. Nevertheless, 90% of the dose was recovered in the erythrocytes after 6 h, suggesting that arsenic be hidden from these compartments immediately after the injection.

In normal rats with the dose of 2.0 mg As/kg body weight, the material balance indicated that 70% of the dose was hidden during the first 1 h, and it was explained by the hepato-enteric circulation, approximately 15% of the dose being estimated not to be recovered from the hepato-enteric circulation, i.e. excreted into feces (Suzuki et al., 2001). On the other hand, the present material balance in EHB rats with a lower dose of 0.5 mg As/kg body weight suggests that the missing arsenic was 35% of the dose. From the balance sheets for the previous normal rats and the present EHB rats, the material balance for normal rats can be estimated, as shown in Figure 6. Because the dose and route of administration alter the distribution, the balance estimated in Figure 6 would change depending on the experimental conditions such as dose even when undertaken with same rats.

Although the material balance depends on a dose, the missing arsenic can be explained by that getting into the hepato-enteric circulation (20% out of 40% of the dose in the liver, 10% of the dose being excreted into feces) and distributing in the muscle (35% of the dose) for normal rats with an administration of 0.5 mg As/kg body weight. The arsenic distributed in the muscle is assumed to be transformed more slowly to DMA than that in the liver, and then excreted into the plasma and taken up by the erythrocytes in rats.

The chemical form of arsenic distributed to each organ/tissue is thought to be the original arsenite because of the immediate distribution from the plasma after the injection. On the other hand, the second distribution to the erythrocytes is thought to be in the form of DMA because of the preferential distribution in rats (Shiobara et al., 2001), as proposed

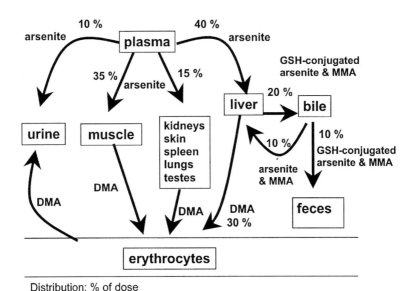

Distribution: % of dose

Figure 6.   Schematic presentation for the distribution of arsenic injected intravenously into normal rats.

in Figure 6. However, alternatively, the arsenic distributed to each organ/tissue may be redistributed to the liver as the second distribution, and then transformed and distributed to the erythrocytes. In animals with low reduction/methylation capacity, the second distribution of arsenic may be in other arsenical forms from DMA because all six possible metabolites were detected in the urine of arsenic-polluted population (Mandal et al., 2001). Although the metabolic pathway for arsenic is quite different in rats from those in other animals owing to the high reduction/methylation capacity and preferential accumulation in erythrocytes, the material balance in rats suggests that arsenic can be delivered to each organ/tissue in the form of arsenite in an organ/tissue-specific manner. The arsenite is, then, transformed to DMA or to the intermediate metabolites in each organ/tissue. The metabolites in each organ/tissue are effluxed into the bloodstream for the second distribution. In rats, the arsenite distributed in any organ/tissue seems to be transformed to the final metabolite DMA, and redistributed in the erythrocytes. On the other hand, in other animals, arsenite is thought to be transformed less effectively and redistributed in the forms of six possible metabolites to other organs or directly to urine.

As for the arsenic not detected in the six organs and bloodstream immediately after the injection, it was proposed to be distributed in the muscle from the following considerations: (1) the arsenic disappears from the plasma within several minutes in the form of arsenite; (2) the amount is 35% of the dose; (3) the arsenic is recovered finally in erythrocytes after being transformed to DMA^{III}. These characteristics suggest that arsenite is taken up by a reserver of big-sized organs/tissues comparable to liver, where arsenite is transformed slowly to DMA^{III}, and then effluxed into the bloodstream. As a possible candidate, muscle and bone can be mentioned, especially the former tissue, and the concentration of arsenic in the muscle was determined in a separate experiment. It amounted to $31.2 \pm 2.3\%$ of the dose on the assumption that muscle is 50% of body weight.

Arsenite is supposed to be transformed more actively to DMA in the liver than in other organs, and then transferred to erythrocytes in rats (Tam et al., 1979; Lerman and Clarkson, 1983; Shiobara et al., 2001; Suzuki et al., 2001). In other words, arsenite is thought to be taken up less efficiently by other organs/tissues, as estimated from the lower concentration of arsenic per organ weight. Furthermore, the efficiency in the transformation to methylated metabolites is also considered to be much lower in muscle and other organs/tissues than in liver, as estimated from the slow accumulation of arsenic in the erythrocytes after the disappearance of arsenic from the liver. Less efficient uptake and less active transformation in muscle than in liver explain the hidden arsenic and the reappearance of arsenic in the erythrocytes in rats after the intravenous injection of arsenite.

The present study pointed out the role of organs/tissues that are not efficient for uptake of arsenic and not active in the transformation compared with liver in the metabolism of arsenic. It was also suggested that arsenite may be transferred directly to organs/tissues in its form, transformed to the methylated metabolites in an organ/tissue-specific manner, and then the arsenic is effluxed into the bloodstream for the second distribution. Alternatively, arsenite may be redistributed without being transformed in muscle and other organs and transferred to the liver for the transformation. In the former case, the arsenite transferred to the target sites is transformed and its metabolites cause the toxicity, while in the latter case the transformed arsenic is transferred to the target sites and causes the toxicity. Although the present results suggest the former possibility in rats, arsenic may be distributed in various organs/tissues and transformed to various intermediates, and then redistributed in a manner specific to animal species.

References

Aposhian, H.V., 1997. Enzymatic methylation of arsenic species and other new approaches to arsenic toxicity. Annu. Rev. Pharmacol. Toxicol., 37, 397–419.

Aposhian, H.V., Gurzau, E.S., Le, X.R., Gurau, A., Haly, S.M., Lu, X., Ma, M., Yip, L., Zakharyan, R.A., Mariorino, R.M., Dart, R.C., Tircus, M.G., Gonzalez-Ramirez, D., Morgam, D.L., Avram, D., Aposhian, M.M., 2000. Occurrence of monomethylarsonous acid in urine of humans exposed to inorganic arsenic. Chem. Res. Toxicol., 13, 693–697.

Charbonneau, S.M., Hollins, J.G., Tam, G.K., Bryce, F., Ridgeway, J.M., Willes, R.F., 1980. Whole-body retention, excretion and metabolism of [^{74}As]arsenic acid in the hamster. Toxicol. Lett., 5, 175–182.

Cherrington, N.J., Hartley, D.P., Li, N., Johnson, D.R., Klaassen, C.D., 2002. Organ distribution of multidrug resistance proteins 1, 2, and 3 (Mrp1, 2, and 3) mRNA and hepatic induction of Mrp3 by constitutive androstane receptor activators in rats. J. Pharmacol. Exp. Ther., 300, 97–104.

Cullen, W.R., Mcbride, B.C., Reglinski, J., 1984. The reaction of methylarsenicals with thiols: some biological implications. J. Inorg. Biochem., 21, 179–194.

Del Razo, L.M., Styblo, M., Cullen, W.R., Thomas, D.J., 2001. Determination of trivalent methylated arsenicals in biological matrices. Toxicol. Appl. Pharmacol., 174, 282–293.

Fischer, A.B., Buchet, J.P., Lauwerys, R.R., 1985. Arsenic uptake, cytotoxicity and detoxification studied in mammalian cells in culture. Arch. Toxicol., 57, 168–172.

Goering, P.L., Aposhian, H.V., Mass, M.J., Cebrian, M., Beck, B.D., Waalkes, M.P., 1999. The enigma of arsenic carcinogenesis: role of metabolism. Toxicol. Sci., 49, 5–14.

Goyer, R.A., 1996. Toxic effects of metals Casarett and Doull's Toxicology. In: Klaassen, C.D. (Ed.), The Basic Science of Poisons. McGraw-Hill, New York, pp. 696–697.

Gregus, Z., Gyurasics, A., Csanaky, I., 2000. Biliary and urinary excretion of inorganic arsenic: monomethylarsonous acid as a major biliary metabolite in rats. Toxicol. Sci., 56, 18–25.

Kala, S.V., Neely, M.W., Kala, G., Prater, C.I., Atwood, D.W., Rice, J.S., Lieberman, M.W., 2000. The MRP2/cMOAT transporter and arsenic–glutathione complex formation are required for biliary excretion of arsenic. J. Biol. Chem., 275, 33404–33408.

Kitchin, K.Y., 2001. Recent advances in arsenic carcinogenesis: modes of action, animal model systems, and methylated arsenic metabolites. Toxicol. Appl. Phamacol., 172, 249–261.

Le, X.C., Lu, X., Ma, M., Cullen, W.R., Aposhian, H.V., Zheng, B., 2000a. Speciation of key arsenic metabolic intermediates in human urine. Anal. Chem., 72, 5172–5177.

Le, X.C., Ma, M., Cullen, W.R., Aposhian, H.V., Lu, X., Zheng, B., 2000b. Determination of monomethylarsonous acid, a key arsenic methylation intermediate, in human urine. Environ. Health Perspect., 108, 1015–1018.

Lerman, S., Clarkson, T.W., 1983. The metabolism of arsenite and arsenate by the rat. Fundam. Appl. Toxicol., 3, 309–314.

Leslie, E.M., Deeley, R.G., Cole, S.P.C., 2001. Toxicological relevance of the multidrug resistance protein 1, MRP1(ABCC1) and related transporters. Toxicology, 167, 3–23.

Lin, S., Cullen, W.R., Thomas, D.J., 1999. Methylarsenicals and arsinothiols are potent inhibitors of mouse liver thioredoxin reductase. Chem. Res. Toxicol., 12, 924–930.

Mandal, B.K., Ogra, Y., Suzuki, K.T., 2001. Identification of dimethylarsinous and monomethylarsonous acids in human urine of the arsenic affected areas in West Bengal, India. Chem. Res. Toxicol., 14, 371–378.

Marafante, E., Vahter, M., 1987. Solubility, retention, and metabolism of intratracheally and orally administered inorganic arsenic compounds in the hamster. Environ. Res., 42, 72–82.

Mitchell, R.D., Ayala-Fierro, F., Carter, D.E., 2000. Systemic indicators of inorganic arsenic toxicity in four animal species. J. Toxicol. Environ. Health, 59, 119–134.

National Research Council Report, 1999. Arsenic in Drinking Water. National Academy Press, Washington, DC.

Petrick, J.S., Ayala-Fierro, F., Cullen, W.R., Carter, D.E., Aposhian, H.V., 2000. Monomethylarsonous acid (MMA^{III}) is more toxic than arsenite in Chang human hepatocytes. Toxicol. Appl. Pharmacol., 163, 203–207.

Reay, P.F., Asher, C.J., 1977. Preparation and purification of ^{74}As-labeled arsenate and arsenite for use in biological experiments. Anal. Biochem., 78, 557–560.

Shibata, Y., Morita, M., 2000. Chemical forms of arsenic in the environment. Biomed. Res. Trace Elements, 11, 1–24.

Shiobara, Y., Ogra, Y., Suzuki, K.T., 2001. Animal species difference in the uptake of dimethylarsinous acid (DMA^{III}) by red blood cells. Chem. Res. Toxicol., 14, 1446–1452.

Styblo, M., Serves, S.V., Cullen, W.R., Thomas, D.J., 1997. Comparative inhibition of yeast glutathione reductase by arsenicals and arsenothiols. Chem. Res. Toxicol., 10, 27–33.

Styblo, M., Del Razo, L.M., Vega, L., Germolec, D.R., LeCluyse, E.L., Hamilton, G.A., Reed, W., Wang, C., Cullen, W.R., Thomas, D.J., 2000. Comparative toxicity of trivalent and pentavalent inorganic and methylated arsenicals in rat and human cells. Arch. Toxicol., 74, 289–299.

Suzuki, K.T., Tomita, T., Ogra, Y., Ohmichi, M., 2001. Glutathione-conjugated arsenicals in the potential hepato-enteric circulation in rats. Chem. Res. Toxicol., 14, 1604–1611.

Tam, G.K.H., Charbonneau, S.M., Bryce, F., Pomroy, C., Sandi, E., 1979. Metabolism of inorganic arsenite (^{74}As) in humans following oral ingestion. Toxicol. Appl. Phamacol., 50, 319–322.

Vahter, M., 1988. Arsenic. In: Clarkson, T.W., et al. (Eds), Biological Monitoring of Toxic Metals. Plenum Press, New York, pp. 303–321.

Vahter, M., 1994. Species differences in the metabolism of arsenic compounds. Appl. Organomet. Chem., 8, 175–182.

Vahter, M., 1999. Methylation of inorganic arsenic in different mammalian species and population groups. Sci. Prog., 82, 69–88.

Vahter, M., 2000. Genetic polymorphism in the biotransformation of inorganic arsenic and its role in toxicity. Toxicol. Lett., 112–113, 209–217.

Vahter, M., Marafante, E., 1983. Intracellular interaction and metabolic fate of arsenite and arsenate in mice and rabbits. Chem. Biol. Interact., 47, 29–44.

Vahter, M., Marafante, E., Dencker, L., 1984a. Tissue distribution and retention of ^{74}As-dimethylarsinic acid in mice and rats. Arch. Environ. Contam. Toxicol., 13, 259–264.

Vahter, M., Marafante, E., Dencker, L., 1984b. Metabolism of arsenobetaine in mice, rats and rabbits. Sci. Total Environ., 30, 197–211.

WHO, 2001. Arsenic Compounds. Environmental Health Criteria 224, 2nd edn. World Health Organisation, Geneva.

Yamauchi, H., Fowler, B.A., 1994. Toxicity and metabolism of inorganic and methylated arsenicals. In: Nriagu, J.O. (Ed.), Arsenic in the Environment, Part II: Human Health and Ecosystem Effects. Wiley, New York, pp. 33–43.

Arsenic Exposure and Health Effects V
W.R. Chappell, C.O. Abernathy, R.L. Calderon and D.J. Thomas, editors
Published by Elsevier B.V.

Chapter 28

Incorporating mechanistic insights in a PBPK model for arsenic

Elaina M. Kenyon, Michael F. Hughes, Marina V. Evans, Miroslav Styblo, Luz Maria Del Razo and Michael Easterling

Abstract

A physiologically based pharmacokinetic (PBPK) model for arsenic provides an integrated framework for addressing issues related to risk assessment, as well as being a tool for hypothesis testing and experimental design. This is because a PBPK model defines the relationship between external exposure and an internal measure of (biologically effective) dose. The arsenic PBPK model is necessarily complex because of the existence of multiple biologically active forms of arsenic and uncertainty concerning their roles in producing toxic effects. Functionally, for arsenic, this requires a minimum of four submodels linked by reduction/oxidation and methylation as well as incorporation of urinary excretion for each metabolite and tissue binding for certain trivalent forms. This is necessary because the availability of any particular arsenical for tissue interactions is a balance between rates of oxidation, reduction, methylation, binding, and excretion. Our current PBPK model structure will be reviewed and unique mechanistic features, together with their experimental basis, will be highlighted. These include dimethylarsinic acid accumulation in lung, inhibition of the second methylation step by arsenite, and assumptions regarding transport and partitioning into tissues. Our modeling experiments and sensitivity analysis have also suggested several important lines of research. Critical information to gather include (1) data on differences in methylation capacity among the human population, and (2) reliable quantitative data collected at the level of individual study participants on arsenic exposure from all sources matched with blood and urinary levels of arsenic metabolites.

Keywords: arsenic, PBPK model, pharmacokinetics, risk assessment

1. Introduction

Internal dosimetry is the bridge to understanding the quantitative relationship between arsenic exposure and health effects. Use of a physiologically based pharmacokinetic (PBPK) model in this context allows one to define the relationship between exposure and an internal measure of biologically effective dose in both experimental animals and humans. Use of PBPK models can account for (1) nonlinear uptake, metabolism and clearance, (2) toxicity associated with products of metabolism rather than parent chemical only and (3) tissue interactions. The underlying assumption here is tissue dose equivalence, i.e. that health effects occur as a result of tissue exposure to the toxic form(s) of the chemical and that equivalent effects will be observed at equal tissue exposure when measured by the appropriate dose metric (Krishnan and Andersen, 1994).

Constructive use of PBPK models in the risk assessment process requires some consensus concerning mode(s) of action and the form of the chemical responsible for the effect of greatest toxicological concern. In this paper, some of the unique challenges encountered in arsenic PBPK model development will be discussed and two specific examples of how pharmacokinetic and mechanistic information have been incorporated in our arsenic model development efforts will be described.

2. Model development issues

Construction of an arsenic PBPK model useful for risk assessment requires one to match the toxic effect of concern with the arsenical species or metabolite responsible for that effect in order to select an appropriate dose metric. Arsenic provides some unique challenges in this respect. Up until approximately a decade ago, trivalent inorganic arsenic (arsenite, AsIII) was believed to be the form of arsenic responsible for the toxic effects observed following exposure to arsenic in drinking water due to its greater acute toxicity compared to pentavalent inorganic arsenic (arsenate, AsV), monomethylarsonic acid (MMA(V)) and dimethylarsinic acid (DMA(V)) (Kaise et al., 1989). However, a growing body of evidence supports the idea that multiple metabolites of arsenic have unique toxicological effects. This implies the need to specifically address issues associated with the incorporation of these metabolites and their associated target tissues in PBPK models.

DMA(V) was the first of the major arsenic metabolites identified as having unique toxicological activity (reviewed by Kenyon and Hughes (2001)). Initiation–promotion studies have demonstrated that DMA(V) acts as a multi-organ tumor promotor in rodents. In initiated rats, DMA(V) administered in drinking water induces tumors in the urinary bladder, kidney, liver, and thyroid (Yamamoto et al., 1995; Wanibuchi et al., 1996). In mice initiated with 4-nitroquinoline 1-oxide, DMA(V) promoted tumor development in lung (Yamanaka et al., 1996). In addition, DMA(V) is a complete bladder carcinogen in rats when administered in diet or drinking water (Van Gemert and Eldan, 1998; Arnold et al., 1999; Wei et al., 1999). It also produces organ-specific damage in the form of DNA-protein crosslinks and single strand scissions in DNA in human lung L132 (type II alveolar) cells (Tezuka et al., 1993; Yamanaka et al., 1993; Kato et al., 1994; Rin et al., 1995; Yamanaka et al., 1995; Yamanaka et al., 1997) and in lung cell DNA of rats and mice exposed orally (Yamanaka et al., 1989a,b; Yamanaka et al., 1991; Yamanaka et al., 1993). Recently, Kitchin and Ahmad (2003) have presented evidence to suggest that DMA(III) is in fact the most likely proximate toxicant in causing many of these effects.

As analytical techniques have improved sufficiently to allow the isolation and toxicological evaluation of the reactive trivalent methylated arsenicals, MMA(III) and DMA(III), they have been identified as having potent cytotoxic and genotoxic effects (Thomas et al., 2001). For example, MMA(III) is a much more potent inhibitor of thioredoxin reductase, an enzyme which protects cellular redox status, compared to either MMA(V) or DMA(V) (Lin et al., 1999). Petrick et al. (2001) found that MMA(III) in solution is a more potent inhibitor of pyruvate dehydrogenase than is arsenite. Trivalent methylated species, particularly DMA(III), are much more genotoxic compared to their pentavalent counterparts and arsenite, as measured by a DNA nicking assay and DNA damage measured in human lymphocytes using the comet assay (Mass et al., 2001). Kitchin and Ahmad (2003) have recently reviewed the mechanistic basis and compelling evidence for biological activity of trivalent organoarsenicals in cancer causation via an oxidative stress mode of action.

The issue of the concordances between metabolites and toxic effects will naturally influence both the functional form of the arsenic model and dose metric selection. The major challenge here is to balance the complexity of the biology with the data available to parameterize the model. Estimation of many parameters from the same data or insufficient data (over parameterization) leads to greater uncertainty in model predictions and limits the utility of the model for risk assessment purposes. One simplifying assumption in both our model and other published models (Mann et al., 1996a,b; Yu, 1999) is that the pentavalent and trivalent organoarsenicals are "lumped," i.e. MMA refers to MMA(III) + MMA(V) and DMA refers to DMA(III) + DMA(V). While the unique and differing toxicological activity of pentavalent and trivalent MMA and DMA suggests the need for separate submodels, it is also possible that "lumped" MMA and DMA may be sufficient for dose–response analysis. Further, given the lack of data to estimate separate chemical-specific parameters to describe partitioning, binding and metabolism for these trivalent and pentavalent organoarsenicals at this time, simplification is necessary. When further mechanistic data become available and analytical techniques improve to the point where these forms can be measured in tissues reliably, the appropriateness of this "lumped" description should be reevaluated.

3. Incorporation of mechanistic data

The arsenic PBPK model is necessarily complex because of the existence of multiple biologically active forms and uncertainty concerning their roles in producing toxic effects. Functionally, for arsenic this requires a minimum of four submodels (As(V), As(III), MMA, DMA) linked by reduction/oxidation and methylation as well as incorporation of urinary excretion for each metabolite and tissue binding for arsenite. This is necessary since the availability of arsenate, arsenite and methylated forms for tissue interactions is a balance between rates of reduction, oxidation, methylation, binding and excretion. In our research, we have used the female B6C3F1 mouse as an animal model for pharmacokinetic studies designed to parameterize and evaluate assumptions used in the model shown in Figure 1. For example, on the basis of pharmacokinetic studies using intravenously injected MMA(V) and DMA(V) in the mouse (Hughes and Kenyon, 1998), we concluded that blood flow limited transport into tissues is a reasonable assumption for most tissues. It is interesting to note that this is contrary to what would be predicted on the basis of *p*Ka values (Mann et al., 1996a).

A unique feature of the DMA submodel is the inclusion of a "deep lung" compartment. This essentially simulates DMA binding to targets in lung and allows DMA to be sequestered in lung (Evans et al., 2001). This was done for two reasons. The first reason is that lung-specific lesions are observed following DMA(V) exposure in both human and rodent tissues as discussed in Section 2. The second reason is that our studies indicate that the lung preferentially accumulates DMA under a variety of conditions. As shown in Figure 2, under conditions of exposure to background levels of arsenic, the concentration of DMA in lung is 3-fold higher compared to levels achieved in the next highest tissue (kidney). In studies in which DMA(V) is administered to mice by the i.v. route, there is

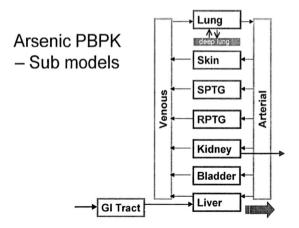

Arsenic PBPK
– Sub models

Figure 1. Basic structure of the *in vivo* arsenic PBPK model. The major circulating organic arsenicals are assumed to be in the pentavalent form; this assumption may need to be modified later, depending on tissue-specific data on the formation of trivalent methylated arsenicals and their transport out of cells and tissues. The structure of all submodels is very similar, except that there is not oral dosing in MMA or DMA submodels and the "deep lung" compartment is found only in the DMA submodel.

a dose-dependent increase in the cumulative dose to lung as measured by area under the curve (AUC). The AUC at a dose of 1.1 mg/kg is 447 nmol min/g, whereas the AUC at a 100-fold higher dose is 2.28×10^6 nmol min/g (Hughes et al., 2000); this observation is consistent with studies reported by Vahter et al. (1984). DMA also accumulates preferentially in lung following an oral dose of arsenic as either sodium arsenate (Fig. 3A and B) or sodium arsenite (Fig. 4) and dose-dependency in DMA accumulation in lung is also observed when mice are dosed with arsenate (Fig. 3A and B).

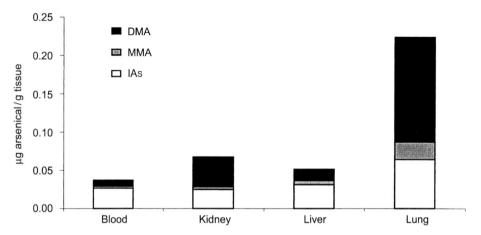

Figure 2. Levels of arsenicals measured in naïve 90-day old female B6C3F1 mice, i.e. those never exposed to experimental doses of arsenicals.

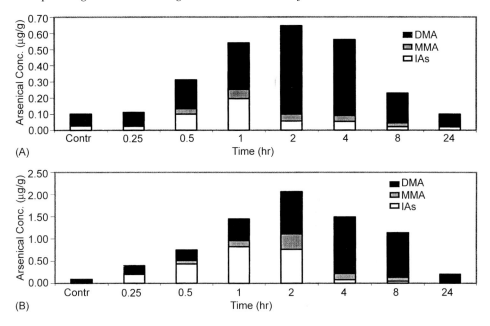

Figure 3. Time course for arsenical species in lung of female B6C3F1 mice following a (A) 10 μmol/kg oral dose of sodium arsenate (0.74 mg/kg as As), (B) 100 μmol/kg oral dose of sodium arsenate (7.4 mg/kg as As).

Another unique mechanistic feature incorporated into our model is arsenite caused inhibition of the formation of DMA from MMA (Kenyon et al., 2001; Easterling et al., 2002). The biological basis for this feature is that in studies using rat liver cytosol or rat and human hepatocytes, as the initial concentration of arsenite is increased, the formation of DMA is delayed, lower amounts of DMA are produced and the ratio of DMA/MMA formation is decreased (Styblo et al., 1996, 1999; Buchet and Lauwreys, 1985). Similar findings have been reported *in vivo* in both mice (Kenyon et al., 2000) and rats

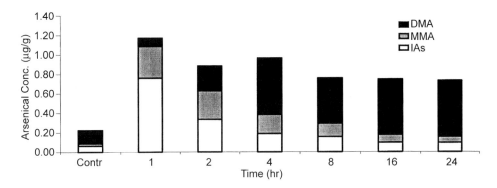

Figure 4. Time course for arsenical species in lung of female B6C3F1 mice following a 100 μmol/kg oral dose of sodium arsenite (7.4 mg/kg as As).

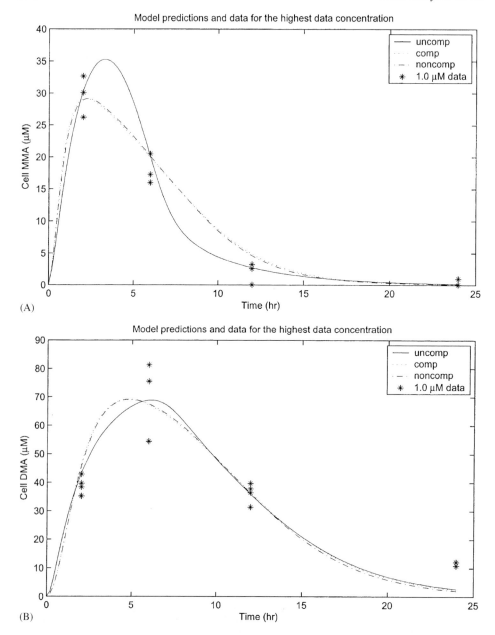

Figure 5. Model predictions of cell (A) [MMA]; (B) [DMA] and data for initial [iAs] = 1.0 μM. The model predictions are for the models with each inhibition type: uncompetitive (solid line), competitive (dotted line), and noncompetitive (dash-dot line).

(Csanaky et al., 2003) administered arsenite orally. In their mechanistic studies of the dose-dependent decrease observed in methylation after administration of arsenite in rats, Csanaky et al. (2003) found an increased concentration of the methyl donor *S*-adenosyl methionine. This indicates that methyl donor (SAM) depletion is not the explanation for the observed decrease in methylation.

We have modeled arsenite methylation in both cytosol and hepatocyte systems considering three possible types of inhibition – competitive, uncompetitive and noncompetitive – as well as no inhibition. When inhibition was not included in the model, the fits were unsatisfactory (Easterling et al., 2002). Although the inclusion of inhibition is necessary, it is not presently possible to definitively differentiate between mechanisms of inhibition using the model and available data alone (Fig. 5A and B). Simulations suggest that uncompetitive inhibition has a very different cellular MMA time course from noncompetitive and competitive inhibition (Fig. 5A) at the highest concentration available (1 μM), whereas this is not the case for DMA (Fig. 5B). The time course of cell MMA concentration peaked at a higher concentration and at a later time when the model used uncompetitive inhibition. The simulations with noncompetitive and competitive inhibition predict almost identical cell MMA concentrations at 1 μM initial AsIII concentration.

4. Conclusions and research needs

The issues and examples discussed in this presentation point out several important findings that suggest further research needs. Pharmacokinetic studies with arsenic need to be directly linked with mechanistic studies in order to match the form(s) of arsenic with specific target tissues and organs. This is essential in order to select appropriate dose metrics. The ultimate application of any PBPK model is human health risk assessment. While pharmacokinetic and mechanistic studies in animals and human cell lines are necessary and highly valuable for model development; ultimately the model must be evaluated using data from exposed human populations. To this end, clinical and field studies in which exposure is quantitatively evaluated at the level of individual study participant and then matched with arsenical blood levels and urinary excretion data for that same individual are essential for model evaluation. The importance of methylation and variation in methylation among humans also needs to be quantitatively evaluated. This is because sensitivity analysis has suggested that methylation parameters are highly influential in determining tissue concentrations of methylated species. The development and use of an arsenic PBPK model that has been adequately evaluated will help define the relationship between exposure and biologically effective dose at the target tissue level. This knowledge will reduce uncertainties in arsenic risk assessment for exposed human populations.

Acknowledgements

This article has been reviewed in accordance with the policy of the National Health and Environmental Effects Research Laboratory, US Environmental Protection Agency, and approved for publication. Approval does not signify that the contents necessarily reflect the views and policies of the agency, nor does the mention of trade names or commercial products constitute endorsement or recommendation for use. Luz Maria Del Razo was a visiting scientist at the US EPA and was supported by a fellowship from Conacyt-Mexico and the Pan American Health Organization.

References

Arnold, L.L., Cano, M., St John, M., Eldan, M., van Gemert, M., Cohen, S.M., 1999. Effects of dietary dimethylarsinic acid on the urine and urothelium of rats. Carcinogenesis, 10, 2171–2179.

Buchet, J.P., Lauwreys, R., 1985. Study of inorganic arsenic methylation by rat in vitro: relevance for the interpretation of observations in man. Arch. Toxicol., 57, 125–129.

Csanaky, I., Nemeti, B., Gregus, Z., 2003. Dose-dependent biotransformation of arsenite in rats – not *S*-adenosylmethionine depletion impairs arsenic methylation at high doses. Toxicology, 183, 77–91.

Easterling, M.R., Styblo, M., Evans, M.V., Kenyon, E.M., 2002. Pharmacokinetic modeling of arsenite uptake and metabolism in hepatocytes-mechanistic insights and implications for further experiments. J. Pharmacokinet. Pharmacodyn., 29, 207–234.

Evans, M.V., Hughes, M.F., Kenyon, E.M., 2001. A novel physiologically based pharmacokinetic (PBPK) model for dimethylarsinic acid (DMA): the lung as a storage compartment. Toxicol., Suppl. Toxicol. Sci., 60 (1-S), 719.

Hughes, M.F., Del Razo, L.M., Kenyon, E.M., 2000. Dose-dependent effects on tissue distribution and metabolism of dimethylarsinic acid in the mouse after intravenous administration. Toxicology, 143, 155–166.

Hughes, M.F., Kenyon, E.M., 1998. Dose-dependent effects on the disposition of monomethylarsonic acid and dimethylarsinic acid in the mouse after intravenous administration. J. Toxicol. Environ. Health, Pt. A, 53, 95–112.

Kaise, T., Yamauchi, H., Horiguchi, Y., Tani, T., Watanabe, S., Hirayama, T., Fukui, S., 1989. A comparative study on acute toxicity of methylarsonic acid, dimethylarsinic acid and trimethylarsine oxide in mice. Appl. Organometal. Chem., 3, 273–277.

Kato, K., Hayashi, H., Hasegawa, A., Yamanaka, K., Okada, S., 1994. DNA damage induced in cultured human alveolar (L-132) cells by exposure to dimethylarsinic acid. Environ. Health Perspect., 102, 285–288.

Kenyon, E.M., Fea, M., Styblo, M., Evans, M.V., 2001. Application of modeling techniques to the planning of in vitro arsenic kinetic studies. ATLA, 29, 15–33.

Kenyon, E.M., Hughes, M.F., 2001. A concise review of the toxicity and carcinogenicity of dimethylarsinic acid. Toxicology, 160, 227–236.

Kenyon, E.M., Del Razo, L.M., Hughes, M.F., 2000. Tissue distribution of arsenite (AsIII) and its methylated metabolites in mice. Toxicol., Suppl. Toxicol. Sci., 54, 258.

Kitchin, K.T., Ahmad, S., 2003. Oxidative stress as a possible mode of action for arsenic carcinogenesis. Toxicol. Lett., 137, 3–13.

Krishnan, K., Andersen, M.E., 1994. Physiologically based pharmacokinetic modeling in toxicology. Principles and Methods in Toxicology, 3rd edn. Raven Press, New York, pp. 149–188.

Lin, S., Cullen, W.R., Thomas, D.J., 1999. Methylarsenicals and arsinothiols are potent inhibitors of mouse liver thioredoxin reductase. Chem. Res. Toxicol., 12, 924–930.

Mann, S., Droz, P.O., Vahter, M., 1996a. A physiologically based pharmacokinetic model for arsenic exposure. I. Development in hamsters and rabbits. Toxicol. Appl. Pharmacol., 137, 8–22.

Mann, S., Droz, P.O., Vahter, M., 1996b. A physiologically based pharmacokinetic model for arsenic exposure. II. Validation and application in humans. Toxicol. Appl. Pharmacol., 140, 471–486.

Mass, M.J., Tennant, A., Roop, B.C., Cullen, W.R., Styblo, M., Thomas, D.J., Kligerman, A.D., 2001. Methylated trivalent arsenic species are genotoxic. Chem. Res. Toxicol., 14, 355–361.

Petrick, J.S., Jagadish, B., Mash, E.A., Aposhian, H.V., 2001. Mono-methylarsonous acid (MMAIII) and arsenite: LD50 in hamsters and in vitro inhibition of pyruvate dehydrogenase. Chem. Res. Toxicol., 14, 651–656.

Rin, K., Dawaguchi, K., Yamanaka, K., Tezuka, M., Oku, N., Okada, S., 1995. DNA-strand breaks induced by dimethylarsinic acid, a metabolite of inorganic arsenics, are strongly enhanced by superoxide anion radicals. Biol. Pharm. Bull., 18, 45–48.

Styblo, M., Delnomdedieu, M., Thomas, D.J., 1996. Mono- and dimethylation of arsenic in rat liver cytosol in vitro. Chem.–Biol. Interact., 99, 147–164.

Styblo, M., Del Razo, L.M., LeCluyse, E.L., Hamilton, G.A., Wang, C., Cullen, W.R., Thomas, D.J., 1999. Metabolism of arsenic in primary cultures of human and rat hepatocytes. Chem. Res. Toxicol., 12, 560–565.

Tezuka, M., Hanioka, K., Yamanaka, K., Okada, S., 1993. Gene damage induced in human alveolar type II (L-132) cells by exposure to dimethylarsinic acid. Biochem. Biophys. Res. Commun., 191, 1178–1183.

Thomas, D.J., Styblo, M., Lin, S., 2001. The cellular metabolism and systemic toxicity of arsenic. Toxicol. Appl. Pharmcol., 176, 127–144.

Vahter, M., Marafante, E., Dencker, L., 1984. Tissue distribution and retention of [74]As-dimethylarsinic acid in mice and rats. Arch. Environ. Contam. Toxicol., 13, 259–264.

Van Gemert, M., Eldan, M., 1998. Chronic Carcinogenicity Assessment of Cacodylic Acid. The Third International Conference on Arsenic Exposure and Health Effects, San Diego, CA.

Wanibuchi, H., Yamamoto, S., Chen, H., Yoshida, G., Endo, G., Hori, T., Fukushima, S., 1996. Promoting effects of dimethylarsinic acid on N-butyl-N-(4-hydroxybutyl) nitrosamine-induced urinary bladder carcinogenesis in rats. Carcinogenesis, 17, 2435–2439.

Wei, M., Wanibuchi, H., Yamamoto, S., Li, W., Fukushima, S., 1999. Urinary bladder carcinogenicity of dimethylarsinic acid in male F344 rats. Carcinogenesis, 20, 1873–1876.

Yamamoto, S., Konishi, Y., Matusuda, T., Murai, T., Shibata, M.-A., Matsui-Yuasa, I., Otani, S., Kuroda, K., Endo, G., Fukushima, S., 1995. Cancer induction by an organic arsenic compound, dimethylarsinic acid (cacodylic acid), in F344/DuCrj rats after pretreatment with five carcinogens. Cancer Res., 55, 1271–1276.

Yamanaka, K., Hasegawa, A., Sawamura, R., Okada, S., 1989a. DNA strand breaks in mammalian tissues induced by methylarsenics. Biol. Trace Elem. Res., 21, 413–417.

Yamanaka, K., Hasegawa, A., Sawamura, R., Okada, S., 1989b. Dimethylated arsenics induce DNA strand breaks in lung via the production of active oxygen in mice. Biochem. Biophys. Res. Commun., 165, 43–50.

Yamanaka, K., Hasegawa, A., Sawamura, R., Okada, S., 1991. Cellular response to oxidative damage in lung induced by the administration of dimethylarsinic acid, a major metabolite of inorganic arsenics, in mice. Toxicol. Appl. Pharmacol., 108, 205–213.

Yamanaka, K., Tezuka, M., Kato, K., Hasegawa, A., Okada, S., 1993. Crosslink formation between DNA and nuclear proteins by in vivo and in vitro exposure of cells to dimethylarsinic acid. Biochem. Biophys. Res. Commun., 191, 1184–1191.

Yamanaka, K., Hayashi, H., Kato, K., Hasegawa, A., Okada, S., 1995. Involvement of preferential formation of apurinic/apyrimidinic sites in dimethylarsenic-induced DNA strand breaks and DNA-protein crosslinks in cultured alveolar epithelial cells. Biochem. Biophys. Res. Commun., 207, 244–249.

Yamanaka, K., Ohtsubo, K., Hasegawa, A., Hayashi, H., Ohji, H., Kanisawa, M., Okada, S., 1996. Exposure to dimethylarsinic acid, a main metabolite of inorganic arsenics, strongly promotes tumorigenesis initiated by 4-nitroquinoline 1-oxide in the lungs of mice. Carcinogenesis, 17, 767–770.

Yamanaka, K., Hayashi, H., Kato, K., Hasegawa, A., Oku, N., Okada, S., 1997. DNA single-strand breaks in L-132 cells resulting from inhibition of repair polymerization shortly after exposure to dimethylarsinic acid. Biol. Pharm. Bull., 20, 163–167.

Yu, D., 1999. A physiologically based pharmacokinetic model of inorganic arsenic. Regul. Toxicol. Pharmacol., 29, 128–141.

PART V
INTERVENTION AND MEDICAL TREATMENT

Arsenic Exposure and Health Effects V
W.R. Chappell, C.O. Abernathy, R.L. Calderon and D.J. Thomas, editors
© 2003 Published by Elsevier B.V.

Chapter 29

Natural history following arsenic exposure: a study in an arsenic endemic area of West Bengal, India

D.N. Guha Mazumder, Nilima Ghose, Kunal Mazumder, Amal Santra, Sarbari Lahiri, Subhankar Das, Arindam Basu and Allan H. Smith

Abstract

The natural history of people who stop drinking arsenic (As) contaminated water is not known. A cohort follow-up study of 1074 people (As-exposed people 623, control population 451) was therefore carried out in 2000, 5 years after the original clinical examination done on the same population. Clinical examination was done on 559 exposed people out of whom 505 people were now found to be consuming safer water (As level $<50 \ \mu g/l$). Clinical examination was also done on a control population of 434 thought to be consuming As-free water throughout.

Out of 199 people with skin lesions among the As-exposed population, who were consuming safe water during the last 5 years, the skin lesions cleared or decreased in 49.7%, according to the responses of participants. However, new skin lesions appeared in 32 (10.5%) out of 306 people who were not diagnosed with such lesions previously.

Cases of chronic lung disease, peripheral neuropathy, dyspepsia and ischaemic heart disease were found in higher numbers in the As-exposed people, compared to the control group during the period of study.

Keywords: Arsenicosis: West Bengal, natural history, safe water, skin lesions

1. Introduction

Elevated As level in drinking water is the major cause of As toxicity in the world. Reports of such contamination are available from Taiwan, Chile, China, Argentina, Mexico, India, Hungary, Bangladesh, USA and Thailand. However, the largest number of people in the world affected from chronic As toxicity due to drinking arsenic-contaminated ground water reside in Bangladesh and India (Chowdhury et al., 2001).

In the past two decades, groundwater in at least nine districts within the Gangetic plain of West Bengal, India has been found to contain elevated concentrations of inorganic arsenic of geologic origin. On the basis of the research carried out in the Department of Medicine and Gastroenterology at the Institute of Post Graduate Medical Education and Research, Kolkata, India, since 1984, the clinical characteristics of chronic arsenic toxicity have been delineated. Previously, major emphasis on chronic As toxicity in man was given to skin manifestations. We have described various non-cutaneous manifestations of chronic As toxicity in West Bengal. On the basis of detailed clinical examination and relevant investigations on 248 cases in the hospital, it was observed that chronic intake of

As in drinking water affects multiple organs of the human body. Over and above the occurrence of spotty pigmentation (94.3%) and keratosis (65.3%), other manifestations are liver enlargement (76.6%), weakness (65.73%), respiratory symptoms (62.5%), anaemia (43.9%), polyneuropathy (29.83%), splenomegaly (29.43%), non-pitting oedema of the legs (9.27%) and peripheral vascular disease (gangrene, 1.2%). Skin, bladder and kidney cancer were seen in a few cases also (Guha Mazumder et al., 1999).

Results of a cross-sectional epidemiological study on a population of 7683 people in one of the worst affected districts of West Bengal in 1995 demonstrated the trends in arsenic-related morbidity in the state (Guha Mazumder et al., 1998a). Keratosis and pigmentation of the skin are two important clinical features of chronic arsenic toxicity. Using individual exposure data, the study showed for the first time that the prevalence of keratosis was strongly related to water arsenic concentration rising from zero in the lowest exposure level (<50 μg/l) to 8.3 per 100 for females drinking water containing >800 μg/l and increasing from 0.2 per 100 in the lowest exposure category to 10.7 per 100 males in the highest exposure level (≥800 μg/l). Findings were similar for hyperpigmentation (Guha Mazumder et al., 1998a). The study further substantiated the results of previous hospital-based data in that the incidences of weakness, liver enlargement, chronic lung disease (CLD) and neuropathy were found to be significantly higher in 4216 people, drinking arsenic-contaminated water, compared to those observed among 3467 people, consuming arsenic-free water (Guha Mazumder et al., 2001).

Exposure assessment is challenging in such studies because families may use several tube wells at any point in time, and the tube wells vary over the years. Subsequent investigation with detailed exposure assessment confirmed the dose response trends for skin lesions. However, no cases with pigmentation changes and/or keratoses occurred without consuming water containing more than 100 μg/l of arsenic, and the large majority had consumed water containing over 200 μg/l at some point in the previous 20 years (Haque et al., 2003).

Information on the natural history of chronic arsenic toxicity following drinking As-free water is scanty. Medical treatment of arsenicosis to modify the health effects is unsatisfactory. Results of long-term intake of arsenic-free water in the affected people in an arsenic endemic area need to be fully ascertained to understand the effect of the intervention programme of providing arsenic-free water to the affected community. A cohort follow-up study was therefore conducted in South 24 Parganas, one of the affected districts of West Bengal, on 1074 people in the year 2000, 5 years after the initial clinical epidemiological study carried out in the year 1995.

2. Methodology

The source population included 7683 individuals who participated in the 1995 population-based cross-sectional survey in South 24 Parganas, West Bengal (Guha Mazumder et al., 1998a). Briefly, two areas of the district were surveyed earlier. The first included 25 villages and was selected because high levels of arsenic were reported in some of the tube wells. In this high-exposure area, which included remote rural areas, a convenience sampling strategy was used. The second area included the remaining part of the district (32 villages in 16 administrative blocks), where people were drinking from shallow tube well not contaminated with arsenic. First, 55–60 people of the participant list of the 1995 study from each of 13 easily accessible villages (by road transport from Kolkata) having relatively high arsenic exposure and 100–150 people from 4 easily accessible villages of 4 blocks of low-exposure areas of the previous survey were identified and an attempt was made to contact them. On the basis of the previous arsenic exposure data from the 1995 study, 672 people who were approached had high arsenic exposure in 1995 (>50 µg/l) and 468 people approached had low arsenic exposure (<50 µg/l). Forty-nine people among the former and 17 of the later population were either absent or could not be contacted during the study.

2.1. Interview and clinical examination

People with previous water As exposure data were re-examined. Regarding skin lesions, they were asked to state whether their pigmentation/keratosis has disappeared, decreased, remained the same or increased. People were also examined and any new appearance of skin lesions (pigmentation/keratosis) was recorded. All study participants were asked about their other ailments, if any, treated in any hospital or health centre. Diagnoses for these conditions were recorded if an authentic record of disease diagnosis was obtained from treatment records that the participants had in their homes. Chronic bronchitis, chronic obstructive airway disease and bronchiectasis are included in the heading of CLD.

2.2. Water analysis

A water history was taken for the period of the last 5 years and water samples were collected from all the tube wells of their home from where the participants were taking water. Total As content of water was estimated by atomic absorption spectrophotometer with hydride generation system in the trace element laboratory of the institute.

2.3. Statistical analysis

Cumulative incidence rates were calculated for the different diseases and reported per 1000 person-years. Person-time at risk was estimated to be half the follow-up time for those with disease outcomes, as the dates of first diagnosis were not available. Based on the cumulative incidences, rate ratios (ratios between incidence among exposed and incidence among non-exposed) were calculated.

3. Results

The number of participants of the study and control population is given in Table 1. 623 participants had a past history of consuming water containing more than 50 µg/l in the last five years, of which 229 (36.7%) cases had skin lesion in 1995. As 64 people had died in this population, the follow-up study could be carried out on 559 people, out of which 210 (39.6%) had skin lesions. On the basis of the analysis of all the water samples collected

Table 1. Distribution of participants of study and control population.

	Study population	Control population
People surveyed in 1995	672	468
People absent	49	17
People surveyed in 2000	623	451
People who had died	64	17
People examined in 2000	559	434

Table 2. Distribution of skin lesion in the study population with data of water history.

	As-exposed subjects			Controls
	Skin lesions present (%)	Skin lesions absent (%)	Total	Total
Cases examined in 1995	229 (36.7)	394 (63.3)	623	451
Cases died	19 (8.29)	45 (11.42)	64 (10.27)	17 (3.77)
Follow-up cases examined in 2000	210 (39.6)	349 (60.4)	559	434
a) Consuming As-free water for 5 yrs	199 (94.76)	306 (87.68)	505	434
b) Consuming As-contaminated water for 5 yrs	11 (5.24)	43 (12.32)	54	X

Table 3. Age and sex distribution of participants of study and control population.

Age	Study cohort ($n = 623$)			Control cohort ($n = 451$)		
	Male (%)	Female (%)	Total (%)	Male (%)	Female (%)	Total (%)
≤20 yrs	52 (15.4)	44 (15.3)	96 (15.4)	71 (34.8)	75 (30.3)	146 (32.3)
21–40 yrs	129 (38.2)	119 (41.6)	248 (39.8)	79 (38.7)	119 (48.1)	198 (43.9)
41–60 yrs	99 (29.3)	84 (29.3)	183 (29.3)	39 (19.1)	40 (16.1)	79 (17.5)
≥60 yrs	57 (16.9)	39 (13.6)	96 (15.4)	15 (7.3)	13 (5.2)	28 (6.2)
Total	337 (100)	286 (100)	623 (100)	204 (100)	247 (100)	451 (100)

from the participants, As levels of 199 and 306 people with and without skin lesions, respectively, were found to be consuming water with As levels less than 50 μg/l (Table 2).

The demographic distribution of the study and control populations is given in Table 3. Sixty-nine percent of people in the As-exposed group belonged to the age groups of 21–60 years while 61.4% in the control group belonged to the same age group. On the other hand, about 15% of the former and 6% of the latter group belonged to the elderly age group (>60 years of age).

Table 4.   Outcome of skin lesions among follow-up patients consuming As-free water (As < 50 μg/l) for five years.

Outcome	Participants consuming safe water (As < 0.05 μg/l) currently (n = 199)	
	No.	%
Cleared	40	20.1
Decreased	59	29.6
Increased	5	2.5
Same	95	47.7
New appearance	32[a]	10.5

[a] Calculated out of 306 participants who were not thought to have skin lesion in 1995.

Table 4 presents data in regard to skin lesions in the follow-up study. Skin lesions were reported to be decreased or could not now be identified (favourable outcome) in 49.7% of cases and increased or remained the same (unfavourable outcome) in 50.3% of cases out of 199 people who were consuming water containing less than 50 μg/l, during the last 5 years. It was interesting to note that new skin lesions appeared in 32 (10.5%) out of 306 people who were not identified to have skin lesions previously and who were thought to have been consuming only water containing less than 50 μg/l.

Among the exposed and control groups, incidences of various diseases as was evident from records of medical treatment of the participant are presented in Table 5. Age and sex distribution of CLD is presented in Table 6. The incidence of CLD was higher among As-exposed people compared to the control population (rate ratio 4.66: 95% CI: 1.04–20.8). Incidences of neuropathy, cerebrovascular accident (CVA) and diabetes mellitus were also found to be higher in arsenic-exposed people compared to the control population.

Table 5.   Distribution of different diseases according to exposure.

Name of disease	Exposed		Non-exposed		Rate ratio
	No.	Incidence	No.	Incidence	(95% CI)
CLD[a]	12	85.99	2	18.4	4.66 (1.04–20.8)
Diabetes	4	28.6	2	18.4	1.55 (0.28–8.48)
Hypertension	3	21.5	5	46.1	0.47 (0.11–1.95)
IHD[b]	0	–	1	9.22	–
CVA[c]	3	21.5	1	9.22	2.33 (0.24–22.4)
Neuropathy	5	35.8	1	9.22	3.88 (0.45–33.2)

[a] Chronic bronchitis, chronic obstructive airway diseases, bronchiectasis.
[b] IHD, ischaemic heart disease.
[c] CVA, cerebrovascular accident.

Table 6. Age and sex distribution of CLD according to exposure.

Gender	Age group	Non-exposed		Exposed	
		Count	Incidence	Count	Incidence
Female					
	21–40	1	1.71	1	1.78
	41–60	1	5	3	7.5
	Total	2	1.64	4	3.1
Male					
	21–40	0	–	3	4.8
	41–60	0	–	3	6.98
	More than 60	0	–	2	9.3
	Total	0	–	8	5.32

From Table 6, it could be seen that the incidence of CLD was higher among men for all age groups above 20 years.

4. Discussion

On the basis of the current available data, it appears that about 31 million people in the Indo-Bangladesh subcontinent are exposed to As due to drinking of contaminated ground water (Chowdhury et al., 2001). Not much information is available in the literature regarding the long-term effect of chronic arsenic toxicity after stoppage of drinking arsenic-contaminated water. Arguello et al. (1938) reported that keratodermia appeared insidiously between the second and third years of intoxication and did not disappear after cessation of exposure. Some individuals were followed up for more than 30 years after termination of exposure.

Data of the present study describe the effect of drinking arsenic-free water in a cohort of 623 previously arsenic-exposed people who have been examined in 1995, and the survivors (559) were re-examined during the year 2000. The morbidity of this population was compared with that of the control population of 434 people examined identically in the same geographical area.

Earlier reported results of cohort follow-up study of 24 patients of chronic arsenic toxicity were re-examined after drinking arsenic-free water for a period varying from 2 to 10 years. Improvement of pigmentation and keratosis was observed in 46% of people studied (Guha Mazumder et al., 1999). Results of the present study also showed that favourable dermatological outcome (clearance or decrement of keratosis and/or pigmentation) occurred in 49.7% of 199 people drinking As-free (<50 μg/l) water for 5 years. However, it was interesting to note that 10.5% of people developed new skin lesions. Observer variation in diagnosis, in particular of mild skin lesions, could explain these findings, and it is also possible that they had arsenic exposure from undetermined sources. The information concerning exposure in this study came only from arsenic concentration in tube wells in the actual homes of participants.

Pulmonary effects due to chronic arsenic toxicity by drinking As-contaminated water have been reported by many workers (Borgono et al., 1977; Chakraborty and Saha, 1987; Guha Mazumder et al., 1988, 1997, 1998b, 2000; Milton et al., 2001). However, reports on the effect of stoppage of drinking As-contaminated water on the respiratory disease are scanty. The results of the present study showed a high incidence of morbidity (incidence 85.99) caused by CLD in spite of drinking water containing less than 50 µg/l. It occurred in all age groups in the exposed population with male being affected more than the female. In an earlier study, we reported new appearance of signs of CLD (shortness of breath and chest signs) in 41.6% of cases, 2–10 years of stoppage of drinking As-contaminated water in the 24 cases studied (Guha Mazumder et al., 1999). The cause of the increased occurrence of lung disease in As-exposed population even after the stoppage of As exposure is an important observation. This may be due to latent effects appearing after some initial cellular damage.

There were five cases of (incidence 35.8) neuropathy among follow-up cases of past arsenic exposure, while there was one among the control population (rate ratio 3.88). Evidence of paraesthesia/peripheral neuropathy due to chronic exposure of arsenic through drinking water has been reported by many workers including us (Hindmarsh et al., 1977; Saha, 1984; Hotta, 1989; Guha Mazumder et al., 1988, 1992, 1997; Kiburn, 1997; Ahmad et al., 1999; Ma et al., 1999). In our previous follow-up study of 24 cases, neuropathy was present initially in 11 cases while in 8 cases after the study period (2–10 years). No new neuropathy developed among the follow-up cases (Guha Mazumder et al., 1999). As we did not have initial neurological data of our study population, we could not ascertain the incidence of recovery of neuropathy among those cases studied.

There were 3 cases of CVA and 4 cases of diabetes mellitus among the exposed population while 1 and 2 cases, respectively, among the control people (rate ratio 2.33 and 1.55, respectively). The relationship between CVA and ingestion of As in drinking water was reported by Chiou et al. (1997), in a cross-sectional study in Taiwan. A significant dose–response relationship was observed between concentration of As in well water and prevalence of CVA after adjustment for age, sex, hypertension, diabetes mellitus, cigarette smoking and alcohol consumption. Association between ingested As and prevalence of diabetes mellitus was reported by Lai et al. (1994) and Tseng et al. (2000) from Taiwan and Rahaman et al. (1998) from Bangladesh. However, these reports were based on studies on the population who were drinking As-contaminated water during previous years. No reports are available regarding incidence of these diseases in a follow-up period after the exposed population had stopped drinking As-contaminated water. The limitation of the present study is that incidences of these diseases were not ascertained in the year 1995. Hence, one does not know whether these diseases were present before the date of discontinuation of As exposure.

There was no case of ischaemic heart disease (IHD) among the exposed population. Occurrence of IHD due to chronic exposure of arsenic has been reported by many workers (Rosenberg, 1974; Chen et al., 1994; Ma et al., 1999). Our current data and results of previous study (Guha Mazumder et al., 1999) suggested that incidence of IHD may be low in West Bengal. However, we need to carry on electrocardiographic study on larger population to find out the actual incidence of morbidity due to IHD in chronic arsenicosis subjects when they take arsenic-free water for long period.

In conclusion, it could be stated that a significant number of arsenicosis patients show evidence of dermatological manifestations, in spite of apparently consuming As-free water for a prolonged period. New skin lesions appeared in a few cases without known exposure. Further, a number of these people show feature of CLD. However it should be noted that there was no monitoring of exposure during the follow-up period, and it is possible that exposure occurred either due to unmeasured or undetermined water sources or perhaps from food contamination with arsenic. Further, more detailed follow-up of larger cohorts is needed to more fully elucidate the health effects of arsenic following termination of exposure.

Acknowledgements

The original epidemiological study was carried out with funding from the "Rajib Gandhi National Drinking Water Mission", Ministry of Rural Development, Govt. of India, Research Grant No.W-1046/2/4/96-TM II (R&D). Further financial support for this study was received from US EPA grant No. R826137-01-0 and NIEHS grants No. P42 E504705 and P30 ES01896-24 through the Center for Occupational and Environmental Health, University of California, Berkeley, USA. Dr. Amal Santra received laboratory training funded by NIH Fogarty D43 TW-000-815-01. The contents of the paper are solely the responsibility of the authors and do not necessarily represent the official views of the funding agencies.

The authors are grateful to the Director and Surgeon Superintendent of the Institute of Post Graduate Medical Education and Research and SSKM Hospital, Kolkata, West Bengal for their help in carrying on this study.

References

Ahmad, S.A., Sayed, M.H.S.U., Hadi, S.A., Faruqua, M.H., Khan, M.H., Jalel, M.A., Ahmed, R., Wadud Khan, A., 1999. Arsenicosis in a village in Bangladesh. Int. J. Environ. Health Res., 9, 187–195.

Arguello, R.A., Cenget, D., Tello, E.E., 1938. Cancer and endemic arsenism in the Cordoba region. Rev. Argent Dermanosifilogr., 22 (4), 461–487.

Borgono, J.M., Vicent, P., Venturino, H., Infante, A., 1977. Arsenic in the drinking water of the city of Antofagasta: epidemiological and clinical study before and after the installation of the treatment plant. Environ. Health Perspect., 19, 103–105.

Chakraborty, A.K., Saha, K.C., 1987. Arsenic dermatosis from tubewell water in West Bengal. Indian J. Med. Res., 85, 326–334.

Chen, C.J., Lin, L.J., Hsuch, Y.M., Chiou, H.Y., Liaw, K.F., Horng, S.F., Chiang, M.H., Tseng, C.H., Tai, T.Y., 1994. Ischaemic heart disease induced by ingested inorganic arsenic. In: Chappell, W.R., Abernathy, C.O., Cothern, C.R. (Eds), Arsenic Exposure and Health. Science and Technology Letters, Northwood, pp. 83–90.

Chiou, H.Y., Huang, W.I., Su, C.L., Chang, S.F., Hsu, Y.H., Chen, C.J., 1997. Dose response relationship between prevalence of cerebro vascular disease and ingested inorganic arsenic. Stroke, 28, 1717–1723.

Chowdhury, U.K., Rahaman, M.M., Mondal, B.K., Paul, K., Lodh, D., Biswas, B.K., Chanda, C.R., Basu, G.K., Saha, K.C., Mukherjee, S.C., Roy, S., Das, R., Kaies, I., Barua, A.K., Palit, S.K., Quamruzzaman, Q., Chakraborti, D., 2001. Ground water arsenic contamination and human suffering in West Bengal, India and Bangladesh. Environ. Sci., 8, 392–415.

Guha Mazumder, D.N., Chakraborty, A.K., Ghosh, A., Das Gupta, J.D., Chakraborty, D.P., Dey, S.B., Chattopadhya, N.C., 1988. Chronic arsenic toxicity from drinking tubewell water in rural West Bengal. Bull. Wld. Health Org., 66, 499–506.

Guha Mazumder, D.N., Das Gupta, J., Chakraborty, A.K., Chatterjee, A., Das, D., Chakraborty, D., 1992. Environmental pollution and chronic arsenicosis in South Calcutta. Bull. Wld. Health Org., 70(4), 481–485.

Guha Mazumder, D.N., Das Gupta, J., Santra, A., Ghosh, A., Sarkar, S., Chattopadhaya, N., Charaboti, D., 1997. Non-cancer effects of chronic arsenicosis with special reference to liver damage. In: Abernathy, C.O., Calderon, R.L., Chappell, W.R. (Eds), Arsenic Exposure and Health Effects, Vol. 10. Chapman and Hall, London, pp. 112–124.

Guha Mazumder, D.N., Haque, R., Ghosh, N., De, B.K., Santra, A., Chakraborty, D., Smith, A.H., 1998a. Arsenic levels in drinking water and the prevalence of skin lesions in West Bengal, India. Int. J. Epidemiol., 27, 871–877.

Guha Mazumder, D.N., Das Gupta, J., Santra, A., Pal, A., Ghosh, A., Sarmar, S., 1998b. Chronic Arsenic Toxicity in West Bengal – the worst calamity in the world. J. Indian Med. Assoc., 96(1), 4–7, 18.

Guha Mazumder, D.N., De, B.K., Santra, A., Das Gupta, J., Ghosh, N., Roy, B.K., Ghosal, U.C., Saha, J., Chatterjee, A., Dutta, S., Haque, R., Smith, A.H., Charaborty, D., Angle, C.R., Centeno, J.A., 1999. Chronic Arsenic Toxicity: Epidemiology, Natural history and Treatment. In: Chappell, W.R., Abernathy, C.O., Calderon, R.L. (Eds), Arsenic Exposure and Health Effects. Elsevier Science, New York, pp. 335–347.

Guha Mazumder, D.N., Haque, R., Ghosh, N., De B.K., Santra, A., Chakraborti, D., Smith, A.H., 2000. Arsenic in drinking water and the prevalence of respiratory affects in West Bengal, India. Int. J. Epidemiol., 29, 1047–1052.

Guha Mazumder, D.N., Ghosh, N., De, B.K., Santra, A., Das, S., Lahiri, S., Haque, R., Smith, A.H., Chakraborti, D., 2001. Epidemiological study on various non-carcinomatous manifestations of chronic arsenic toxicity in a district of West Bengal. In: Chappell, W.R., Abernathy, C.O., Calderon, R.L. (Eds), Arsenic Exposure and Health Effect IV. Elsevier Science, New York, pp. 153–164.

Haque, R., Guha Mazumder, D.N., Samanta, S., Ghosh, N., Kalman, D., Smith, M.M., Mitra, S., Santra, A., Lahiri, S., Das, S., De, B.K., Smith, A.H., 2003. Arsenic in drinking water and skin lesions: dose-response data from West Bengal, India, Epidemiology., 14, 174–182.

Hindmarsh, J.T., McLetchie, O.R., Heffernan, L.P.M., Hayne, O.A., Eltenberger, H.A., McCurdy, R.F., Thiebaux, H.J., 1977. Electromyographic abnormalities in chronic environmental arsenicalism. J. Anal. Toxicol., 1, 270–276.

Hotta, N., 1989. Clinical aspects of chronic arsenic poisoning due to environmental and occupational pollution in and around a small refining spot [in Japanese]. Nippon Taishitsugaku Zasshi [Jpn. J. Const. Med.], 53 (1/2), 49–70.

Kiburn, K.H., 1997. Neurobehavioral impairment from long-term residential arsenic exposure. In: Abernathy, C.O., Calderon, R.L., Chappell, W.R. (Eds), Arsenic Exposure and Health Effects, Vol. 14. Chapman and Hall, London, pp. 159–177.

Lai, M.S., Hsneh, Y.M., Chen, C.G., 1994. Ingested inorganic arsenic and prevalence of diabetes mellitus. Am. J. Epidemiol., 139, 484–492.

Ma, H.Z., Xia, Y.J., Wu, K.G., Sun, T.Z., Mumford, J.L., 1999. Human exposure to arsenic and health effects in Bayingnormen, Inner Mongolia. In: Abernathy, C.O., Calderon, R.L., Chappell, W.R. (Eds), Arsenic Exposure and Health Effects. Elsevier Science, London, pp. 127–131.

Milton, A.H., Hasan, Z., Rahman, A., Rahman, M., 2001. Chronic arsenic poisoning and respiratory effects in Bangladesh. J. Occup. Health, 43, 136–140.

Rahaman, M.M., Tondel, M., Ahmad, S., et al., 1998. Diabetes mellitus associated with arsenic exposure in Bangladesh. Am J. Epidemiol., 148, 198–203.

Rosenberg, H.G., 1974. Systemic arterial disease and chronic arsenicism in infants. Arch. Pathol., 97 (6), 360–365.

Saha, K.C., 1984. Melanokeratosis from arsenic contaminated tube well water. Indian J. Dermatol., 29, 37–46.

Tseng, C.H., Tai, T.Y., Chong, C.K., et al., 2000. Long term arsenic exposure and incidence of non-insulin dependent diabetes mellitus: a cohort study in arsenic hyperdemic villages in Taiwan. Environ. Health Perspect., 108, 847–851.

Arsenic Exposure and Health Effects V
W.R. Chappell, C.O. Abernathy, R.L. Calderon and D.J. Thomas, editors

Chapter 30

Saha's grading of arsenicosis progression and treatment

Kshitish C. Saha

Abstract

Arsenicosis in West Bengal, India and Bangladesh was discovered by the author in 1982 and 1984, respectively. Since 1983, periodical field surveys showed increasing severity of arsenicosis. The disease due to arsenicosis was confirmed by high arsenic level in consumed water, urine, nails, hair, and skin scales. The arsenic content was initially estimated by silver diethyldithiocarbazine method at the School of Tropical Medicine and much later by flow injection hydride generation atomic absorption spectrometry method at the School of Environmental Studies, Jadavpur University, Kolkata, India.

According to severity, progression of arsenicosis has been classified by the author into 4 stages, 7 grades, and 20 sub-grades. The 4 stages are I) pre-clinical, II) clinical, III) complication, and IV) malignancy. Each stage is further graded as follows. Pre-clinical (stage I) is graded 0 with 2 sub-grades; 0-a (labile or blood phase) and 0-b (stable or tissue phase). Clinical stage (stage II) has been sub-divided to four grades, 1) melanosis, 2) spotted keratosis on palms or soles, 3) diffuse keratosis on palms and soles, and 4) dorsal keratosis. Each of the four grades has been further sub-divided into three sub-grades, a, b, and c according to severity. Complications (stage III) and malignancy (stages IV) are graded 5 and 6, respectively; each of the grades has been further sub-divided to a, b, and c.

The features of different sub-grades are as follows, diffuse melanosis on palms (1-a), spotted melanosis on trunk (1-b), generalized melanosis (1-c), number of keratotic nodules 0–6 (2-a), number of keratotic nodules more than 6 (2-b), large keratotic nodules (2-c), diffuse keratosis on palms or soles (3-a), diffuse keratosis on both palms and soles (3-b), diffuse keratosis complete on whole palms and soles (3-c), nodular lesions on hands or feet (4-a), nodular lesions on hands and feet (4-b), nodular lesions on hands and feet along with extension of keratosis all over the body (4-c), palpable liver (5-a), jaundice (5-b), ascites (5-c), malignancy with single lesion (6-a), malignancy having two lesions (6-b), and malignancy having more than two lesions (6-c).

The various stages of arsenicosis can be treated as follows. Grade 0 can be eliminated by replacing arsenic contaminated with arsenic-free water, high protein diet, and fresh fruits. In addition, this treatment can be applied to the patients of advanced grades along with the supply of arsenic-free water. Bronchitis and obstructive asthma can be treated by antibiotics and bronchodilators, respectively. The chelating agent, dimercaptopropane sulfonate, therapy is carried out at grade 2-b onwards; complications may be prevented however, there is little improvement of keratosis. The prognosis of grade 5-b onwards is poor. At the stage of malignancy, i.e. stage 4 (grades 6-a, 6-b, and 6-c), surgical removal can only help patients of grade 6-a and 6-b provided glands are not affected. The role of antioxidants is under trial.

Such differentiation of various stages of arsenicosis is helpful to detect asymptomatic cases in pre-clinical or sub-clinical phase and to find the severity of the disease in order to prevent its further progress to complication and malignancy stages.

Keywords: arsenicosis, stages and grades, arsenical dermatosis, symptomatology, malignancy

1. Introduction

Historically, arsenic has been used for various purposes (Chisholm, 1970; La Dou, 1983; Hughes et al., 1988; Nriagu and Azcue, 1990), e.g. medicinal, domestic, agricultural, industrial etc. (Jhaveri, 1959; Flemming, 1964; Liessella et al., 1972; Hughes et al., 1988). Arsenical dermatosis (ASD) was known to occur from prolonged use of medicines for the treatment of syphilis (Moore, 1933), amebiasis, tropical eosinophilia (Macgrath, 1966), trypanosomiasis (Nash, 1960; Most, 1972), psoriasis (Goodman and Gilman, 1942), anemia (Leslie and Smith, 1978), asthma (Tay and Scale, 1975), and leukemia (Forknar and Scott, 1931). Now these are replaced by non-toxic drugs and as a consequence, ASD was practically unknown for the last 50–60 years. In the early 20th century, natural calamities like volcanic eruptions, scattered incidences of agricultural and domestic use of arsenic containing pesticides, rodenticides (Chisholm, 1970), wall paints, arsenic-contaminated beer (Luchtsath, 1972; Wolf, 1974), and industrial calamities in the western world were identified as factors causing arsenicosis.

West Bengal, a state in eastern India and the closest neighbor of Bangladesh, is one of the most densely populated places of the world having a population of over 68 millions with an area of 55,769 sq. miles (89,230 km^2.). West Bengal lies in the flood plain and is part of the River Ganges delta, while Bangladesh occupies the other half of the delta. Arsenicosis in West Bengal as well as Bangladesh was discovered by the author in 1982 and 1984, respectively. The first patient of West Bengal came to the out-patient department of the Calcutta School of Tropical Medicine (CSTM) from Gangapur village of North 24 Parganas district of the state (Saha, 1984), and the first case of Bangladesh came from Khulna district (Saha, 1985) of the country. The diagnosis of arsenicosis was confirmed by observing high arsenic content in nails, hair, and skin scales of the patients. Subsequent research and a field study in West Bengal and Bangladesh revealed the greatest arsenical calamity of the world.

The ground water in the villages is acquired for consumption by hand-operated pumps, locally known as "tube wells". Arsenic enters the body through intake of arsenic-contaminated ground water containing more than 0.05 mg/l of arsenic. Arsenic-contaminated ground water is found in arsenic affected areas, if the water is pumped by shallow tube wells of depth less than 300 ft. (commonly 50–150 ft.) (Saha and Poddar, 1986; Chakrabarty and Saha, 1987; Saha, 1995). Soil is the source of arsenic in most of the cases of arsenicosis. Leaching of arsenic from soil occurs due to lowering of underground aquifer level by excessive water withdrawal for agricultural and domestic purposes. Two hypotheses, "pyrite oxidation theory" and "iron oxy-hydroxide reduction theory", are put forward to explain leaching of arsenic from soil (Kinneburgh et al., 1944; Bhattacharya et al., 1997; Nickson et al., 1998). Probably both the mechanisms play a role in arsenic leaching leading to arsenicosis of geological cause. However, the author came across two instances of arsenicosis of exogenous origin in West Bengal (Saha, 1999, 2000a,b). Increasing number of arsenicosis cases are found from more than 2600 villages of 76 police station covering areas (locally known as "blocks") in 9 districts of West Bengal (Saha, 1997, 1998, 2000b) and about 30 districts of Bangladesh (Chowdhury et al., 2000).

2. Materials and methods

4865 patients of ASD in 1206 villages were thoroughly examined for clinical features. A field survey was initiated in 1983 (Saha, 1984) and it included clinical investigation, collection of tube well water samples, nails, hair and skin scales of patients. Measurement of arsenic content in skin scales was originally proposed by the author for easier processing of scales relative to nails. The cases of mild complications like asthmatic bronchitis, pyoderma, anemia, and avitaminosis received treatment in the field. However, serious cases with complications like jaundice or ascites were admitted in the hospital of the CSTM, Calcutta, India during 1983–1987 and other hospitals after that period, for further studies and chelation drug therapy. Arsenic content in tube well water, nails, hair and skin scales of the samples had been measured by SDDC (silver diethyldithiocarbazine) method at CSTM until 1987 (Saha, 1984, 1985, 1995; Saha and Poddar, 1986; Chakrabarty and Saha, 1987). A collaborative study was initiated with the School of Environmental Studies (SOES), Jadavpur University, India, in 1989 following the author's retirement from CSTM in 1987, and arsenic content has been measured by FI-HGAAS (flow injection hydride generation atomic absorption spectrometry) (Saha, 1997, 1998, 1999, 2000b; Chowdhury et al., 2000).

3. Results

The patients were evaluated for various symptomatology including dermatological as well as non-dermatological features (Table 1). Arsenicosis has been classified (Saha et al., 1999; Saha, 2000c) by the author into 4 stages, 7 grades and 20 sub-grades (Table 2). The four stages are pre-clinical (I), clinical (II), complications (III) and malignancy (IV).

3.1. Stage I (pre-clinical or asymptomatic or occult stage)

In the pre-clinical stage, there are no clinical signs on the patient's skin. This stage, also referred as grade 0, has further been sub-divided into two. The labile or blood phase has been designated as 0-a and the stable or tissue phase has been designated as 0-b. In labile phase, arsenic is present in both blood and urine but it disappears on withdrawal of arsenic-contaminated water. Arsenic is excreted in urine as monomethyl arsonic acid (MMA) and dimethyl arsinic acid (DMA). Although skin signs were absent in pre-clinical stage, nails, hair and skin scales showed high arsenic content in stable or tissue phase (0-b). Normal level of urine arsenic is 5–40 μg/day (water intake 1.5 l per day). Normal nails and hair arsenic content are 0.43–1.0 mg/kg (toxicity >1 mg/kg) and 0.08–0.25 mg/kg (toxicity >1 mg/kg), respectively. In 0-b or stable, sub-clinical stage, nails, hair, skin scales or other tissues contained high arsenic levels, though clinical signs were absent. Unaffected members of the affected family often belonged to this stage. Duration of pre-clinical stage varied between 6 months and 10 years depending on the concentration of arsenic in tube well water, volume of arsenic-contaminated water consumed per day and nutritional status. This stage lasted for 10 years or more if the arsenic content in water is within 0.06–0.1 mg/l, volume of arsenic-tainted water

Table 1. Analysis of clinical symptomatology (total – 4,865 cases).

	No. of cases	%
(I) Dermatological (external or cutaneous) features		
(A) Major dermatological features		
(1) Diffuse melanosis	4,865	100.0
(2) Spotted melanosis (rain drop pigmentation)	2,792	57.4
(3) Leucomelanosis (pigmented/depigmented)	1,542	31.7
(4) Palmo-planter keratosis (spotted/diffuse/hyperkeratosis)	2,987	61.4
(5) Dorsal keratosis	1,467	30.0
(B) Minor dermatological features		
(6) Mucous membrane pigmentation (in tongue, lips)	345	7.1
(7) Non-pitting edema	248	5.1
(8) Conjunctival congestion	194	4.0
(II) Non-dermatological (internal or systemic) features and complications		
(A) Major		
(1) Lungs – asthmatic bronchitis	1,576	32.4
(2) Liver – hepatomegaly	908	27.7
(3) Spleen – splenomegaly	72	1.5
(4) Liver fibrosis – ascitis	24	0.5
(B) Minor		
(5) Weakness and anemia	698	14.35
(6) Myalgia/burning body, neuropathy (sensory)	82	1.68
(7) Laryngitis	14	0.3
(8) Hypothyroidism	14	0.3
(9) Suprarenal deficiency	4	0.1
(10) Myopathy	1	0.02
(11) Ischemic gangrene (black feet)	1	0.02
(III) Malignancy	250	5.14
(A) Skin malignancy	212	4.35
(B) Internal malignancy	38	0.78

consumed was less than 1.5 l in 24 h, and the nutritional status was good. Skin features were often mild if these developed at all. Cases were found where patients of good nutritional status showed mild or no clinical signs despite consuming arsenic-contaminated water. Conversely, patients having malnutrition showed classical signs of arsenicosis if they consumed water containing borderline concentration of arsenic, i.e. 0.06–0.1 mg/l. In addition, children were affected quite frequently if malnutrition was observed. However, this stage lasted for only 6 months to 2 years, if the arsenic content in water was higher than 1 mg/l, volume of arsenic-contaminated water consumed per day was higher than 1.5 l/day, and nutritional status was poor, i.e. diet poor in protein, selenium, and vitamins. The features of clinical phase did not appear at all if the unsafe water having more than 0.05 mg/l of arsenic [above permissible limit recommended by WHO (World Health Organization (WHO) Environment Health Criteria, 1981)] was withdrawn in grade 0.

Table 2. Saha's grading of arsenicosis.

Stages	Grades	Inference
I. Pre-clinical	0	Pre-clinical
	0-a	Labile–blood phase
	0-b	Stable–tissue phase
II. Clinical	1	Melanosis
	1-a	Diffuse melanosis on palms
	1-b	Spotted melanosis on trunk
	1-c	Generalized melanosis
	2	Spotted keratosis on palms and soles
	2-a	Mild – 1–6 nodules
	2-b	Severe – more than 6 nodules
	2-c	Large nodules
	3	Diffuse keratosis on palms and soles
	3-a	Partial – only soles
	3-b	Severe – soles and palms
	3-c	Complete
	4	Dorsal keratosis
	4-a	Hands or legs
	4-b	Hands and legs
	4-c	Generalized
III. Complications	5	Hepatic disorder
	5-a	Palpable liver
	5-b	Jaundice
	5-c	Ascitis
IV. Malignancy	6	Malignancy
	6-a	Single lesion
	6-b	Two lesions
	6-c	More than two lesions

3.2. Stage II (clinical or symptomatic stage)

This symptomatic stage appeared after 6 months to 10 years (commonly 2–5 years). This stage has been divided into 4 grades.

Grade 1: melanosis
Grade 2: spotted keratosis in palms/soles
Grade 3: diffuse keratosis in palms/soles
Grade 4: dorsal keratosis

3.2.1. Grade 1 (melanosis)

Melanosis was the earliest sign to develop after 6 months to 10 years (usually 2–5 years). This phase was characterized by diffuse pigmentation of palms followed by gradual spreading over body. Melanosis was reversible as it disappeared in a few months after withdrawal of unsafe water and/or treatment with chelating agents such as British anti-Lewisite (BAL), D-penicillamine, dimercaptosuccinic acid (DMSA), or dimercaptopropane sulfonate (DMPS), etc. This grade has been further sub-divided into 3 sub-grades. Grade 1-a (Fig. 1) showed melanosis that was restricted to palms and soles. Mild melanosis could be revealed by comparing with normal palm. Spotted melanosis or rain drop pigmentation that appeared in trunk and limbs has been referred as grade 1-b (Fig. 2). Generalized melanosis, where whole body gradually darkens, has been designated as grade 1-c (Fig. 3). Chelating drugs were not usually necessary at this stage or grade, especially at grade 1-a.

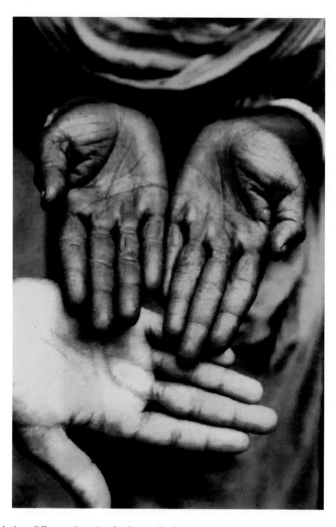

Figure 1.   Grade 1-a: diffuse melanosis of palms and soles.

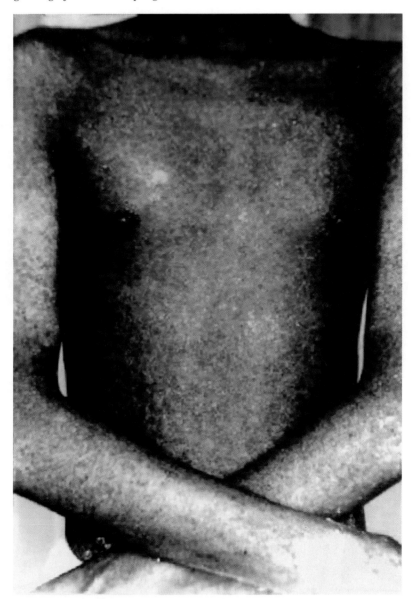

Figure 2.   Grade 1-b: spotted melanosis in trunk/limbs.

3.2.2. Grade 2 (spotted keratosis in palms/soles)

Keratotic spots gradually appeared on palms and soles. Small spots were hardly visible but palpable. The keratotic spots increased in number and size gradually, and became visible. Grade 2 was sub-divided into 3 sub-grades. Grade 2-a (Fig. 4) showed small keratotic spots (1–6 spots) on palms and soles. Grade 2-b (Fig. 5) showed more than 6 keratotic

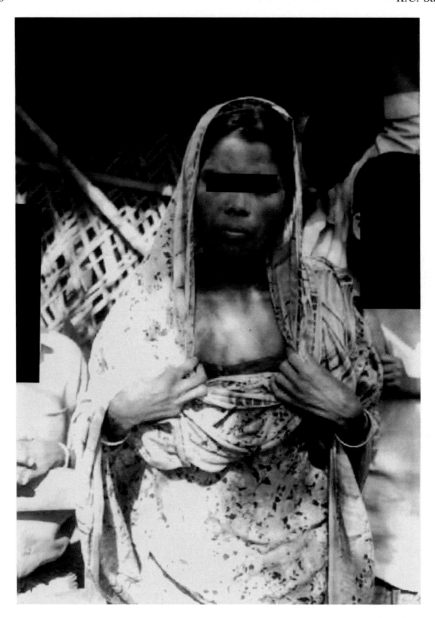

Figure 3.   Grade 1-c: generalized melanosis.

spots on palms/soles. Grade 2-c (Fig. 6) showed large keratotic nodules in palms/soles. The combination of melanosis and keratotic spots in an adult is almost a sure clinical sign of arsenicosis. This grade and all other subsequent grades were irreversible, i.e. they did not disappear despite the withdrawal of arsenic-contaminated unsafe water or chelating drug therapy.

Figure 4.   Grade 2-a: spotted keratosis – palms/soles (1–6 nodules).

3.2.3. Grade 3 (diffuse keratosis of palms/soles)

Grade 3 was characterized by gradual thickening of palms and soles. Spotted keratosis still remains at this stage and subsequent stages. This grade has been further sub-divided into 3 sub-grades. Grade 3-a (Fig. 7) showed partial thickening of palms and soles. Grade 3-b showed greater partial thickening of palms and soles. Grade 3-c (Fig. 8) showed complete thickening. The skin of palms and soles could not be pinched in complete, diffuse hyperkeratosis.

3.2.4. Grade 4 (dorsal keratosis)

This grade was characterized by visible nodular spots on dorsum of feet and hands. This feature may or may not be associated with complete diffuse keratosis of palms and soles; however, some partial diffuse keratosis was always present. Diffuse and spotted melanosis in body were also present.

This grade has been further sub-divided into 3 sub-grades. Grade 4-a (Fig. 9) showed nodular lesions on hands or feet. Grade 4-b (Fig. 10) showed nodular lesions on hands and feet and keratosis extended to body in grade 4-c (Fig. 11). The keratosis in this phase has also been distributed over limbs and trunk. Spotted and diffuse melanoses on body and nodular palmo-plantar keratosis persisted at this grade.

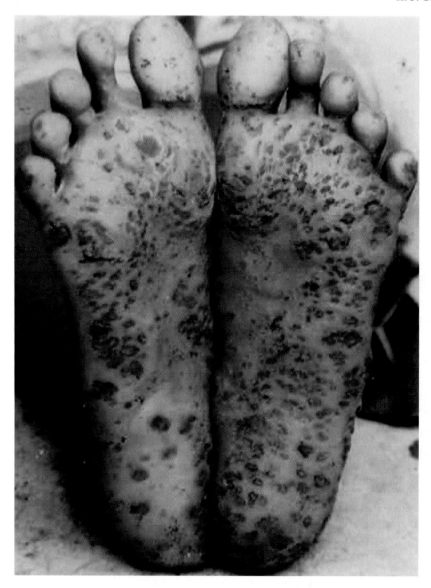

Figure 5.   Grade 2-b: spotted keratosis of palms/soles (>6 nodules).

3.3. Stage III (complication stage)

This stage was also characterized as grade 5 and associated with liver damage. Lung complication with asthmatic bronchitis could be seen in earlier stages, i.e. grade 2, 3, and 4. The stage has been further sub-divided into 3 sub-grades. Grade 5-a was associated with palpable liver. Grade 5-b was associated with jaundice. Grade 5-c was associated with ascitis that developed from portal hypertension due to non-cirrhotic portal fibrosis (NCPF).

Figure 6. Grade 2-c: spotted keratosis of palms/soles (large nodules).

Liver damage due to arsenicosis could be inferred, if the associated features of arsenicosis such as melanosis and keratosis were present. Isolated hepatomegaly, jaundice or ascitis without skin signs of arsenicosis should not be inferred to result from arsenicosis. Jaundice or ascitis was hardly recoverable. Chelating agents (DMPS/DMSA) were of little benefit at this stage.

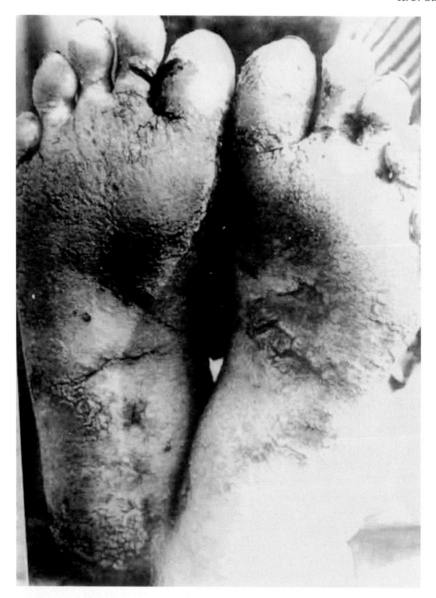

Figure 7. Grade 3-a: diffuse keratosis partial – palms or soles.

3.4. Stage IV (malignancy)

The stage of malignancy (Hutchinson, 1887; Hill and Faneng, 1948; Tseng et al., 1968; Morton et al., 1976; International Agency for Research on Cancer (IARC), 1980; Tsuda et al., 1995; Saha, 2001), also classified as grade 6, usually developed 10–20 years after the onset of symptoms (Saha, 2000c, 2001). There was hardly any recovery once this stage was attained. Again, cancer due to arsenicosis should only be inferred if the associated

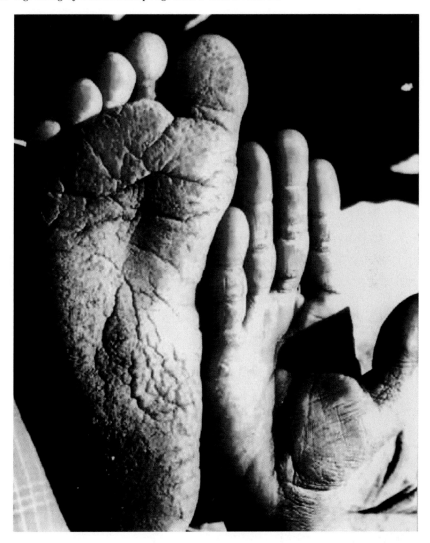

Figure 8. Grade 3-c: diffuse keratosis of complete palms/soles.

signs of arsenicosis, i.e. melanosis and keratosis, were present. 250 cases of malignancy consisting of both skin (212 cases) as well as internal organs (38 cases) were found. In skin malignancy, 161 cases of squamous cell carcinoma and 51 cases of Bowen's disease were found. Internal malignancy included lung carcinoma (25 cases), gastrointestinal tract carcinoma (3 cases), bladder carcinoma (5 cases), and genital tract carcinoma (5 cases) (Table 3).

This stage has further been sub-divided into 3 sub-grades. Grade 6-a (Fig. 12) was associated with single lesion. Grade 6-b (Fig. 13) was associated with two lesions. Grade 6-c (Fig. 14) was associated with more than 2 lesions. All cases of arsenicosis did not advance to the malignant stage. However, this stage was likely to appear if unsafe water

Figure 9.   Grade 4-a: dorsal keratosis of hands or feet.

intake was not withdrawn or chelating therapy was not instituted in keratotic stage. The
author observed that the malignancy continued to develop 15 years after cessation of
unsafe water (>0.05 mg/l) intake and chelating drug, e.g. 2,3-dimercaptopropanol and
D-penicillamine, therapy.

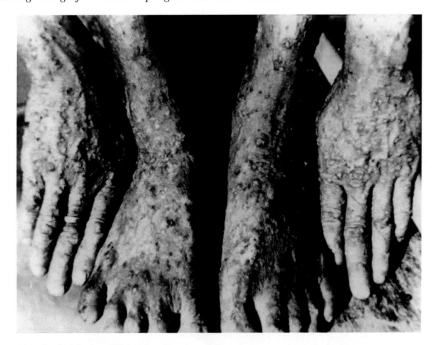

Figure 10.   Grade 4-b: dorsal keratosis of hands and feet.

3.5. Other unclassified features

Apart from melanosis and keratosis, leucomelanosis (parallel appearance of depigmented and pigmented spots), non-pitting edema, conjunctival congestion, mucous membrane melanosis, and sensory neuropathy were present in some cases, which were not included in the classification (Saha, 1995, 1999, 2000a,b). Black foot disease (Tseng et al., 1961) or ischemic gangrene, which is common in Taiwan, was observed in only one case in the author's series of finding of arsenicosis in West Bengal, India. In Taiwan, perhaps high concentration of arsenic in water acted on the wall of the vessels producing sclerotic ischemic necrosis. Relatively low concentration of arsenic in West Bengal with prolonged action might result in proliferation of melanocytes and keratocytes rather than acting on the wall of the vessels.

4. Treatment

There is no satisfactory treatment for arsenicosis. Avoidance of arsenic-contaminated water is the best option, if possible. Arsenic-free water along with the maintenance of nutritious diet supplemented with protein and vitamins, especially vitamin A, C and selenium are required for all stages (Table 4). The melanosis and mild keratosis (grade 1 and 2-a) were often cleared in 2–3 months by the above treatment.

Chelating drugs, such as DMPS, DMSA, etc., were administered on cases of grade 2-b having more than six keratotic nodules and further advanced grades to prevent progression

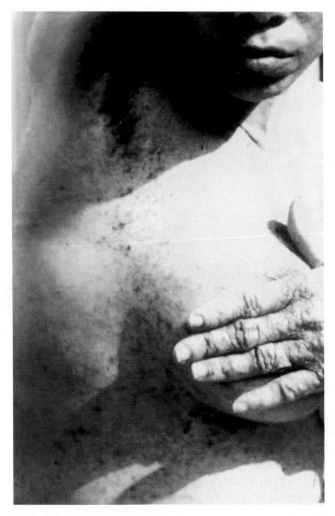

Figure 11.   Grade 4-c: generalized keratosis – body.

of disease (Table 5). BAL was effective but not administered currently because of toxicity. However, D-penicillamine was not found to be effective at all.

Keratolytics, such as 6% salicylic acid with preferably 20% urea, was given to soften hard keratotic nodules on palms and soles.

Mechanical scrapping of thickened soles, as proposed by the author, often relieved the patients' discomfort due to thick soles.

Surgery was carried out for large keratotic nodules, ischemic gangrene, and non-metastatic (isolated) carcinoma.

Antibiotics were given for bronchitis, ulcers, gangrene, and other infections. Bronchodilators and mucolytics were given for asthmatic bronchitis. Other symptomatic treatment such as hematinics (iron, folic acid, etc.) were often given for anemia and vitamin B1 for neuropathy.

The antioxidants are under trial.

Table 3. Malignancy in arsenicosis.

District	TSM	SCC	BD	Lung	GIT	Blad	GT	Total	Percent
N. 24 parganas	65	45	20	5	1	2	2	75	30.00
Murshidabad	43	29	14	6	2	1	2	54	21.60
Nadia	39	34	5	5	0	0	1	45	18.00
S. 24 parganas	21	15	6	4	0	1	0	26	10.40
Malda	19	18	1	3	0	1	0	23	9.20
Bardhaman	21	16	5	1	0	0	0	22	8.80
Calcutta	4	4	0	1	0	0	0	5	2.00
Hugli	0	0	0	0	0	0	0	0	0.00
Haora	0	0	0	0	0	0	0	0	0.00
Total	212	161	51	25	3	5	5	250	100.00

TSM, total skin malignancy (SCC + BD); SCC, squamous cell carcinoma; BD, Bowen's disease; GIT, gastrointestinal tract; Blad, bladder; GT, genital tract.

5. Discussion

Both under-diagnosis and over-diagnosis of arsenicosis should be avoided. The melanosis of arsenic origin (grade 1) needs to be differentiated from other causes of pigmentation. *Sun melanosis* is restricted to exposed parts. In *Addison's disease*, the patient is prostrated with hypotension and pigmentation of palmar creases with low serum cortisol. In *hemochromatosis*, the patients have diabetes and high serum iron. In *ephelids*, spotted

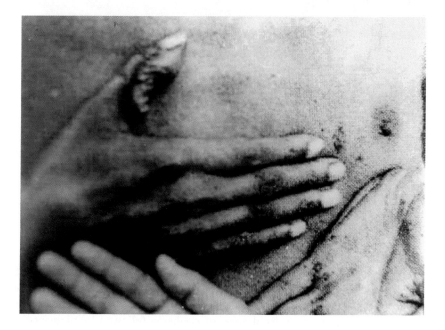

Figure 12. Grade 6-a: malignancy (single lesion).

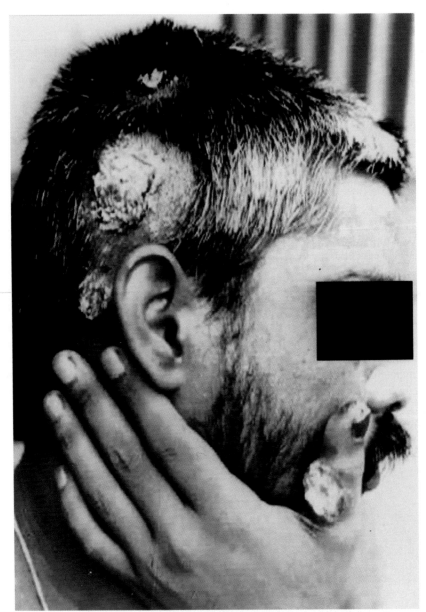

Figure 13.   Grade 6-b: malignancy (2 lesions).

pigmentation is restricted to face, and is present since childhood. In *xeroderma pigmentosa*, spotted and generalized pigmentation or leucomelanosis is found since birth, with horizontal pedigree pattern. However, palms are not pigmented. *Seborrheic dermatitis* is associated with itching, alopecia, acne, and seborrheic distribution. In *fixed drug rash*, fixed areas of erythema followed by pigmentation are found. The pigmentation is due to intake of a fixed drug, which an individual is sensitive

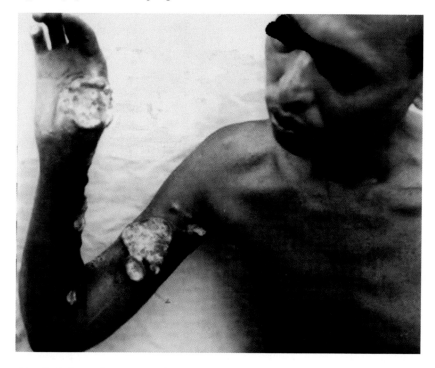

Figure 14. Grade 6-c: malignancy (>2 lesions).

to. *Nevus* is a birthmark present since birth. *Contact dermatitis* is associated with history of medicated oil application or other contactants with itching and scaling. *Pityriasis versicolor* is associated with small scales on small grain-like lesions that are present around large coalescent patches in the center of the trunk. *Epidermo dysplasia virruciformis* is associated with multiple, small, flat, dark-colored warts, of mostly the same size. These warts are non-pruritic having Koebner's phenomenon often. *Ichthyosis* is associated with fish scale appearance of skin. *Chronic porphyria* (erythropoetic porphyria or porphyria cutanea tarda) is associated with hyperpigmentation, hypertrichosis, blistering, scarring, red urine, and sclerodermoid changes. In addition, red fluorescence of RBC and urine are observed along with excess porphyrin excretion in urine in erythropoetic porphyria.

Grade 2 having spotted keratosis in palms and soles has to be differentiated from genetic keratoderma, verruca vulgaris (warts), and pitted keratolysis. *Genetic keratoderma* is present at birth and familial with no melanosis. *Verruca vulgaris* and *verrucous tuberculosis* are associated with warty lesions (chronic papules with rough surface) usually unilateral with no melanosis of year's duration. In *verrucous tuberculosis*, tuberculin test and ELISA for *Mycobacterium tuberculosis* are positive. Pitted keratolysis is associated with multiple and small lytic depressions on both soles that result from water-logged condition.

Grade 3, i.e. diffuse keratosis on palms/soles, should be differentiated from *tylosis*, *scleroderma*, and *mycetoma*. *Tylosis* is associated with bilateral keratosis, present since

Table 4. Plan of treatment at different stages and grades in arsenicosis.

Stages	1 Pre-Clinical	2 Clinical				3 Complication	4 Malignancy
Grades	0 0-a (labile), 0-b (stable)	1 Melanosis	2 Spotted keratosis	3 Diffuse keratosis	4 Dorsal keratosis	5 Lungs, liver, others	6 Skin internal
Treatment							
Arsenic water avoidance	+	+	+	+	+	+	+
Nutrition	+	+	+	+	+	+	+
Iron, vit. A, C, Se, suppl.	+	+	+	+	+	+ (bronchitis,	+
Antibiotics	−	−	−	−	−	+ (bronchitis, ulcers, gangrene)	(sec. infection)
Chelation therapy	−	−	− (2-a) + (2-b, c)	+	+	+	+
Symptomatic Tr.	−	−	+ (2-c)	+	+	+ (gangrene)	+
Surgery	−	−	+ (2-b, c)	−	−	−	+
Keratolytics	−	−	+ (2-b, c)	+	+/−	−	−
Mechanical	−	−	+ (2-b, c)	+	+/−	−	−
Antioxidants	−	−	?	?	?	−	−

+, effective; −, ineffective; ?, under trial.

Table 5. Chelating agents used in arsenicosis.

Chelating drugs	Initial dose	Subsequent dose
BAL (2,3 dimercaptopropanol)	3–5 mg/kg/i.m. every 4 h	B.D. till urine As <50 mg/daily
D-Penicillamine	250 mg T.D. or Q.D.	For 4–6 weeks
DMSA (dimercaptosuccinic acid)	10 mg/kg/T.D. for 7 days	10 mg/kg/B.D. 14 days
DMPS (dimercaptopropane sulfonate)	100 mg T.D. or Q.D.	Alternate weeks 3 courses

i.m., intra-muscular; T.D., three times daily; Q.D., four times daily; B.D., two times daily.

birth with no melanotic spots in the body, and hereditary in origin. *Scleroderma* is associated with thick, non-elastic, and hard skin. The symptoms get aggravated on cold exposure with painful hardened finger or toe tips (Raynaud's phenomenon). Mycetoma is associated with chronic swelling of one ankle or foot or other part of the body with multiple sinuses. The lesions last for months or years. The fungus can be identified by culture of crushed granules.

Grade 4, i.e. dorsal keratosis, should also be differentiated from seborrheic keratosis, Darier's disease, epidermo dysplasia verruciformis, pityriasis rubra pilaris, sporotrichosis, acne, rosacia, syringoma, trichoepithelioma, lichen amyloidosus, follicular lichen planus, tuberculosis verrucosa cutis, post Kala Azar dermal leishmaniasis (PKADL), adenoma sebeceum, lepromatous leprosy, and neurofibromatosis. *Seborrheic keratosis* is associated with pigmented, waxy, raised, non-pruritic papules on the surface of the back of an elderly person. *Darier's disease* is associated with generalized papules with scales, non-pruritic, present since childhood in seborrheic distributions (scalp, face, sides of ear, nose, neck, chest, axilla, and limbs*). Pityriasis rubra pilaris* is associated with generalized erythema with fine scales and small papules present since childhood affecting dorsal aspect of fingers (particularly, index finger) also. *Sporotrichosis* is associated with chronic nodules followed by ulceration in linear arrangement due to lymphatic spread and usually found in the fore-arm of an adult. *Acne* has associated papules and pustules with comedones, i.e. blackheads in the center of papules that usually appear on the face, chest, and back in adults. Cheesy material is expressed from papules with black top. It is often associated with dandruff in scalp and seborrheic dermatitis. *Rosacia*is associated with papules and pustules on face without comedones. Flushing of face is observed on exposure to sunlight. *Syringoma* is associated with multiple small papules below the eyes and present since childhood. *Trichoepithelioma*, a familial disease, is associated with multiple papules around nose and present since birth. *Lichen amyloidosus* is associated with chronic pruritic papules on both legs of adults. *Follicular lichen planus* is very pruritic and associated with flat-topped papules of light violet color and present in legs or forearms. *Tuberculosis verrucosa cutis* is a warty lesion present on one foot and lasts for years. Tuberculin test is highly positive. Biopsy confirms the diagnosis. *PKADL* is associated with a history of Kala Azar in the past. All three types of lesions, papular, hypopigmented, and erythematous, are present in the same patient. Axilla, neck, and waist are spared from the disease. Slit smear from nodular lesion shows many LD bodies on Leishman stain. *Adenoma sebaceum* is associated with multiple and small papules on central face around nose and present since childhood. This is often associated with lesions in brain resulting in convulsive fits, mental deficiency, subungual fibroma (Kinen's tumor) and Shagreen's plaque. *Lepromatous leprosy* is associated with multiple nodules on body, ear lobules, maderosis, i.e. loss of hair on outer half of eyebrows, and glossy skin. Slit smear shows many AFB on ZN stain. *Neurofibromatosis* is associated with multiple, non-pruritic skin nodules having caffe-au-leit spots and present since childhood.

6. Conclusions

The following conclusions can be drawn:

a) Features of arsenicosis are classified by the author into 4 stages, 7 grades, and 20 sub-grades.
b) Grading of arsenicosis is helpful to know the severity of the disease and plan the treatment.
c) Grade 1, i.e. melanosis, is reversible by withdrawal of arsenic-contaminated water. Chelating agents therapy is usually not necessary at this grade.
d) Further progression to subsequent stages (II, III, and IV) and grades (1–6) can be prevented if arsenic-contaminated water intake is ceased and/or chelating agents therapy is instituted, especially from grade 1-b onwards.
e) Grades 2–6 are irreversible.
f) Chelating agents are helpful to prevent further progression of disease, if they are used in grades 2, 3, and 4.
g) The use of chelating agents cannot rule out the possibility of further advancement of the disease to the stage of malignancy.

Acknowledgements

1. Poddar, G., Professor of Chemistry, The School of Tropical Medicine, Calcutta, West Bengal, India.
2. Chakrabarty, D., Director, School of Environmental Studies, Jadavpur University, West Bengal, India.
3. The Indian Council of Medical Research (ICMR) for grant supporting research during 1984–1986.

References

Bhattacharya, P., Chatterjee, D., Jacks, G., 1997. Int. J. Water Res. Dev., 13, 79–92.
Chakrabarty, A.K., Saha, K.C., 1987. Indian J. Med. Res., 85, 326–334.
Chisholm, J.J., Jr., 1970. Arsenic poisoning from rodenticides. Pediatr. Clin. N. Am., 17, 591.
Chowdhury, U.K., Biswas, B.K., Chowdhury, T.R., et al., 2000. Arsenic contamination in Bangladesh and West Bengal, India. Environ. Health Perspect., 108 (5), 393–397.
Flemming, A.J., 1964. Industrial hygiene and medical control procedures. Arch. Environ. Health, 8, 266–270.

Forknar, C.E., Scott, T.F.M., 1931. Arsenic as therapeutic agent in chronic myelogenous leukemia – preliminary report. J. Am. Med. Assoc., 97, 3–5.

Goodman, L., Gilman, A., 1942. The pharmacological basis of therapeutics, 4th printing. Macmillan, New York.

Hill, A.B., Faneng, E.L., 1948. Studies in the incidence of cancer in a factory handling inorganic compounds of arsenic. I. Mortality experience in the factory. Br. J. Ind. Med., 5, 1–15.

Hughes, J.P., Polissar, L., VanBelle, G., 1988. Evaluation and synthesis of health effects studies of communities surrounding arsenic producing industries. Int. J. Epidemiol., 17, 407–413.

Hutchinson, J., 1887. Arsenic cancer. Br. Med. J., 2, 1280–1281.

International Agency for Research on Cancer (IARC), 1980. Monographs on the Evaluation of the Carcinogenic Risk of Chemicals to Humans: Some Metals and Metallic Compounds, Vol. 23. Lyon, France, pp. 39–141.

Jhaveri, S.S., 1959. A case of cirrhosis and primary carcinoma of liver in chronic industrial arsenic intoxication. Br. J. Ind. Med., 16, 242–248.

Kinneburgh, D.G., Gale, J.N., Smedley, P.L., et al., 1944. Appl. Geochem., 9, 175–195.

La Dou, J., 1983. Potential occupational health hazards in the microelectronics industry. Scand. J. Environ. Health, 9, 42–46.

Leslie, A.C.D., Smith, H., 1978. Self poisoning by the abuse of arsenic containing tonics. Med. Sci. Law, 18, 159–162.

Liessella, F., Long, K.R., Scott, H.G., 1972. Health aspects of arsenicals in the environment. J. Environ. Health, 34, 511.

Luchtsath, H., 1972. Cirrhosis of liver in chronic arsenic poisoning of vinters. German Med. Monthly, 2, 127.

Macgrath, B., 1966. Tropical eosinophilia. Prices Textbook of Medicine, Oxford University Press, Bombay.

Moore, J.E., 1933. Modern Treatment of Syphilis. Charles C. Thomas, Springfield III.

Morton, W., Starr, G., Pohl, D., Wagner, S., Weswig, D., 1976. Skin cancer and water arsenic in Lane county, Oregon. Cancer, 37, 2523–2532.

Most, H., 1972. Drug therapy – treatment of common parasitic infections of man encountered in the United States. N. Engl. J. Med, 287, 698.

Nash, T.A., 1960. A review of the African trypanosomiasis problem. Trop. Dis. Bull., 57, 973–1013.

Nickson, R., McArthur, J., Burgess, W., et al., 1998. Nature, 395, 398.

Nriagu, J.O., Azcue, J.M., 1990. Food contamination with arsenic in the environment. Adv. Environ. Sci. Technol., 23, 121–143.

Saha, K.C., 1984. Melanokeratosis from arsenic contaminated tube well water. Indian J. Dermatol., 29 (4), 37–46.

Saha, K.C., 1985. Skin disease from arsenic contaminated tube well water in West Bengal. Found. Souvenir Calcutta School Trop. Med., 8 (24th Feb).

Saha, K.C., 1995. Chronic arsenical dermatoses from tube well water in West Bengal during 1983–1987. Indian J. Dermatol., 40 (1), 1–12.

Saha, K.C., 1997. Arsenic tainted water in West Bengal. Environ. Rev., 14 (2), 6–9.

Saha, K.C., 1998. Arsenic poisoning from ground water in West Bengal. Breakthrough, 7 (4), 5–14.

Saha, K.C., 1999. A bird's eye view on arsenical calamity in West Bengal. Indian J. Dermatol., 44 (3), 116–118.

Saha, K.C., 2001. Arsenicosis. Continued education series. Indian J. Dermatol., 46 (1), 8–17.

Saha, K.C., 2000. 17 years experience in arsenicosis (abstract). Fourth International Conference on Arsenic Exposure and Health Effects, SEGH, San Diego, California, June 18–22, p. 53.

Saha, K.C., 2000c. Stages and grades of arsenicosis (abstract). Fourth International Conference on Arsenic Exposure and Health Effects, SEGH, San Diego, California, June 18–22, p. 147.

Saha, K.C., 2001. Cutaneous malignancy in arsenicosis. Br. J. Dermatol., 145/1, 185.

Saha, K.C., Poddar, J., 1986. Further studies on chronic arsenical dermatosis. Indian J. Dermatol., 31, 29–33.

Saha, J.C., Dikshit, A.K., Saha, K.C., 1999. A review of arsenic poisoning and its effects on human health. Crit. Rev. Environ. Sci. Technol., 29 (3), 281–313.

Tay, C.H., Scale, C.S., 1975. Arsenic poisoning from anti-asthmatic herbal preparations. Med. J. Aust., 2, 424–428.

Tseng, W.P., Chen, W.Y., Sung, J.L., Chen, J.S., 1961. A clinical study of black foot disease in Taiwan. Mem. Coll. Med. Natl Taiwan Univ., 7, 1–17.

Tseng, W.P., Chu, H.M., How, S.W., et al., 1968. Prevalence of skin cancer in an endemic area of chronic arsenicism in Taiwan. J. Natl cancer, 40, 453–463.

Tsuda, T., Babazono, A., Yamamoto, E., Kurumatani, N., Mino, Y., Ogawa, T., Kishi, Y., Aoyama, H., 1995. Ingested arsenic and internal cancer: a history cohort study followed for 33 years. Am. J. Epidemiol., 141 (3), 198–209.

Wolf, R., 1974. On the question of occupational arsenic poisoning in vineyard workers. Berufsdermatosen, 22 (1), 34–37.

World Health Organization (WHO) Environment Health Criteria, 1981. Arsenic. WHO, Geneva.

Arsenic Exposure and Health Effects V
W.R. Chappell, C.O. Abernathy, R.L. Calderon and D.J. Thomas, editors
415

Chapter 31

Painting tube wells red or green alone does not help arsenicosis patients

Quazi Quamruzzaman, Mahmuder Rahman, M.A. Salam,
A.I. Joarder, M. Shahjahan and S.U. Mollah

Abstract

The problem of arsenic contamination of hand-pump shallow tube wells in Bangladesh became known in 1993. As 97% of the population drink tube well water, various studies have predicted that nearly 85 million people are at risk of arsenic poisoning.

The experience of working at the field level in Bangladesh shows that the present mitigation programmes may not be effective due to faulty design. These programmes are mainly centred on identifying contaminated tube wells by field test kits and then painting the tube well red or green according to an unsafe or safe arsenic level. Some programmes include models of alternative water supply options and some include field-level patient identification with the provision of vitamins and ointments for lesions.

The Dhaka Community Hospital's experience of treating hundreds of affected patients demonstrates that investigation, diagnosis, appropriate treatment, physical and socio-economic rehabilitation, cost analysis and burden should be essential components of any arsenic mitigation programme.

The current emphasis of programmes and funds on safe water supply only, ignores the human dimension of the arsenic problem. Safe water supplies are essential – but the management of those already affected by arsenicosis should be the main focus of any arsenic mitigation programme.

Keywords: arsenic mitigation programme design, patients

1. Introduction

Arsenic contamination of groundwater in Bangladesh has been recognised since 1993. Ninety-seven per cent of the Bangladeshi population has relied on an estimated 11 million shallow tube well hand pumps for their daily water supply and it now appears that 85 million people could be at risk of exposure to arsenic. After a slow start to mitigation activities the problem now attracts many millions of dollars in pursuit of a solution. Projects have been undertaken by major donors and the Government of Bangladesh under the "Bangladesh Arsenic Mitigation Water Supply Project" (BAMWSP), which has also been co-ordinating all the work.

It is proposed here that the present projects whose main outcome is the painting of tube wells either red (contaminated) or green (safe) will not produce effective mitigation because of their design limitations and that the current emphasis of the projects ignores the human dimension of the disaster, so funding is not being used to achieve maximum benefits for the people of Bangladesh.

2. Existing project designs

The main component of the existing projects is the testing of shallow tube well water for the presence or absence of arsenic. Testing is carried out by field test kit, the owner is advised of the result and implications for the future use of the well and the well painted red or green.

Most of the projects advise users to seek alternative water supplies when the tube well is marked red and some include the construction and donation of a demonstration model for an alternative water supply.

Another component of these projects has been the identification of people already suffering from symptoms of chronic arsenic poisoning. Identification is made through visual diagnosis of skin manifestations, sometimes by doctors through health worker referrals and sometimes through health worker diagnosis only. Once identified, the patients are then sometimes given a three week supply of vitamins and salicylic ointment for skin lesions and advised to seek further help from the available government health service providers. The testers carry out some awareness campaigning during testing. Only one project has a research component and that is concerned with the development of domestic filters.

3. Problems with existing designs

When it became established that the aquifers of Bangladesh contained arsenic it was a laudable ambition to make sure that people did not drink contaminated water but the effort was directed at looking for contaminated water rather than looking for safe water and almost no attempt was made to help those people whose lives were already affected by chronic poisoning.

To mobilise enough teams to seek and test and then retest every tube well would require a mammoth undertaking in a developed country where communication, transportation and recording are well established. In areas where it can take 2 days to reach a single village the time to cover the entire country is prohibitive. To date, the BAMWSP project has only tested 41 of the 460 targeted upazilas (administrative areas) and only 1.3% have received help with establishing alternative safe water supplies, even when some villages have almost 100% contaminated wells.

All these projects use a one-off testing procedure but it is acknowledged that the action of arsenic in groundwater is dynamic; "green" tube wells may become contaminated as arsenic levels fluctuate. Ideally tube wells need regular re-testing but facilities for this do not exist, as the testing teams move on, so do the test kits.

Few advances have been made in the field test kits in terms of safety and reliability since the problem was first detected in Bangladesh. The tester continues to be exposed to toxic gas inhalation and the test result reliability is subject to the tester's consistency and result interpretation (especially in matching degrees of colour-shade). Few of the safer kits test at the 0.05 mg/l level of the Bangladeshi Government's safety limit.

A further difficulty is the problem of locating all tube wells and maintaining their status. Those sunk inside properties are sometimes not tested at all and those outside risk losing their paint. With no constant awareness campaign, motivation to seek "safe" water declines, especially when the tube well is used for commercial purposes.

While water testing goes on, arsenicosis patients receive virtually no funding from any mitigation project apart from the ointment and vitamins, valued at $0.27 and $0.90. The efficacy of the vitamin treatment is not established. Meanwhile there is no provision for training any medical staff in managing arsenicosis patients at the centres to which they are referred. There are no rehabilitation plans.

Once the extent of contamination became apparent, it would have been appropriate to re-evaluate project impacts and re-examine the project designs. This is a human disaster but little attention has been given to the effects already being felt by some users or the potential health damage that could occur in the future.

4. Proposed design

The design proposed here is based on the Dhaka Community Hospital (DCH) field experience in arsenic exposed areas of identifying, treating and managing over 13,500 chronic arsenicosis patients suffering from melanosis, keratosis or carcinogenic changes as well as the usual debilitating effects produced by chronic poisoning of the human body. It centres on the investigation, diagnosis, appropriate treatment and physical and socio-economic rehabilitation of patients, taking into account the cost analysis and burden to make the projects sustainable in the long term.

By focussing on the victims, a sustainable project design emerges that is both practical and relevant to local conditions. Costs become affordable when the design is engineered to the local context.

5. Investigation and diagnosis

Diagnosis of arsenicosis should not take place without full investigation that should include a history of water consumption and the testing of water supplies for arsenic. It should be based on entire households assuming they are sharing a common water supply. Family discrepancies should be recorded for primary care follow-up as these people may be at increased risk.

Definite diagnosis of arsenicosis necessitates biological testing of urine, hair and/or nails and skin biopsies, where relevant. Patients have the right to correct diagnosis. The cost of these tests (approximately $10.50) can be borne either by government for those of limited income or by the patient for those preferring to use private facilities.

6. Management

Patient management (currently a one-off activity) must be ongoing for the chronic condition suffered. It involves primary, secondary and tertiary health care and the existing health care facilities can be utilised to provide this. (In the long term, training in arsenicosis management should be introduced at all levels of medical training). The cost responsibility of patient management will be reflected in the same way as current practices.

The obvious first step in primary management is to stop consumption of contaminated water and this necessitates water testing and the introduction of alternative safe water supplies. As long as there is no central government policy of water management replacing tube wells then it makes sense to have a reliable water testing method based at local level. Portable spectrophotometers give more accurate results and are friendlier to use than field test kits and could be used for other testing purposes as well, if required. Although the test cost is higher (field kit testing costs between $0.20 and $0.76 while AAS costs $6), it is the tube well owner's responsibility to maintain the tube well and field experience shows that anyone who has privately invested in a tube well is willing to pay maintenance costs for that supply. The government has licensing systems in practice (e.g. radios) and a similar system could be used for drinking water with regular renewal dates required. Government, government regulated private organisations or companies could carry out testing.

Currently funded alternative safe water options take little account of the natural water resources of the country, a deltaic region with an annual monsoon where historically each village has had its own central water tanks and communal water sharing points, sometimes ponds, sometimes dug wells. Culturally, without a nationally organised welfare system, small communities have developed their own mutual support systems and field experience shows that community water supplies are possible and people are prepared to pay jointly for water, e.g. rainwater harvesting or piped reservoir water. Filtration units may be a valuable solution in some contexts but how it will benefit the rural poor of Bangladesh is yet to be financially analysed and the sludge disposal problem has still to be solved.

Primary level patient management conducted at the household level with health worker and community support is observed to work well but requires continual awareness building reinforcement; a task that DCH has found to be efficiently performed by young people of the communities as well as health workers.

Advanced arsenicosis symptoms, e.g. malignancies require secondary and tertiary intervention and these need to be integrated within the current health system so that consistent care is achieved.

Post-secondary and tertiary intervention can be followed up by the health workers with community support, particularly necessary in cases of permanent handicap and chronic sickness.

7. Rehabilitation

There are no rehabilitation activities in any of the current projects but patients require both health and socio-economic rehabilitation. Most rural work is labour intensive and symptoms of arsenicosis such as keratosis make labouring impossible. Those who suffer carcinogenic changes have even greater need for rehabilitation. The impact of awareness building and primary health care helps to enable health and social rehabilitation. The government already has a vulnerable group support of food and housing scheme that could be extended to include arsenicosis victims to assist in economic rehabilitation (another reason why diagnostic techniques must be accurate). It is also DCH's experience that the change to alternative shared water supplies and intensive community health work has generated some employment for patients and their families, assisting in their economic rehabilitation.

8. Research

By focussing mitigation projects on the patients, it is possible to incorporate research components that may be conducted by national institutions under their present research facilities. This should include applied and community need investigations, development of water options suitable for Bangladesh, patient management and epidemiology. Little is known of the long-term effects of arsenic exposed populations in Bangladesh and the early establishment of a government recording and monitoring system provides an opportunity to increase understanding of the nature and long-term outcome of the disease and so provide knowledge for future economic planning, particularly important in a country that relies on labour intensive activities for its economic survival.

Project mitigation activities for arsenic contamination of groundwater must be directed towards the human aspect so that solutions that are appropriate and relevant will emerge.

Arsenic Exposure and Health Effects V
W.R. Chappell, C.O. Abernathy, R.L. Calderon and D.J. Thomas, editors
© 2003 Published by Elsevier B.V.

Chapter 32

Arsenic mitigation in Bangladesh
Progress of the UNICEF–DPHE Arsenic Mitigation Project 2002

Colin Davis

Abstract

In Bangladesh, the magnitude of the task and the consequent dimension of the effort needed to mitigate the problem have both being growing rapidly as the level of understanding increases. The UNICEF–DPHE Project of arsenic measurement and mitigation has developed in four phases so far, with 10% of the nation's upazilas (sub districts) being allocated to the Project, most of these being in the known "hot spot" areas. The 1st phase was the national random survey. The 2nd phase was the Five Upazila Action Research Project. The 3rd phase was an additional 15 upazila expansion and the 4th phase, a further 25 upazilas, will be implemented in 2002 making a blanket tube well testing project area of 45 of the "hot spot" upazilas. In addition to this, UNICEF and DPHE have completed plans to carry out a rapid survey in the 200 upazilas, which are not yet allocated to any Project and this work has begun in the second half of 2002. This means that the UNICEF–DPHE Project is testing in almost half of the country and this work will be completed by the end of 2002.

Working in partnership with government and NGOs, over 450,000 tube wells have already been tested by the UNICEF–DPHE Project. In 2002, it is estimated that an additional 650,000 wells will be tested making a total of over 1,100,000. Surprisingly, preliminary analysis of testing data indicates that approximately 50% of the contaminated handpumps have actually only been installed for less than 5 years. This work has discovered hundreds of villages and schools that have no safe source at all. More than 2700 new arsenicosis patients have been diagnosed in the 3rd phase upazilas, equating to a prevalence rate of 0.92/1000 for those people using contaminated wells. Extrapolated for the total 45 upazila Project area this would be approximately 8000 people with arsenicosis. In addition to the testing programme, the Project must simultaneously divert some resources and ingenuity to helping people to obtain a safe water source.

Keywords: testing, arsenicosis, upazila, hot-spot

1. Introduction

Beginning in the 1970s, groundwater has been exploited in development programmes by governments and development agencies in many parts of the world, as a means of providing safe drinking water. The exploitation of groundwater for drinking water purposes has been instrumental in decreasing the incidence of waterborne disease and has made an important contribution to improving public health.

Because the geology of Bangladesh allows for a simple and low cost manual technology for the drilling of tube wells, the private sector has been active in installing shallow tube wells and handpumps for many years before government and development agencies began to exploit groundwater for developmental purposes. Most tube wells in the country, estimated to be between 6 and 10 million in number, have been installed privately, which is why there are no exact figures on how many tube wells there are in use.

There have been two large-scale national random surveys so far implemented in Bangladesh, independently by UNICEF during 1996–1998 and the British Geological Survey during 1998. These indicated that approximately 29 and 26%, respectively, of all tube wells may be contaminated with arsenic above the current national acceptable level, set at 0.05 mg/l.

This is a report of the progress on arsenic mitigation in a 45 upazila Project area developed and implemented by the UNICEF Water & Environmental Sanitation Programme, as set out in the Project "Arsenic measurement and mitigation" (hereinafter called the Project) for the planned period 2001–2005. This Project is implemented in partnership with the Government of Bangladesh – Department of Public Health Engineering (DPHE).

With a quoted coverage rate of 97% (Progotir Pathey 2000) access to safe water in Bangladesh, together with up to 43.4% (Progotir Pathey 2000) access to a sanitary latrine (in itself far advanced compared to many other developing countries) the achievements made in water supply and sanitation in Bangladesh are significant and noteworthy. While recognising and not negating these impressive achievements the Project must now deal quickly with the very serious problem of arsenic contamination. The problem, however, is complex. It is important to understand the various dimensions of the issue, which are as follows.

1.1. The magnitude

With 61 out of 64 districts affected and 264 out of 493 upazilas (BAMWSP – Bangladesh Arsenic Mitigation Water Supply Project) being the most affected it is estimated that at least 26 million people are at risk from contracting arsenicosis at the level of 0.05 mg/l.

1.2. The complexity

There is no pattern to arsenic contamination of groundwater. One well in a village may be safe while another well 100 yards away may be contaminated and the one after that may again be safe. There is a growing understanding that arsenic may be entering the food chain through contaminated irrigation water.

1.3. The severity

While arsenicosis is not a contagious disease it often appears to be to the affected rural communities. There are reports of children who have developed the characteristic hyperkeratosis symptoms being asked to leave school. Marriage prospects for the affected are severely limited.

1.4. The urgency

Preliminary analysis of testing data from the UNICEF–DPHE 3rd phase 15 upazila Project seems to indicate that many wells have actually only been installed during the last 4–6 years. This could explain why the country is not reporting many more cases of arsenicosis since exposure time for many people at risk has been relatively short. This, however, also implies that the programme is in a window of opportunity where if mitigation can be effected as soon as possible, then many thousands of people can be saved from contracting arsenicosis. Given the time exposure already lapsed; this window of opportunity may be a short one.

2. Methodology

2.1. Scale of project development

Since 1996, the UNICEF–DPHE Arsenic Mitigation Project has had a phased evolution, which is explained below:

Phase 1; During 1996–1998: A national random testing programme for arsenic contamination, wherein 51,000 tube wells were tested at random across the country and it was determined for a national average that approximately 29% of tube wells are contaminated above 0.05 mg/l.

Phase 2; During 1999–2000: A "Five Upazila Action Research Project" in which UNICEF–DPHE worked in partnership with four major NGOs: BRAC, Dhaka Community Hospital (DCH), Grameen Bank and Rotary/ISDCM. In this phase over 105,000 tube wells were tested and many hundreds of safe water options were installed. Over 750 arsenicosis patients were diagnosed and assisted with palliative treatment. The prevalence rate for the people using contaminated sources was 0.98/1000.

Phase 3; During 2000–2001: A 15 upazila Project area, where over 298,000 wells were tested and over 2700 arsenicosis patients were diagnosed. The prevalence rate for those using contaminated sources was 0.92/1000. A total of 1600 doctors and 15,000 medical health workers were trained on diagnosis. More than 800 safe water sources were installed benefiting over 140,000 people. Of those wells tested, a total of 75,000 wells in three upazilas were plotted by GPS so as to develop a database to be used to plan and monitor progress of mitigation activities.

Phase 4; During 2002: At the time of reporting, UNICEF–DPHE have completed the logistics to test wells, disseminate communication and identify patients in a further 25 "hot spot" upazilas and this work is ongoing. Additional water supply options will be installed and GIS will be further developed as a management and planning tool. Plans have also been completed to use the existing DPHE field structure to carry out an additional rapid survey at the rate of 1000 wells per upazila in 200 upazilas, which have not yet been allocated to any project. This will be an attempt to quickly survey and identify further hot spots that may then, need blanket testing. This will represent a Project area of 245 Upazilas; almost half of the country benefiting approximately 65 million people. In this phase, it is estimated that approximately 650,000 tube wells will have been tested during 2002.

Phase 5; Post 2002: The Project will concentrate more on the provision of safe water options. More efforts will be made in the encouragement and support to the private sector to provide services such as testing and provision of safe water options. The focus will be on safe water options for schools. The health system will be encouraged and supported to diagnose and care for arsenicosis patients.

3. Strategy

UNICEF–DPHE continues to implement a four part integrated strategy that was developed during the Five Upazila Action Research Project, which is:

1. *Blanket testing of all handpump tube wells and shallow irrigation wells to identify both safe and unsafe sources*: The complexity of the problem means that it is not possible by random testing to determine if a specific tube well is contaminated or not. The only way in which this can be achieved is by blanket testing of each and every well. Using field test kits it is possible to achieve this by locally trained health workers. It is important, in order to ensure maximum impact in terms of sensitising the community to the risks that the testing is carried out in the presence of the owner/use of the well.
2. *Informing communities about the risks and the need to avoid consuming arsenic contaminated water*: Even now, many people are reluctant to believe that the water they are consuming from handpumps may be putting their health at risk; people will always want to believe that the problem will affect someone else not themselves. It is important, therefore, to ensure that all communities are continuously exposed to the communication materials and messages. A communication campaign was developed during 1998 and 1999, through a process of study, field-testing and consensus building amongst all stakeholders. The campaign was officially released as the national communication campaign by the Minister for Health in December 1999.
3. *The identification and management of those people affected with arsenicosis*: While the epidemiology of arsenicosis is not yet well understood, it is suggested that the onset of the disease may occur in adults after 9–15 years consumption of contaminated water, depending on the level of contamination. Children may develop the disease after a shorter exposure time. In the advanced stages the disease leads to gangrene, internal disorders and possibly cancer. It is therefore essential to locate possible patients for them to be properly diagnosed to either rule out or confirm arsenicosis and to then assist any patient to be referred for hospital treatment.
4. *The provision of alternative and safe water options*: Having identified contaminated tube wells, communities need to be assisted to obtain access to a safe water source. Any use of groundwater in Bangladesh carries with it long-term obligations to test periodically, and not only for arsenic contamination. The British Geological Survey Technical report WC/100/19, volume 1, February 2001, mentions that "while arsenic is the single greatest problem in Bangladesh ground waters, other elements of concern from a health point of view are manganese, boron and uranium". For this reason, options which avoid groundwater such as a pond with a sand filter, dug wells and rainwater harvesting (RWH) are being promoted at this time until more is known about the safety of deeper aquifers and until suitable testing practices can be developed.

In addition to the "avoidance" technologies, adsorbent filters with imported or locally manufactured media are being tested.

4. Area of operation

During 2001, the Project area which has been assigned to UNICEF–DPHE by the Government of Bangladesh BAMWSP Project (the national co-ordinating Arsenic Mitigation Project) was expanded to cover a total of 45 of the identified "hot spot" upazilas (including the original five Action Research Project upazilas). This Project area represents approximately 10% of the total upazilas in the country, and 20% of the currently known "hot spot" upazilas. Approximately 13.3 million people live within this Project area. In addition to blanket testing in the already identified "hot spots", UNICEF–DPHE recognised that so far only the hot spot upazilas have actually been allocated to Projects, whereas all of the upazilas in the country have to be tested. Consequently, a plan to utilise the organisational capacity of DPHE to carry out rapid testing of 200 of the remaining upazilas has been completed and testing is underway at the rate of 1000 tube wells per upazila. This makes a total Project area of 245 upazilas which will benefit approximately 65 million people.

5. Field work

For the 3rd phase work, the Project continued to build upon the partnership between government and NGO that began in the 2nd Phase "Five Upazila Action Research Project", building partnerships with local NGOs so as to better facilitate sustainability. The number of partner NGOs was expanded from 4 to 8. For the 4th phase work, the number of NGOs will expand to 15, in order to include more local level NGOs. In addition, the DPHE itself will directly implement activities in six upazilas using its field level presence, institutional capacity and influence with local government. The partner NGOs and assigned upazilas are listed in Table 1 by phase of work:

Building from the experience gained on the "Five Upazila Action Research Project", the first task was to put in place a structure so as to ensure co-operation at field level, build confidence and to make clear the intention to help the local populace to help themselves.

Contact was established with the district and upazila local government structure through the Deputy Commissioner (DC) and Upazila Nirbahi Officer (UNO). This process, was aided by the issuance of a government cabinet level proclamation in November 2000, which sets out the terms of reference and scope of work for Arsenic Mitigation Committees to be established at District, Upazila, Union and Ward (village) levels.

Following on from this process the NGOs then organised arsenic management and co-ordination committees at district, Upazila and union levels, trained the various field workers and volunteers in how to use the communication materials, how to use the field test kits and how to record the results accurately. The number of committees formed, meetings held and people trained in this 3rd phase work (provisional data) are presented in the Table 2.

Table 1. List of NGOs contracted to work on the project by allocated Upazila and by phase.

No.	Upazila	District	NGO
2nd Phase			
1	Sonorgaon	Narayanganj	BRAC
2	Kachua	Chandpur	Grameen Bank
3	Manikganj Sadar	Manikganj	Rotary/ISDCM
4	Jikorgacha	Jessore	BRAC
5	Bera	Pabna	DCH
3rd Phase			
6	Nabinagar	Brahmanbaria	CDIP
7	Muradnagar	Comilla	Grameen
8	Haimchar	Chadpur	BRAC
9	Bhanga	Faridpur	BRAC
10	Manirampur	Jessore	BRAC
11	Shibchar	Madaripur	GUP
12	Serajdikhan	Munshiganj	DCH
13	Homna	Comilla	ISDCM
14	Damurhuda	Chuadanga	ISDCM
15	Shahrasti	Chadpur	Grameen
16	Babuganj	Barisal	NGO Forum
17	Bancharampur	Brahmanbaria	ISDCM
18	Barura	Comilla	BRAC
19	Rajair	Madaripur	GUP
20	Kalia	Narail	EPRC
4th Phase			
21	Dhamrai	Dhaka	DPHE
22	Keranigonj	Dhaka	DPHE
23	Jamalpur Sadar	Jamalpur	SSKS
24	Bakshigonj	Jamalpur	NGO Forum
25	Saturia	Manikganj	GUP
26	Ghior	Manikganj	NGO Forum
27	Itna	Kishoreganj	BRAC
28	Astogram	Kishoreganj	SOHAC
29	Nikli	Kishoreganj	BRAC
30	Munshiganj Sadar	Munshiganj	DPHE
31	Mymensingh Sadar	Mymensingh	SWORD
32	Muktagacha	Mymensingh	DCH
33	Nandail	Mymensingh	Grameen Shikka
34	Guiripur	Mymensingh	Grameen Shikka
35	Monohardi	Narsindi	DPHE
36	Palash	Narsindi	DPHE
37	Shibpur	Narsindi	DPHE
38	Sherpur Sadar	Sherpur	UPACOL
39	Nakla	Sherpur	HELP
40	Bashail	Tangail	ISDCM
41	Mirzapur	Tangail	CDIP

Table 1. (continued)

No.	Upazila	District	NGO
42	Tangali Sadar	Tangail	Unnayan Shamunnay
43	Bandur	Narayanganj	DPHE
44	Jessore Sadar	Jessore	EPRC
45	Kaligonj	Jhenaidha	Asia Arsenic Network

6. Blanket testing of wells

To test the level of contamination in the wells, the Project has continued to use field test kits designed around the Gutzeit method, which are easily used by trained field workers. The Project has so far used the Merck field test kit, but in the 4th phase will also use the HACH 2 stage field test kit, which is expected to deliver a clearer and more accurate result for lower levels of contamination.

It is believed that the field kit result is accurate enough for the rapid assessment that must be carried out to inform people as quickly as possible whether they are at risk or not. Accuracy is checked by a small sample of tests being revalidated by laboratory analysis. At the time of reporting, preliminary results indicate an average of 5.75% false positive and 3% false negative. It is noted that the field test kit provides a more accurate reading

Table 2. Number of people trained and committees formed.

No.		Male	Female	Total
Category trained				
1	Ward committee members	1958	600	2558
2	Union committee members	957	233	1190
3	Upazila committee members	245	20	265
4	Persons trained in communication	3336	1352	4688
5	Tube well testers trained	396	793	1189
	Total	6892	2998	9890
Committees formed				
6	Ward arsenic mitigation committee			1517
7	Union arsenic mitigation committee			193
8	Upazila arsenic mitigation committee			15
9	District arsenic mitigation committee			10
	Total			1735
Planning/sensitisation meetings held				
10	Ward arsenic committee			782
11	Union arsenic committee			203
12	Upazila arsenic committee			52
	Total			1037

where the contamination levels are higher. The field kits are able to detect to 0.05 mg/l with satisfactory accuracy.

7. Patient identification

So far all cases have been detected by active case search. In the 3rd phase, three methods of searching for and diagnosing patients were used. These were:

1. House to house searches carried out by DCH in Haimchar, Monirampur, Bhanga, Shibchar, Muradnagar, Nabinagar and Sirajdikund,
2. Health camps organised jointly by the DCH and the Directorate General of Health Services (DGHS) in Bancharampur, Damurhuda, Homna, Barura, Babuganj, Kalia and
3. Using the DGHS network of community health workers in Shahrasti Upazila.

8. Results

8.1. Testing of tube wells

Analysing the results of the 3rd phase 15 upazila area, it is seen that the situation is indeed grave.

This 3rd phase is the largest blanket testing Project area so far implemented by UNICEF–DPHE and it has yielded the most significant data so far:

- More than 2700 arsenicosis patients have been diagnosed, equating to a prevalence rate of 0.92/1000.
- A total of 298,000 wells were tested between September 2001 and January 2002. Of these, the average number of wells contaminated was 68%.
- Over 580 villages have been identified as having no safe source at all, affecting approximately 70,000 families. Many villages have more than 60% contamination. Many schools have contaminated wells.
- In the 3rd phase 15 upazila Project area alone it is estimated that almost 3 million people are at risk of contracting arsenicosis by consuming water from contaminated sources.

As a result of the Project testing programme, all people in the Project area have been alerted to the extent of the problem. This will help to minimise the number of people who might otherwise fall ill from arsenicosis. It is however also clear that there is an urgent need to help these affected people to obtain a safe water source; as soon as possible. The complete testing and patient identification results for the 2nd and 3rd phase testing areas are presented in Tables 3 and 4.

There is estimated to be over 13.3 million people who will eventually be assisted in the 45 "hot spot" upazilas of the Project area. Using the contamination and prevalence rates determined from the 3rd phase area, it can be extrapolated that an estimated 8.9 million people are at risk and an estimated 8000 people may develop arsenicosis.

Table 3. Results of 2nd phase testing programme and list of 3rd phase "no source" villages.

Sl. no.	Upazila	No. of union	No. of village	Tube well surveyed			% of contaminated TW	Total population surveyed	Arsenicosis patient identified
				Total wells	Safe wells	Contaminated wells			
(a) Results of 2nd phase Action Research Project									
1	Kachua	12	243	17,747	363	17,424	98.17%	293,683	202
2	Manikganj	11	371	21,300	15,062	6338	29.75%	237,771	19
3	Sonargaon	11	476	25,048	8877	16,171	64.56%	261,881	252
4	Jhikorgachha	11	187	26,637	10,989	15,658	58.78%	237,711	151
5	Bera	10	169	14,407	7011	7396	51.33%	208,897	120
	Total	55	1446	105,139	42,302	62,987	59.90%	1,239,943	744

	Upazila	No. of villages 100% contaminated	No. of families affected	Estimated number of people affected
(b) 3rd phase 15 upazila Project – no safe source villages (provisional)				
1	Bancharampur	21	3474	20,844
2	Barura	89	9290	55,740
3	Bhanga	63	7315	43,890
4	Haimchar	21	2442	14,652
5	Homna	55	6375	38,250
6	Kalia	55	5068	30,408
7	Monirampur	6	437	2622
8	Muradnagar	149	21,368	128,208
9	Shaharasti	108	13,826	82,956
10	Shibchar	22	944	5664
	Total	589	70,539	423,234

From the 3rd phase work a number of villages were identified as having no safe source at all, which will impact on the strategy to promote well sharing. These villages are listed in Part (b).

Table 4. Results of 3rd phase testing programme and analysis of identified arsenicosis patients.

Area		Results of blanket tube well testing								Analysis			
Name of NGOs	Name of upazila	Population (1999 est.)	Total No. of TW tested	Total No. of TWs out of order	Found conta-minated	%	Found-safe	%	Arsenicosis patients identified	Average popu-lation per well	Estimated exposed population	Pre-valence per 1000 exposed	Prevalence per 1000 all wells
ISDCM	Homna	246,000	15,605	796	10,805	69.2	4800	30.8	69	15.76	170,232	4.5	2.8
	Bancharampur	310,000	17,628	537	12,464	70.7	5164	29.3	68	17.59	219,170	3.1	2.1
	Damurhuda	250,000	25,263	893	6835	27.0	18,428	73.0	199	9.90	67,500	29.48	7.9
BRAC	Monirampur	384,000	35,170	472	17,954	51.0	17,216	49.0	50	10.92	195,840	2.5	1.3
	Bhanga	248,000	21,147	1329	19,172	90.6	1975	9.4	488	11.73	224,688	21.72	19.6
	Haimchar	135,000	3665	256	3100	84.6	565	15.4	101	36.83	114,210	8.8	7.4
	Barua	260,000	25,986	1729	18,482	71.1	7504	28.9	28	10.01	184,860	1.5	1.0
GUP	Rajoir	236,000	13,022	887	9893	76.0	3129	24.0	85	18.12	179,360	4.7	3.6
	Shibchar	353,000	31,964	88	14,501	45.4	17,463	54.6	281	11.04	160,262	17.5	7.9
GS	Muradnagar	614,000	30,199	390	28,234	93.5	1965	6.5	440	20.33	574,090	7.6	7.1
	Shahrasti	214,000	16,380	470	16,221	99.0	159	1.0	411	13.06	211,860	19.4	19.2
CIDP	Nabinagar	449,000	30,122	864	26,181	86.9	3941	13.1	212	14.91	390,181	5.4	4.7
NGOF	Banuganj	156,000	5956	252	3987	66.9	1969	33.1	106	26.19	104,364	10.1	6.7
DCH	Serajdikhan	264,000	15,058	542	6623	43.9	8462	56.1	134	17.53	115,896	11.5	5.0
EPRC	Kalia	220,000	11,307	734	8201	72.5	3106	27.5	40	19.46	159,500	2.5	1.8
Totals	15	4,339,000	298,472	10,239	202,653	67.9	95,846	32.1	2712	14.54	2,946,043	9.2	6.2

8.2. Experiences of the field workers

The NGOs that carry out the fieldwork have reported many experiences as they implement the testing work for the communities and business houses in the Project area. They have to build relations with local government, reassure the communities that they have good intent, deal with businesses and landlords who might not want their well tested for fear of losing income. Some experiences have been positive, others less so! The experiences recounted below are from CDIP who implemented the Nabinagar Project upazila:

- Our relation with the government administration was good. The Upazilla Nirbahi Officer accorded all support during the Project period.
- People, especially the women and children, were very keen to have the tube wells tested and watched the process closely. If the well tested positive, there was always great sadness, if negative great joy.
- Our field testers requested all safe tube well owners to help the neighbours by supplying drinking water to them and almost all agreed. When the school tube wells were marked red, it was observed that the children carried water from the green tube wells in bottles.
- The villagers took the communication posters with great care. They asked the children not to destroy the posters. Many shopkeepers requested for posters to hang in their shops.
- Some people refused to paint their tube wells. A few people were angry with the tube well testers and refused to test their tube wells. They said they had been using the tube wells for a long time and they were not affected by any disease. Some landlords and restaurant owners refused to paint their tube well red.
- Some people spread the rumour that the surveyors were from the "tube manufacturing companies" and after declaring the tube wells "red" they will be selling tubes to install the tube wells deeper.
- The fear of drinking from the red tube wells generally remained for few days only and due to non-availability of safe water, people started drinking again from their red tube wells and blamed their fate.

This has been some of the experiences as recounted by one of the NGOs working in the field.

8.3. Water supply options

Perhaps the greatest challenge yet, is to help people to obtain a safe and sustainable water supply. To obtain an arsenic free source of water, the choice is to use technologies that either *avoid* arsenic contamination by exploiting surface or other non-groundwater sources, or those that *remove* arsenic from groundwater by filtration, chemical coagulation or adsorption. Exploitation of deeper aquifers may also offer a source of potable water, but this will be developed with care because of the risk from other contaminants such as uranium, boron, manganese, etc. in addition to arsenic and also with caution that there may be geo-chemical processes which are still ongoing that might affect groundwater quality in the future. Each technology choice has positive and negative aspects.

It has to be recalled that the exploitation of groundwater was a development choice for a number of reasons:

- It was in order to provide a bacterially safe water source; in which it has been mostly successful.
- It was to reduce drudgery and workload of the water carrier, usually the women and girl child.
- It was, by reduction of the workload, to improve the nutritional status of the water carrier.
- It was to make available time for the water carrier, which could be used for education, rest and for taking better care of the younger children.
- It provided a livelihood for many thousands of people as construction, sales and maintenance service providers.

The contribution made by the handpump option towards helping people to attain a better health and well-being, is evidenced in the reduction in under five mortality rate from 151 per thousand live births in 1990 to 102 in 1999 and the reduction in the infant mortality rate from 94 in 1990 to 57 in 1999 (Progotir Pathey 2000). Any alternative sources to tube wells and handpumps, will need to ensure that there is no reversal to those gains, which have been made.

With the large amount of data being generated by the Project and the need to be able to assess the local magnitude of the problem, as well as progress with regard to communities obtaining a safe water source, the Project has utilised the GPS to log all well sites in three upazilas. This is an experiment to try to develop a computerised database that can be used for Project development and monitoring. These upazilas were Bhanga in Faridpur district, Sirajdikhan in Munshiganj district and Muradnagar in Comilla district. In all, 75,000 well sites were plotted by using six low-cost hand-held GPS receivers in each of the upazilas. The teams of field workers were supervised by two technical advisers. The exercise took 3 months to complete. The Project now has a GIS database for three upazilas which will be used to determine the use and desirability of continuing the work to plot all wells in the Project area.

The use of GPS had two objectives. The first was to check the quality and accuracy of the field well testing work. The GPS survey indicated that the field testing work had been completed satisfactorily with 98% of the recorded wells existing and correctly marked. The second objective was to use the database to determine the extent of the problem in each upazila, to map the contaminated areas and to attempt to utilise the data to plan safe water options.

The data is still being assessed, but preliminary findings indicate that in Muradnagar, 69% of people live within 300 m of a safe tube well, using the current safe level of contamination of 0.05 mg/l. Using the WHO GV level of 0.01 mg/l, the percentage within 300 m decreases to 55%. The issues are whether the owners of safe wells are willing to share the water on a continual basis, and whether people are willing to walk the 300 m to obtain safe water (some informal indications are that they are not!)

Those options which UNICEF–DPHE and its partners have installed so far on are listed in Table 5 together with the approximate cost ranges for each technology.

Table 5.   Safe water options installed as of May 2002.

Type	Quantity	Number of beneficiaries	Approximate total cost (USD)
Rain water harvesting	368	3196	43,592
Pond sand filter	70	35,000	4900
Protected dug well	134	40,200	80,400
Imported adsorbent	127	31,650	110,000
Indigenous adsorbent	2100	12,600	8700
Rapid surface water treatment	2	2000	17,000
Piped scheme	1	600	30,000
Deep tube well	44	15,400	34,100
Totals	2846	140,646	319,692
		Average cost per head	2.27
		Average cost per installation	112.00
		Average beneficiaries per installation	49

There follows below a description of each technology so far utilised to provide a safe water source:

8.4.   Avoidance of groundwater

8.4.1.   Rainwater harvesting, a household option

A total of 368 units have been installed so far, benefiting approximately 3200 people. This system is usually a household option ranging from a simple 500 l bladder or clay container to a 3200 l cement storage jar. The cost for the cement jar is USD70 (Taka3900) which has been reduced by designing the cement jar without steel reinforcing. Bangladesh has the most rainfall of all its neighbouring countries, at 1500–4000 mm/year. The issue is whether or how well people are able to conserve the stored water during the dry season.

8.4.2.   Pond sand filtration (PSF), a community option

The Project has installed 70 units so far, at an approximate cost of USD700 (Taka40,000) each. These units have benefited approximately 35,000 people. The main issue is that the community has to agree to work together to carry out regular maintenance otherwise it quickly falls into disrepair. It is also important that the unit is well constructed otherwise there can be infiltration by insects and helminthes.

8.4.3.   Protected dug well (PDW), a community option

Costing approximately USD160 (Taka9280) for renovation of an old well to USD600 (Taka35,800) for installation of a new one, this is another traditional technology that the Project has reverted to, with 134 being installed to date benefiting approximately 40,000 people. This solution avoids arsenic by being installed above the zone where arsenic usually occurs. However, it must be stated that arsenic has also been found at shallower depths in some areas and so it cannot be assumed that all dug wells would be arsenic free.

8.4.4. Rapid surface water treatment, a community option

The Project has procured and installed two rapid surface water treatment plants at a cost of USD17,000 (Taka986,000) and benefiting approximately 2000 people. This technology utilises a system of sedimentation, pre-chlorination and filtration with activated carbon elements and deliver approximately 1000 l of treated water per hour. While the technology is effective, it does require considerable maintenance and operational procedures and the capital cost is very high at this time.

The technologies described above represent the main traditional systems and a new rapid filtration system, which can be used as options for the provision of safe water by avoiding arsenic contaminated sources. It should be borne in mind that the PSF and PDW add to the workload of the water carrier and may therefore be less popular as a first choice.

8.4.5. Removal/filtration of arsenic, a household or community option

In addition, UNICEF–DPHE are testing filtration systems, which employ a column filter filled with a synthetic media (ferric hydroxide, activated alumina) or non-synthetic media, which adsorbs arsenic and other metalloids. Once adsorbed, the arsenic should be contained on the media indefinitely.

8.4.5.1. Imported adsorbent media
A total of 127 units have been installed to date benefiting approximately 32,000 people. The cost for this technology ranges from USD35 (Taka2030) for a household filter to USD8875 (Taka514,750) for a large capacity community model. The types of media used so far are ferrous hydroxide and activated alumina. While the capital cost for the imported synthetic media may at first seem prohibitive, the cost per treated litre of water is in fact very small at only a few paise per litre as calculated against the initial capital cost. The Project will develop ways in which the spent media can be disposed of safely from an environmental perspective. With the amount of arsenic being only approximately 5 g per media change, disposal of spent media should not be an insurmountable problem and even simple burial within the village could be a feasible option.

8.4.5.2. The Shapla filter
As well as imported adsorbent media, there are alternatives made from indigenous materials such as brick chips treated with ferrous sulphate, which produces a low cost option for household filtration. A single unit will filter approximately 3000–4000 l of contaminated water. The cost is low at approximately USD7 (Taka406) for a new unit and only USD1.70 (Taka100) for subsequent replacement media. Even if several units were needed each year it would still be a viable option. A total of 100 of these adsorbent units have been installed to date which has benefited approximately 600 people.

8.4.5.3. The 3 Kolshi filter
During the Five Upazila Action Research Project, another low cost adsorption technology was tried out, of which approximately 2000 were installed which would have benefited approximately 12,000 people. At a cost of only USD4 each (Taka232) this was the 3 Kolshi filter, a system designed with three clay pots positioned one above the other. This simple filter works acceptably in removing arsenic but it was found to be susceptible to bacterial growth and did not last long. This system is no longer utilised on the Project.

It is a simple process to remove arsenic from groundwater by simply passing the water to be cleaned over an adsorbent media. The drawback is the cost and availability of imported media and the fact that the beneficiary community will have to pay repeatedly to replace the media when spent. The issue of safely disposing of waste media is not thought to be an obstacle to using this media.

8.5. Groundwater exploitation

8.5.1. Piped scheme, a community option

At a cost of USD30,000 (Taka1,740,000) the Project has installed one piped scheme so far which has benefited approximately 600 people. One of the Project partner NGOs, BRAC, together with the Rural Development Academy (RDA), designed a borehole and small scale piped scheme which was constructed at Pakundar village in Sonargaon. This piped scheme is novel in that it aims to provide water for irrigation as well as domestic use, the supplied irrigation water subsidising the cost for the domestic users. This installation is providing some valuable experiences, but it would need to be much cheaper and would need to serve many more people if it is to be replicable.

8.5.2. Deep tube wells, a community option

Earlier theories were that deep tube wells (below 200 m) would be arsenic free, but this has not yet been substantiated. The Project has installed 44 deep wells in the Jikorgacha area with some success at an approximate cost each of USD775 (Taka45,000). These wells have benefited approximately 15,500 people. However, attempts to exploit the deeper aquifers in Manikgonj were less successful when many of the deep wells there began to produce saline water. With regard to exploitation of deep aquifers, the issue of the presence of a wider range of contaminants will need to be fully discussed and a policy adopted with regard to long-term testing obligations. At that time it might be possible to design deep wells to serve as the source for extensive local area piped scheme networks.

In summary, the Project has so far installed almost 3000 safe water options, benefiting over 140,000 people at an approximate cost of USD320,000 (Taka18,560,000) this calculates to approximately USD2.27 (Taka132) per head. The aim of the Project is to reach a stage where affected communities can be informed of a range of options which might be feasible for each given situation; where the community will then decide which is the option they want to choose.

9. Discussion

The Project will continue to evaluate experiences gained so far before making decisions about which technologies will be mainly promoted and supported. At this time one likely technology that may provide a solution seems to be RWH, perhaps augmented by an adsorption system during the dry season months.

However, with rainwater systems situated at the household level, coverage will be slow and the cost high. The traditional systems of PSF and dug wells may offer one way of serving a larger area rapidly in the short term until RWH or other systems can be installed in the medium to long term.

Some predictions have been that up to 80 million people are at risk from contracting arsenicosis in Bangladesh. However, the figures that are resulting from the UNICEF–DPHE tube well testing programme so far indicate a fairly low prevalence rate and using extrapolated data from the 3rd phase work it would seem that less than 9000 people may have actually contracted the disease in the 45 upazila area. Having said that, given that the prevalence rate indicated is from one time active case searches in each Project area there could be more cases which have not yet been detected and diagnosed. Given also that a fairly large proportion of wells installed seem to be relatively young, the Project could be in a window of opportunity to save many people from contracting arsenicosis, if the current speed of work in the UNICEF–DPHE Project area can be maintained.

Even though the risk associated with arsenic contamination has been regularly publicised for a number of years, people are still installing shallow tube wells with handpumps without having them tested; and this is another dimension to the mitigation work which is proving difficult to handle.

The extent of contamination, which may be affecting the food chain is not yet well understood at the time of reporting and this will surely be another aspect of mitigation which will have to be dealt with; adding further to the already formidable task.

10. Conclusion

The UNICEF–DPHE Project is moving very quickly to test all of the wells in its allocated 45 "hot spot" upazila Project area and this, work will be completed by the end of 2002. In addition, the Project is on track by the end of 2002 to complete the rapid survey of 200 upazilas which were previously not allocated to any Project. Since the early Project work of the national random testing programme which began in 1996 and including the projected wells that will be tested this year, 2002, more than 1,100,000 wells will have been tested at an average of more than 183,000 wells per year. This rapid output has served more than 1 million people every year on average; by informing them as to whether their well is contaminated or not and by advising them of what to do if it is. By so providing this advice and service the Project may well have been instrumental in preventing many thousands of people from contracting arsenicosis by the act of testing alone.

The challenge is to continue to advise and in addition to work as quickly as possible to help people to obtain a safe and sustainable water supply. This will involve continued communication for awareness so that people will not consume arsenic contaminated water, so that they may share safe sources and so that people will not install new handpumps without due care.

Adoption of the handpump option has been successful and has contributed to the fall in U5MR and IMR in Bangladesh. So as not to lose these gains, it is crucial that the testing and mitigation Project continues to move with all speed to tackle the wider issue of all possible groundwater contaminants and not only arsenic.

The UNICEF–DPHE Project is on track to meet its commitments to blanket test all wells in 45 upazilas, in the known "hot spot" areas, within the current GoB-UNICEF five year Master Plan of Operations, which represents 10% of the country and 20% of the "hot spot" upazilas. In addition, the Project is carrying out a rapid survey of an additional 200 upazilas at the rate of 1000 wells per upazila, so as to quickly identify if other "hot spot" upazilas exist which would then have to be blanket tested. This represents almost half of the upazilas in the country; extremely significant on a national scale. Even then, it will be important for other Projects to pick up speed if this "window of opportunity" is not to be missed. It will also be important to have a much better co-ordination amongst all actors involved in arsenic mitigation, which is something UNICEF can facilitate if the government and Donors so desire.

For its part, DPHE must take the lead in developing and implementing testing protocols for ground waters and for advising and assisting those people who are still desirous of utilising groundwater. For continued monitoring of arsenicosis patients and in order to obtain a more accurate status of the number of people falling ill from contaminated water, the health system must recognise arsenicosis as a reportable disease and treat it accordingly within its existing health infrastructure.

Arsenic Exposure and Health Effects V
W.R. Chappell, C.O. Abernathy, R.L. Calderon and D.J. Thomas, editors
439

Chapter 33

Normative role of WHO in mitigating health impacts of chronic arsenic exposure in the South-east Asia region

Deoraj Caussy

Abstract

Groundwater contamination in excess of the World Health Organization (WHO) guideline value of 0.01 mg/l has been observed in many parts of the world, including five member states from the South-east Asia region of WHO, namely, Bangladesh, India, Nepal, Myanmar and Thailand. Over 10 million tube wells are in use in the region thereby exposing between 40 and 50 million persons to unsafe level of water. The associated disease burden is projected to be around 12.5 million subjects within 10 years. To mitigate this problem, the WHO Regional Office for South-east Asia has launched a new initiative to provide a normative role in training manpower for using harmonized case detection and surveillance and clinical management of patients, formulation of a national policy for arsenic testing and removal, research on co-factors and strengthening of the infrastructure for laboratory diagnosis of arsenic.

Keywords: WHO normative role, arsenic mitigation, chronic arsenicosis, south-east region, case definition, case management

1. Arsenic problem in South-east Asia

One of the major burdens of disease caused by the exposure to arsenic worldwide is from contamination of groundwater. For instance, groundwater contamination in excess of the World Health Organization (WHO) guideline value of 0.01 mg/l or the respective prevailing national standard has been observed in USA, Canada, Mexico, Argentina, Chile, Hungary, China, Taiwan and Vietnam and many countries of the South-east Asia region of WHO (WHO, 2001a and WHO, 2001d). The South-east Asia region encompasses a population of almost 1.9 billion spanning several geographic zones and comprises 10 member countries. Groundwater from tube wells is a predominant source of drinking water in many of the member countries. For instance, some 10 million such tube wells have been dug during the green revolution in Bangladesh, and the arsenic concentrations in some of these tube wells can be sometimes 25-fold in excess of the WHO guideline value. Well-protected groundwater is safer in terms of microbial quality than water from open dug wells and ponds. However, groundwater is notoriously prone to chemical contaminants, including arsenic from geological or anthropogenic sources.

Many studies support the contention of the existence of a natural arsenic-rich eco-region formed by the alluvial/deltaic sediment in the Gangetic delta region. As shown in Figure 1, countries of South-east Asia that fall in this belt include Bangladesh, parts of India, Myanmar and Nepal. Contamination of groundwater from anthropogenic activities of mining arsenic-containing iron ores has also been detected in one province of Thailand.

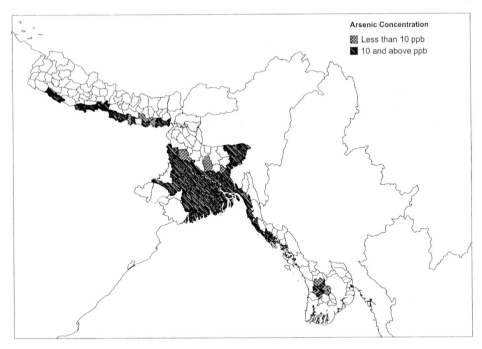

Figure 1.   Arsenic concentration in Bangladesh, India, Myanmar and Nepal.

1.1. Historical background

In environmental health, an unusual clustering of health event often leads to the quest for cause and exposure. Thus, the appearance of a cluster of clinical cases of arsenic disease in West Bengal – India (Chakraborty and Saha, 1987), Bangladesh (Dhar et al., 1997) and Thailand (Choprapawan, 1994) led to the hypothesis that substantial exposure to arsenic was occurring in these populations. Specialists from different fields applied their knowledge in a common effort to determine the size and characteristic of the exposed population, the environmental distribution and exposure pathway of arsenic, including the species of metal involved. Once it was established that the clinical cases presenting with skin lesions were due to arsenic exposure, a systematic search for the source was undertaken in these countries.

1.2. Size of the disease burden

Precise estimates of the number of persons exposed in the south-east region are difficult to obtain because of uncertainty about the number of tube wells, the accuracy of field test kits and the limitation of resources for conducting large-scale surveys. Systematic surveys conducted by the British Geological Survey and others have shown that in Bangladesh alone nearly 37 million people are consuming water that are in excess of their national arsenic standard or the WHO guideline value (BGS, 1999). The estimated figure for the South-east Asia region is between 45 and 50 million exposed subjects, comprising

37 million from Bangladesh, 5 million from West Bengal, 2.5 million from Nepal and 2.5 million from Myanmar. Assuming a 25% disease rate for the exposed subjects developing clinical diseases, based on observation from other parts of the world, the South-east Asia region will experience a disease burden of at least 12.5 million afflicted subjects with various clinical consequences arising out of the 50 million exposed subjects.

1.3. Health impacts of chronic arsenic poisoning

The health impacts of exposure to arsenic depend on the dose, the modality and duration of exposure as well as the source and type of arsenic. Prolonged exposure to non-lethal doses of 0.005–0.09 mg/kg-body weight/day results in arsenic diseases called arsenicism, also referred to as arsenicosis, black foot disease, black skin fever etc. in various parts of the world (ATSDR, 1999). The hallmarks of chronic arsenicism are dermal changes concomitantly characterized by increased pigmentation and hardening of the skin (ATSDR, 1990). Some subjects may also show systemic disease of the central and peripheral nervous systems as well as the liver and kidney. Other complications after a mean latency period of about 10 years include the development of various types of cancers, such as squamous cell carcinoma, and cancer of lungs, bladder, and kidney.

The evidence linking arsenic to diseases has been adequately summarized elsewhere (WHO, 2001c). Cancer of the lungs, bladder, kidney and skins has been consistently observed in subjects drinking arsenic-contaminated water. The evidence linking arsenic to cardiovascular disease is inconclusive, and diabetes and negative reproductive outcomes are suggestive. However, a consistent hallmark of chronic poisoning is the manifestation of non-cancerous dermal changes known as hyperkeratosis.

2. Critical epidemiological assessment of arsenic poisoning

Hyperkeratotic lesions are the first dermal sign of arsenic contamination and consequently deserve a systematic analysis. The Regional Office of WHO undertook a meta-analysis of published reports to assess the nature of the epidemiological association of arsenic with dermal lesions. A total of 18 studies from Bangladesh, 16 from India and 23 from elsewhere were systematically reviewed for coherency of observation using set criteria of exposure and outcome measurements (WHO, 2002). An analysis of the prevalence rates and their confidence intervals enabled the following corollaries to be drawn: first, the factors predicting the onset of dermal lesions or progression of dermal lesions from moderate to severe forms are poorly understood (Oshikawa et al., 2001), but not all exposed subjects in a given population progress to disease onset: at the most, 25% of the exposed subjects progress to a disease state. Secondly, considerable variation in the prevalence rates of arsenicosis patients was observed.

These variations are open to biologic and epidemiologic interpretations: biologically, one can postulate inherent population variability in their response to arsenic exposure. From an epidemiologic standpoint, one can also explain the results in terms of study design and measurement errors. Because the study subjects are not from inception cohorts, they have different durations of exposure and thus reach the disease end-points at different times. Hence they cannot be accounted in the average figure at a single point in time.

Varying prevalence rates can also result from errors in: (1) the measurement of exposure to arsenic because of lack of validity of testing kits, (2) the measurement of the outcome of exposure to arsenic due to lack of objective criteria for defining a clinical case and (3) estimating the prevalence of arsenicosis due to uncertainty in measuring the exposure and the outcome. All these uncertainties are interlinked and will propagate together to compound the errors in measurement. If a case is not clearly defined, it will be misclassified leading to erroneous counts. By the same token, if exposure to arsenic is not accurately measured, the exposure status will be misclassified. Both these errors will lead to biased estimates of true prevalence.

3. Normative role of WHO in mitigating health impacts of arsenic exposure

Realizing the serious health impacts of arsenic contamination, the WHO Regional Office for South-east Asia has provided policy and technical support to national governments since 1997 (WHO, 1997). The main activities have been intercountry exchange of experts, skills and knowledge. These activities have been consolidated and formalized by the establishment of a WHO initiative on arsenic mitigation in 2002. This initiative has the highest level of policy support from the WHO Regional Office. First, a high-level task force consisting of representatives from member countries recommended that intercountry collaboration on arsenic poisoning should be intensified in the region. Secondly, the Regional Committee for South-east Asia, a high-level policy making body, adopted a resolution on arsenic poisoning urging WHO and member countries of the South-east Asia region to intensify collaborative efforts in this field (WHO, 2001b). This was followed by the WHO Advisory Committee on Health Research adopting a recommendation for health research in arsenic mitigation (WHO, 2002).

As illustrated in Figure 2, the arsenic mitigation initiative is implemented through a strategic plan focusing on three main goals: (1) responding to arsenic hazards through exposure assessment, risk determination and risk management; (2) strengthening infrastructure for arsenic mitigation; and (3) capacity-building through human resource development.

3.1. Exposure assessment

The validity of any prevalence estimate is contingent on accurate laboratory measurement of arsenic in groundwater. Laboratory assays must distinguish between forms and species of arsenic, including markers of exposures under field conditions. Our objectives in exposure measurement are to: validate all field test kits for sensitivity and specificity; establish a standard operating procedure (SOP) for arsenic testing; introduce quality control by the use of internal or external standards (QA/QC); and support the formulation of a national policy for the import and testing of kits. It is also anticipated to use the validated test for a comprehensive exposure assessment from all environmental media, including water, air, soil and food. As both organic and inorganic arsenic may be present in the food chain, proper research is needed to distinguish these species of arsenic in the food chain. The results will have tremendous public health significance because potentially

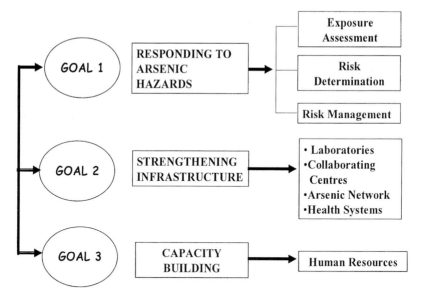

Figure 2. WHO strategic goals for arsenic mitigation in the South-east Asia region.

erroneous exposure assessment will result if one measures exposure only in drinking water in instances where concurrent exposure by food consumption is predominant.

3.2. Risk determination

Our aims in risk determination are to: (1) establish the magnitude of the risk for developing arsenicosis after exposure to arsenic-contaminated water, (2) establish the population attributable risk and (3) characterize dose–response relationship, particularly for long-term exposure to low doses of arsenic.

The bulk of the evidence linking arsenical dermatitis to arsenic exposure is derived from cross-sectional studies. While such studies provide the first line of evidence, the interpretation of the conclusion is limited by confounding due to other exposures. Thus, the extent to which unexposed or exposed persons develop clinical dermatological conditions mimicking chronic arsenical dermatitis is not known. Similarly, the extent to which exposed persons do not develop chronic arsenical dermatitis is not known. These limitations can only be overcome by a case-control study in which both exposed and non-exposed subjects are investigated for arsenic-related diseases. In order to elucidate these possibilities, WHO has provided technical and financial support to the International Center for Diarrhoeal Disease and Research in Bangladesh for a case-control study on the role of nutritional co-factors in the genesis of dermal lesions.

There is a wide range in the projected number of arsenic-affected patients in our region. This is partly due to active case searches being conducted in the proximity of a contaminated well and generalizing the results to the whole population. A reliable estimate of occurrences of arsenicosis can be made only by using valid laboratory assays, harmonized case definition and sound epidemiological design.

3.3. Risk management

Our aims in risk management are to set the norm, standards and guidelines for harmonized protocol on case detection and management as well as to formulate a protocol for verification of technologies for providing arsenic-safe water.

The accurate detection of arsenic cases is the cornerstone for good case management and reporting. Until now, no uniform case definition of arsenicosis has been developed or validated regionally or internationally, with the consequence that there is a wide discrepancy in the prevalence of arsenicosis. In disease surveillance, a case is defined by clinical signs, symptoms or laboratory measures. In applying this approach for arsenic case definition, two practical difficulties arise. First, there are several skin conditions that share major features with arsenic-defining conditions and, secondly, the use of laboratory measures is not uniformly available under all local conditions. The selection of case definition for arsenicosis ultimately depends on the objective of the public health program. Our objectives in formulating a case definition are to: (1) achieve consistent case detection and reporting, (2) provide an objective way to evaluate the efficacy and effectiveness of any interventions, (3) attain consistency in training of health care workers and (4) enable valid comparison of studies.

In order to achieve all these goals and in keeping with standard public health practice, the WHO Regional Office for South-east Asia has formulated criteria for classifying cases into the categories of *suspected, probable* and *confirmed*. This allows flexibility to identify cases in the field with limited laboratory facilities as well as unambiguously confirming cases at the tertiary level where adequate laboratory facilities are available. Figure 3 illustrates the fundamental approach in developing an algorithm for case definition. First, the literature was reviewed for all explicit and implicit definitions used so far and the most commonly occurring features were selected. An expert committee consisting of dermatologists, oncologists, internists, toxicologists and epidemiologists was convened nationally and regionally to arrive at a consensus. A consensus version, shown in Figure 4, has been formulated and field-testing is now in progress.

The lack of currently available proven therapy for clinical management of chronic arsenic poisoning has led to a number of unsubstantiated therapeutic measures being used for treating arsenicosis. The type of treatment that has been used can be broadly classified as chelators, retinoids, vitamins, anti-metabolites and surgical intervention (Kosnett, 1999). Only clinical trials or cohort studies using objective criteria were selected for further evaluation. There are only three such studies for chelators and the results are inconclusive. For retinoids, one study showed some efficacy in Bowen's disease. For vitamins, one study found improvement of the skin lesions depending on the stage of the disease and for surgical intervention or anti-metabolite, one study found moderate success in Bowen's disease (Mazumder Guha et al., 1998; Huq et al., 2000; Sikder et al., 2000; Oshiwaka et al., 2001). The dearth of information for clinical management warrants an expert committee to critically review the state-of-the-art therapy for chronic arsenicosis and makes evidence-based recommendations for the proper management of patients suffering from arsenicosis. This work is in progress to formulate a protocol.

The prohibitive cost of arsenic removal technologies has prompted the search for alternative options. These include: water treatment using a packet of chemicals for household treatment, use of deeper aquifers, promotion of rainwater use and consideration

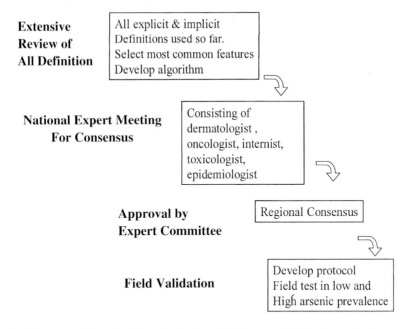

Figure 3. Overview of the methodology involved in validating case definition for arsenicosis.

of piped schemes based on central supply of surface. Researchers, NGOs, and private sectors have developed numerous community and point-of-use arsenic removal technologies. Since the efficacy of these technologies varies, the regional office is convening an expert meeting for the verification of the different protocols. Research will be supported to empower the community and strengthen the national capacity to independently evaluate such technologies.

3.4. Strengthening of infrastructure

Realizing the role of infrastructure in arsenic mitigation, the regional office is seeking donor support for strengthening key existing infrastructures in India, Bangladesh, Thailand and Nepal. Two aspects are being addressed: strengthening of reference laboratories and setting up of an arsenic network. In order to monitor and validate arsenic testing and arsenic removal technologies, centres will be supplied with equipment and reagents and imparted training in the use of the protocols. A regional network will be created by linking existing national and regional centres of excellence with international centres.

3.5. Human resource development

The goal of human resource development is to produce a critical mass of trained manpower that can effectively handle laboratory and clinical diagnosis of arsenic diseases

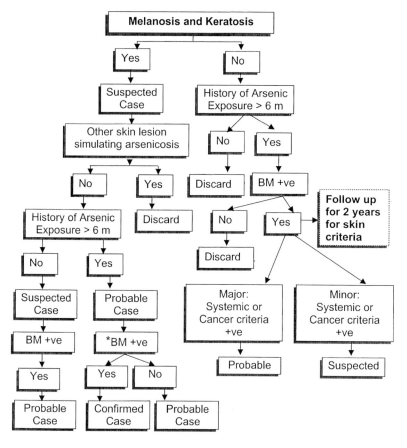

* BM – Bio-marker – Arsenic or metabolites in Urine/Hair/Nail

Figure 4. A proposed case definition for arsenicosis.

as well as clinically manage affected patients. This will be implemented by conducting training workshops for trainers and supporting regional and international exchange of experts in this area. WHO has sponsored a number of scientists in the intercountry exchange of expertise and skills.

4. Partnership, the way forward

The strength of WHO is the partnership it forges with other agencies and institutions in fulfilling its normative functions in health. In order to provide the best evidence-based options in these goals addressing the above research gaps, the partnership of the research community is essential in creating a regional network for exchanging ideas, methods and experience. We already work in partnership with UNICEF with whom we are jointly implementing a United Nations Fund (UNF) project in Bangladesh. The regional office will also work with other agencies, such as the International Atomic Energy Commission (IAEC), on defining regional policy and guidelines for arsenic testing and validation of test kits. Other agencies and donors include the Centers for Disease Control, Atlanta, and the Department for International Development, UK.

References

ATSDR, Agency for Toxic Substance and Disease Registry, 1990. Case Studies in Environmental Medicine, Arsenic Toxicity. U.S. Department of Health and Human Services, June.

ATSDR, Agency for Toxic Substances and Disease Registry, 1999. Toxicological Profile for Arsenic (Update). U.S. Department of Health and Human Services, Atlanta.

British Geological Survey/Department of Public Health Engineering, Government of Bangladesh, 1999. Groundwater Studies for Arsenic Contamination in Bangladesh. Final Report, Vol. I: Summary, Department of Public Health and Engineering, Government of Bangladesh, Dhaka, www.bgs.ac.uk/hydro/B_find.htm

Chakraborty, A.K., Saha, K.C., 1987. Arsenical dermatosis from tubewell water in West Bengal. Indian J. Med. Res., 85, 326–334.

Choprapawan, C., 1994. Arsenic poisoning problem at Ronpiboon district, Nakorn Si Thammarat province. Division of Environmental Health, Ministry of Public Health, 1992. Conclusion of arsenic poisoning situation at Ronpiboon district, Nakorn Si Thammarat province.

Dhar, R.K., Biswas, B.K., Samanta, G., Mandal, B.K., Chakraborti, D., Roy, S., Jafa, A., Islam, A., Ara, G., Kabir, S., Khan, W., Ahmed, S.K., Hadi, S.A., 1997. Groundwater arsenic calamity in Bangladesh. Curr. Sci., 73, 48–59.

Huq, M.A., Misbahuddin, M., Choudhury, S.A.R., 2000. Spirulina in the treatment of chronic arsenic poisoning, research studies on health impact of arsenic exposure. Bangladesh J. Physiol. Pharmacol., 16 (1), 15–16.

Kosnett, J.M., 1999. Clinical approaches to the treatment of chronic arsenic intoxication: from chelation to chemoprevention. In: Chappel, W.R., Abernathy, C.O., Calderon, R.L. (Eds), Arsenic Exposure and Health Effects, Elsevier, Oxford, pp. 349–353.

Mazumder Guha, D.N., Ghoshal, U.C., Saha, J., Santra, A., De, B.K., Chatterjee, A., Dutta, S., Angel, C.R., Centeno, J.A., 1998. Randomized placebo-controlled trial of 2,3-dimercaptosuccinic acid in therapy of chronic arsenicosis due to drinking arsenic contaminated subsoil water. Clin. Toxicol., 36 (7), 683–690.

Oshikawa, S., Geater, A., Chongsuvivatwong, V., Piampongsan, T., Chakraborti, D., Samanta, G., Mandel, B., Hotta, N., Kojo, Y., Hironaka, H., 2001. Long term changes in severity of arsenical skin lesions following intervention to reduce arsenic exposure. Environ. Sci., 8 (5), 435–448.

Sikder, M.S., Maidul, A.Z.M., Karim Khan, M.A., Asadul Huq, M., Choudhury, S.A.R., Misbahuddin, M., 2000. Effects of spirulina in the treatment of chronic arsenicosis. Bangladesh J. Dermatol. Venereal. Leprol., 17 (1), 9–13.

World Health Organization for South-east Asia, (1997). Arsenic in Drinking Water and Resulting Arsenic Toxicity in India and Bangladesh – Report of a Regional Consultation, New Delhi, India, 29 April–1 May, SEA/EH/507.

WHO, 2001a. World Health Organization for South-east Asia, Arsenic Contamination in Groundwater Affecting Some Countries in the South-east Asia Region, New Delhi, July, SEA/RC54/83.

WHO, 2001b. World Health Organization for South-east Asia, Resolution of the WHO Regional Committee for South East Asia on Arsenic Contamination in Groundwater Affecting Some Countries in the South-east Asia Region, New Delhi, 3 July, SEA/RC54/R3.

WHO, 2001c. World Health Organization and International Programme on Chemical Safety Environmental Health Criteria 224, Arsenic and Arsenic Compounds, World Health Organization, Geneva.

WHO, 2001d. World Health Organization, Arsenic in Drinking Water, Fact Sheet no. 201, World Health Organization, Geneva.

WHO, 2002. World Health Organization for South-east Asia Health Research in Arsenic Poisoning. Deliberations of the WHO South-east Asia Advisory Committee on Health Research, New Delhi, 7 May, SEA/ACHR/27-Report.

PART VI
WATER TREATMENT AND REMEDIATION

Arsenic Exposure and Health Effects V
W.R. Chappell, C.O. Abernathy, R.L. Calderon and D.J. Thomas, editors
© 2003 Published by Elsevier B.V.

Chapter 34

"Arsenic Solutions" Web platform of >50 options for developing countries: collaborative design and innovation for the common good

Susan Murcott

Abstract

Knowledge of the extent of arsenic occurrence and its effect on public health has not kept pace with remediation efforts. Arsenic-contaminated tubewells were discovered in West Bengal in the early 1980s and in neighboring Bangladesh in the early 1990s. Arsenic-contaminated water is also found in the Argentina, Australia, Brazil, Chile, Hungary, Mexico, Mongolia, Nepal, Nicaragua, Taiwan, Thailand, US, Vietnam, and elsewhere. Since then, adequate solutions have not become widely available. This delay in providing remediation and safe alternatives to arsenic-contaminated water has enormous public health consequences, especially for the poor. The affected populations in only four countries illustrate this need:

- *Bangladesh: 25–77 million (WHO, 2002)*
- *West Bengal: 4.5–6.9 million*
- *Nepal: 0.5 million*
- *Inner Mongolia: 0.3 million.*

A first step to action is to know the alternatives. Our approach is to combine a knowledge base and a collaborative design platform. The knowledge base is the "Arsenic Solutions" Web site http://web.mit.edu/murcott/www/arsenic; the "ThinkCycle" collaborative design platform is located at http://www.thinkcycle.org. The "Arsenic Solutions" site discusses the universe of arsenic remediation methods, beginning with alternatives to contaminated tubewells, then covering over 50 tubewell treatment technologies. After briefly introducing the site and its main features, we show how the Web site knowledge base has been used in the course of one invention and technology implementation process. We then discuss the features of "ThinkCycle" that enable new forms of distributed design collaboration, based on the model of "Open Source" software development. Finally, we explore the next steps in collaborative innovation for the common good.

Keywords: arsenic remediation technologies, developing countries, collaborative innovation, South Asia, Open Source, ThinkCycle

1. Introduction

Arsenic contamination of drinking water could be described as the first 21st century water crisis and an unprecedented humanitarian disaster. But unlike natural disasters, famine or war, it has not received the same level of media attention, public awareness and action, perhaps because arsenic is a slow killer. The statistics on the affected people in just four South Asian countries are mind numbing:

- Bangladesh: 25–77 million (WHO, 2002)
- West Bengal: 4.5–6.9 million
- Nepal: 0.5 million
- Inner Mongolia: 0.3 million.

Ten to twenty years have passed since the arsenic crisis was first discovered in South Asia and still adequate safe water solutions have not yet become generally available. How do we move from information to action? In our approach, we make use of the World Wide Web because of its knowledge-sharing capabilities and collaborative design potential. This paper describes how we might take the Web beyond information dissemination and current forms of Web-based research collaboration – e-mail, chat rooms, e-conferences – to interactive collaboration among distributed users, based on the "Open Source" model of software development.

Water for health is a human right. On December 4, 2002, the United Nations Committee on Economic, Cultural and Social Rights took the unprecedented step of agreeing on a General Comment on water as a human right, saying

> Water is fundamental for life and health. The human right to water is indispensable for leading a healthy life in human dignity. It is a prerequisite to the realization of all other human rights.

Also in 2002, the United Nations Summit on Environment and Development in Johannesburg committed itself to halving, by 2015, the number of people without access to safe drinking water and sanitation. If we are to achieve these lofty goals, the Web is one critical tool. To turn these principles and goals into reality calls for cooperative innovation on a new and unprecedented scale.

2. Approach

Our approach in this paper is to combine a knowledge base and a collaborative design platform. The knowledge base is the "Arsenic Solutions" Web site http://web.mit.edu/murcott/www/arsenic; the "ThinkCycle" collaborative design platform, developed by Sawhney (2003) is located at (http://www.thinkcycle.org).[1]

The "Arsenic Solutions" Web site:

- focuses exclusively on arsenic remediation solutions, i.e. the alternative safe water supply options and arsenic removal technologies through tubewell treatment and provides a comprehensive knowledge base;
- organizes data according to a common format;

---

[1] Thinkcycle is an academic, non-profit initiative supporting distributed collaboration on challenges facing underserved communities and the environment. Thinkcycle seeks to create a culture of open source design and ongoing collaboration among a distributed global community of researchers, domain experts and stakeholders.

- presents data on performance, cost, laboratory and field tests applications, contact persons, references, links,[2] etc.;
- provides a comprehensive knowledge base, which is one foundation for a collaborative network.

When the "Arsenic Solutions" knowledge base is combined with the "ThinkCycle" platform, a workspace is generated to implement product design collaboration that can:

- provide capability for file management and archiving, up- and down-loading of files with versioning features and search capabilities;
- index and makes searchable all content to the site;
- allow access control so that content owners can set permissions on any contribution to enable edit privileges and other inter-active options in order to maintain up-to-date information on arsenic removal technologies and safe water alternatives information;
- feature an online discussion forum, image and file-sharing repository, as well as a digital library for archiving publications.
- notify subscribers to the topic by e-mail whenever new content is posted.

Through these mechanisms, the "Arsenic Solutions" knowledge base acts as a vehicle for collaborative solution design, research dissemination and action.

3. Results and discussion

3.1. Knowledge base

In this section, we briefly describe the content of the knowledge base, then, we walk through one specific example showing how the knowledge base has been applied to obtain a viable arsenic treatment solution.

http://web.mit.edu/murcott/www/arsenic presents the current universe of arsenic remediation methods appropriate for low income and developing countries – alternative safe water options and tubewells remediation technologies. Applying a common template, each technology in the database is described in detail, including its performance, the analytic method used to determine performance, and the sites of laboratory, pilot, or full-scale tests. All tubewell treatment options are categorized according to the dominant

---

[2] The following LINKS constitute the "essential background reading" on arsenic remediation options:
BUET/United Nations University "Technologies for Arsenic Removal from Drinking Water Workshop," May 5–7, 2001 (http://www.unu.edu/env/water/arsenic/Workshop-Report.html);
United Nations Synthesis Report on Arsenic in Drinking Water (http://www.who.int/water_sanitation_health/Arsenic/ArsenicUNRep6.htm);
Arsenic Crisis Information Center (W.S. Atkins Phase I&II Report, March 2001) (http://www.bicn.com/acic/resources/arsenic-on-the-www/safewater.htm#ENV);
Ontario Centre for Environmental Technology Assessment's Web site devoted to Environmental Technology Verification for their Arsenic Mitigation Project (http://www.oceta.on.ca/programs/etv-am.htm) currently identifies no specific technologies on that site;
Harvard arsenic site: (http://phys4.harvard.edu/~wilson/arsenic_project_main.html);
EPA: Technologies and Costs for Removal of Arsenic from Drinking Water (http://www.epa.gov/safewater/arsenic.html).

treatment process – oxidation, coagulation/precipitation, filtration, adsorption, ion exchange, membrane processes, biological – so that similar technologies and approaches can be compared. We learn what equipment is needed, step-by-step procedures used to carry out laboratory and field tests, capital and O&M costs, sludge issues, pros/cons and which options are promoted by which agencies, companies, and academic institutions. Published references are listed and e-mail addresses of key contact people enable the possibility of networking.

"How to obtain arsenic-free drinking water, especially low cost solutions for developing countries?" Broadly speaking, there are two kinds of solutions:

(I) *Alternative safe water options*, which may or may not include the need for treatment in order that the water be safe to drink. Options included so far on this Web site are

1. shallow ("dug") wells;
2. deep wells;
3. sharing existing safe (arsenic-free) tubewells;
4. pond sand filters;
5. river bank infiltration;
6. distillation;
7. rainwater harvesting.

(II) *Arsenic remediation options* through tubewell treatment. Over 50 options have been included so far on the "Arsenic Solutions" Website – too numerous to include here. The reader is referred to http://web.mit.edu/murcott/www/arsenic for a complete list.

3.2. Example of use of the "Arsenic Solutions" knowledge base

In order to demonstrate the role of a comprehensive knowledge base in the process of design, we now take the reader though one example of invention and implementation.

Step 1: Problem definition. The author's concern with the development of arsenic solutions came about because of a presentation on the human impact of the arsenic crisis by a delegation of Bangladeshi women doctors and social workers at the September 1998 Second International Conference on Women and Water. At that time, the arsenic crisis had not yet been covered by the US news media. (*The New York Times* published its first article on the arsenic crisis as a headline story on November 10, 1998).

Step 2: Idea generation. The author undertook research and began to develop a database of arsenic solutions. She presented a preliminary set of 22 remediation technology options at the February 27–28, 1999 "Arsenic in Bangladesh Groundwater" conference, organized by the Bangladesh Chemical and Biological Society of North America. The author's paper "Appropriate Remediation Technologies for Arsenic-Contaminated Wells in Bangladesh" synthesized the then present state-of-knowledge by providing a "preliminary comprehensive list" of available and proposed remediation technologies for treating arsenic-contaminated tubewells in Bangladesh. Fundamental engineering process categories: sedimentation, coagulation/precipitation, adsorption, etc. provided an organizing framework under which to discuss specific options. Also proposed were three, and later four, major screening criteria: treatment performance, cost, social acceptability and overall sustainability (technical, economic, environmental, social). The paper concluded that there were already a number of available technologies that performed well and were affordable, even for very low income countries such as Bangladesh, but information on field testing and implementation in Bangladesh or other developing countries was, at that stage, limited.

Step 3: Database → knowledge base. Efforts to convert the database of arsenic solution options into a knowledge base involved taking a sea of information and organizing it. This required analysis and evaluation, which in turn required some expert judgment. This process began at the Fourth International Conference on Arsenic Exposure and Health Effects in June 2002, sponsored by the Society for Environmental Geochemistry and Health, the University of Colorado and the US EPA. The author's presentation "A Comprehensive Review of Low-cost Tubewell Water Treatment Technologies for Arsenic Removal" included a "Fact Sheet" template for an arsenic remediation options database. The request went out to the participants – national and international experts on arsenic – to provide feedback on the template so that it would ask all the right questions and, when filled out, present concise, but complete, knowledge on each option. The "Fact Sheet" template draft was also sent out electronically to a network of colleagues who were arsenic experts engaged in providing technical solutions to the arsenic crisis. From this process, there emerged a final template. This was published both in the conference proceedings (Chappell et al., 2001) and on the "Arsenic Solutions" Web site. The final Fact Sheet template was then widely distributed.

Step 4: Experimentation and analysis. Once the Fact Sheets on many different options were completed and turned in, that information was posted on the Arsenic Solutions Web site. From that knowledge base, a subset of "most promising options" was selected for in-depth research. Under the supervision of the author, three promising technologies from the "Arsenic Solutions" Web site, all of which were in the process category "Adsorption," were studied as a Master of Engineering thesis by Jessica Hurd in the 2000–2001 academic year. This research took place both at MIT and in a remote field site in Parasi, Nepal, where high concentration (250 µg/l) of arsenic had been identified in area tubewells. Hurd fabricated two systems: the iron filings in a jerry can technology (A6) and the 3-kolshi/3-gagri system (A7) and she obtained, by donation, at an activated alumina metal oxide technology (Apryon; A1). The iron filings used in her 3-kolshi/3-gagri system were obtained from the same US source as the iron filings used by Khan et al. for their 3-kolshi technology laboratory studies in the US (A7) and by Nickolaidis for his AsRT Technology (A5). Awareness of the criteria of social acceptability and project sustainability led us to look for a local Nepali supply of iron filings. The iron filings Hurd transported in her luggage from the US to Parasi, Nepal, were not available locally. The author suggested trying iron nails, which we were able to find in the local Parasi marketplace. Although the short duration of the field testing for that technology did not allow a breakthrough curve to be established, in the comparative tests we did perform, the local iron nails gave as effective an arsenic removal as the iron filings brought from the US. Hurd went on to determine that the 3-kolshi/3-gagri was the most promising technology of the three she had selected from the "Arsenic Solutions" knowledge base in terms of the criteria: performance, cost and social acceptability.

Step 5: Innovation. In the next academic year, three different technologies from the "Arsenic Solutions" knowledge base were studied by three new MIT Master of Engineering students, again under the author's supervision. The three systems were the 2-kolshi system (CP13), iron oxide coated sand (A22) and another activated alumina metal oxide system (Aquatic Treatment Systems; A21). Separately, in another Master of Engineering project, Heather Lukacs, was studying the CAWST biosand filter as a promising option for treatment of microbially contaminated water. One of the arsenic students, Tommy Ngai, who was well versed in the subject of arsenic treatment, combined ideas from Hurd and Lukacs to come up with an innovation, a biosand arsenic filter using iron nails as the adsorption medium. Ngai submitted this innovation to an MIT design competition "IDEAS,"[3] won the top prize, and used his prize money to implement his invention, first in a laboratory set-up in at RWSSSP in Butwal, Nepal, and then, in collaboration with RWSSSP, in field trials in Sarawal and Devadaha, Nepal.

Step 6: Field testing. To date, our MIT team has performed Phase I testing of seven arsenic remediation technologies from the "Arsenic Solutions" knowledge base in the US and Nepal. Screening of these seven technologies by the criteria of performance, cost, social acceptability and overall sustainability (technical, economic, environmental, social) has now led to concurrent field testing of the three most promising options: 2-kolshi (CP13), 3-kolshi (A7), and the biosand arsenic filter (A25). Research and analysis of results is still underway, but preliminary results in the context of Nepal indicate that, in terms of performance, the biosand arsenic filter is the best, while in terms of cost, the 3-kolshi is the best among the options tested. We are now seeking ways to bring down the cost of the biosand arsenic filter so that this best-performing option is available to all who might benefit from it. Concurrently, our partner organizations in Nepal, Environment and Public Health Organization (ENPHO)/Red Cross and RWSSSP are engaged in additional pilot tests of these same three most promising options.

From this example, we can see that the Arsenic Solutions Web site has been used principally as a knowledge base and a networking tool in the design process. In order to expand its capabilities, we are now engaged in combining the Arsenic Solutions Web site with the Thinkcycle collaborative design platform.

3.3. ThinkCycle's collaborative design platform

Direct uploading of new information. The "Arsenic Solutions" knowledge base has been maintained by a single domain expert, and this has slowed down the process of information dissemination. A person or group with a new technology, or with updated information for an existing technology, e-mails the domain expert, she replies by sending a template, the template is filled out and returned, then the new information is posted to the Web site – a multiple-step, tedious process. One obvious modification is to provide the option to freely upload as well as download information from the Web site without having to go through the domain expert bottleneck. The down-side is that junk might be posted. The up-side is that it should be relatively obvious to sort out the junk from reliable information.

Distributed design teams. So far, the "Arsenic Solutions" Web site has been used by a design team that is largely co-located in space (i.e. Cambridge, MA). The next step is to explore whether it is possible and useful to design from multiple (distributed) locations. Product design teams have traditionally worked in a hands-on, face-to-face environment and have used the Web overwhelmingly for browsing and searching for relevant

---

[3] http://web.mit.edu/~ideas/

information. As regards Web-based distributed (as opposed to co-located) research collaboration, e-mail has enabled the exchange of ideas and documents, e-conferences have linked up people around the world and chat rooms have occasionally brought researchers together around a common theme, a prerequisite for collaboration. However, distributed collaboration is just beginning to be possible, when the workspaces of individual researcher and their computers can be shifted to common workspaces on a platform such as www.thinkcycle.org.

Our partner organizations in Nepal. ENPHO/Red Cross, RWSSSP and the International Buddhist Society are each working on different facets of the arsenic crisis in Nepal, and collaboration has, so far, only been through conventional channels. (We go to Nepal and do joint or parallel field testing, they come to the US and present at conferences, we write e-mails and share documents via attachments). Yet, we are a team working in two different parts of the world – Massachusetts and Nepal – on one common problem. In agricultural societies, when there is a common need – for example – harvest time – the entire community takes part. In our global village, we have a common need – safe, clean drinking water. Can the global community work cooperatively? More concretely, assuming we have a "common need" (for safe drinking water) and a "common field" (Web based common space as on www.thinkcycle.org), could we bring these distributed team members together to work on technology design and implementation in a new, more powerful way by expanding Web capabilities? If yes, then one next step might be to begin to share a common work space with our partner organizations in Nepal and/or to partner academic institutions, for example, Cambridge University.

Design collaboration using thinkspaces. Thinkspaces are online design notebooks for design teams to access and contribute resources and design ideas. Design teams may choose to set up their ThinkSpace as private-access only, selectively sharing their ongoing designs with subscribed members or gradually making specific project contributions publicly accessible. This allows flexibility in both dissemination and peer-review for rapid design iterations, while enabling private exchange among team members in the early stages of a design project. Currently, the online design tools are being extended to provide visualizations of the design process and better support for requirements of collaborative engineering design among remote users. We will explore and report on this in future publications.

Collaborative design and intellectual property. Cooperative design raises challenging and, as yet, unanswered questions about intellectual property. A variety of intellectual property rights approaches are currently supported on ThinkCycle:

- digital "paper trail" of design contributions from distributed participants, i.e. registering all contributions in the database with a timestamp and Internet host address;
- history of member contributions from all topical areas and ThinkSpaces;
- moderated access for online projects with private access to specific members and gradual public disclosure of selective content;
- replication of all public domain content on multiple mirror servers worldwide to ensure accessibility and redundancy of data.

In the future, several general intellectual property rights agreements will be provided on the site to support diverse forms of disclosure, dissemination and licensing for cooperative innovation.

4. Summary

"Technology" is typically defined from a science/engineering and an anthropological perspective, as follows:

1. the application of science and engineering, especially to industrial, military and commercial objectives;
2. the body of knowledge that is available to a civilization.

We urgently need new technologies today to meet the basic human need for safe, clean drinking water for those in our global community who lack this fundamental security and sufficiency. While these new technologies are primarily intended to provide water for health as a basic human right, and only secondarily for industrial, military or commercial objectives, they certainly contribute to the well-being and security of us all, broadly defined. Through an examination of one collaborative design process spanning two different continents, and through a review of how the Web has been used in that process to date, this paper explores ways to bring "collaborative design and implementation for the common good" one next step closer to fruition.

Acknowledgements

This work acknowledges the collaboration and support of and the Master of Engineering (ME) Program of the Department of Civil and Environmental Engineering, Massachusetts Institute of Technology and specifically Dr Eric Adams, the director, along with ME students who have worked on arsenic technologies or the "Arsenic Solutions" Web site: Patricia Halsey, Shaheer Hussam, Teresa Yamana, Jessica Hurd, Soon Kyu Hwang, Tommy Ngai, Barika Poole, Georges Tabbal. We also acknowledge Roshan Shrestha, Arinita Maskey and their colleagues of the Environment and Public Health Organization (ENPHO), Kalawati Pokharel and the Rural Water Supply and Sanitation Support Program, Bhikkhu Maitri and the International Buddhist Society, our partner organizations in Nepal.

References

Chappell, W., Abernathy, C.O., Calberon, R.L. (Eds), 2001. Arsenic Exposure and Health Effects IV. Elsevier Science, London.
Sawhney, N., 2003. Cooperative innovation in the commons: rethinking distributed collaboration and intellectual property for sustainable design innovation. PhD Thesis. Massachusetts Institute of Technology. Media Arts and Sciences, School of Architecture and Planning, February 2003.
WHO, 2002. Fact Sheet. http://www.who.int/inf-fs/en/feature206.htm; March 2002.

Arsenic Exposure and Health Effects V
W.R. Chappell, C.O. Abernathy, R.L. Calderon and D.J. Thomas, editors
© 2003 Published by Elsevier B.V.

Chapter 35

Investigation of arsenic removal technologies for drinking water in Vietnam

Pham Hung Viet, Tran Hong Con, Cao The Ha, Hoang Van Ha, Michael Berg, Walter Giger and Roland Schertenleib

Abstract

Severe and widespread contamination by arsenic in groundwater and drinking water has been recently revealed in rural and suburban areas of the Vietnamese capital of Hanoi with similar magnitude as observed in Bangladesh and West Bengal, India. This fact has prompted the need to develop simple, rapid and low-cost techniques for reducing arsenic concentrations in supplied water. In the present study, laboratory and field tests were conducted to assess the suitability of using oxidation processes by activated hypochlorite in water treatment plants in Hanoi city and naturally occurring minerals as sorbents in household-based systems to reduce arsenic concentrations in drinking water. Sorption experiments indicated that coprecipitation of arsenate [As(V)] in ferric hydroxide is much more efficient than of arsenite [As(III)]. With Fe concentration of 5 mg/l, As(V) can be efficiently reduced from concentrations of 0.5 mg/l levels lower than the Vietnam standard of 0.05 mg/l. Activated hypochlorite was additionally introduced after the aeration tank in the conventional water treatment process that is currently used in the water treatment plants of Hanoi city. This modified process was able to lower arsenic concentrations below the standard level with relatively low Fe concentration (5 mg/l). Investigations on pilot scale apparatus indicated that the removal efficiency of As in this system was much higher than that in the laboratory experiments. To reduce As concentrations to the levels lower than the standard level of 0.05 mg/l, initial Fe/As concentration ratios used in the pilot system and laboratory experiment were 16 and 50, respectively. Laterite and limonite, which are naturally and widely occurring minerals in Vietnam, can be used as potential sorbents for arsenic removal in smaller scale water treatment systems. The sorption capacities of laterite and limonite for As(V) were estimated to be 1100 and 900 mg/kg, respectively. Initial results of field tests indicated that arsenic concentrations decreased to levels < 0.05 mg/l. The household system based on an adsorption column packed with these minerals seemed to be a suitable technique for small-scale groundwaters remediation in rural and sub-urban areas.

Keywords: arsenic removal, co-precipitation, sorption, chlorine oxidation, naturally occurring minerals, laterite, limonite

1. Introduction

The concern over arsenic contamination in drinking water and groundwater has increased in recent years and has become a worldwide problem. Severe contamination has been reported for a decade in Bangladesh and West Bengal, India, where millions of people are consuming arsenic-poisoned groundwater (Nickson et al., 1998). Serious arsenicosis has been observed for a large population in these areas (Chowdhury et al., 2000). Arsenic problems have also been observed in developed nations. In the US, the Environmental Protection Agency has recently announced to lower the maximum contamination level for arsenic in drinking water from 50 to 10 μg/l. The increasing awareness of arsenic toxicity and the regulation changes have prompted considerable attention towards developing suitable methods for lowering arsenic levels in drinking water.

Natural occurring contamination by arsenic has also been observed in the Red River delta of Northern Vietnam. A recent comprehensive survey conducted in our laboratory has revealed elevated arsenic concentrations over a large rural and suburban area of the Vietnamese capital of Hanoi (Berg et al., 2001). In four districts of the rural Hanoi area, arsenic concentrations in about 48% of the investigated groundwaters exceeded the Vietnam guideline of 50 μg/l, and hence, point to a high risk of chronic arsenic poisoning. This fact has prompted the need to investigate suitable methods for lowering arsenic concentrations in drinking water by rapid, simple and low-cost techniques.

A number of recent studies have proposed the use of zerovalent iron filings as filter medium for removing arsenite [As(III)] and arsenate [As(V)] from groundwater (Farrell et al., 2001; Su and Plus, 2001a,b). The process is based on the adsorption and co-precipitation of As(III) and As(V) onto Fe(III) oxides (Melitas et al., 2002). Adsorption capacity of arsenic in the form of arsenite and arsenate onto various ferric clay minerals has been well investigated (Farquhar et al., 2002). In Bangladesh, several efforts have been made to develop household filtration systems with effective low-cost technologies. Coprecipitation with ferric chloride is an effective and economic technique for removing arsenic from water, because iron hydroxides formed from ferric salt have a high sorption capacity for arsenate (Meng et al., 2001). However, the applicability of such methods depends largely on the geological characteristic of the groundwater. For example, in Bangladesh, elevated concentrations of phosphate and silicate may enhance the mobility of As(V) in soils contaminated with arsenate (Peryea and Kammereck, 1997; Hug et al., 2001). In addition, recent studies have suggested that silicate may disturb the removal of As(III) and As(V) by coprecipitation with ferric chloride (Meng et al., 2000).

In Vietnam, our recent investigations showed that the current arsenic contamination in the Red River delta area has been as serious as observed in Bangladesh and West Bengal (Berg et al., 2001). Furthermore, the chemical composition of groundwater in Vietnam is similar to that in Bangladesh. In the present study, we investigated the applicability of a simple and economic technique for removing arsenic in groundwater during the treatment process in water treatment plants of urban Hanoi. Furthermore, we have evaluated laterite and limonite, which occur very widely in Vietnam, as potential sorbents for arsenic. The sorption kinetics of these minerals for As(III) and As(V) were investigated and their applicability in household adsorption and filtration system for arsenic removal was assessed.

2. Materials and methods

2.1. Experiments for arsenic removal by adsorption onto Fe hydroxide and oxidation by hypochlorite

Raw groundwater samples were collected from water supplies of Hanoi city. Appropriate Fe(II) chloride amounts were added and the pH was maintained at 7.0 ± 0.2. Fe(II) was oxidized to Fe(III) by air purging until Fe(II) could not be detected by the orthophenanthroline method. As(III) and As(V) in the form of AsO_3^{3-} and AsO_4^{3-} at concentration of 0.5 mg/l were added. Solutions were stirred gently for 10 min and settled for 15 min for precipitation. The precipitate was discarded and the solution was analyzed for As and Fe concentrations. Chlorine in the form of hypochlorite was added to a series of Fe(II) solutions with concentrations of 1, 5, 10, 15, 20, 25 and 30 mg/l and arsenic at a constant concentration of 0.5 mg/l. For arsenic analysis, an on-line hydride generation device coupled with atomic absorption spectroscopy (HVG-AAS) (Shimadzu, Kyoto, Japan) was used. Further details on the determination of As can be found in our earlier article (Berg et al., 2001).

2.2. Sorption capacity of laterite and limonite for As(III) and As(V)

Laterite and limonite were first treated (see below) and then subjected to determination of their chemical composition as well as naturally occurring arsenic contents (see Table 3). Arsenic possibly present in these minerals was removed by washing in an alkali solution (10 M NaOH) and by heating to 900°C for 2 h. Isothermal sorption experiments were carried out using treated laterite and limonite as sorbents, with initial As(III) and As(V) concentrations of 2, 5, 10, 20, 30, 40, 50 and 100 mg/l and under atmospheric pressure and at 28°C. The suspensions were centrifuged and the supernatant solutions were filtered through 0.45 μm membrane filters prior to arsenic determination.

Treated laterite and limonite were packed into a column and applied as filtration device in a household water treatment system. Raw groundwater was pumped through the column. Raw groundwater and filtered water samples were collected periodically (about 3–4 times a week) and were analyzed for total arsenic concentrations.

3. Results and discussion

3.1. Removal of arsenic in the form of As(III)

In anoxic groundwater, arsenic is present in the form of arsenite (products of H_3AsO_3) due to the reductive conditions. After aeration in the Hanoi water treatment plants, most Fe(II) is oxidized to Fe(III). After Fe is completely oxidized, the dissolved oxygen increases and then facilitates the oxidation of As(III). In treated water of the water treatment plants, As(V) concentration after aeration varied substantially with a maximum level of about 20% of total As concentration. However, the coprecipitation and sorption mechanism is much more efficient for As(V) as compared to As(III). To clarify this, we investigated the sorption capacity of As(III) and As(V) onto iron (III) hydroxide under the conditions of the water treatment plants in Hanoi.

Figure 1 shows the arsenic sorption capacity of iron (III) hydroxide in the sorption experiment. Fe(II) concentrations of 1, 5, 10, 15, 20, 25 and 30 mg/l were used, and the As(III) concentration was kept constant at 0.5 mg/l. The sorption of As(III) increased with increasing Fe(II) concentration. As shown in Figure 1, to reduce the As concentration to the level below the Vietnamese standard (0.05 mg/l), a minimum Fe(II) concentration of 25 mg/l was required. If this technique is applied for water treatment plants in Hanoi, it is

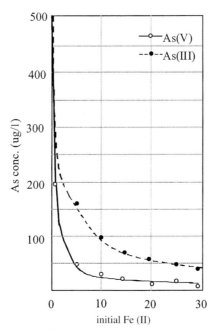

Figure 1. As (III) and As (V) removal efficiency by coprecipitation with iron(oxy)hydroxides formed from aerated Fe(II) solutions (initial As concentration = 500 μg/l).

difficult to reduce arsenic concentrations to the WHO standard level (10 μg/l). Therefore, we have further investigated the possibility of lowering arsenic concentrations in supplied water in the form of As(V).

3.2. Removal of arsenic in the form of As(V)

In this experiment, As(III) was oxidized to As(V) using hypochlorite. In this experiment, the active chlorine solution was added in excess (0.5 mg/l) for complete oxidation of As(III) to As(V). The sorption isotherm for As(V) onto iron (III) hydroxide showed that the adsorption capacity for As(V) is much more efficient than that of As(III) (Fig. 1). For example, with a relatively low Fe concentration of 5 mg/l, the arsenic concentration can be substantially reduced to a level below 50 μg/l. If treated water contains As concentrations of less than 500 μg/l, the required Fe concentration for lowering such As levels should be more than 5 μg/l.

3.3. Influence of active chlorine to improve arsenic removal efficiencies

In this experiment, chlorine concentrations ranging from 0.25 to 1.25 mg/l were used and the initial arsenic (III) concentration was kept constant at 500 μg/l. The capacity for total inorganic arsenic removal (%) was examined with different Fe concentrations: 1, 5, 15 and 25 mg/l (Fig. 2). Interestingly, the removal efficiency remained constant at more than

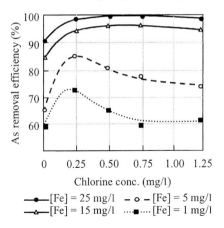

Figure 2. Influence of active chlorine concentrations on As removal efficiency (initial As(III) concentration = 500 μg/l).

80% for relatively high concentrations of Fe. However, for lower Fe concentrations, the removal efficiency curve had a maximum and the efficiency decreased thereafter with increasing chlorine concentrations (Fig. 2). This phenomenon may be due to the oxidation of other compounds or/and the formation of other Fe species (Meng et al., 2000). Fortunately, the Fe(II) concentration is quite high (average 15–20 mg/l) in groundwater of the Red River Delta. The effect of other compounds such as silicate and phosphate was not investigated in this study.

Based on the results of the arsenic removal efficiency in the form of As(V), we proposed to add hypochlorite right after the aeration step in the conventional process for water treatment in the urban Hanoi water treatment plants (Fig. 3). After aeration, Fe(II) was fully oxidized to Fe(III), and As(III) was also oxidized to As(V). The removal of As(V) was efficient and the hypochlorite can also act for water sanitation purposes. We suggest that this process can be applied for As concentrations in the city water treatment plants. In this process, the added amount of ClO^- depends on the chemical composition of the groundwater and a residual 0.5 mg/l chlorine must be achieved.

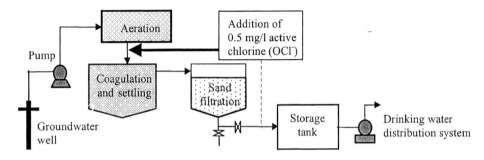

Figure 3. Proposed schematic diagram for additional oxidation by active chlorine in the water treatment process of the urban Hanoi water treatment plants.

3.4. Evaluation of As removal efficiency in a pilot equipment consisting of floating filtration and sand filtration

To further investigate As removal, we also tested the removal efficiency on the pilot equipment for groundwater treatment that is currently installed in our centre (Fig. 4). Groundwater is pumped from a 40 m deep well (1) to an ejector (3) placed in a floating filtration tank (4). The oxidation of Fe(II) to Fe(III), precipitation of iron(oxy)hydroxides and co-precipitation of As(V) took place in this tank. After coagulation and floating filtration, the water was transferred through the sand filtration system (5) and finally to the reservoir (6) (Fig. 4). In order to evaluate the quality of the raw groundwater, samples were taken and were analyzed for total Fe, As, phosphate, soluble silicate concentrations, dissolved oxygen and pH continuously for two weeks. The composition of the groundwater before the pilot scale installation is given in Table 1.

Because the initial Fe(II) concentration is quite high, Fe(II) was not added into the pilot system. To assess the potential for As removal, As(III) was continuously added in the form of AsO_3^{3-} with a series of concentrations from 0.15 to 1.7 mg/l. The results are presented in Table 2 and Figure 5 will be changed into "It is recognised that arsenic was removed in the floating filtration tank by coprecipitation of As(V) and ferric hydroxide. For the given iron content of ~ 25 mg/l, the maximum arsenic inlet concentration was 1.0 mg/l to reach arsenic in levels below 0.05 mg/l after the floating filter (sampling point S_3). This corresponded to a Fe/As ratio of ≥ 23 (S_3). After passing the sand filter, As was

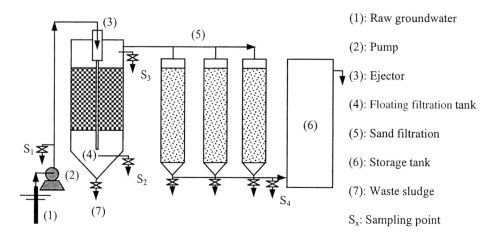

(1): Raw groundwater

(2): Pump

(3): Ejector

(4): Floating filtration tank

(5): Sand filtration

(6): Storage tank

(7): Waste sludge

S_x: Sampling point

Figure 4. Schematic diagram of the water treatment pilot system.

Table 1. Composition of groundwater before the pilot water treatment system.

Composition	Total Fe (mg/l)	Total As (μg/l)	DO (mg/l)	pH	PO_4^{3-} (mg/l)	Soluble Si (mg/l)
Level	25.5	20.1	1.2	6.8	0.12	4.36

Table 2. Arsenic removal efficiency at different sampling points in the pilot water treatment system (see Figure 4).

Spiked As (mg/l)	Fe/As ratio	As (mg/l) and Fe (mg/l) at sampling points							
		S_1		S_2		S_3		S_4	
		Fe	As	Fe	As	Fe	As	Fe	As
0.00	1221	25.64	0.021	22.36	0.020	1.42	0.004	0.53	0.003
0.15	153	26.54	0.173	–	–	2.86	0.012	0.32	0.008
0.35	66	24.56	0.372	–	–	2.61	0.015	0.11	0.009
0.55	53	30.41	0.574	–	–	1.34	0.021	0.43	0.011
0.65	34	23.32	0.677	–	–	1.86	0.028	0.08	0.012
1.00	23	23.43	1.024	–	–	1.67	0.043	0.12	0.014
1.30	20	26.52	1.319	–	–	2.06	0.066	0.01	0.018
1.50	18	27.04	1.522	–	–	4.32	0.151	0.01	0.027
1.60	16	26.02	1.621	–	–	4.22	0.177	0.08	0.043
1.70	15	26.05	1.725	–	–	3.75	0.191	0.21	0.068

further lowered to 0.014 mg/l (sampling point S_4). By this way, the desired As standard (0.05 mg/l) could even be reached with an initial Fe/As concentration ratio of 16, which corresponded to an inlet As concentration of 1.6 mg/l."

These results indicate that As removal in the pilot system was much more efficient than in the laboratory experiments. To reduce the As concentrations to levels <50 µg/l, the initial Fe/As concentration ratio in the laboratory experiment was 50, while this value was 16 for the pilot system.

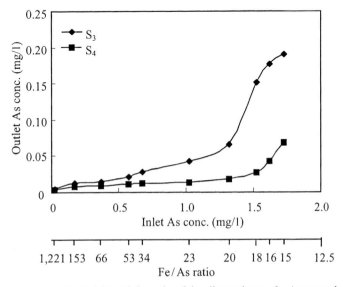

Figure 5. As concentrations in the inlet and the outlet of the pilot equipment for As removal.

3.5. Household sorption and filtration system

In Vietnam, private wells have been used for a long period of time for rural and sub-urban areas. In 1990s, UNICEF's pumped tube well systems have been widely developed and throughout the country. UNICEF wells play a very important role and are the main source of water supply for many people in Vietnam. However, as mentioned above, recent findings of the unexpected severe arsenic pollution in groundwater raised a serious concern that millions of people living in rural and suburban areas are consuming arsenic-rich groundwater and are at risk of arsenic poisoning (Berg et al., 2001). Due to the lack of knowledge and education, the risk of arsenic exposure for people in rural areas may be more serious. In this study, we therefore, also investigated the applicability of naturally occurring iron minerals having a high sorption capacity for some inorganic ions, including As(III) and As(V). Such minerals, nam2ely laterite and limonite, are abundant in midland areas (e.g. Ha Tay, Vinh Phu province in Northern Vietnam) and are often relatively clean. We anticipated that these minerals could be used as potential sorbents for a household sorption and filtration system to remove arsenic from tubewell water.

Laterite and limonite minerals were collected, treated, sieved and subjected to the determination of composition as well as naturally occurring arsenic contents. The results of the analysis of laterite and limonite compositions and arsenic contents in these minerals is shown in Table 3. Sorption isotherms and breakthrough curves of limonite and laterite are shown in Figures 6, 7 and Figures 8, 9, respectively. A Langmuir sorption isotherm was able to describe the sorption kinetics of As(III) and As(V) onto laterite and limonite.

Table 3.   Laterite and limonite composition and arsenic content.

Material	SiO_2 (%)	Al_2O_3 (%)	Fe_2O_3 (%)	CaO (%)	MgO (%)	As_2O_3 (mg/kg)		
						Initial	After washing by alkali solution	After heating at 900°C
Laterite	40.96	14.38	32.14	0.14	0.18	41.83	33.77	5.36
Limonite	11.25	4.12	84.24	0.25	0.16	16.25	14.27	1.29

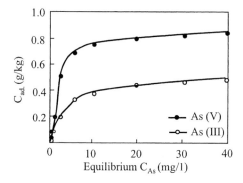

Figure 6.   Sorption isotherm of As (III) and As (V) onto limonite (initial As concentration = 500 μg/l).

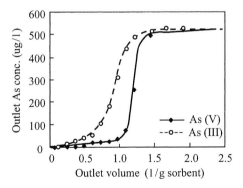

Figure 7.   Breakthrough curves of sorption of As(III) and As (V) for limonite (initial As concentration = 500 µg/l).

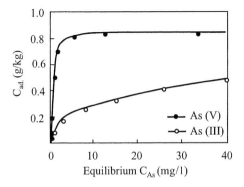

Figure 8.   Sorption isotherm of As (III) and As (V) onto laterite (initial As concentration = 500 µg/l).

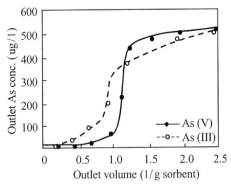

Figure 9.   Breakthrough curves of sorption of As (III) and As (V) for laterite (initial As concentration = 500 µg/l).

It is obvious that the sorption capacity of As(V) is higher than that of As(III), suggesting the suitability of using these materials to remove arsenic in the form of As(V) from groundwater. Based on the sorption isotherm, the sorption capacity of limonite for As(III) and As(V) was calculated as 500 and 900 mg/kg, respectively. For laterite, the sorption capacity was slightly higher for As(III) [600 mg/kg] suggesting a more effective sorption ability of this mineral for lowering arsenic concentrations in groundwater using household-based filtration and adsorption system. We also tested the arsenic concentrations before and after the sorption column. Our initial results showed that this system was able to reduce arsenic concentrations below the Vietnam standard of 50 mg/l. In addition, manganese was also efficiently removed and there was no contamination by sorbent-originated elements. Further investigations are necessary to provide detailed information on the efficiency and capacity of arsenic removal of this household water treatment system.

4. Conclusions

The investigations of suitable techniques for lowering arsenic concentrations in water treatment plants of Hanoi and household adsorption and filtration system for rural and suburban areas indicated that arsenic could be efficiently removed from drinking water in the form of arsenate. In the water treatment plants, hypochlorite (NaClO) for oxidizing As(III) to As(V) was added to the conventional process applied in the plants. With an Fe concentration of 5 mg/l, As concentrations can be decreased to the level below the Vietnam standard of 0.05 mg/l from the initial concentration of 0.5 mg/l. The investigation of the pilot scale apparatus installation indicated that removal of As in this system is more effective than that in the laboratory experiments. For small scale water treatment systems in rural and suburban areas, naturally occurring minerals such as laterite and limonite, can be used as sorbents for arsenic in adsorption and filtration columns. The relatively high sorption capacity of these minerals for arsenite and arsenate suggests the feasibility of using them in household-based water treatment systems.

Acknowledgements

The authors acknowledge the excellent cooperation and technical support of coworkers Bui Van Chien, Luyen Tien Hung of CETASD and colleagues at EAWAG. Funding was provided by the Albert Kunstadter Family Foundation (New York) and SDC (Swiss Agency for Cooperation and Development) for ESTNV Project.

References

Berg, M., Tran, H.C., Nguyen, T.C., Pham, H.V., Schertenleib, R., Giger, W., 2001. Arsenic contamination of groundwater and drinking water in Vietnam: a human health threat. Environ. Sci. Technol., 35, 2621–2626.
Chowdhury, U.K., Biswas, B.K., Chowdhury, T.R., Samanta, G., Mandal, B., Basu, G.C., Chanda, C.R., Lodh, D., Saha, K.C., Mukherjee, S.K., Roy, S., Kabir, S., Quamruzzaman, Q., Chakraborti, D., 2000. Groundwater arsenic contamination in Bangladesh and West Bengal, India. Environ. Health Perspect., 108, 393–397.
Farquhar, M.L., Charnock, J.M., Livens, F.R., Vaughan, D.J., 2002. Mechanism of arsenic uptake from aqueous solution by interaction with goethite, lepidocrocite, mackinawite, and pyrite: an X-ray adsorption spectroscopy study. Environ. Sci. Technol., 36, 1757–1762.
Farrell, J., Wang, J., O'day, P., Conklin, M., 2001. Electrochemical and spectroscopic study and arsenate removal from water using zerovalent iron media. Environ. Sci. Technol., 35, 2026–2032.

Hug, S.J., Canonica, L., Wegelin, M., Gechter, D., von Gunten, U., 2001. Solar oxidation and removal of arsenic at circumneutral pH and iron containing waters. Environ. Sci. Technol., 35, 2114–2121.

Melitas, N., Wang, J., Conklin, M., O'day, P., Farrel, J., 2002. Understanding soluble arsenate removal kinetics by zerovalent iron media. Environ. Sci. Technol., 36, 2074–2081.

Meng, X., Bang, S., Korfiatis, G.P., 2000. Effects of silicate, sulfate, and carbonate on arsenic removal by ferric chlorine. Water Res., 34, 1255–1261.

Meng, X., Korfiatis, G.P., Christodoulatos, C., Bang, S., 2001. Treatment of arsenic in Bangladesh well water using a household co-precipitation and filtration system. Water Res., 35, 2805–2810.

Nickson, R., McArthur, J., Burgess, W., Ahmed, K.M., 1998. Arsenic poisoning of Bangladesh groundwater. Nature, 395, 338.

Peryea, F.J., Kammereck, R., 1997. Phosphate-enhanced movement of arsenic out of lead arsenate contaminated topsoil and through uncontaminated sub-soil. Water Air Soil Pollut., 93, 117–136.

Su, C., Puls, R.W., 2001a. Arsenate and arsenite removal by zerovalent iron: kinetics, redox transformation, and implications for *in situ* groundwater remediation. Environ. Sci. Technol., 35, 1487–1492.

Su, C., Puls, R.W., 2001b. Arsenate and arsenite removal by zerovalent iron: effects of phosphate, silicate, carbonate, borate, sulfate, chromate, molybdate and nitrate, relative to chlorine. Environ. Sci. Technol., 35, 4562–4568.

Arsenic Exposure and Health Effects V
W.R. Chappell, C.O. Abernathy, R.L. Calderon and D.J. Thomas, editors
© 2003 Published by Elsevier B.V.

Chapter 36

Removing arsenic from drinking water: a brief review of some lessons learned and gaps arisen in Chilean water utilities

Ana María Sancha

Abstract

Theoretically, arsenic removal from water can be achieved by using different technologies. Application at full scale of each one might not be as successful as at laboratory tests. At full scale some variables are difficult to control and these may interfere in the removal process. These interferences may not be detected when working at laboratory tests with analito (As) solutions in distilled water.

The selection of the best available technology for arsenic removal should be based on some key factors, such as removal goals, quality of the water matrix, water quantity, operator skill requirements, operational water treatment costs, arsenic speciation, availability of analytical methods, sludge management, and others.

Some of these issues may play an important role in the feasability of arsenic removal practices, and therefore, must be considered and assesed before the selection of any arsenic removal technology. Small water utilities may face more challenges than larger systems. The situation for family systems is even worse.

This paper addresses a critical review of some of these issues and the lessons learned about arsenic removal in Chilean water utilities in the last 30 years.

Keywords: arsenic, arsenic removal, water treatment, Chile

1. Background

1.1. Arsenic occurrence in water sources

High arsenic levels are generally associated with specific geochemical environments. Arsenic may be naturally present in both organic and inorganic species in natural ground water and surface water. Methylated arsenic species may be sometimes found in eutrophic surface water but is rarely found in ground water. The organic species may originate from bacterial action (biomethylation) on inorganic species or agriculture use of arsenic pesticides. In the absence of man-made pollution and biological activity, water will contain only inorganic arsenic species. In general, for drinking water treatment the organic species are considered irrelevant if compared to inorganic species. (Ferguson and Garvis, 1972; Cullen and Reimer, 1989; Anderson and Bruland, 1991).

1.2. Water sources and arsenic speciation

In surface water, arsenic is present predominantly in its pentavalent form As(V). In groundwater, arsenic may be present as As(III) and As(V). Bacterial action may facilitate the presence of methylated species. The geochemical actions of soil may favor formation of As–S complexes. The meteoric water (rain water) in general should not contain arsenic, save for those regions influenced by smelting emissions or other As emitting industrial activities.

1.3. Water quality and its implications

Little is known about the "real" quality of water that contains arsenic. "Real" water – water that are not solutions of the analyte arsenic in distilled water, but contains other substances subsumed under the term "matrix" – corresponds to solutions of ionized and non-ionized compounds which, as a group or separately, may have antagonistic or synergistic effects with respect to arsenic and its implications on health, toxicity and/or removal processes. The water matrix may influence the speciation of As in water: inorganic or organic; soluble or particulate; ion, complex ion, chelated ion or molecule; colloidal, precipitated or adsorbed. Also pH and redox potential may influence to a large extent arsenic speciation.

Among the best documented cases of As in water are those of Taiwan, Argentina and Chile, which show enormous differences with respect to "water matrix" although they might contain similar As concentration ranges (Table 1). In the case of Taiwan and Inner Mongolia, As(III) accounted for 70–90% of the total As and the water

Table 1. Characterization of water matrix.

	Chile[a]	Argentina[b]	Mongolia[c]	Taiwan[c]
pH	8.0–8.4	8.5–8.0	6.89–8.37	7.45–7.52
Total dissolved solids	700–800	1655–2595	228.2–2343.3	733.1–801.6
Arsenic	0.40–0.60	0.80–1.25	0.024–1.12	0.10–0.13
Fluorescence intensity			2.30–69.40	26.84–32.57
Sulfate	80.0–100.0	145–173	0.00–3024.9	
Chloride	120.0–140.0	78–213	21.30–1701.52	
Chemical oxygen demand			4.63–66.63	
Alkalinity (CaCO$_3$)	100.0–120.0	1020–1615		
Hardness (CaCO$_3$)	130.0–150.0	36–44		
Silica (SiO$_2$)	40–50			
Boron (B)	3.0–4.0			
Fluoride (F)	0.30–0.50	24.0–10.5		
Nitrate (mg/l)		15–41		
Vanadium (mg/l)		4.50–3.00		

[a] Sancha et al. (1998).
[b] Cusimano et al. (1992).
[c] Xiaoying (2001).

contains natural organic matter in large amounts which suggests the feasibility of organoarsenical complexes. Some recent studies show that well samples from Inner Mongolia and Taiwan are quite different in the structure of natural organic matter (Xiaoying, 2001). In the case of Chile our experience, based on surrogate parameters, indicates that the Chilean water matrix is characterized by the absence of natural organic matter, oxidant conditions, hardness, silica, boron and high salinity (Sancha, 2000). In the case of Argentina, the water matrix contains high alkalinity, fluoride and vanadium levels (Cusimano et al., 1992).

For Taiwan, some researchers consider that blackfoot disease may be caused by high arsenic content in drinking water (Tseng, 1977). But other researchers consider that the cause may be a combination of high As and organic matter (Lu, 1990). For Chile, with drinking water of high content of arsenic blackfoot disease has rarely been reported. Only skin hyperpigmentation is reported.

The following proposes some questions that need to be studied:

- What is the role of the water matrix in As removal processes and/or toxicity?
- Are all species of arsenic removed with the same processes?
- Are all species of arsenic equally toxic?
- Antagonism vs synergism among water matrix components?

1.4. Analytical methods for arsenic in water

The specialized literature contains numerous methods to determine total As in water (Irgolic, 1994). The methods that can be easily used in treatment plants and which actually are the most widely used are silver diethyldithiocarbamate colorimetric procedure and atomic absorption spectometry with hydride generation.

At present there are several questions about the subject of As quantification in waters that are crucial to choose a treatment option. Among the questions to be answered are

- Are the analytical methods available – classical or advanced – enough to provide sustained control of As removal processes?
- How practical and reliable are the analytical methods at low As concentrations?
- What is the practical and true quantification levels at routinary scale?
- Will the laboratories existing at the treatment plants be able to measure accurately and consistently the As concentrations with the available methods?
- Are there reference materials for analytical quality control?
- Is there a good knowledge of all that is related to stability, contamination avoidance and water sample preservation?
- What is the cost of the analyses and what are the analyst's training requirements?

1.5. Maximum arsenic content in drinking water

Given the carcinogenic character of arsenic (IARC, 1980), the "ideal goal" for the maximum content of this element in drinking water should be 0 mg/l. In 1993, the World Health Organization (WHO, 1993) set a guide value of 0.01 mg As/l in drinking water. This guideline value is non-enforceable, but is of vital importance for all those countries that have this contaminant in their drinking water supplies because the

compliance with this standard is required by the international organizations that grant
financial support to countries that need to implement arsenic removal technologies in
order to supply safe water to their people. Many times these requirements are difficult
to comply with due to the complexity of the water quality matrix and the local socio-
economic conditions.

The reality of this situation is that there still exist many questions and uncertainties to
have complete information that may allow to establish national standards about As content
in drinking water. Among these subjects are

- Which is the extent of exposure to As through drinking water relative to other routes of
  exposure (e.g. food, inhalation or dermal contact)?
- How to manage the risk of people exposed to As through different ways?
- Is it possible to monitor compliance with the MCL using routine analytical methods?
- What is the cost of arsenic removal technologies?
- What are the real effects of arsenic on health?

2. Removing arsenic from water

2.1. Water quality vs arsenic removal technologies

The inorganic arsenic speciation is an important factor in the efficiency of arsenic removal
processes. (Sorg and Logsdon, 1978). Pentavalent arsenic As(V) is more effectively
removed than trivalent arsenic As(III). The former consists of arsenic acid (H_2AsO_4,
$H_2AsO_4^-$, $HAsO_4^{2-}$, AsO_4^{3-},) and its conjugate bases and the latter consists of
arsenious acid (H_3AsO_3) and its conjugate bases.

The technologies available to remove As from water include conventional
technologies: adsorption–coprecipitation with hydrolyzing metals, such as aluminum or
iron; adsorption on activated alumina; ionic exchange and processes with membranes such
as reverse osmosis, electrodialysis and nanofiltration. Activated carbon, is not selective for
arsenic. Emerging technologies are Fe–Mn oxidation; porous media sorbent; sorption on
reduced metals; in situ arsenic immobilization and biological treatment.

Most of the conventional and emerging technologies have not been tried at the plant
level with "real" water. The existing experiences have been obtained only at laboratory
scale with "non-real" water (analyte dissolved in distilled water). So, when testing some of
these methods in a real scenario, some of them might present important limitations
originated in the water matrix and in the loss of control of some important parameters that
are relevant in the treatment processes, but are difficult to keep under control when
working at real scale or in the field, as is the case of pH and temperature.

For all these reasons, it is important to conduct bench-scale testing using the actual
water ("real water") to be treated when designing full-scale or field-scale arsenic removal
systems.

The most common treatment technologies for removing As from drinking water are
conventional coagulation and filtration, lime softening and iron removal. All are well
suited for large-scale treatment plants. Coagulation with iron salts has been chosen as the
most appropriate technology for centralized arsenic removal (WHO, 2002). Coagulant
sludge is generally safe for disposal in municipal landfills.

The coagulation processes may be affected, according to the Chilean experience at plant-scale, by temperature, pH and water alkalinity as well as by the oxidation state of arsenic. Therefore, oxidation pretreatment of Arsenic (III) using chlorine is used (Sancha et al., 1992a,b). The efficiency of arsenic removal by coagulation will depend on the formation of insoluble products, arsenic adsorption in them and later separation from the water mass. Experience shows that any problem arising in these processes will hinder the success of arsenic removal. Some recent studies (Holn, 2002) show that some anions (particularly phosphate, carbonate and silicate) may compete with $AsO_4{}^{3-}$ for hydrous ferric oxide sorption sites and interfere with As removal. Arsenic sorption from water containing $HCO_3{}^-$, Si and $PO_4{}^{3-}$ may be less than arsenic sorption in which the only sorbate is As(V).

Ion exchange (IX) and activated alumina (AA) processes may be effective for As removal provided that these systems are regenerated or recharged properly and in a timely manner. If the IX system unlike AA, is operated beyond breakthrough, arsenic effluent concentration could exceed the influent levels (Gurian and Small, 2002). The efficiency will be strongly affected by the ionic species of As and water matrix. Because both IX and AA processes remove primarily anionic species, As(III) must be oxidized to As(V) to achieve the desired removal. Because of the competition of some ions (matrix water) for exchange sites, IX systems are not applicable (economically attractive) for waters containing high total dissolved solids (TDS > 500 mg/l) or sulfate (>150 mg/l). Activated alumina adsorption of arsenate is not as greatly affected by competing ions, however, it is highly dependent on water pH. Therefore, water pH values often need adjustment for an optimal run length (Wang et al., 2000; Holn, 2002).

Disposal of residuals generated by IX and AA processes is another important issue in their full-scale application. It must pass the toxicity characteristic leaching procedure (TCLP) test for disposal as a non-hazardous waste.

In the case of reverse osmosis and nanofiltration technologies, the presence of calcium salts (hardness) as well as of natural organic materials in suspension will require a previous conditioning of the water to be treated. The suspension material removal may be very important for surface waters during certain times of the year. The situation is different in the case of groundwater due to the absence of suspension material. For groundwater, the great interferences may be the presence of humic substances or other organic or inorganic materials (methane, sulfide, amino nitrogen and nitrite-nitrogen).

Iron or manganese-based adsorption media (such as granular ferric hydroxide or greensand) have been shown to possess high arsenic removal capacities in bench-and pilot-scale tests. However, their full-scale applications are still limited.

Besides all these considerations (Table 2), it is necessary to consider issues related to the skill that the human resources who operate the different systems must have (Table 3). In some cases it is necessary to have highly trained people to operate the system. For this reason, the staff selection is of the greatest importance in order to get a sustained efficiency in the arsenic removal process.

Another important factor to be considered when choosing the technology to be used in the arsenic removal is water availability. Reverse osmosis systems, for instance, recover 50–80% of the raw water. This loss of water resources in addition to the intensive energy use probably makes the use of this technology unfeasible in arid zones.

476

Table 2.   Water quality requirements for selected arsenic removal processes.

Technology	Remarks
Coagulation and filtration	As(V) removal is easier than As(III) Phosphate or silicate may reduce arsenic removal rates Chloride, carbonate and sulfate have little effect on removal rates pH adjustment Calcium and magnesium can enhance arsenic removal
Ion exchange resins (IX)	As(V) removal is easier than As(III) Sulfate, nitrite, nitrate and chromate are also removed by most arsenic removal resins Run lengths are largely determined by sulfate levels, IX must not be used in water with > 120 mg/l sulfate High levels of dissolved solids (TDS > 500 mg/l) will shorten run times High levels of dissolved iron can precipitate out and clog the filter
Activated alumina (AA)	pH adjustment Precipitation of iron can clog AA As(V) removal is easier than As(III) Phosphate, sulfate, chromate and fluoride are also removed by activated alumina, but nitrate is not
Membrane methods (reverse osmosis and nanofiltration)	Chloride, sulfate, nitrate and heavy metals are also removed Low water recovery rates Operate at high pressures High-quality influent water is required Colloidal organic matter may foul membranes Iron and manganese can lead to scaling and membrane fouling
Lime softening	Use only with very hard waters pH of treated water is in the range of 10–12 Phosphate may reduce arsenic removal, especially below pH 12 Large volume of sludge As(V) removal is easier than As(III)
Iron-coated sand, Greensand iron removal, Biological removal processes	Technologies are being tested

Table 3.   Recommended skill levels for operators of arsenic removal processes.

Process	Operator skill
Activated alumina	Advanced
Coagulation–filtration	Advanced
Direct filtration	Intermediate
Ion exchange	Advanced
Reverse osmosis	Advanced
Lime softening	Intermediate
Oxidation (chlorination)	
Gas	Intermediate
Solution	Basic
Granules	Basic

All these issues raise some questions that need to be studied:

- What is the role of the water matrix in As removal processes?
- Are all species of arsenic removed with the same processes?
- Is it possible to use in practice any technology that has proved successful in the laboratory?
- Volume and handling of the generated sludge?
- Are there low cost conditions to dispose off the sludge safely?

2.2. Arsenic in drinking water sources and perspectives of water industry

Arsenic removal can be done at a centralized treatment level or at a family level. Centralized arsenic removal can be used when there is only one water source and there exists a water distribution system. For those systems that use groundwaters, it may be difficult to implement a treatment to remove As from water, for, in general, these systems are not currently required to treat the water. Because of this, many of these systems may have minimal operational infrastructure. Another factor of concern is that groundwater supply systems are generally smaller than those of surface waters and many times are privately owned. For this reason, their owners may be limited in the financial resources available for them to meet regulatory requirements. The compliance with these regulatory requirements will have a higher cost per treated water unit due to the losses of the scale economy. The treatment at family level should be used when there does not exist a water distribution system through pipes or when the water supply sources are very sparsely distributed. Such is the case of rural settlements that obtain their water supply from groundwater that is pumped manually.

These two scenarios (only one source or numerous sparse sources) require very different solutions to remove arsenic from water. The former can be solved with a treatment plant that could use any of the currently known technologies. Each of these technologies offers, as pointed out before, advantages and disadvantages that are necessary to know and balance at the time of choosing the treatment (Table 2).

The most important factors to consider when making a decision are water matrix, removal goal and feasibility to sustain the treatment. An important aspect that must not be neglected is the ability to control the efficiency of arsenic removal by means of analytical methods that are within reach of the communities they serve. The specialized literature offers many technologies, but the ones that could really be applied at routine scale are very few because of the requirement of special equipment and skilled personnel. Small water systems often do not have engineers or other technical people on staff and may be lucky to have a certified operator who stays on the job for any length of time.

The second scenario, the one of the sparsely located sources that supply family groups, is more difficult to solve. In these cases, the solution may pose a significant impact due to the fact that it is a non-centralized location problem. For this reason, the treatment to be used for removing arsenic from groundwater must be done at individual wells. Additionally, we must bear in mind that groundwaters are usually low in turbidity so it is not necessary to treat these water sources. Therefore, they do not have an installation of conventional coagulation–sedimentation–filtration systems in places where it is possible to remove arsenic in conjunction with other contaminants. An exception to this situation could be the current fluorine removal programs (Susheela and Ghosh, 2000).

Only a few examples of arsenic removal at community level have been documented. Most of these have been preliminary level or pilot experiments in Argentina, Chile and India. The processes used were coagulation–sedimentation–filtration or similar. The success of these methods depends fundamentally on a proper maintenance and operation of the equipment, as well as timely reagent supply. It is also necessary to have the constant financial support of regional and local authorities not only for start up, but also for operation and periodic maintenance and repair. (Cusimano et al., 1992; Sancha et al., 1992a,b; Chakraborti et al., 1995; Esparza et al., 1998; WHO, 2002).

Specialized literature also suggests that for removing arsenic at family scale it would be possible to use activated alumina, reverse osmosis and ionic exchange. However, the co-occurrence of arsenic with high levels of silica, hardness, selenium, sulfate and/or other ions in groundwater may limit the usefulness of these treatments (Table 2). An additional factor to be considered is that some of these methods, such as ionic exchange and osmosis, produce a large wastage of water and contaminated brines or residuals which will be difficult to handle at the family level.

Another issue to consider in the second scenario is that technologies have to be able to provide water at the required quantity and quality, even under suboptimal conditions. So it must be a technology that is socially acceptable and which offers little chance of errors.

Given the relatively limited availability of field-proven technologies, analytical capabilities and social acceptability, there are still great doubts about arsenic removal at all levels.

At the centralized treatment-level:

- Will it be possible to reach – at the treatment plants – sustained removal efficiency and residual arsenic levels like the ones recommended by the World Health Organization?
- Will the community that uses treated water be able to defray the treatment cost?
- What can be the acceptable tolerance margin in residual arsenic concentration?

At the family level:

- Will small systems and/or a family system be able to put arsenic removal programs in operation?
- Will small systems and/or a family system be able to comply with the regulation?
- Do arsenic removal technologies which are economically feasible and socially acceptable exist?

2.3. Goals in the removal of arsenic from water

The goals of arsenic removal that should be reached in the different countries have been recently pointed out by the World Health Organization (WHO, 1993). The principal aim of this recommendation is the consumers' health protection. Factors such as the real feasibility to achieve and control this aim as well as its associated costs do not seem to have been considered.

The goal of maximum arsenic content in potable water of 0.010 mg/l recommended by the World Health Organization is very difficult to achieve sustainably with conventional methods such as coagulation/filtration. In order to achieve this goal it will be necessary to use more expensive technologies such as reverse osmosis which, in general, requires a previous conditioning of the water to be treated and produces a significant loss of water. This goal also requires the use of advanced analytical methods and, of course highly trained personnel.

For the water treatment utilities, the impact of the actions that are necessary to take to reach the goals of maximum arsenic content in water are economic as well as technical and arise largely from the limitations of available technologies to easily achieve the desired arsenic level. Current problems include

- need for highly qualified personnel
- high removal plant operational costs
- removal efficiency vs water matrix
- water quality temporal variations
- limited water resources vs high water loss during the treatment;
- disposal of the residues generated in the treatment;
- power availability
- high operation/maintenance costs associated with small, geographically isolated systems.

3. Uncertainties and challenges in arsenic removal

The aforementioned shows some of the challenges and uncertainties that national, regional and local authorities as well as the water industry will have to face in order to solve the serious problems derived from the presence of arsenic in drinking water. Removal of arsenic from water raises many questions that are difficult to answer, due to the lack of data. Among issues requiring more studies we must point out

- occurrence and speciation
- water matrix influence
- exposure and effects on human health
- analytical methods at laboratory and at the field levels
- removal methods at full-scale water utilities and at point-of-use treatment device
- costs of treatment technologies
- waste management.

All these subjects present important knowledge gaps. It will be necessary to count on a joint effort from everybody in order to orient the resources available to carry out research that may find solutions that are feasible to apply – at a reasonable cost – in the affected countries. The implications of these uncertainties are significant for the public that will ultimately bear the costs.

References

Anderson, L.C.D., Bruland, K.W., 1991. Biochemistry of As in natural waters: the importance of methylated species. Environ. Sci. Technol., 25, 39.

Cullen, W.R., Reimer, K.J., 1989. Arsenic speciation in the environment. Chem. Rev., 89–713.

Cusimano, N.O., Deambrosi, N.E., Albina, L.C., Callegaro, R.S., 1992. Arsenic in potable water of Republica Argentina. Techniques for removal. International Seminar Proceedings (51–58). Arsenic in the Environmental and its Incidence on Health, Universidad de Chile, Santiago, Chile.

Chakraborti, D., Roy Chowdhury, T., Samanta, G., Mukherjee, D.P., Chanda, C.R., Saha, K.C., Mandal, B.K., 1995. A simple household device to remove arsenic from groundwater making it suitable for drinking and cooking. International Conference on Arsenic in Ground Water: Cause, Effect and Remedy, Jadavpur University, India.

Esparza, M.L., Wong, M., 1998. Abatimiento de arsénico en aguas subterráneas para zonas rurales. XXVI Congreso Interamericano de Ingenieria Sanitaria y Ambiental, Brasil.

Ferguson, J.F., Garvis, F., 1972. A review of the arsenic cycle in natural water. Water Res., 6, 1259.

Gurian, P., Small, M., 2002. Point-of-use treatment and the revised arsenic MCL. J. AWWA, 94, 3.

Holn, T., 2002. Effects of CO_3^{2-}/bicarbonate, Si and PO_4^{3-} on arsenic sorption to HFO. J. AWWA, 94, 4.

International Agency for Research on Cancer, IARC, 1980. Monograph on the evaluation of carcinogenic risk of chemicals to humans. Some Metals and Metallic Compounds, Vol. 23 Lyon, France.

Irgolic, K., 1994. Determination of total arsenic and arsenic compounds in drinking water. In: Chappel, W.R., Abernathy, C.O., Cothern, C.R. (Eds.), Arsenic Exposure and Health. Environ. Geochem. Health, Vol. 16, Special issue, 51–60.

Lu, F.J., 1990. Blackfoot disease: arsenic or humic acids? Lancet, 336, 115–116.

Sancha, A.M., et al., 1988. Remoción de Arsénico en Aguas de Abastecimiento para Pequeñas Comunidades. XXI Congreso Interamericano de Ingeniería Sanitaria y Ambiental, Brasil.

Sancha, A.M., Rodriguez, D., Vega, F., Fuentes, S., Lecaros, L., 1992a. Arsenic removal by direct filtration. An example of appropriate technology. In: International Seminar Proceedings (165–172). Arsenic in the Environment and its Incidence on Health, Universidad de Chile, Santiago, Chile.

Sancha, A.M., Vega, F., Fuentes, S., 1992b. Speciation of arsenic present in water inflowing to the Salar del Carmen treatment plant in Antofagasta, Chile and its incidence on the removal process. International Seminar Proceedings (165–172). Arsenic in the Environment and its Incidence on Health (183–186), Universidad de Chile, Santiago, Chile.

Sancha, A.M., 2000. Removal or arsenic from drinking water supplies: Chile experience. Water Supply, 18 (1), 621–625.

Sorg, T.J., Logsdon, G.S., 1978. Treatment technology to meet the interim primary drinking water regulations for inorganics. Part 2. J. AWWA, 70, 379.

Susheela, A.K., Ghosh, G., 2000. Fluorosis management programme in India: the impact due to net working between health and rural water supply implementing agencies. Interdisciplinary Perspectives on Drinking Water Risk Assessment and Management. IAHS Publication No. 260, (159–165).

Tseng, W.P., 1977. Effects and dose–response relationships of skin cancer and blackfoot disease with arsenic. Environ. Health Perspect., 19, 109–119.

Wang, L., Chen, A., Sorg, T., Fields, K. 2000. Field evaluation of As removal by IX and AA. J. AWWA, 94, 4.

World Health Organization, WHO, 1993. Guidelines for Drinking-Water Quality. Recommendations, Vol. 1, 2nd edn.

World Health Organization, WHO, 2002. Arsenic in drinking water. Safe Water Technology, Chapter 6.

Xiaoying, Y., 2001. Humic acids from endemia arsenicosis areas in Inner Mongolia and from the blackfoot disease areas in Taiwan: a comparative study. Environ. Geochem. Health, 23, 1.

Arsenic Exposure and Health Effects V
W.R. Chappell, C.O. Abernathy, R.L. Calderon and D.J. Thomas, editors
© 2003 Published by Elsevier B.V.

Chapter 37

Disposal of wastes resulting from arsenic removal processes

Michael J. MacPhee, John T. Novak, Rodney N. Mutter and David A. Cornwell

Abstract

Due to the reduction in the USEPA arsenic MCL 50 μg/L to 10 μg/L, many systems will need to add treatment to remove arsenic. Since all arsenic removal processes will produce arsenic-containing residuals, management and disposal of those wastes is a critical consideration. This paper focuses on the quality of arsenic-laden residuals solids produced by full-scale coagulation, lime softening, and iron removal treatment processes from across the U.S. Extensive laboratory testing was performed to assess the conditions under which arsenic leaches from those residuals over time, with particular focus on redox effects. The fresh and aged residuals were characterized in paired TCLP (federal) and Ca WET (California only) tests to develop a database comparing those two hazardous residuals characterization procedures. Arsenic leaching under the Ca WET test was dramatically higher than that under the TCLP test conditions. Leaching characteristics with both tests varied with residuals percent solids, with dry solids resulting in consistently lower arsenic concentrations than residuals dewatered to 20 percent. The findings have important cost and operational implications for utilities in California that will likely produce hazardous residuals solids if they use coagulation or iron removal processes. Utilities in the rest of the country will not produce hazardous residuals with these treatment processes, unless the TCLP test procedure is altered in the future, or if the TCLP regulatory limit for arsenic is lowered.

Keywords: residuals, Ca WET, leaching, landfill, redox

1. Overview

The management of arsenic (As)-laden residuals is an important issue due to the potential re-release of arsenic into the environment as a result of the various handling and disposal techniques commonly used by utilities. Because arsenic removal is sensitive to both pH and oxidation state, any process that changes pH or results in a reducing environment may release arsenic from the solid phase. These processes include chemical conditioning during dewatering, storage and lagooning, and ultimate disposal options such as landfilling, land application, and indirect discharge to a sanitary sewer. With the arsenic maximum contaminant level (MCL) of 10 μg/l, modifications of the treatment processes are needed to meet these lower levels. Residual handling and disposal will be further impacted and there are many questions regarding differences in environmental conditions and the retention of arsenic in residuals.

The objective of this chapter was to better understand the factors that may cause the release of arsenic from solid residuals and, thereby, allow arsenic to re-enter the environment. The project tasks were structured in order to determine arsenic release from residuals due to chemical conditioning for dewatering, thickening, short- and long-term storage, as well as to determine innovative methods for preventing arsenic mobility from residuals into the environment. These research efforts will help provide the data necessary to develop recommendations for regulators and utilities to minimize arsenic release during handling, storage, and disposal of water treatment plant (WTP) residuals.

Characterization and laboratory treatability work were used to develop a database comparing paired toxicity characteristic leaching procedures (TCLPs) and California waste extraction test (Ca WET) results for the same residuals. These residuals were collected from full-scale coagulation, lime softening, and iron/manganese removal processes.

2. Utilities sampled

The utilities sampled during the project are summarized in Table 1. Eight settled residual samples and six spent filter backwash (SFBW) samples were collected. One of the lime softening residuals was collected near the end of the study, and consequently was used only for characterization.

All of the solid samples passed both the TCLP and Ca WET tests with supernatant arsenic levels less than 5.0 mg/l. The LADWP solid sample, however, was only slightly below the arsenic limit using the Ca WET test. Based on these one-time analyses, the samples evaluated would be considered non-hazardous. Because the test is subject to technique, any Ca WET level greater than 1 mg/l may be of concern. These data also indicate significant differences between the TCLP arsenic and Ca WET arsenic concentrations measured from the solid residual samples. These data indicated (using 100% dry solids) that the Ca WET extraction increased the leached arsenic concentration from eight to 80 times higher than the TCLP arsenic extraction. This increase is expected due to the use of citric acid buffer, 48 h extraction period, and anaerobic test conditions for the Ca WET extraction, which is more aggressive than acetic acid extraction for TCLP analysis.

3. Results – arsenic leached in lagoon simulations versus hazardous potential leaching tests

The laboratory treatability research tasks were conducted to assess the factors influencing the release of arsenic from solid-phase residuals. Lagoon storage simulations were

Table 1. Sample collection locations.

Utility name	Sample type	Chemical treatment
Louisiana Water Company, New Iberia, LA	Settled solids/SFBW	Lime softening
City of Great Falls, MT	Settled solids/SFBW	Alum
City of Helena, MT	Settled solids/SFBW	Alum
Dept. of Public Utilities, Billings, MT	Settled solids/SFBW	Ferric chloride
Lockwood Water Users Association, Lockwood, MT	Settled solids/SFBW	Ferric chloride
Los Angeles Dept. of Water & Power (LADWP), CA	Settled solids/SFBW	Ferric chloride
Indiana-American Water Company, IN	Settled solids	Ferric chloride
Heath, OH	Settled solids	Lime softening

performed using settled residuals (most collected from sedimentation basins) from each of the utilities to determine under what conditions and to what extent arsenic is released. Sample sludges were placed in a controlled environment in closed, airtight containers, similar to conditions that might be expected in thickeners, lagoons, sedimentation basins, and monofills. Conditions would likely be much different in a landfill due to the presence of organics, low pH, and redox conditions. Sludge was analyzed for total arsenic, iron, and aluminum initially, and after 2 and 6 months of storage. TCLP and Ca WET were conducted on dry (100%) sludge and 20% solid concentration sludges. Lagoon supernatant, pH, dissolved oxygen, and redox potentials were measured at the same intervals, along with arsenic, iron, and aluminum.

3.1. Coagulation residuals

Lagoon supernatant arsenic concentration was measured over time to demonstrate the effect of lagoon conditions on arsenic release from the solid to liquid phase. The data demonstrate that arsenic leaching from the lagoon solids occurred soon after lagoon simulation testing was initiated. All residuals, except for Lockwood, MT, demonstrated increased levels of arsenic in the supernatant. The total arsenic concentration in the supernatant very closely followed the trend of total iron release from the lagoon solids, and generally followed a lowering of the redox potential.

3.2. Lime softening residuals

The only lime softening residual used for lagoon simulation testing was from New Iberia, LA. The lagoon test results demonstrated that there was no release of arsenic to the lagoon supernatant under lagoon storage conditions. The lagoon pH remained at 11.7 and the redox potential after storage increased to as high as 178 mV. There were only small concentrations of Fe and Al present in the lagoon supernatant residual stream initially, and no leaching over time was noted for either of these metals. Due to the chemistry of lime softening residuals, arsenic released under lagoon storage conditions is not expected to be a problem.

3.3. Redox potential

Figure 1 presents arsenic concentrations in lagoon supernatant collected at 0, 2, 4, and 6 months into lagoon simulation testing versus redox potential. For the majority of data points, redox potential became negative quickly, and was on the order of those found under reducing conditions in landfills. A key exception was the softening solids. Trends were similar for iron release (see Figure 2).

3.4. TCLP and Ca WET results

Solid samples were collected from the lagoon simulations analyzed after 2 and 6 months of storage time for TCLP and Ca WET arsenic analyses. These data (using 100% solid samples) are shown in Table 2 along with the results from the fresh residual sample analyzed prior to lagoon testing. Ca WET arsenic levels increased after 2 months of

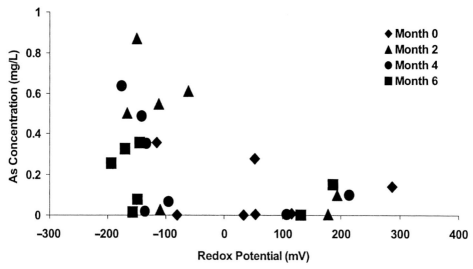

Figure 1.   Arsenic in lagoon simulation supernatant versus redox potential.

storage for all residuals except the lime softening residuals. TCLP arsenic levels were very low regardless of aging. Both tests were also conducted on sludge dried to a 20% solid concentration. Ca WET arsenic concentrations were generally higher for the 20% solid concentration samples than the 100% solid concentration samples.

The results demonstrate that each of the lagoon residuals was able to easily meet the TCLP arsenic limit. The maximum arsenic concentration noted was approximately 0.15 mg/l. The Ca WET arsenic data, however, present a different picture in terms of arsenic leaching. These data indicate that the fresh residuals would only be slightly below

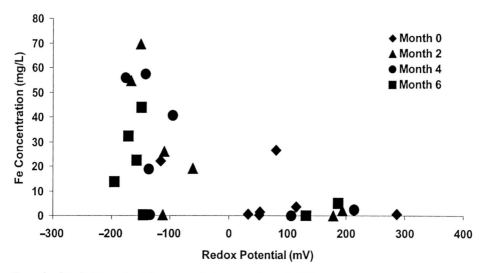

Figure 2.   Iron in lagoon simulation supernatant versus redox potential.

Table 2. Lagoon simulation TCLP and Ca WET results.

Utility name	TCLP As (mg/l)			Ca WET As (mg/l)		
	Fresh	Aged 2 months	Aged 6 months	Fresh	Aged 2 months	Aged 6 months
Louisiana Water Company, New Iberia, LA	0.012	0.01	0.01	0.096	0.05	0.108
City of Great Falls, MT	0.158	0.048	0.109	4.83	6.11	7.18
City of Helena, MT	0.125	0.13	0.144	1.88	5.61	10.3
Dept. of Public Utilities, Billings, MT	0.013	0.01	0.009	0.863	2.26	3.19
Lockwood Water Users Association, Lockwood, MT	0.011	0.009	0.013	0.224	0.26	0.266
Los Angeles Dept. of Water & Power (LADWP), CA	0.162	0.079	0.093	4.931	15.24	13.1
Indiana-American Water Company, IN	0.031	0.01	0.02	2.47	10.23	5.38

Note: TCLP and Ca WET test results in this table were conducted for 100% solid concentration.

the 5.0 mg/l limit of the Ca WET test, and after lagoon storage for 6 months, four of the six residuals would exceed the 5.0 mg/l Ca WET arsenic limit.

3.5. Comparison of extraction tests to lagoon release

The Ca WET test caused a much higher arsenic release than the TCLP test for the lagoon simulations for almost all of the residuals tested. Both tests are designed to give some sense of leaching potential. TCLP and Ca WET arsenic concentrations in 20% residuals collected at 2 months are compared with measured lagoon release in Figure 3. Ca WET arsenic levels were much higher (in some cases by more than two orders of magnitude) than both TCLP arsenic levels and measured arsenic concentrations in the lagoon supernatant. TCLP arsenic levels tracked much more closely than Ca WET concentrations

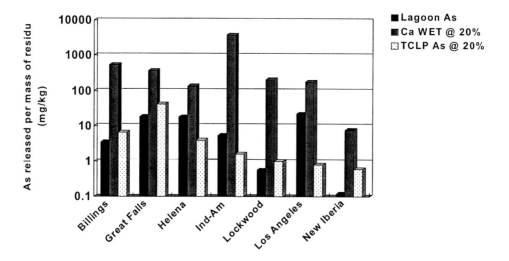

Figure 3. Ca WET and TCLP test results compared to lagoon simulation arsenic releases.

Table 3. Ratio of TCLP and Ca WET to arsenic lagoon release at 2 months.

Utility	Ca WET:lagoon		TCLP:lagoon	
	20%[a]	100%[a]	20%[a]	100%[a]
New Iberia, LA	63	3.7	5.00	15.8
Great Falls, MT	20	3.3	2.30	0.05
Helena, MT	8	3.1	0.20	0.14
Billings, MT	156	6.6	1.90	0.06
Lockwood, MT	362	4.6	1.80	0.35
Los Angeles, CA	8	6.9	0.04	0.01
Indiana-American Water Company, IN	694	19.0	0.30	0.04

[a] Refers to percent solid concentration of the residuals used in the test.

with lagoon supernatant concentrations. At 6 months of lagoon storage, the results were similar.

The ratios of Ca WET and TCLP arsenic to measured lagoon arsenic are provided in Table 3. The Ca WET test arsenic concentration for the 20% solid concentration residuals was as much as 700 times greater than the lagoon arsenic release. The Ca WET test (100% solid concentration) measured arsenic release was 3–7 times higher than the lagoon measured arsenic release. Overall, the Ca WET data measured using 100% solids demonstrated a much closer relationship to the lagoon release than did the 20% solid Ca WET results.

4. Summary

Based on the results of this research, several important conclusions can be drawn, including:

- U.S. utilities should have no difficulty meeting the TCLP arsenic limit of 5 mg/l if there is no change made to the test procedure.
- California utilities will have difficulty meeting the Ca WET test limit of 5 mg/l.
- Reducing conditions in water plant residuals can develop quite quickly and result in releases of arsenic back into the liquid phase.
- Such releases may not be predicted by the TCLP test.

These findings suggest that sludge accumulating in settling basins or sludge thickeners will contribute arsenic to the water phase, which could represent an important secondary source of arsenic in water plants. The consequences are especially important for plants that recycle SFBW water or clarifier blowdown, which is a very common practice in the U.S. treatment facilities. Moreover, on-site stockpiling of residuals could result in further contamination of the source groundwater. Long-term environmental contamination due to arsenic leaching from landfills and monofills could be an important issue facing municipalities in certain areas of the world if they receive large quantities of arsenic-laden water plant residuals.

Arsenic Exposure and Health Effects V
W.R. Chappell, C.O. Abernathy, R.L. Calderon and D.J. Thomas, editors
© 2003 Elsevier B.V. All rights reserved.

Chapter 38

Development of a low-waste technology for arsenic removal from drinking water

József Hlavay, Klára Polyák, János Molnár, Kornél Gruber,
Pál Medgyesi and Márta Hódi

Abstract

*The purification of drinking water containing inorganic arsenic compounds causes important problems in Hungary. Arsenic ions are accompanied by high concentrations of ammonium-, Fe-, and Mn-ions, humic acids (about 5–10 mg/l), dissolved gases, and the water has a high temperature, >30°C. This contamination arises from natural leaching of arsenic rocks by the percolating water. New low-waste technology was developed by a combination of ion exchange and adsorption methods. It is appropriate for selective removal of ammonium, iron, manganese, and arsenic ions, as well as humic acids from drinking water. Processes were applied in laboratory and field experiments. Natural ion exchangers and adsorbents were used as sodium-form natural clinoptilolite (Na-Cli), manganese-form natural clinoptilolite (Mn-Cli), granulated activated carbon (GAC), and granulated $Al_2O_3/Fe(OH)_3$. Natural zeolite is mined in Hungary and the clinoptilolite content was found to be 65–70 m/m% by XRD analysis. Optimal exhaustion–regeneration cycles were estimated and a pilot-plant set-up was designed. The Na-form of clinoptilolite was produced by 20 BV 20 g NaCl/l solution, then washed with distillated water. The Mn-form was prepared from the Na-form with 20 BV of 1 mol/l $MnSO_4$ and 20 BV of 10 g/l $KMnO_4$. $Al_2O_3/Fe(OH)_3$ adsorbent was prepared from granules of 0.3–1.0 mm of activated Al_2O_3 and $Fe(OH)_3$ was freshly precipitated onto the surface of particles. Laboratory and field experiments were carried out by 3.2 cm i.d.*15 cm and 8 cm i.d.*90 cm columns. Adsorption and ion exchange capacities were estimated for all materials. In the model experiments, up to the 10 μg/l As, the adsorption capacities were as follows: $Al_2O_3/Fe(OH)_3$, 86.8 μg/g; GAC, 66.3 μg/g; Mn-Cli, 15.3 μg/g. The experimental set-up proved to be efficient for the removal of all analytes with concentrations higher than the maximum contaminant level (MCL).*

Keywords: arsenic ions, drinking water, novel type adsorbents, combined purification technology, environmentally friendly materials

1. Introduction

In Hungary it was well known 20 years ago that more than 80 settlements with 400,000 inhabitants were served with tap water containing As-ions >50 μg/l concentration. This contamination arises from the natural leaching of arsenic rocks by the percolating water. Arsenic ions are accompanied by high amounts of ammonium-, Fe-, and Mn-ions, humic acids (about 5–10 mg/l), dissolved gases, and the water has a high temperature, >30°C. In a recent survey, arsenic exposure, besides its carcinogenic effect, has been found to be associated with elevated risk of some adverse pregnancy outcomes such as stillbirth and spontaneous abortion. In a survey, seven villages (with a total population of 25,000) and two towns (with 22,000 and 25,000 inhabitants, respectively) supplied with drinking water containing arsenic levels above 100 µg/l were selected as exposed settlements and six villages (with a total population of 21,000) and one town (25,000 inhabitants) supplied with drinking water <10 µg/l were chosen as controls (Rudnai et al., 2002). Data on pregnancy outcomes between 1970 and 2001 were collected from the district nurses' registries. Linear regression models were used for statistical analysis and significant associations between arsenic level and spontaneous abortion ($p = 0.0275$) or stillbirth ($p = 0.0087$) were found. Today more than 1.6 million inhabitants on 400 settlements in Hungary are served with tap water containing arsenic greater than 10 µg/l.

In soils and natural water inorganic arsenic is usually found in the form of As(III) as arsenite, or As(V) as arsenate. In the pH range of natural water, the dominant arsenite species is neutral ($H_3AsO_3^0$; $pK_{a,1} = 9.20$) and the negatively charged anions ($H_2AsO_4^-$ and $HAsO_4^{2-}$; $pK_{a,1} = 2.91$ and $pK_{a,2} = 6.94$) are the principal arsenate species. Both arsenate and arsenite are subjected to chemically and/or microbially mediated redox and methylation reactions in soils and natural water (Andreae and Klumpp, 1979; Freeman et al., 1986).

Naturally occurring dissolved arsenic usually arises from a sulfide complex formed with iron, nickel, and cobalt. The most calamitous occurrence is that of Bangladesh and West Bengal where arsenic is derived from natural deposits in the Ganges River delta (Dhar et al., 1997). Arsenic concentrations greater than the World Health Organization (WHO) recommended standard (10 µg/l) have been reported in water supplies throughout the world in places like Bangladesh (Dhar et al., 1997), West Bengal (Chatterjee et al., 1995; Das et al., 1995; Mandal et al., 1996), Argentina (Borgono et al., 1977), Taiwan (Chen et al., 1994), Mexico (Del Razo et al., 1990), and the United States (Feinglass, 1973; Morton et al., 1976; Matisoff et al., 1982; Boudette et al., 1985; Welch et al., 1988).

An American Water Works Association study estimated in United States alone, the compliance costs for a lower arsenic standard ranging from $330 million/year for 20 µg/l maximum contaminant level (MCL) to more than $4.1 billion/year for a 2 µg/l MCL (Frey et al., 1998). These estimates represent a 10- to 20-fold increase in the USEPA's preliminary cost estimates. The National Research Council, Subcommittee on Arsenic in Drinking Water has estimated that the 50 µg/l is associated with lifetime excess cancer risks of between 1 in 100 and 1 in 1000, which is well above the EPA's target risk range for drinking water standards of between 1 in 10,000 and 1 in 1,000,000. Arsenic exposure is widespread as roughly half of U.S. public supplies have detectable arsenic concentration (>0.5 µg/l) in their finished water. Hence, there is a compelling need to find cost-effective technologies for arsenic removal. A variety of methods have been developed to remove arsenic from drinking water but these are usually only effective for arsenate and require a pre-oxidation step for arsenite removal. This is necessary because these technologies operate best if arsenic is in the oxidized form. Oxidants include chlorine, ferric chloride, permanganate, ozone, and hydrogen peroxide. Frey et al. (1998) have surveyed treatment plants and found that the most prevalent treatment used was coagulation or lime softening with some enhancements. Several other advanced treatment technologies have been applied to smaller scale facilities. These include anion exchange, activated alumina, and reverse osmosis (Frey et al., 1998). Ferric iron hydroxides have been widely studied and applied to arsenic removal in both natural and engineered systems (Davis and Kent, 1990;

Fuller et al., 1993; Waychunas et al., 1993; Edwards, 1994; Hering et al., 1996; Wilkie and Hering, 1996; Raven et al., 1998).

As was stressed earlier, the special character (chemical composition, physicochemical features) of the confined subsurface water resources of Hungary with elevated As concentration clearly requires the development of unique technology. Problems of high concentrations of ammonium ion, iron, manganese, as well as high concentrations of humic substances together represents a complex challenge to technology development of arsenic removal. This unique problem clearly covers a large region of a candidate member state of the EU involving about 2.2 million people. Recently new absorbents were developed from environmentally friendly raw materials (Hlavay and Polyák, 2001). The main features of absorbents were established. The pH value corresponding to the zero point of charge of the adsorbent was determined in equilibrium solutions in the presence of As(III)- and As(V)-ions (pH_{iep}). The amounts of surface charged groups, surface charge, surface potential, stability constants of compounds formed on the surface of the adsorbent and the effect of pH on the adsorbed amount of arsenic ions were established. Results were evaluated by chemical and electrical double layer models. In the following, the development of a new efficient arsenic removal technology to provide the 10 μg/l concentration is discussed. Adsorbents and ion exchangers are produced from environmentally friendly raw materials and combined adsorption and ion exchange processes are used.

2. Experimental

Laboratory and field experiments were carried out by 3.2 cm i.d.*15 cm (bed volumes, BV = 80 ml) and 8 cm i.d.*90 cm columns (BV = 4 l), respectively. The average exhaustion flow rate was about 5 BV/h.

2.1. Materials

Natural ion exchangers and adsorbents were used in the experimental work as sodium-form natural clinoptilolite (Na-Cli), manganese-form natural clinoptilolite (Mn-Cli), granulated activated carbon (GAC), and granulated $Al_2O_3/Fe(OH)_3$. Natural zeolite was mined at Tokaj Mountain in the northern Hungary. The clinoptilolite content of zeolite was found as 65–70 m/m% by XRD analysis. The ion exchange column was filled with the 0.5–1.0 mm grain size fraction of the clinoptilolite. The Na-form of clinoptilolite was produced by a 20 BV of 20 g NaCl/l solution, then washed with distilled water to remove the excess chloride ions. The Mn-form was prepared similarly, at first the clinoptilolite was brought to the Na-form, and then a 20 BV of 1 mol/l $MnSO_4$ and next a 20 BV of 10 g/l $KMnO_4$ solutions were introduced to the zeolite. Finally, the ion exchanger was dried at room temperature. Granulated activated carbon (Filtrasorb 300) was produced by the Chemviron Carbon (Belgium). It was stored in plastic and paper box up to prevent the contamination.

2.2. Preparation of the $Al_2O_3/Fe(OH)_3$ adsorbent

The support of the adsorbent was granulated from Al_2O_3 (COMPALOX AN/V-801) made by the Alusuisse-Martinswerk GmbH (Germany). Fractions of the grain size ranged from

0.3 to 1.0 mm were activated at 450°C over 4 h. The surface of the Al_2O_3 was covered by $Fe(OH)_3$ by an *in situ* precipitation method. Al_2O_3 was placed in an exsiccator and a 10 g/l $FeCl_3$ solution was added in excess. The air was eliminated by an aspirator pump from the pores. After bubbling, the excess $FeCl_3$ was poured off and $Fe(OH)_3$ was formed on the surface of the granules by NH_4OH solution. The resulting $Fe(OH)_3$ impregnated porous adsorbent was dried at room temperature, packed into an ion exchange column and washed with distillated water to remove the excess reagent.

2.3. Laboratory experiments

The laboratory system was set up as follows: ion exchange columns were filled with 80 ml (1 BV) ion exchanger and adsorbent. The columns were put together and model water was flowed through the system with a 5 BV/h (400 ml/h) flow rate, 5 days weekly and 6–8 h daily. Model water was prepared similarly to the raw water composition at field experiments: As(V): 70 μg/l, Fe(III): 0.2 mg/l, Mn(II): 0.05 mg/l, and NH_4^+: 2 mg/l.

2.4. Analytical procedures

Samples, taken in all cases from raw and effluent waters, were preserved with cc. HNO_3 and cooled till analysis. Before and after measurements all glassware and plastic vessels were treated with a solution of 10 m/m% HNO_3 for 48 h, and washed with doubly distilled ion exchanged (DDI) water. The ammonium ion concentration was measured by the Hungarian Standard (MSZ ISO 7150-1:1992). Iron, manganese, and arsenic concentrations of samples were determined by electrothermal atomic absorption spectrometry (Perkin Elmer 5100 PC GEM Software, deuterium background correction, AS-60 Autosampler).

2.5. Field experiments

Ion exchange and adsorption technologies were operated at a deep well of the Water Company of Makó for 4 months. The columns were filled with 4 l (1 BV) ion exchangers and adsorbents. Raw water was stored up in a plastic barrel, at first it was charged to the top of the first column and the prepurified water was led from the bottom of this column to the top of the second column, and so on (Fig. 1). The experiment was carried out 5 days/week and 10 h/day. Average concentrations of investigated components in water of the well (No. 10) were: As: 60–70 μg/l, NH_4^+: 1.4–1.9 mg/l, Fe: 0.16–0.2 mg/l, Mn: 0.05–0.07 mg/l, and humic acid: 2.5–4.0 mg/l. During the experimental period, samples were taken once a week from the influent water and effluent water of all columns. Analytical measurements were performed in the laboratory of Hydra Ltd. (Csongrád) by the Hungarian Standards. Complete water chemical analysis was performed monthly from the collected samples.

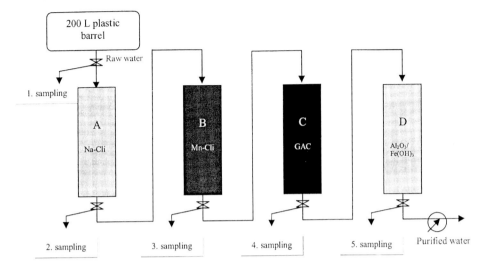

Figure 1. Experimental set-up at the field experiments in the 1st stage.

3. Results and discussion

The $Al_2O_3/Fe(OH)_3$ adsorbent had been found to be suitable for removal of arsenic ions from the drinking water. The Fe content of the adsorbent was found as 0.313 ± 0.003 m/m% (56.1 mmol/g). Its mechanical and chemical stability proved to be appropriate in solutions. The total capacity of the adsorbent was 0.115 mmol/g, the pH of zero point of charge, $pH_{zpc} = 6.9 \pm 0.4$. Depending on the pH of solutions the adsorbent can both be used for binding of anions and cations. If $pH_{eq} < pH_{zpc}$, anions are sorbed on the surface of adsorbent through $\{S-OH_2^+\}$ and $\{S-OH\}$ groups (Hlavay and Polyák, 2001). In the preliminary laboratory experiments model water was prepared as: As(V): 70 µg/l, Fe(III): 0.2 mg/l, Mn(II): 0.05 mg/l, and NH_4^+: 2 mg/l. The results showed that the concentration of the NH_4^+, iron, and manganese ions in the treated water was <LoD (limit of detection). Thus, the aim of further experiments was to evaluate the ability of arsenic removal of ion exchangers and adsorbents. Mn-Cli and $Al_2O_3/Fe(OH)_3$ adsorbents were studied separately and Na-Cli was combined with GAC. Breakthrough curves for all columns were prepared and the adsorption capacity of each adsorbents and ion exchangers was determined. The breakthrough concentration was chosen at 10 µg As/l as MCL and 30 µg As/l for practical and economical purposes. The results are summarized in Table 1.

It was found that the $Al_2O_3/Fe(OH)_3$ adsorbent was the most effective for arsenic removal from drinking water since 1530 BV (122 l) of model water was treated with this adsorbent until the breakthrough concentration (10 µg/l) was reached. GAC was the second most effective adsorbent, and Mn-Cli was found to be the least efficient one. The advantage of this experimental set-up is that the adsorption capacity of GAC and Mn-Cli to bound As-compounds can be applied to decrease significantly the concentration of As in the influent water fed to the $Al_2O_3/Fe(OH)_3$ adsorbent. Further, in the case of continuous

Table 1. Arsenic removal in laboratory experiments, efficiency and adsorption capacity (10 μg/l) of investigated adsorbents.

Adsorbent	Volume of treated water		Amount of initial As (mg)	Amount of adsorbed As (mg)	Efficiency (%)	Adsorption capacity (μg/g ads.)
	BV	l				
Al$_2$O$_3$/Fe(OH)$_3$	1530	122	8.0	7.7	96	86.8
GAC	650	52	3.2	3.0	93	66.3
GAC[a]	755	60	3.6	3.2	88	72.2
Mn-Cli	260	21	1.2	1.1	90	15.3
Mn-Cli[a]	415	33	1.9	1.5	77	21.1

[a] Applies to 30 μg As/l.

arsenic removal technology, depending upon the composition of the raw water, columns can be accordingly changed for efficient operation.

3.1. Field experiments

In the field experiments the efficiency of the combined ion exchange and adsorption technologies, as well as the adsorption capacities were determined. The exhaustion experiments were cyclically carried out and the regeneration of spent adsorbents was also studied. The Na-form clinoptilolite was applied as an ion exchanger for removal of ammonium and iron ions from the raw water. Average concentrations of analytes in the raw water ranged as follows: As: 65.4–68.6 μg/l, NH$_4^+$: 1.55–1.73, Fe: 0.15–0.18, Mn: 0.06–0.08, and humic acid: 3.33–3.74 mg/l. Since the manganese content of the water was low the Mn-Cli has been applied for removal of the arsenic ions.

During the field experiments the efficiency of Na-Cli was found as 67–74% for ammonium ions (1135 BV, 4538 l, and capacity: 0.8–0.95 mg/g ads.) (Table 2).

Regeneration of the spent Na-form clinoptilolite ion exchanger was accomplished by a 20 BV of 20 g NaCl/l solution, then washed with distilled water. After regeneration, the Na-Cli was found to be appropriate to remove ammonium ions. In the plastic barrel

Table 2. Ammonium ions removal by Na-Cli.

Na-Cli ion exchanger	Volume of treated water		Amount of initial NH$_4^+$ (g)	Amount of exchanged NH$_4$ (g)	Efficiency (%)	Capacity (mg/g ads.)
	BV	l				
1st regeneration	545	2177	4.0	3.0	74	0.95
2nd regeneration	590	2361	3.8	2.6	67	0.8

the iron ions in the raw water forms an iron hydroxide precipitate and this can block the ion exchange sites during the purification technology. Therefore, removal of iron is also of importance and efficiency of removal has been found as 69% (treated water: 1160 BV, 4640 l, amount of initial Fe: 845 mg, amount of adsorbed Fe: 584 mg, and adsorption capacity: 0.19 mg/g). Along with the removal of ammonium and iron ions the concentration of arsenic ions was also monitored. In the first period it was found that 26% of arsenic content of the raw water was removed by Mn-Cli (Table 3).

Poor adsorption capacity of the GAC for As was obtained (3.6%), since the removal of humic acids was the main purification process. The volume of treated water was 1358 BV (5433 l) up to the humic acid concentration in effluent water reached the MCL of the humic acid (0.5 mg/l). The amount of the initial humic acid was 18.5 mg, while the amount of the adsorbed humic acid was found as 16.9 mg. So the efficiency of the purification was 91%, which meant an adsorption capacity of 8.4 mg/g adsorbent. Consequently, the arsenic loading of $Al_2O_3/Fe(OH)_3$ adsorbents was mainly decreased by Mn-form clinoptilolite and, in a small degree, by activated carbon column about one-third ratio.

Since the raw water did not contain high concentrations of Mn, the set up shown in Figure 1 was changed as follows: A column: Na-Cli, B column: GAC, C column: $Al_2O_3/Fe(OH)_3$, and D column: $Al_2O_3/Fe(OH)_3$. With the two columns filled by $Al_2O_3/Fe(OH)_3$ adsorbent a more efficient removal was achieved.

Furthermore, the regenerated $Al_2O_3/Fe(OH)_3$ adsorbent was again set to the technological chain and the cycles of exhaustion–regeneration processes have been estimated. Results of first experiments showed that arsenic concentration of purified water continuously decreased, and after 275 BV, it reached the 10 µg/l MCL, and with the treatment of 675 BV it reached the 30 µg/l. The spent $Al_2O_3/Fe(OH)_3$ adsorbent was regenerated as follows: the adsorbent was dried at 60°C, after that $Fe(OH)_3$ was freshly precipitated onto the surface of particles of the alumina granules and dried at room temperature. During the regeneration a new column filled with $Al_2O_3/Fe(OH)_3$ adsorbent was operated in the water purification system as column D (Fig. 2).

After regeneration the adsorbent was again found to be appropriate to remove arsenic ions. The regenerated $Al_2O_3/Fe(OH)_3$ adsorbent was set up into the system as a third column. Adsorption capacities and efficiency of $Al_2O_3/Fe(OH)_3$ adsorbent are summarized in Table 4 and the breakthrough curve prepared for the exhaustion cycle is shown in Figure 3.

Table 3. Arsenic removal by Mn-Cli and GAC.

Adsorbent	Volume of treated water		Amount of initial As (mg)	Amount of adsorbed As (mg)	Efficiency (%)	Adsorption capacity (µg/g ads.)
	BV	l				
Mn-Cli	1026	4104	311	80	26	25.6
GAC column	2195	8780	503	18	3.6	9.0

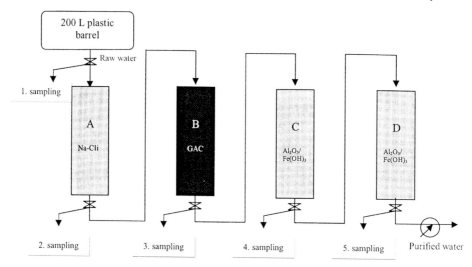

Figure 2. Experimental set-up at the field experiments in the 2nd stage.

After regeneration the adsorption capacity proved to be even higher than in the first experimental period. It means that this regeneration process is efficient. The frequency of exhaustion–regeneration cycles can only be estimated with longer operational time.

During the regeneration of the adsorbent, a newly prepared $Al_2O_3/Fe(OH)_3$ adsorption was set into the purification technology (Fig. 2, column D). In this period, the prepurified water from column C (filled with $Al_2O_3/Fe(OH)_3$ adsorbent) was led to the fresh adsorbent. As it can be seen in Figure 3, up to 2195 BV, no breakthrough concentration of As was found, so the adsorption capacity of new $Al_2O_3/Fe(OH)_3$ could not be estimated yet. With the new adsorbent the volume of treated water was 638 BV (2551 l), the amount of initial As was 50.7 mg, while the amount of adsorbed As was as 35.6 mg, so the efficiency up to 6.7 µg/l As was found as 70%. Further experiments are necessary to calculate the adsorption capacity either to 10 or 30 µg/l of As.

Table 4. Arsenic removal by $Al_2O_3/Fe(OH)_3$ adsorbent.

$Al_2O_3/Fe(OH)_3$ adsorbent (µg/l)	Volume of treated water		Amount of initial As (mg)	Amount of adsorbed As (mg)	Efficiency (%)	Adsorption capacity (µg/g ads.)
	BV	l				
10	294	1177	60.5	55.4	92	13.62
30	626	2504	125.0	91.7	73	22.55
Regeneration						
10	365	1461	98.9	87.6	89	21.5
30	851	3403	211.1	154.6	73	38.0

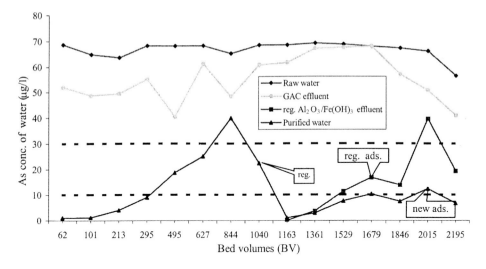

Figure 3. Breakthrough curve for arsenic ions (July 9–October 25, 2001).

The aim of the newly developed water purification technology is a complex, efficient, and selective process for removal of As, Fe, Mn, NH_4^+, and humic acid. Selectivity of the adsorption and ion exchange procedures can be proved by the analysis of all analytes in the purified water that are set into the drinking water standards. In Table 5 the complete chemical analysis is summarized. The analysis was performed every month during the experimental period.

As can be seen there is no significant difference between the raw and purified water concerning the basic components of drinking water (Ca, Mg, Na, K, pH, and COD). Only the analytes of concentrations near or greater than the MCL were removed to meet the standards.

Table 5. Complete chemical analysis of raw and purified water between August and October 2001.

Component (mg/l)	August 16, 2001		September 27, 2001		October 25, 2001	
	Raw	Purified	Raw	Purified	Raw	Purified
COD	3.0	0.7	1.0	1.2	3.0	0.6
pH	7.72	7.82	7.83	7.90	7.80	8.16
Ca	11.7	10.0	12.9	11.6	12.1	10.7
Mg	6.1	5.3	6.6	6.7	6.8	5.1
Na	115	109.5	115	113.0	118.5	104.0
K	1.0	5.0	1.1	0.5	1.05	0.24
NH_4^+	1.73	0.95	1.55	0.15	1.56	0.13
As (μg/l)	**67.9**	**18.6**	**68.6**	**8.3**	**65.4**	**6.7**
Fe	0.18	0.04	0.15	0.04	0.16	0.03
Mn	0.08	0.02	0.05	0.06	0.06	0.04
Humic acid	3.74	<0.25	3.33	<0.25	3.60	0.68

4. Conclusion

A water purification technology was developed by ion exchange and adsorption methods. Laboratory and field experiments were carried out to investigate the applicability of the process. In the laboratory experiments model water was used and efficiencies of the adsorbents for removal of As were determined as follows: $Al_2O_3/Fe(OH)_3 >$ $GAC > Mn-Cli$ adsorbents. In the field experiments deep-well water containing concentrations of pollutants greater than the MCL was purified. A continuous system was operated with different adsorbents and ion exchanger columns. Ammonium and iron ions were effectively removed. However, it should be noted that these two components were not identified in high concentration. During field experiments the efficiency of the GAC was found as 91% for humic acid. Removal of As was carried out by $Al_2O_3/Fe(OH)_3$ adsorbents with an efficiency of 70–92%. The adsorption capacities of adsorbents and ion exchangers were also studied during laboratory and field experiments. As it was expected, the capacities of environmentally friendly adsorbents and ion exchanger were higher in the laboratory experiments than in the field ones. The $Al_2O_3/Fe(OH)_3$ adsorbents and the Na-Cli ion-exchanger were regenerated several times during the experimental period. Optimal exhaustion–regeneration cycles cannot be estimated yet in this short operation of the purification technology.

In the future, laboratory and field experiments are needed to estimate the effective and economical purification technologies. The evaluation of technologies by the pollutant removal efficiency, simplicity, production of technological wastes, investment, and operational costs are of importance.

Acknowledgements

The research was supported by OTKA 043220 (Hungarian Science Foundation).

References

Andreae, M.O., Klumpp, D., 1979. Biosynthesis and release of organoarsenic compounds by marine algae. Environ. Sci. Technol., 6, 738–741.

Borgono, J.M., Vincent, P., Venturino, H., Infante, A., 1977. Arsenic in the drinking water of the city of Antofagasta: epidemiological and clinical study before and after installation of a treatment plant. Environ. Health Perspect., 19, 103–105.

Boudette, E.L., Canney, F.C., Cotton, J.E., Davis, R.I., Ficklin, W.H., Motooka, J.M., 1985. High levels of arsenic in the ground waters of southeastern new Hampshire: a geochemical reconnaissance, U.S. Geological Survey, Open-File Report 85-202.

Chatterjee, A., Das, D., Mandal, B.K., Chowdhury, T.R., Samanta, G., Chakraborti, D., 1995. Arsenic in ground water in six districts of West Bengal, India: the biggest arsenic calamity in the World. Part 1. Arsenic species in drinking water and urine of the affected people. Analyst – R. Soc. Chem., UK, 120, 643–650.

Chen, S.L., Dzeng, S.R., Yang, M.H., Chiu, K.H., Shieh, G.M., Wai, C.M., 1994. Arsenic species in ground waters of the blackfoot disease area, Taiwan. Environ. Sci. Technol., 28, 877–881.

Das, D., Chatterjee, A., Mandal, B., Samanta, G., Chanda, B., Chakraborti, D., 1995. Arsenic in ground water in six districts of West Bengal, India: the biggest arsenic calamity in the World. Part 2. Arsenic concentration in drinking water, hair, nails, urine, skin-scale and liver tissue (biopsy) of the affected people. Analyst – R. Soc. Chem., UK, 120, 917–924.

Davis, J.A., Kent, D.B., 1990. Surface complexation modeling in aqueous geochemistry. In: Hochella, M.F., White, A.F. (Eds), Mineral–Water Interface Geochemistry. Mineral Society of America, pp. 177–260.

Del Razo, L.M., Arellano, M.A., Cebrian, M.E., 1990. The oxidation states of arsenic in well water from a chronic arsenicism area of Northern Mexico. Environ. Pollut., 64, 143–153.

Dhar, R.K., Biswas, B.Kr., Samanta, G., Mandal, B.K., Chakraborti, D., Roy, S., Jafar, A., Islam, A., Ara, G., Kabir, S., Khan, A.W., Ahmed, S.A., Hadi, S.A., 1997. Groundwater arsenic calamity in Bangladesh. Curr. Sci., 73, 48–58.

Edwards, M.A., 1994. Chemistry of arsenic removal during coagulation and Fe–Mn oxidation. J. Am. Water Works Assoc., 86 (9), 64–78.

Feinglass, E.J., 1973. Arsenic intoxication from well water in the United States. N. Engl. J. Med., 288 (16), 828–830.

Freeman, M.C., Aggett, J., O'Brien, G., 1986. Microbial transformations of arsenic in Lake Ohakuri, New Zealand. Water Res., 20, 283–294.

Frey, M.M., Owen, D.M., Chowdhury, Z.K., Raucher, R.S., Edwards, M.A., 1998. Cost to utilities of a lower MCL for arsenic. J. Am. Water Works Assoc., 90 (3), 89–102.

Fuller, C.C., Davis, J.A., Waychunas, G.A., 1993. Surface chemistry of ferrihydrite: Part 2. Kinetics of arsenate adsorption and coprecipitation. Geochim. Cosmochim. Acta, 57, 2271–2282.

Hering, J.G., Chen, P.Y., Wilkie, J.A., Elimelech, M., Liang, S., 1996. Arsenic removal by ferric chloride. J. Am. Water Works Assoc., 88 (4), 155–167.

Hlavay, J., Polyák, K., 2001. Surface properties of an advanced adsorbent developed for arsenic removal from drinking water. In: Chappell, W.R., Abernathy, C.O., Calderon, R.L. (Eds), Arsenic Exposure and Health Effects IV. Elsevier, Amsterdam, pp. 431–440.

Mandal, B.K., Chowdhury, T.R., Samanta, G., Basu, G.K., Chowdhury, P.P., Chanda, C.R., Lodh, D., Karan, N.K., Dhar, R.K., Tamili, D.K., Das, D., Saha, K.C., Chakraborti, D., 1996. Arsenic in ground water in seven districts of West Bengal, India – the biggest arsenic calamity in the World. Curr. Sci., 70 (11), 976–986.

Matisoff, G., Khourey, C.J., Hall, J.F., Varnes, A.W., Strain, W.H., 1982. The nature and source of arsenic in northeastern Ohio ground water. Ground Water, 20, 446–455.

Morton, W., Starr, G., Pohl, D., Stoner, J., Wagner, S., Weswig, P., 1976. Skin cancer and water arsenic in Lane County, Oregon. Cancer, 37, 2523–2532.

Raven, K.P., Jain, A., Loeppert, R.H., 1998. Arsenite and arsenate adsorption on ferrihydrite: kinetics, equilibrium, and adsorption envelopes. Environ. Sci. Technol., 32, 344–349.

Rudnai, P., Farkas, I., Csanády, M., Kiss, V., 2002. Antagonism of arsenic and iodine in pregnancy outcomes. Epidemiology, 13 (4), S247.

Waychunas, G.A., Rea, B.A., Fuller, C.C., Davis, J.A., 1993. Surface chemistry of ferrihydrite: Part 1. EXAFS studies of the geometry of coprecipitated and adsorbed arsenate. Geochim. Cosmochim. Acta, 57, 2251–2269.

Welch, A.H., Lico, M.S., Hughes, J.L., 1988. Arsenic in ground water of the western United States. Ground Water, 26, 333–347.

Wilkie, J.A., Hering, J.G., 1996. Adsorption of arsenic onto hydrous ferric oxide: effects of adsorbate/adsorbent ratios and co-occurring solutes. Colloids and Surf. A: Physicochem. Eng. Aspects, 107, 97–110.

PART VII

AN OVERVIEW OF SOME US EPA AND NIEHS PROGRAMS ON ARSENIC

Arsenic Exposure and Health Effects V
W.R. Chappell, C.O. Abernathy, R.L. Calderon and D.J. Thomas, editors
Published by Elsevier B.V.

Chapter 39

An update on some arsenic programs at the US EPA[*]

Charles O. Abernathy, Mike Beringer, R.L. Calderon,
T. McMahon and E. Winchester

Abstract

Exposure to arsenic (As) has been reported to cause adverse health effects in humans, including internal and skin cancers, vascular, neurological, and dermal manifestations. Several offices of the US Environmental Protection Agency (EPA) have As programs, and selected activities will be described. The Office of Water (OW) is concerned with the Safe Drinking Water and Clean Water Acts. After reviewing the As database, the EPA established a maximum contaminant level goal of zero and a maximum contaminant level of 10 μg/l on January 22, 2001. The effective date for the rule was extended by the administrator and the As database was re-examined. After carefully considering the new reports, the EPA announced on October 31, 2001 that the rule would go into effect without delay. The Office of Pesticides Program (OPP) has been working with chromated copper arsenate (CCA). It was realized that humans, especially children, were exposed to CCA from playground equipment and soils. OPP negotiated with the wood treatment industry about CCA and they announced a voluntary phase-out program of CCA for residential uses. Since As is found in two-thirds of contaminated waste sites, it presents potential hazards. The Office of Solid Waste and Emergency Response (OSWER) must make remediation decisions based on human and environmental risk assessments. To support these decisions, OSWER is developing toxicity values for short-term As exposures and improving bioavailability data. These two activities will enable OSWER to make more rational decisions. The Office of Research and Development has developed and implemented a research framework. The primary themes are identifying key events in As-induced health effects and developing dose–response relationships in As-exposed human populations.

1. Introduction

Although arsenic (As) can exist in four valence states (-3, 0, $+3$, and $+5$; Welch et al., 1988), the major environmental forms of interest are arsenate ($+5$) and arsenite ($+3$). Both natural and anthropogenic sources contribute to the environmental As load. Natural sources include the geologic formations (e.g. rocks, soil, and sedimentary deposits), geothermal and volcanic activity; As concentrations in the earth's crust vary, but are generally reported to range from 1.5 to 5 mg/kg (Abernathy et al., 1997; NAS, 1977; ATSDR, 2000).

Anthropogenic sources include wood preservatives, pesticides, industrial, mining, and smelting wastes. Their impact depends on factors such as the level of human activity, distance from the pollution sources, and dispersion and fate of the released As. In recent years, about 90% of the US annual industrial As use is for chromated copper arsenate

---

[*] The opinions expressed in this manuscript are those of the authors and do not necessarily reflect the opinions or policies of the US Environmental Protection Agency.

(CCA; Reese, 1998, 1999). CCA is used in wood treatment for construction of decks, fences, or other outdoor residential applications, but under a voluntary agreement reached with the wood treatment industry, these uses are being phased out (FR, 2002). Agricultural uses of As included pesticides, herbicides, insecticides, defoliants, and soil sterilants. Inorganic As pesticides are no longer used (the last agricultural application was canceled in 1993) and are presently only used in sealed ant bait and wood preservatives, although organic arsenicals are still a constituent of a few pesticides. The most widely used is monosodium methanearsonate, applied to cotton to control broadleaf weeds (Jordan et al., 1997). Small amounts of disodium methanearsonate are also applied to cotton fields as herbicides. Roxarsone, an organic arsenical, has been approved by the Food and Drug Administration for use as a feed additive for poultry and it undergoes little or no degradation before excretion (Calvert, 1975; NAS, 1977; Garbarino et al., 2003).

Since arsenicals vary greatly in their toxicity, it is important to consider the form and valence state in assessing toxicity. (When As is reported as total As, the data are often of little use in understanding As toxicity or exposure.) Although, we do have some data on various forms of As, it is limited. In marine life, arsenobetaine (AsB) and arsenocholine (AsC) are the main forms of As and appear to have little or no toxicity (Sabbioni et al., 1991; Donohue and Abernathy, 1999). Since early reports stated that the inorganic forms of As were more toxic than organic forms (NAS, 1977; NRC, 1999), the early research focus was on arsenite ($+3$), which was more toxic than the arsenate ($+5$) form. Recently, however, it has been reported that the trivalent ($+3$) monomethylated and dimethylated metabolites of inorganic As are more toxic than arsenite and are likely candidates for the putative active forms (Thomas et al., 2001; Nesnow et al., 2002; Styblo et al., 2002; Abernathy et al., 2003).

Inorganic As can exert many adverse health efforts after acute or chronic exposures. At non-lethal, but high doses, As can cause gastroenterological (GI) effects, shock, neuritis, and vascular effects in humans (Buchanan, 1962; Abernathy et al., 1997). Although acute exposures to high doses of As can cause adverse effects, the EPA is mainly concerned with the chronic effects of exposure to low concentrations of As (North et al., 1997).

2. Health effects

2.1. Non-cancer

Many non-carcinogenic effects have been reported in humans after exposure to drinking water contaminated with As. The most commonly observed signs are skin effects, including alterations in pigmentation and palmar–planter keratoses on the hands, the soles of the feet, and the torso. Their presence on parts of the body not exposed to the sun is characteristic of As exposure (Yeh, 1973) and similar alterations have been reported in patients treated with Fowler's solution (1% potassium arsenite; Cusick et al., 1982).

Chronic exposure to As is often associated with alterations in GI function (Morris et al., 1974; Nevens et al., 1990). Physical examination may reveal spleen and liver enlargement and histopathological examination may demonstrate periportal fibrosis (Nevens et al., 1990; Guha Mazumder et al., 1997). There have been a few reports of cirrhosis after As exposure, but alcohol consumption is a confounding factor (NRC, 1999).

Peripheral vascular changes after As exposure have also been reported. In Taiwan, blackfoot disease (BFD) has been the most severe manifestation of this effect. BFD is a peripheral vascular insufficiency, which may result in gangrene of the feet and other extremities. Other vascular effects, e.g. Reynaud's disease, have been described in Chile (Zaldivar, 1974) and Mexico (Cebrian, 1987). In the US, increased SMRs for hypertensive heart disease were noted in both males and females from Utah after As exposure from drinking water (Lewis et al., 1999) and indicate that As affects the cardiovascular system.

2.2. Cancer

The first reports that As in drinking water was associated with cancer came from southwest Taiwan (Tseng et al., 1968; Tseng, 1977). A dose- and age-related increase in skin cancer was found. Limitations included those caused by the grouping of villages into wide exposure groups and the lack of experimental detail. However, these studies were corroborated by Albores et al. (1979) and Cebrian et al. (1983). Exposure to As in medicines (Cusick et al., 1982) and pesticides (Roth, 1956) caused similar effects. These studies were the basis for EPA's classification of inorganic As as a known human carcinogen (Group A) for skin cancer by the oral route (US EPA, 1988).

Exposure to As in drinking water has also been associated with the development of internal cancers. Reports from Taiwan (Chen et al., 1985; Wu et al., 1989) observed increases in mortality for bladder, kidney, lung, liver, and colon cancers. Corroborating studies have come from Argentina (Borgono, Greiber, 1972; Hopenhayn-Rich et al., 1996, 1998) and Chile (Smith et al., 1998).

There have only been a few studies of As exposure via drinking water in the US. Most of the reports have generally been small and have not considered cancer as an endpoint (Southwick et al., 1983; Valentine et al., 1979; Vig et al., 1984). Bates et al. (1995) reported no association of bladder cancer and As in non-smokers, but positive trends in smokers, especially in the 30–39 year old group. More recently, Lewis et al. (1999) used SMRs to examine a cohort of 4058 people from Utah. They reported an increase in prostate cancer in males, but no significant association between As exposure and bladder cancer. However, Utah has a low incidence of bladder cancer and also a low population of smokers and smoking is a well-known risk factor for bladder cancer.

The following discussions will present concise overviews of some work in the Office of Water (OW), the Office of Pesticides Program (OPP), the Office of Solid Waste and Emergency Response (OSWER) and the Office of Research and Development (ORD). For additional information, specific web addresses are listed in Section 8.

3. Safe drinking water act (OW)

The road to the regulation of As in the US began in 1942 with the US Public Health Service standard of 0.05 mg/l (i.e. 50 μg/l or 50 ppb; US PHS, 1943). In 1975, the EPA reviewed the available database and established an interim drinking water regulation for As of 50 μg/l US. Over the next 20 years, there were several EPA, Science Advisory Board (SAB) and National Research Council (NRC) meetings and reports on As (see Abernathy et al., 2000 for a detailed review of the process). There was scientific uncertainty over the health effects of As and in 1996, Congress specifically included a schedule for the As drinking water regulation in the reauthorization of the Safe Drinking Water Act (SDWA; PL 104-182). This law required the EPA to outline a research program that would support rule making and consult with the NRC, federal agencies, private and public entities. The EPA was also required to propose a new drinking water regulation by January 1, 2000 and issue a final regulation no later than January 1, 2001.

To accomplish this goal, the EPA contracted with NRC to review the As database and issue a report (NRC, 1999). The report used data from Taiwan and stated that As in drinking water caused bladder, lung, and skin cancers in humans and the risks calculated using the Taiwan data were similar to those calculated using data from studies in Chile and Argentina. Since As in drinking water caused cancer and there were insufficient data to prove a non-genetic mechanism of action, the EPA proposed a health-based, non-regulatory maximum contaminant level goal (MCLG) of zero. In addition to the health risks, the Agency also considered the practical quantitation limit (3 µg of As/l), the occurrence of As in water systems across the US, the costs for various As treatments affordable for several size systems, the analyses of health benefits and proposed an enforceable maximum contaminant level (MCL) of 5 µg/l for As in drinking water on June 22, 2000. This proposal also asked for comment on 3, 10, and 20 µg/l as potential MCLs (FR, 2000). After consideration of public comment on the proposal, on January 22, 2001, EPA established an MCLG of zero and an MCL of 10 µg/l (FR, 2001). In March and May, 2001, the administrator extended the effective date for the January 22, 2001 rule, but not the 2006 compliance data, in order to conduct a review of the Agency's risk assessment, new health effects' studies and costs/benefits used to support the As rule. The NRC was asked to review the new data on the health effects of As. In addition, the National Drinking Water Advisory Council (NDWAC) was requested to review the EPA's cost calculations and the SAB to evaluate the benefits analysis. The NRC (2001) concluded that the Taiwanese data remained the most appropriate database for estimating risk, the methylated As metabolites might directly react with DNA, and the risk might be greater than the risk estimated by the previous NRC (1999) report. The NDWAC panel stated that EPA had produced a credible estimate of costs considering the constraints of the available models and data, while the SAB recommended that the EPA consider a "cessation lag" (i.e. the time between reduction in As exposure and decrease in adverse health effects), identify ages of cancer incidence and quantify some of the unquantified health endpoints in their analyses. After carefully considering the comments and recommendations of NRC, NDWAC and the SAB (note – see the OW website for NDWAC and SAB reports), the EPA announced on October 31, 2001 that the final As regulation of 10 µg/l would go into effect without any further delay and that the Agency will consider the impacts of the NRC, NDWAC, and SAB reports, and all new information, on As as part of the next 6-year review in 2006.

4. Arsenic: bioaccumulation and ambient water quality criterion (OW)

EPA first published the Clean Water Act (CWA) Section 304(a) ambient water quality criterion (AWQC) document for As in 1980 (US EPA, 1980). This included a criterion for consumption of both water and organisms (i.e. fish/shellfish) and one for organisms only.

In 1992, the As criterion was revised as part of the final National Toxics Rule (FR, 1992) and subsequently adopted nationally as CWA 304(a) criteria. Development of 304(a) AWQC to protect human health requires data on health effects and exposure (US EPA, 2000a). In this derivation, intake is expressed in mg/kg-day and includes intake from both drinking water and fish/shellfish. Since aquatic organisms may accumulate high levels of certain chemicals in their bodies through a process called bioaccumulation, this pathway must be considered. In deriving an AWQC, EPA's latest exposure methodology accounts for intake of a given chemical from fish/shellfish through the use of national bioaccumulation factors (BAFs). The national BAF (l/kg) relates the concentration of a chemical in water to its expected concentration in fish/ shellfish at a specified trophic level. This AWQC methodology provides guidance for assessing exposure and a procedural framework for deriving national BAFs (US EPA, 2000a).

The earliest As AWQC were derived using the 1980 Agency guidance. Then, the intake obtained through consumption of fish/shellfish was often estimated using a bioconcentration factor (BCF) rather than a BAF due to lack of bioaccumulation data or of understanding about the bioaccumulation process. Unlike a BAF, a BCF only accounts for accumulation of a chemical through the water medium. Thus, if a compound biomagnifies through the food chain or if it is metabolized in the food chain, a BCF may not accurately represent actual exposure potential. The data available in 1980 for deriving As BCFs were limited and the final BCF used in the AWQC was derived from the weighted-average of two BCFs, resulting in a final BAF of 44 for total inorganic As.

A substantial amount of new information on As has been published since 1980, and the Agency is seeking to update the current As AWQC. However, the As AWQC is for freshwater and estuarine systems, not marine and most of the published As occurrence and bioaccumulation information is for marine fish (Edmonds and Francesoni, 1993). This presents a challenge for the Agency in locating useful data for updating the AWQC. EPA recently conducted an extensive literature search and direct inquiries to persons involved in As research, to obtain exposure and bioaccumulation data on various forms of As. Evaluation of the collected literature and data sources to determine what may be useful to the update is currently underway, although some general findings have been noted.

The general consensus is that 85 to >90% of the As found in edible portions of marine fish and shellfish is organic in nature, e.g. AsB, AsC, dimethylarsinic acid (DMA) (Donohue and Abernathy, 1999; NRC, 1999). However, less is known about the forms of As in freshwater fish. In addition, recent toxicological data indicate that two organic metabolites [monomethylarsonic acid, MMA (+3)] and DMA (+3) are the putative toxic moieties (Thomas et al., 2001; Nesnow et al., 2002; Styblo et al., 2002). Furthermore, DMA is a tumor promoter and low toxicity cannot be assumed (Wanibuchi et al., 1996). Thus, it is important to determine if there is any data on the valence state of As, MMA, and DMA in the BAF literature.

Although some organic forms of As can pose toxicological concerns, very little organic As is present in surface waters and groundwaters (NRC, 1999). Accordingly, the revised AWQC will likely be developed for inorganic As as in 1980; however, EPA has sought information on all species of As to ensure that all relevant species are being considered. Depending on the data limitations, which initial reviews indicate are considerable, the Agency seeks to derive BAFs for as many forms of As as possible, including total As, total inorganic As, dissolved inorganic As, the valence state of each As species, AsB, AsC,

MMA, and DMA. Data permitting, BAFs will be derived for trophic levels two through four for lakes, rivers/streams, and estuarine systems using the appropriate methods as presented in the 2000 AWQC Methodology (US EPA, 2000a).

5. Pesticides (OPP)

CCA is a wood preservative mixture that is manufactured as three different types. Types A, B, and C of CCA contain As (as the pentoxide) in varying percentages (16, 45, and 34%, respectively). They can be applied to wood via pressure treatment, brush, spray, low-pressure injection, soak, or bandage treatment, but the predominant method is pressure treatment of lumber intended for outdoor use in constructing a variety of residential landscape and building structures, as well as home, school, and community playground equipment. This CCA-treated wood, predominantly of southern yellow pine, represents the majority of pressure-treated dimensional lumber marketed to the general consumer via lumberyards/hardware stores and other retailers. Major commercial installations include utility poles, highway railings, roadway posts/barriers, bridges, bulkheads, and pilings (ATSDR, 2000).

In 1978, EPA's OPP issued an FR notice initiating a rebuttable presumption against registration (RPAR) for three wood preservatives, including CCA, to consider whether their registrations should be cancelled or modified due to adverse risk criteria identified from animal toxicity testing (US EPA, 1978). The Agency determined that, for CCA, unacceptable risks existed with respect to carcinogenicity, mutagenicity, and reproductive and developmental toxicities. The conclusion of the RPAR process in 1984 and settlement agreements with stakeholders in 1986 established CCA as a restricted use pesticide and modified the terms and conditions of registration of this pesticide. Specifically, CCA-treated wood was no longer allowed under circumstances where it may come into contact with food or animal feed, be used for cutting boards or countertops, or used where it would come into direct or indirect contact with public drinking water except for incidental contact such as docks and bridges. It could only be used for patios, decks, and walkways if visibly clean and free of surface residue.

Although the CCA manufacturing use product solution used to treat wood was a restricted use pesticide, the CCA-treated wood itself was not classified as restricted use based on exemption of treated articles from registration under 40 CFR 152.25(a). Thus, CCA-treated wood has been widely used for residential decking and fencing as well as for wooden play structures. However, the Agency continued to voice concern for human exposure to arsenic from contact with CCA-treated wood even after the RPAR agreement reached in 1986. The Agency considered requiring warnings on the labels of the treated wood itself, or promulgation of a rule under the Toxic Substances Control Act (TSCA). However, the Agency realized that labeling data could not be expected to reach the consumer at the point of purchase, and promulgation of a rule under TSCA would take several years. As the Agency desired to develop a more immediate means of educating the public about the use of CCA-treated wood, it adopted a program called " the consumer awareness program" suggested by the American Wood Preservers Institute. Under this voluntary program on industry's part, a consumer information sheet would be distributed at the point of purchase describing safe use and handling practices of CCA-treated wood as well as disposal practices in effect at that time (US EPA, 1984).

OPP was made aware in recent years of concerns raised by the public as to the potential hazards associated with inorganic As as a component of CCA-treated lumber, especially from playground equipment to which infants and children may be exposed through dermal contact with the treated wood and/or soil around the treated wood structure or through oral ingestion of chemical residue from touching of wood and/or soil and subsequent hand-to-mouth behaviors.

In addition, EPA determined that there were more effective ways to communicate information to consumers regarding safe handling and use precautions of CCA-treated wood than the methods included in the initial consumers awareness program. As a result of these events, the wood treatment industry put forth a proposal in June 2001, which strengthened the nature of the consumer awareness program. The EPA approved this proposal in July 2001. The updated proposal, as implemented, includes placement of individual end tags on each piece of treated lumber with specific safe handling and use information. End tag information includes consumer information on the nature of the pesticide in the wood (As), use-site precautions, and safe handling precautions. The proposal as implemented also includes other measures to inform consumers about safe use and handling of CCA-treated wood, such as display of in-store stickers and signs, and the establishment of a toll-free hotline and a web site.

OPP also entered into negotiations with the wood treatment industry regarding the continued use of CCA-treated wood in playground settings and in residential settings. As a result of these negotiations, the industry announced a program in which a voluntary phase-out was agreed to regarding use of CCA-treated wood for residential uses (US EPA, 2002). The agreement affects virtually all residential uses of CCA-treated wood, including wood used in play structures, decks, picnic tables, landscaping timbers, residential fencing, patios, and walkways/boardwalks. As of January 1, 2004, EPA will not allow CCA products to be used to treat wood intended for any of these residential uses. EPA has not concluded that there is unreasonable risk from exposure of consumers to CCA-treated wood, but believes that any reduction in exposure to As from this use is desirable.

6. Issues in establishing cleanup goals for arsenic contaminated waste sites (OSWER)

Arsenic is found at about two-thirds of contaminated waste sites on the National Priorities List, those sites deemed to present the greatest potential threat to human health and the environment (ATSDR, 2000). The main sources of As at contaminated sites are mining, smelting, and commercial/industrial wastes, and pesticide formulation facilities. Since remediation decisions are based on human health and ecological risk assessments, OSWER is engaged in two main activities to improve the As assessments. They include developing a toxicity value for evaluating short-term exposures to As and drafting guidance on the use of bioavailability data.

6.1. Derivation of an acute/subchronic oral reference dose (RfD)

Human health risk assessments and site-specific cleanup goals are typically based on long-term (i.e. chronic) exposure scenarios as toxicity values (e.g. oral RfD) are usually lower for chronic exposure than for short-term exposures. With As, adverse effects in humans have been noted at similar doses regardless of the exposure duration (NRC, 1999). In addition, since children ingest relatively large amounts of soil over a short time period (US EPA, 1997), there is a need to evaluate short-term exposure scenarios for As. EPA has derived an oral RfD of 0.0003 mg/kg-day for evaluating chronic or long-term exposure to As in environmental media such as soil or water (US EPA, 2002). Because, there has not historically been a toxicity value available for evaluating short-term As exposure, a work group was formed to derive a short-term toxicity value. (Acute exposures were considered to be between 1 and 14 days and subchronic was between 15 days and 7 years.) The work group focused its evaluation on poisoning incidents, case reports, and epidemiological investigations with a similar exposure duration and that provide sufficient information to estimate a daily exposure. In most cases, the actual dose of As was not known and was estimated.

The present data show that exposure to As at 0.05–0.06 mg/kg-day from drinking water or other sources for 1 day to 10 years will cause adverse effects following acute and subchronic exposure, a result that may partially be due to dose estimation uncertainties. Hyperpigmentation and hyperkeratosis are the most consistent finding at this exposure (Silver and Wainman, 1952; Zaldivar and Ghai, 1980; Huang et al., 1985; Guha Mazumder et al., 1998). These effects were generally of minimal severity and the prevalence was relatively low. More severe effects (electrocardiogram abnormalities and peripheral neuropathy) have been documented at exposures of 0.11–0.12 mg/kg-day (Barbey et al., 2001; Soignet et al., 2001; Wang, 2001). Mizuta et al. (1956) also reported electrocardiogram abnormalities and mild peripheral neuropathy at a dose of 0.05 mg/kg-day, but there is considerable uncertainty regarding this exposure level. There is less robust evidence suggesting a no-observed-adverse-effect level (NOAEL) was 0.015–0.040 mg/kg-day from subchronic exposure (Tseng et al., 1968; Tseng, 1977; Cebrian et al., 1983; Guha Mazumder et al., 1998).

A lowest-observed-adverse-effect level (LOAEL) of 0.05 mg/kg-day was selected as the point of departure for deriving an acute RfD because there is consistent evidence across numerous studies that adverse effects occur at or slightly above this level, while the data to establish a definitive NOAEL is less conclusive. Appropriate uncertainty factors to account for human variability, sensitive individuals, and database limitations are still under discussion. Once this decision is reached, an acute/subchronic RfD will be calculated and considered a provisional toxicity value until a complete consensus review is completed and merits addition to EPA's Integrated Risk Information System. It is important to note that external peer review of the toxicological assessment for arsenic was conducted on two separate occasions.

6.2. Use of bioavailability information

Bioavailability is an important factor in assessing potential risks. Metals in contaminated soils may not be absorbed as well as the fraction absorbed in the studies used to establish toxicity values. For As, EPA's current oral toxicity values are derived from drinking water studies where As is soluble and virtually completely absorbed. The default assumption is that the same fraction of As is absorbed from ingestion of contaminated soil as from drinking water. However, the relative bioavailability (RBA) of As in soil is almost certainly less than from water. Characterizing the RBA is significant because making such adjustments in the risk assessment to account for the differences in bioavailability between soil and water may affect the cleanup decision and/or may have a substantial impact on the cleanup goal itself.

Methods for assessing bioavailability can be separated into three broad categories: mineralogical studies, *in vivo*, and *in vitro* methodologies. While mineralogical data can provide useful information to characterize bioavailability, it cannot be used *per se* for making quantitative adjustments in risk assessment. Many different *in vivo* methods and several laboratory species have been used to assess oral bioavailability (Hrudey et al., 1996; Diamond et al., 1997). These involve measuring chemical concentrations in blood, feces, urine, or other body tissues after dosing. It has been common practice to determine the bioavailability of As by quantifying the fraction present in urine because the majority of bioavailable As is excreted in urine within a relatively short period of time.

Recently, there has been a significant effort expended on developing *in vitro* methods for assessing bioavailability due to their ease of use and potential cost savings as compared to more traditional *in vivo* methods. Several researchers have investigated physiologically based extraction tests, which attempt to simulate the conditions in the GI tract (Davis et al., 1992; Ruby et al., 1996, 1999). These assays provide a measure of bioaccessibility or the amount solubilized in the GI fluid and available for potential absorption.

The bioavailability of As in soil has been studied in several laboratory species. For example, the RBA was about 20% at a site in Montana, using *cynomolgus* monkeys (Freeman et al., 1995). At the Vasquez Boulevard and I-70 site located in Colorado, the RBA of As was determined to be 42% in juvenile swine (US EPA, 2000b). Lorenzana et al. (1996) reported a RBA of 78% for arsenic in soil at a smelter site in Washington. Several *in vitro* studies have also been conducted showing reduced RBA in soil as compared to the default of 1.0 (Ruby et al., 1996; Rodriguez et al., 1999). Studies in several species demonstrate that As' RBA in soils can vary significantly and is often substantially less than the standard default of 100%.

EPA's "Risk Assessment Guidance for Superfund" (US EPA, 1989) supports the consideration of bioavailability in evaluating site-specific risks. However, the use of RBA adjustments have not been widespread due to several reasons, including limited data, uncertainty regarding methodologies for assessing bioavailability, and lack of specific guidance on collecting information to support risk assessments. As a result, OSWER is leading a cross-Agency work group to develop additional guidance, which will address several issues such as when it is appropriate to consider RBA adjustments. Other areas include validation of alternative methods to assess bioavailability, adequate site characterization, and incorporation into site-specific risk adjustments. Ultimately, this guidance will facilitate more widespread use of bioavailability data which has the potential to reduce uncertainty in cleanup decisions at waste sites.

7. Health effects research (ORD)

Research on the health effects of chronic exposure to As supports the scientific basis to support the Safe Drinking Water Act. ORD's overall strategy for research on the health effects of As was initially outlined in an arsenic research plan (US EPA, 1998) which provided a guide for evaluating the value of research in terms of the agency's goal of reducing uncertainty in the risk assessment for As. The arsenic research plan placed special emphasis on identifying areas in which research could be directed to reduce uncertainties in the risk assessment for As. ORD's strategy posed three broad questions about the health effects of exposure to As. These questions provide a useful framework to classify research now underway at NHEERL.

1. What are the health effects of and dose–response associated with As exposure?
2. What are the dose–response relationships at low doses?
3. What are the modifiers of As susceptibility?

The five-year implementation plan for research on the health effects of chronic exposure to As is based on the ORD research strategy, the NRC reports, and NHEERL's ongoing As health effects research program. In addition, discussions with risk assessors and regulators in OW identified the types of health-effect data that might have the greatest impact on the risk assessment or the cost/benefit analysis. To emphasize the close relationships among the various areas in which NHEERL researchers were working and to acknowledge the potential value of this research in the risk assessment process, two broad thematic areas describe ORD's health effects research program for As.

Theme 1: Identification of key events or mode of action for cancer, non-cancer effects, and metabolism. Research on the mode of action for As will attempt to predict the shape of the dose–response curve (e.g. tumor formation, formation of skin keratoses) at very low levels of As exposure. For example, if metabolism of inorganic As to methylated forms renders this metalloid carcinogenic, then understanding the mechanistic basis of metabolism and its control are absolutely essential to improve the risk assessment of As. Because the methylation process may be inherently non-linear, one might expect non-linear dose–response relationship at low levels of exposure to As. Factoring these aspects of metabolism and mode of action into the risk assessment may materially assist EPA in evaluation of this metalloid.

A distinctive feature of the metabolism of As is its conversion to methylated species (predominantly methyl and dimethyl As) which are excreted in urine. Research from ORD and elsewhere has confirmed the critical role of oxidation state of As in the control of its reactivity; trivalent As, not pentavalent As, is the preferred substrate for the reaction that methylates As. Indeed methylated arsenicals that contain As in the trivalent oxidation state are more potent cytotoxins, genotoxins, and enzyme inhibitors than inorganic As (reviewed by Thomas et al., 2001). Furthermore, methylated arsenicals, especially DMA, are carcinogens and/or tumor promoters (reviewed by Kitchin, 2001). Recently, the enzyme that methylates As to both the MMA and DMA forms has been purified (Lin et al., 2002). This information provides an impetus to examine the role of methylation in the action of As as a toxin and carcinogen.

As noted in the 1999 NRC report on As in drinking water, understanding variability in the metabolism of inorganic As in humans is likely critical to understand the response of humans to chronic exposure to inorganic As. Wide variation in response in humans chronically exposed to inorganic As in occupational and environmental settings suggest that as-yet-identified factors modify the response of individuals to As. The health effects research program is investigating whether this inter-individual variation in response is related to differences in the capacity to metabolize inorganic As, these differences in metabolic capacity are primarily determined by the kinetic properties of the enzyme that catalyzes the methylation reaction, and the kinetic properties of this enzyme are determined by its primary sequence and, thus, by the genotype for this protein. Differences in susceptibility to the toxic or carcinogenic effects of chronic As exposure might also be related to other genetic polymorphisms. For example, the occurrence of As-induced skin cancer in Taiwanese chronically exposed to As in drinking water might reflect a genetically determined inability to repair DNA damage that initiates the development of tumors in skin. The absence of skin cancer in US populations chronically exposed to As might be explained by a different pattern of genotypes for those enzymes that are involved in DNA repair. Determining the frequency of the genotypes for specific DNA repair enzymes in populations that differ in the occurrence of specific adverse effects of chronic As exposure (e.g. hyperkeratosis) would provide a test of this hypothesis.

Arsenic Metabolism

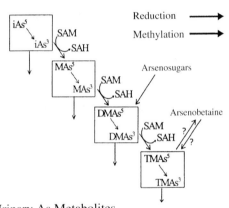

Urinary As Metabolites

Environmental factors may also influence the toxicity and carcinogenicity of As. Selenium (Se) is an essential nutrient that shares many chemical properties with As. A complex pattern of interactions, including both antagonism and potentiation, exists between these metalloids. EPA's health effects research program has shown that Se status affects the rate of As methylation and clearance in mice. Some Se compounds affect the rate of As methylation and clearance in cultured cells and exacerbate the cytotoxicity of arsenicals. It is widely suspected that Se deficiency exists in some of the populations worldwide that are chronically exposed to As and this deficiency exacerbates the effects of As exposure.

Theme 2: Determination of health effects or dose–response relationships at low exposures/doses with a focus on population studies, the development and validation of biomarkers of effect and exposure (dose). The fundamental issue in the risk assessment of As remains the shape of the dose–response curve, particularly for the low levels of exposure that may occur in the United States. If As is not a direct-acting genotoxin, then the suitability of an assumption of linearity in the dose–response relationship is, at best, questionable. Non-linearity in the dose–response relation for As may be a consequence of its reductive metabolism and methylation in which the oxidation state (and reactivity) of As changes as it passes through sequential methylation reactions. Data are lacking on the relative potencies of various arsenicals as initiators of carcinogenesis and on the additivity of As exposure from sources other than drinking water including As in food, and air exposure in areas with mining activity. Indeed, the absence of a reliable animal model in which any arsenical is an initiator has impeded research on the molecular basis of As carcinogenesis. The issue of dose–response is further complicated by recent evidence that at least one organic arsenical, DMA, may act as a tumor initiator, promoter, or progressor. This finding suggests that the metabolism of As is a critical factor in its carcinogenesis.

In the proposed research, two complementary approaches will be taken for understanding the effects of chronic exposure to As. In the first, population-based studies will ascertain the extent of As exposure in the US populations and in other populations worldwide. This approach permits a much-needed assessment of the extent of As exposure in the US populations and allows exposure comparisons between the US and non-US populations. In addition, these studies will focus on the evaluation of biomarkers of exposure and effect in chronically exposed populations. The development and validation of biomarkers for assessing exposure and effects of chronic exposure to As in drinking water at concentrations below 200 μg/l will enhance our ability to conduct low-exposure studies and will provide insight into mechanistic aspects of As toxicity. Biomarkers may potentially increase the efficiency of epidemiologic studies by increasing the sensitivity with which we measure exposure and effects. In addition, these studies will assist in elucidating the nature of the dose–response relationship

8. US EPA arsenic websites

1. OW/drinking water: http://www.epa.gov/safewater/arsenic
2. OW/arsenic water quality criteria: http://www.epa.gov/waterscience/criteria/arsenic/
   [All ambient water quality criteria: http://www.epa.gov/OST/standards/wqcriteria.html]
3. OPP/pesticides (CCA): http://www.ccasafety.com
4. OSWER (general): http://www.epa.gov/superfund/programs/risk/humhlth.html

References

Abernathy, C.O., Calderon, R.L., Chappell, W.R. (Eds), 1997. Arsenic: Exposure and Health Effects. Chapman and Hall, London, 429 pp.

Abernathy, C.O., Dooley, I.S., Taft, J., Orme-Zavaleta, J., 2000. Arsenic: moving toward a regulation. In: Salem, H., Olajos, E.J. (Eds), Toxicology in Risk Assessment. Taylor and Francis, Philadelphia, pp. 211–222.

Abernathy, C.O., Thomas, D.J., Calderon, R.L., 2003. Health effects and risk assessment of arsenic. J. Nutr., 133, 1536S–1538S.

Albores, A., Cebrian, M.E., Tellez, I., Valdez, B., 1979. Comparative study of chronic hydroarsenicism in two rural communities in the Region Laguna of Mexico. Bol. Oficina Sanit. Panam., 86, 196–205.

ATSDR (Agency for Toxic Substances and Drug Registry), 2000. Toxicological profile for arsenic, prepared for the US Department of Health and Human Services, ATSDR by the Research Triangle Institute, 349 pp.

Barbey, J.T., Singer, J.W., Unnikrishnan, D., Dutcher, J.P., Varshneya, N., Lucariello, R., Wiernik, P.H., Chiaramida, S., 2001. Cardiac toxicity of arsenic trioxide. Blood, 98, 1632–1634.

Bates, M.N., Smith, A.H., Cantor, K.P., 1995. Case-control study of bladder cancer and arsenic in drinking water. Am. J. Epidemiol., 141, 523–530.

Borgono, J.M., Greiber, R., 1972. Epidemiological study of arsenicism in the city of Antofagasta. Trace Subst. Environ. Health, 5, 13–24.

Buchanan, W.D., 1962. Toxicity of Arsenic Compounds. Elsevier Scientific Publishers, Amsterdam, 155 pp.

Calvert, C.C., 1975. Arsenicals in animal feeds and waste. In: Woolson, E. (Ed.), Arsenical Pesticides. American Chemical Society Series 7, American Chemical Society, Washington, DC, pp. 70–80.

Cebrian, M., 1987. Some potential problems in assessing the effects of chronic arsenic exposure in North Mexico [preprint extended abstract]. Amer. Chem. Soc, New Orleans, LA.

Cebrian, M.E., Albores, A., Aguilar, M., Blakely, E., 1983. Chronic arsenic poisoning in the north of Mexico. Hum. Toxicol., 2, 121–133.

Chen, C.J., Chuang, Y.C., Lin, T.M., Wu, H.Y., 1985. Malignant neoplasms among residents of a blackfoot disease endemic area in Taiwan. Cancer Res., 45, 5895–5899.

Cusick, J., Evans, S., Price Evans, D.A., 1982. Medicinal arsenic and internal malignancies. Br. J. Cancer, 45, 904–911.

Davis, A., Ruby, M.V., Bergstrom, P.D., 1992. Bioavailability of arsenic and lead in soils from the Butte, Montana mining district. Environ. Sci. Technol., 26, 461–468.

Diamond, G.L., Goodrum, P.E., Felter, S.P., Ruoff, W.L., 1997. Gastrointestinal absorption of metals. Drug Chem. Toxicol., 20, 345–368.

Donohue, J.M., Abernathy, C.O., 1999. Exposure to inorganic arsenic from fish and shellfish. In: Chappell, W.R., Abernathy, C.O., Calderon, R.L. (Eds), Arsenic Exposure and Health Effects. Elsevier, Amsterdam, pp. 89–98.

Edmonds, J.S., Francesoni, K.A., 1993. Arsenic in seafoods: human health aspects and regulations. Marine Pollut. Bull., 26, 665–674.

FR (Federal Register), 1992. Part II. Environmental Protection Agency. Water quality standards; establishment of numeric criteria for priority toxic pollutants; states' compliance final rule. 40 CFR part 131, Tuesday, December 22, 1992.

FR (Federal Register), 2000. Part II. Environmental Protection Agency. National primary drinking water regulations; arsenic and clarifications to compliance and new source contaminants monitoring; proposed rule. 40 CFR parts 141 and 142, Thursday, June 22, 2000.

FR (Federal Register), 2001. Part VIII. Environmental Protection Agency. National primary drinking water regulations; arsenic and clarifications to compliance and new source contaminants monitoring; final rule. 40 CFR parts 9, 141 and 142, Thursday, January 22, 2001.

FR (Federal Register), 2002. Part XXX. Environmental Protection Agency. 67 CFR part, February, 2002.

Freeman, G.B., Schoof, R.A., Ruby, M.V., Davis, A.O., Dill, J.A., Liao, S.C., Lapin, C.A., Bergstrom, P.D., 1995. Bioavailability of arsenic in soil and house dust impacted by smelter activities following oral administration of cynomolgus monkeys. Fundam. Appl. Toxicol., 28, 215–222.

Garbarino, J.R., Bednar, A.J., Rutherford, D.W., Beyer, R.S., Wershaw, R.L., 2003. Environmental fate of roxarsone in poultry litter. I. Degradation of roxarsone during composting. Environ. Sci. Technol., 37, 1509–1514.

Guha Mazumder, D.N., Das Gupta, J., Santra, A., Pal, A., Ghose, A., Sarkar, S., Chattopadhaya, N., Chakraborty, D., 1997. In: Abernathy, C.O., Calderon, R.L., Chappell, W.R. (Eds) Arsenic: Exposure and Health Effects. Chapman and Hall, London, pp. 112–123.

Guha Mazumder, D.N., Haque, R., Ghosh, N., De, B.K., Santra, A., Chakraborty, D., Smith, A.H., 1998. Arsenic levels in drinking water and the prevalence of skin lesions in West Bengal, India. Int. J. Epidemiol., 27, 871–877.

Hopenhayn-Rich, C., Biggs, M.L., Fuchs, A., Bergoglio, R., Tello, E.E., Nicolli, H., Smith, A.H., 1996. Bladder cancer mortality associated with arsenic in drinking water in Argentina. Epidemiology, 7, 117–124.

Hopenhayn-Rich, C., Biggs, M.L., Smith, A.H., 1998. Lung and kidney cancer mortality associated with arsenic in drinking water in Cordoba, Argentina. Int. J. Epidemiol., 27, 561–569.

Hrudey, S.E., Chen, W., Rousseaux, C.G., 1996. Bioavailability in Environmental Risk Assessment. CRC Press, Boca Raton, FL, 296 pp.

Huang, Y., Qian, X., Wang, G., Xiao, B., Ren, D., Feng, Z., Wu, J., Xu, R., Zhang, F., 1985. Endemic chronic arsenism in Xinjiang. Chin. Med. J., 98, 219–222.

Jordan, D., McClelland, M., Kendig, A., Frans, R., 1997. Monosodium methanearsonate influence on broadleaf weed control with selected post-emergence-directed cotton herbicides. J. Cotton Sci., 1, 72–75.

Kitchin, K.T., 2001. Recent advances in arsenic carcinogenesis: modes of action, animal model systems, and methylated arsenic metabolites. Toxicol. Appl. Pharmacol., 172, 249–261.

Lewis, D.R., Southwick, J.W., Ouellet-Hellstrom, R., Rench, J., Calderon, R.L., 1999. Drinking water arsenic in Utah: a cohort mortality study. Environ. Health Perspect., 107, 359–365.

Lin, S., Shi, Q., Nix, F.B., Styblo, M., Beck, M.A., Herbin-Davis, K.M., Hall, L.L., Simeonsson, J.B., Thomas, D.J., 2002. A novel S-adenosyl-L-methionine: arsenic (III) methyl transferase from rat liver. J. Biol. Chem., 277, 10795–10803.

Lorenzana, R.M., Duncan, B., Ketterer, M., Lowry, J., Simon, J., Dawson, M., Poppenga, R., 1996. Bioavailability of arsenic and lead in environmental substrates. Document Control No. EPA/910/R-96-002, U.S. Environmental Protection Agency, Region 10, Seattle, WA.

Mizuta, N., Mizuta, M., Itâ, F., Itâ, T., Uchida, H., Watanabe, Y., Akama, H., Murakami, T., Hayashi, F., Nakamura, K., Yamaguchi, T., Mizuia, W., Oishi, S., Matsamura, H., 1956. An outbleak [sic] of arsenic poisoning caused by arsenic contaminated soy-sauce (shâyu): A clinical report of 220 cases [sic]. Bull. Yamaguchi Med. School, 4, 132–149.

Morris, J.S., Schmid, M., Newman, S., Scheuer, P.J., Sherlock, S., 1970. Arsenic and noncirrhotic portal hypertension. Gastroenterology, 66, 86–94.

NAS (National Academy of Sciences), 1977. Arsenic. Medical and Biological Effects of Environmental Pollutants. National Academy of Sciences Press, Washington, DC, 332 pp.

Nesnow, S., Roop, B.C., Lambert, G., Kadiiska, M., Mason, R.P., Cullen, W.R., Mass, M.J., 2002. DNA damage induced by methylated trivalent arsenicals is mediated by reactive free radicals. Chem. Res. Toxicol., 15, 1627–1634.

Nevens, F., Fevery, J., Van Steenbergen, W., Sciot, R., Desmet, V., De Groote, J., 1990. Arsenic and noncirrhotic portal hypertension: a report of eight cases. J. Hepatol., 11, 80–85.

North, D.W., Gibb, H.J., Abernathy, C.O., 1997. Arsenic: past, present and future considerations. In: Abernathy, C.O., Calderon, R.L., Chappell, W. (Eds), Arsenic: Exposure and Health Effects. Chapman and Hall, London, pp. 406–423.

NRC (National Research Council), 1999. Arsenic in Drinking Water. National Academy Press, Washington, DC, 310 pp.

NRC (National Research Council), 2001. Arsenic in Drinking Water: 2001 Update. National Academy Press, Washington, DC, 225 pp.

Reese, R.G., Jr., 1998. Arsenic in United States Geological Survey Minerals Yearbook, 1998. Fairfax, VA.

Reese, R.G., Jr., 1999. Arsenic in Mineral Commodity Summaries, January, 1999.

Rodriguez, R.R., Basta, N.T., Casteel, S.W., Pace, L.W., 1999. An *in vitro* gastrointestinal method to estimate bioavailable arsenic in contaminated soils and solid media. Environ. Sci. Technol., 33, 642–649.

Roth, F., 1956. Concerning chronic arsenic poisoning of the Moselle wine growers with special emphasis arsenic carcinomas. Z. Krebsforschung, 61, 287–319.

Ruby, M.V., Davis, A., Schoof, R., Eberle, S., Sellstone, C.M., 1996. Estimation of lead and arsenic bioavailability using a physiologically-based extraction test. Environ. Sci. Technol., 30, 422–430.

Ruby, M.V., Schoof, R., Brattin, W., Goldade, M., Post, G., Harnois, M., Mosby, D.E., Casteel, S.W., Berti, W., Carpenter, M., Edwards, D., Cragin, D., Chappell, W., 1999. Advances in evaluating the oral bioavailability of inorganics in soil for use in human health risk assessment. Environ. Sci. Technol., 33, 3697–3705.

Sabbioni, E., Fischbach, M., Prozzi, G., Pietra, R., Gallorini, M., Piette, J.L., 1991. Cellular retention, toxicity and carcinogenic potential of seafood arsenic. I. Lack of toxicity and transforming activity of arsenobetaine in the BALB/3T3 cell line. Carcinogenesis, 12, 1287–1291.

Silver, A.S., Wainman, P.L., 1952. Chronic arsenic poisoning following use of an asthma remedy. J. Am. Med. Assoc., 150, 584.

Smith, A.H., Goycolea, M., Haque, R., Biggs, M.L., 1998. Marked increase in bladder and lung cancer mortality in a region of northern Chile due to arsenic in drinking water. Am. J. Epidemiol., 147, 660–669.

Styblo, M., Drobna, Z., Jaspers, I., Lin, S., Thomas, D.J., 2002. The role of biomethylation in toxicity and carcinogenicity of arsenic: a research update. Environ. Health Perspect., 110 (Suppl. 5), 767–771.

Thomas, D.J., Stable, M., Lin, S., 2001. Review: the cellular metabolism and systemic toxicity of arsenic. Toxicol. Appl. Pharmacol., 176, 127–144.

Tseng, W.P., 1977. Effects and dose–response relationships of skin cancer and blackfoot disease with arsenic. Environ. Health Perspect., 19, 109–119.

Tseng, W.P., Chu, H.M., How, S.W., Fong, J.M., Lim, C.S., Yeh, S., 1968. Prevalence of skin cancer in an endemic area of chronic arsenicism in Taiwan. J. Natl Cancer Inst., 40, 453–463.

US EPA (US Environmental Protection Agency), 1978. Fed. Reg., 43 (202), 48299–48617.

US EPA (US Environmental Protection Agency), 1980. Ambient Water Quality Criteria for Arsenic. EPA 440/5-80-021.

US EPA (US Environmental Protection Agency), 1984. Wood Preservative Pesticides: Creosote, Pentachlorophenol, Inorganic Arsenicals. Position Document 4.

US EPA (US Environmental Protection Agency), 1988. Special Report on Inorganic Arsenic: Skin Cancer; Nutritional Essentiality. EPA 625/3-87/013, 124 pp.

US EPA (US Environmental Protection Agency), 1989. Risk assessment guidance for superfund, volume I: human health evaluation manual (part A), interim final. US Environmental Protection Agency, Office of Research and Development, Washington, DC. EPA/540/1-89/002.

US EPA (US Environmental Protection Agency), 1998. As research plan.

US EPA (US Environmental Protection Agency), 2000a. Methodology for Deriving Ambient Water Quality Criteria for the Protection of Human Health (2000). EPA-822-B-00-004, 165 pp.

US EPA (US Environmental Protection Agency), 2000b. Relative Bioavailability of Arsenic in Soils from the VB170 Site. Report prepared for USEPA Region VIII by Syracuse Research Corporation, February, 2001.

US EPA (US Environmental Protection Agency), 2002. IRIS File on Arsenic, verified November 15, 1990. Available online at http://www.epa.gov/iris

US PHS (US Public Health Service), 1943. Public Health Service drinking water standards. Public Health Reports, 58, 69–111.

Valentine, J.L., Kang, H.K., Spivey, G., 1979. Arsenic levels in human blood, urine, and hair in response to exposure via drinking water. Environ. Res., 20, 24–32.

Vig, B.K., Figueroa, M.L., Cornforth, M.N., Jenkins, S.H., 1984. Chromosome studies in human subjects chronically exposed to arsenic in drinking water. Am. J. Ind. Health, 6, 325–338.

Wang, Z.-Y., 2001. Arsenic compounds as anticancer agents. Cancer Chemother. Pharmacol., 48, S72–S76.

Wanibuchi, H., Yamamoto, S., Chen, H., Yoshida, K., Endo, G., Hori, T., Fukushima, S., 1996. Promoting effects of dimethylarsinic acid on N-butyl-N-(4-hydroxybutyl)nitrosamine-induced urinary bladder carcinogenesis in the rat. Carcinogenesis, 17, 2435–2439.

Welch, A.H., Lice, M., Hughes, J., 1988. Arsenic in groundwater of the western United States. Ground Water, 26, 333–347.

Wu, M.M., Kuo, T.L., Hwang, Y.H., Chen, C.J., 1989. Dose–response relation between arsenic concentration in well water and mortality from cancers and vascular diseases. Am. J. Epidemiol., 130, 1123–1132.

Yeh, S., 1973. Skin cancer in chronic arsenicism. Hum. Pathol., 4, 469–485.

Zaldivar, R., 1974. Arsenic contamination of drinking water and food-stuffs causing endemic chronic poisoning. Beitr. Pathol., 151, 384–400.

Zaldivar, R., Ghai, G.L., 1980. Clinical epidemiological studies on endemic chronic arsenic poisoning in children and adults, including observations on children with high- and low-intake of dietary arsenic. Zentralbl. Bakt. Hyg. I. Abt. Orig. B., 170, 409–421.

Subject index